RESEARCHES ON FUNGI

VOLUME VII

RESEARCHES ON FUNGI

VOLUME VII

THE SEXUAL PROCESS IN THE UREDINALES

BY THE LATE

A. H. REGINALD BULLER, F.R.S.

B.Sc. (LOND.); D.Sc. (BIRM.); PH.D. (LEIP.)
LL.D. (MAN., SASK., AND CALCUTTA)
D.Sc. (PENN.); F.R.S.C., ETC.

FORMERLY PROFESSOR OF BOTANY AT THE
UNIVERSITY OF MANITOBA

WITH ONE HUNDRED AND TWENTY-FOUR FIGURES IN THE TEXT

PUBLISHED FOR

THE ROYAL SOCIETY OF CANADA

BY

THE UNIVERSITY OF TORONTO PRESS

1950

Toronto: University of Toronto Press
London: Geoffrey Cumberlege
Oxford University Press

TO THE STUDENTS AND COLLEAGUES
OF THE LATE A. H. REGINALD BULLER

FOREWORD

STUDENTS of cryptogamic botany, plant pathologists, and biologists in general will welcome the posthumous publication of this last volume of Professor A. H. Reginald Buller's classical *Researches on Fungi*. In this work he brings together and interprets a wealth of material, much of it the result of his own experiments, dealing with the sexual process in the rust fungi.

After thirty-two years of distinguished service as Professor and Head of the Department of Botany at the University of Manitoba, Dr. Buller voluntarily retired in 1936 in order to devote himself to a number of researches on which he was engaged. His work on the present volume was well advanced, when early in 1944, illness prevented its completion. In his will, he left all his scientific drawings, photographs, illustrations and unpublished manuscripts to the Royal Botanic Gardens at Kew. Upon their arrival in July 1947, Dr. G. R. Bisby found that this particular manuscript was nearly ready for the press; he made a few minor alterations in the text, supplied footnotes and references, and prepared blocks for a number of illustrations.

Longmans, Green and Co., London, who published Buller's six preceding volumes, were unable to undertake the printing due to the current paper shortage in England. The Council of the Royal Society of Canada has gladly accepted this responsibility and is pleased to publish this valuable research under the joint imprimatur of the Society and the University of Toronto Press. The late Dr. Buller made substantial bequests to the Winnipeg Foundation, and to the Royal Society of Canada, of which he was a distinguished past president, and thus it is a duty, as well as a privilege, for the Society to assure the publication of his last work.

It is very appropriate that this volume has been dedicated to Dr. Buller's former students and colleagues. Undoubtedly this would have pleased him. He loved his students and never failed to give them credit for their assistance. He dedicated the previous volumes to teachers who had inspired him, and to eminent mycologists with whom he had been associated.

Acknowledgements are here made to Dr. J. H. Craigie, and Dr. F. L. Drayton of the Dominion Experimental Farm, Ottawa, and especially to Dr. W. F. Hanna, Director of the Dominion Laboratory of Plant Pathology, Winnipeg, for their invaluable assistance in reading the proofs and in offering suggestions, as well as to the Honorary Editor of the Society, Dr. G. W. Brown. The index, so essential to a treatise of this character, has been compiled under the supervision of Dr. W. H. Hanna, and every effort has been made to have it complete. Dr. Buller was ever a perfectionist, and he devoted much care to the preparation of his indexes.

Among the manuscripts left by Dr. Buller was a History of Mycology, which Dr. Bisby considers should be published. He had projected also an eighth volume entitled "The Sexual Process in Pyrenomycetes, Discomycetes and Lichens," of which the first five chapters were tentatively written. These manuscripts and miscellaneous notes in the library at Kew Gardens are available to interested students.

The sizeable collection of poems, rhymes, and limericks which Dr. Buller left will probably never be published. Many of them possess real merit, and all bespeak the good sense of humour and kindly nature of this really great scientist. Two of them have been selected for inclusion in this volume: "Pond Life," written in 1903 when he was in Birmingham, vividly depicts the wondrous universe of a drop of water; and "The Sporobolomycetologist," composed forty years later in Winnipeg, was his last poem, which he regarded as his requiem.

The enthusiasm of this inspiring teacher, his devotion to his beloved science, and his passion for research were naturally communicated to his students. To the Dominion Laboratory of Plant Pathology at Winnipeg, in which he was always deeply interested and to which his inspiration and advice through the years meant so much, he left his exceedingly valuable scientific library of over two thousand volumes which, unfortunately, is unavailable to the staff for lack of space, and of necessity is stored in a fire-proof vault in Winnipeg. It is hoped that in the near future the "Buller Library" will be suitably established in a new building where the donor's work and influence will survive. This would constitute a fitting monument to the memory of this eminent biologist, whose many discoveries brought honours to him, and to his adopted country.

JOSEPH A. PEARCE, *President*
The Royal Society of Canada

A. H. REGINALD BULLER

SCIENTIFIC AFFILIATIONS AND HONOURS

ARTHUR HENRY REGINALD BULLER, 1874-1944

Societies British Association; British Mycological Society (President 1913); American Phytopathological Society; Botanical Society of America (President 1928); Canadian Phytopathological Society (President 1920); associate member Société Royale de Botanique de Belgique; corresponding member Société Botanique Neerlandaise; honorary member Indian Botanical Society; Fellow Royal Society of Canada 1909 (President 1927-28); Royal Society of London 1929; member International Botanical Congress (President Section VI, 1935).

Medals Royal Society of Canada, Flavelle Medal, 1929; Natural History Society of Manitoba Medal, 1936; Royal Society of London, Royal Medal, 1937.

Honorary Degrees University of Manitoba, LL.D., 1924; University of Saskatchewan, LL.D., 1928; University of Pennsylvania, D.Sc., 1933; University of Calcutta, LL.D., 1938.

Publications A list of his many publications are recorded in *Canadian Who's Who 1938-39* and *Who's Who 1944*. Biographical sketches of his life and work were published in *Nature* August 5, 1944, by Bisby, *Science* October 5, 1944, by Brodie and Lowe, *Phytopathology* August 1945, by Hanna, Lowe and Stakman, and in the *Proceedings of the Royal Society of Canada 1945*, by R. B. Thompson.

POND LIFE

By A. H. Reginald Buller

Upon the slide, with best of light
  And lenses "5" or "7",
I see a sight as wondrous as
  The Milky Way in heaven.

A myriad creatures live and move
  Within the little ocean;
Atlantic fishes never were
  In livelier commotion.

A gauzy Stephanoceros
  Now glows with silvery sheen,
And draws within its tiny mouth
  Some tinier cells of green.

An emerald Volvox rolls in sight,
  Ah! What co-ordination!
A thousand willing oarsmen help
  To row the globy nation.

A Diatom in armour clad,
  Now glides into the scene:
It seems to be in miniature
  A perfect submarine.

Lo! Vorticella on a stalk,
  The bell a whirlpool making.
What change o'ercomes thee, when I give
  Thy little world a shaking?

Thy stalk was straight, but in a trice
  A spring-like coil appears,
That slowly lengthens out once more,
  When calm allays thy fears.

A Paramecium hastens by
  With countless cilia beating.
Its body is transparent: I
  Can see what it's been eating!

To Spirogyra's restful threads
  My vision now is guided;
Each segment is with spiral bands
  Of chlorophyll provided.

Two hurrying swarm-spores meet and touch,
  Rebound and pass for ever;
Or else—they fuse and form one cell
  And then—there's parting never.

There slowly flows a plasmic mass,
  An ancient form of life.
Thou art a bachelor all thy days
  Nor ever needest wife,

For when descendents thou wouldst have
  (To have them is thy mission),
Some pseudopodia are cut off
  By act of simple fission.

Amoeba! Learned men have said,
  From thee I am descended,
And all the mysteries of life
  In thy small frame are blended.

The fiery orbs of silent night,
  Sunk deep in stellar space,
Astronomers can measure and
  Their proper courses trace.

Their flaming gas is analysed
  By spectroscopic prism:
But still holds fast its secret, an
  Amoeboid organism.

How wondrous is a mighty sun,
  That lights a boundless chasm!
More wondrous still I deem a speck
  Of living protoplasm.

The University, Birmingham.   December, 1903

THE SPOROBOLOMYCETOLOGIST

On little yeasts he rose to fame,
  He will be sadly missed:
He was an ardent Spor-o-bol-
  O-my-cet-o-log-ist!

Chorus

  He was an ardent Spor-o-bol-
    O-my-cet-o-log-ist!
      Spor-o-bol-o,
      My-cet-o-lo,
  Spor-o-bol-o-my-cet-o-log-ist;
  He was an ardent Spor-o-bol-
    O-my-cet-o-log-ist!

No golf or billiards would he play,
  And maidens never kissed:
He was an ardent Spor-o-bol-
  O-my-cet-o-log-ist!

He kept no dog, he gave no ear
  To Beethoven or Liszt:
He was an ardent Spor-o-bol-
  O-my-cet-o-log-ist!

"I have no days for Shakespeare's plays"
  He often would insist:
He was an ardent Spor-o-bol-
  O-my-cet-o-log-ist!

Perhaps in heaven, where angels are,
  His yeast-thoughts will persist:
He was an ardent Spor-o-bol-
  O-my-cet-o-log-ist!

Set up a monument to him,
  Say on a slab of schist:
He was an ardent Spor-o-bol-
  O-my-cet-o-log-ist!

A. H. Reginald Buller
Jan. 17, 1943.

EDITOR'S NOTE

DR. BULLER'S last illness came upon him suddenly at the end of 1943 and prevented him from making the last alterations to this book. In due course his manuscripts, drawings, notes, and other data were forwarded to me, arriving in July, 1947. I was glad to find that this volume was nearly ready for the press; I have made practically no alteration to the text except that of deleting a few lines here and there that seemed to me superfluous or repetitive, and adding a very few foot-notes within square brackets. Most of the blocks had been made, and these were forwarded to me by his block-makers, Wallage and Gilbett of Birmingham. The figures for Chapters III, V, and XI had been prepared, but the blocks not yet made, and there were fifteen figures for other chapters prepared but without blocks; all these have now been included in the text. His notes show that Chapter X was the last to be revised (in November, 1943). Unfortunately he had not managed to prepare his projected figures for Chapter X, and I have been able to add only three figures from amongst his miscellaneous photographs.

Conditions have made it impracticable for Longmans, Green and Co., who published his six preceding volumes of Researches, to undertake this one; but the Royal Society of Canada—of which he was long a prominent member—has fortunately been able to take charge.

Amongst the manuscripts left by Dr. Buller was a History of Mycology, which will doubtless soon be published. He had projected an eighth volume of Researches entitled "The Sexual Process in Pyrenomycetes, Discomycetes, and Lichens, and a Further Contribution to Our Knowledge of the Sporophores of the Hymenomycetes," of which the first five chapters were written up tentatively as follows: I, Historical and Critical Review of the Problems of Sex in the Lichenes; II, The Function of the Spermatia of the Non-lichenous Pyrenomycetes and Discomycetes; III, Sex, Inheritance, and the Functions of the Conidia in *Neurospora sitophila*; IV, Sex and Inheritance of Lethal Factors in *Neurospora tetrasperma*; V, The Sexuality of *Pleurage anserina*. Other chapters, mostly written some years ago and with many blocks prepared, were listed by Buller as a projected volume VII (before his interests switched primarily to sex in Rust Fungi) as follows:

Fungi in Some of their Relations with Water.

Fungi as Hydrophytes, Mesophytes, and Xerophytes.

Drop-excretion and Insects.

Absorption and Storage of Water by the Pilei of Certain Hymeno-mycetes.

The Vitality of Desiccated Fruit-bodies.

The Vitality of Desiccated Spores.

Grafting Phenomena with Special Reference to the Grafting of Fruit-bodies on Mycelia.

The Attachment of the Lamellae to the Stipe in Agaricaceae.

The Polyporaceae and Boletaceae Compared and Contrasted.

Polyporus betulinus and its Spore-discharge Period.

The Hexagonal Hymenial Tubes of Hexagona.

The Form and Texture of the Smallest and Largest Polyporaceae.

Various Observations on Polyporeae and Agaricaceae.

Cryptoporus volvatus and Its Relations with Insects.

The Relations of the Sporophore with the Mycelium.

The Connexion of Fruit-bodies with Buried Roots and Sticks.

The Upward-growing Stromatous Strand.

The Mycorrhizal Cord.

The Blocking Layer of the Mycelium of Wood-destroying Hy-menomycetes and the Luminosity of the Mycelium of Armil-laria mellea.

Further Observations on the Coprini.

Marine Fungi and Lichens.

Dr. Buller's will bequeathed his miscellaneous manuscripts to the Library of the Royal Botanic Gardens, Kew. Since it seems unlikely that anyone could now adequately edit these various chapters into a volume VIII of Buller's Researches, it seems best to deposit them at Kew together with the blocks, illustrations, and notes. A student working on any of the topics mentioned above can consult the manuscripts at Kew, and could ask the authorities there for permission to publish Buller's observations or figures.

G. R. BISBY

IMPERIAL MYCOLOGICAL INSTITUTE
KEW, *November*, 1947

TABLE OF CONTENTS

PAGE

FOREWORD BY THE PRESIDENT OF THE ROYAL SOCIETY OF CANADA vii

A. H. REGINALD BULLER (PORTRAIT) ix

SCIENTIFIC AFFILIATIONS AND HONOURS. xi

Pond Life xii

The Sporobolomycetologist xiv

EDITOR'S NOTE xv

CHAPTER I

THE DISCOVERY OF THE FUNCTION OF THE PYCNIDIA OF THE UREDINALES

Introduction — The Discovery of Pycnidia — The Pycnidia (Spermogonia) and Pycnidiospores (Spermatia) of the Lichens — The Pycnidia and the Spermogonia of the Non-lichenous Discomycetes and Pyrenomycetes — The Antheridia and Spermatia (Antherozoids) of the Laboulbeniales — Discovery of Pycnidia (Spermogonia) and Pycnidiospores (Spermatia) in the Uredinales — A Century of Observation and Speculation — Entomological Background — Cytological Background — Experimental Background — Discovery of Heterothallism in the Uredinales — Discovery of the Function of the Pycnidia of the Uredinales — Discovery of Flexuous Hyphae in the Pycnidia of the Uredinales and of their Fusion with Pycnidiospores of Opposite Sex — Species of Rust Fungi Proved to be Heterothallic or Homothallic 3

CHAPTER II

THE PYCNIDIA AND PROTO-AECIDIA OF THE UREDINALES

Species Studied — Preparation of Material for Inoculation — Inoculation Method — The Formation of Pycnidia and Proto-aecidia in a Haploid Pustule — Conversion of a Proto-aecidium into an Aecidium — Nuclear Migration, Cell Fusions, Septal Pores, and the Diploidisation of a Proto-aecidium — The Structure of a Mature Pycnidium — Terminology — The Development of a Pycnidium of *Puccinia graminis* — Periphyses and their Functions — The Pycnidiosporophores and the Pycnidiospores — Horns of Pycnidiospores produced abnormally by *Puccinia graminis* — Oil-drops in Extruded Masses of Pycnidiospores — The Drop of Nectar — The Nectar is Antiseptic — The Odour of Rust Fungi including *Puccinia graminis* — Insects in their Relations with Rust Fungi — Effect of Rain on Pycnidia — The Flexuous Hyphae — Conversion of Mature Periphyses into Flexuous Hyphae 89

CHAPTER III

FLEXUOUS HYPHAE IN THE UREDINALES IN GENERAL

Introduction — Previous Observations — New Observations — Cronartium — *Melampsora lini* — *Melampsorella cerastii* — Milesia — *Gymnoconia peck-*

iana — Gymnosporangium — Phragmidium — *Tranzschelia pruni-spinosae* —
Puccinia and Uromyces — Ashworth's Observations on *Melampsoridium betulinum* and *Melampsora larici-capraearum* — Conclusion 187

CHAPTER IV

THE UNION OF PYCNIDIOSPORES AND FLEXUOUS HYPHAE IN PUCCINIA GRAMINIS AND THREE OTHER RUSTS

1. *Puccinia graminis.* Introduction — Methods — Fusions between Pycnidiospores and Flexuous Hyphae described and discussed — Time elapsing
between Mixing the Nectar and the Appearance of Fusions — Pycnidiospores
and Pointed Periphyses — Pycnidiospores and· Hyphae emerging between
Epidermal Cells — Pycnidiospores fuse with Flexuous Hyphae only — Pycnidiospores incapable of Independent Germination — Pycnidiospores, Oidia,
and Pollen Grains — Passage of a Nucleus of a Pycnidiospore down a Flexuous
Hypha to a Proto-aecidium — Multiple Fusions and Hybridisation. 2. *Puccinia helianthi*. 3. *P. coronata avenae*. 4. *Gymnosporangium clavipes* . . 215

CHAPTER V

THE PRESENCE OR ABSENCE OF PYCNIDIA AND ASSOCIATED SPORE-FORMS IN CERTAIN UREDINALES

Microcyclic Rust Fungi without Pycnidia — A Microcyclic Rust with Pycnidia only on Certain Hosts — Desirability of Experiments on Microcyclic
Rusts provided with Pycnidia — Vestigial Pycnidia in *Coleosporium pinicola*
and *Calyptospora goeppertiana* — A Variety of *Puccinia coronata* with Vestigial
Pycnidia — Correlation of Facts concerning the Presence or Absence of Pycnidia and Sexuality — Pycnidia and the Production of Uredospores and Teleutospores of *Puccinia graminis* on Barberry Leaves — Short-cycling in *Uromyces
fabae* — Supposed Occasional Association of Pycnidia and Uredospore Sori in
Puccinia helianthi — *Uromyces hobsoni* and its Pycnidia. 241

CHAPTER VI

MODES OF INITIATING THE SEXUAL PROCESS IN THE RUST FUNGI

The Sexual Process — Modes of Initiating the Sexual Process — Modes I and
II, the Normal Modes — Mode III: Fusion of a Diploid Mycelium with a
Haploid Mycelium — Critical Remarks concerning Mode I — Mode IV, in
Homothallic Rusts — Mode V: Annual Self-diploidisation in Certain Rusts
having a Systemic Mycelium — De-diploidisation and Self-diploidisation in
Puccinia minussensis — Defect of Homothallism — Rust Fungi in which the
Sexual Process has become Imperfect or Inoperative — Uninucleate Rusts
considered as Self-propagating Haploid Strains derived from Heteroecious
Species 264

CHAPTER VII

COMPARISON OF THE SEXUAL PROCESSES IN THE UREDINALES WITH THOSE IN THE HYMENOMYCETES

Dikaryotic and Synkaryotic Diploid Cells in the Hymenomycetes and Uredinales — Uredinales and Hymenomycetes as Closely Related Groups —
Puccinia graminis and *Coprinus lagopus* as Types for Comparison — Modes
of Initiating the Sexual Process — Mode I — Mode II — Mode III — Mode
IV — Mode V — Appendix 297

CHAPTER VIII

A REVIEW OF CYTOLOGICAL WORK ON THE SEXUAL
PROCESS IN HETEROTHALLIC UREDINALES

Introduction — Period I — Period II — Period III — Living Fungi and Dead
Preparations — Conclusion   313

CHAPTER IX

CRONARTIUM RIBICOLA AND ITS SEXUAL PROCESS

The White Pine Blister Rust Disease — Mode of Infection — Incubation
Period for Canker Formation — Successive Formation of Pycnidia and Aecidia — Morphology and Cytology of an Old-established Canker — Comparison of Systemic Mycelia — Simple and Compound Cankers — Heterothallism of *Cronartium ribicola* — Initiation of Diploidisation in a Simple
Canker — Initiation of Diploidisation in a Compound Canker — Cell-fusions
in the Basal Cells of the Proto-aecidia — Conclusion   321

CHAPTER X

PUCCINIA SUAVEOLENS AND ITS SEXUAL PROCESS

Introduction — The Name *Puccinia suaveolens* — Host-plant — Can *Puccinia
suaveolens* parasite the Dandelion? — Geographical Distribution of Host and
Parasite — Cause of Absence of *Puccinia suaveolens* from Central Canada and
its Artificial Introduction — Growth and Reproduction of *Cirsium arvense* —
The Effect of *Puccinia suaveolens* on the Structure and Physiology of its Host
— Economic Value of *Puccinia suaveolens* — Effect of External Conditions on
the Production of Secondary Uredospores and Teleutospores — Source of
Fungus Material — Primary and Secondary Uredospores — Localised and
Systemic Mycelia — Forms of Localised Mycelia — Forms of Systemic Mycelia — Localised Mycelia derived from Uredospores — Localised Mycelia
derived from Basidiospores — Establishment of Systemic Mycelia in *Endophyllum euphorbiaesylvaticae*, *Aecidium leucospermum*, and *Puccinia minussensis* in Herbaceous Perennials — Establishment of Systemic Mycelia of
Puccinia suaveolens in *Cirsium arvense* Seedlings — Experiments with Root-
buds — Systemic Mycelium in Thistle Roots — Binucleate Condition of the
Systemic Mycelium in Stems and Roots — Systemic Mycelia in Leaves and
the Production of Pycnidia, Primary Uredospores, and Teleutospores — De-
diploidisation, Re-diploidisation, Self-diploidisation, and Cross-diploidisation
—Biological Significance of Pycnidia on Systemic Mycelia — Systemic Mycelia bearing Secondary Uredospores and Teleutospores only — Systemic Mycelia bearing Pycnidia only — *Puccinia suaveolens* is Heterothallic — Experiments proving that Systemic Mycelia may arise from Uredospores under
Natural Conditions — Uredospores and the Persistence of *Puccinia suaveolens*.   344

CHAPTER XI

THE GENUS GYMNOSPORANGIUM

Introductory Remarks — The Genus Described — Heteroecism — Geographical Distribution — Species in Central Canada — Cytology and Heterothallism — Appearance and Origin of Cornute Aecidia — Proto-aecidia and their
Development into Cornute Aecidia — Two Groups of Species, one with Cupulate the other with Cornute Aecidia — Cornute Aecidia and Puff-balls —
Hygroscopic Movements of the Peridium of Cornute Aecidia — Non-violent
and Violent Discharge of Aecidiospores in Rust Fungi in general — Evolution
of Gymnosporangium — Pores and Pore-plugs in the Aecidiospores of *Gymnosporangium ellisii* and other Rust Fungi   389

CHAPTER XII

THE GEOLOGICAL TIME DURING WHICH THE PYCNIDIA
OF THE UREDINALES ATTAINED THEIR PRESENT FORM
AND FUNCTION, WITH SOME REMARKS ON THE
EVOLUTION OF OTHER ENTOMOPHILOUS FUNGI

Introduction — Rock-systems — Land Plants of the Past — Cambrian and
Ordovician Periods — Silurian Period — Devonian Period — Carboniferous
and Permian Periods — Triassic Period — Jurassic Period — Cretaceous
Period — Tertiary and Quaternary Periods — Insects of the Past — Flowers
and Insects — Fungi and Insects — Fungi dependent on Insects and the Time
of their Evolution — Geological Time of Evolution of the Pycnidia of the
Uredinales — Time of Evolution of the Phallales, etc. — Conclusion   .   .   .   415

INDEX   429

RESEARCHES ON FUNGI

VOLUME VII

RESEARCHES ON FUNGI

CHAPTER I

THE DISCOVERY OF THE FUNCTION OF THE PYCNIDIA
OF THE UREDINALES

Introduction — The Discovery of Pycnidia — The Pycnidia (Spermogonia) and
Pycnidiospores (Spermatia) of the Lichens — The Pycnidia and the Spermogonia
of the Non-lichenous Discomycetes and Pyrenomycetes — The Antheridia and
Spermatia (Antherozoids) of the Laboulbeniales — Discovery of Pycnidia
(Spermogonia) and Pycnidiospores (Spermatia) in the Uredinales — A Century
of Observation and Speculation — Entomological Background — Cytological
Background — Experimental Background — Discovery of Heterothallism in
the Uredinales — Discovery of the Function of the Pycnidia of the Uredinales —
Discovery of Flexuous Hyphae in the Pycnidia of the Uredinales and of their
Fusion with Pycnidiospores of Opposite Sex — Species of Rust Fungi Proved
to be Heterothallic or Homothallic.

Introduction.—The pycnidia (pycnia, spermogonia)[1] of the Rust
Fungi (Fig. 1) have long been known; but, for many years and until
recently, they were supposed to be functionless.

De Bary, by means of experiments made in 1864 and 1865, proved
conclusively that *Puccinia graminis* can pass from its graminaceous
host-plants to Barberry bushes by means of basidiospores derived

---

[1]As I do not regard a pycnidium of a Rust Fungus as a male organ, I have
avoided referring to it as a "spermogonium" producing "spermatia." The term
pycnidium has been employed by me for the "spermogonia" of the Lichenes, the
Pyrenomycetes, and the Uredinales. It seems probable that the "spermatia" of
the Ascomycetes and of the Uredinales originated from conidia. Those of many
of the Lichenes and non-lichenous Ascomycetes have been found to germinate. The
inability of the "spermatia" of certain Pyrenomycetes (Pleurage) and Discomycetes
(Sclerotinia) and of the Laboulbeniales and Uredinales to germinate may have
resulted from a physiological specialisation which, in the course of evolution,
appeared as a conidial mutation and was retained because it proved to be advantageous for promoting the initiation of the sexual process. Arthur's term *pycnium*
(producing *pycniospores*) is restricted to the Uredinales and has attained wide usage
among uredinologists (*cf.* J. C. Arthur, *The Plant Rusts*, New York, 1929, p. 9).

from teleutospores[1] and back again from Barberry bushes to graminaceous host-plants by means of its aecidiospores;[2] but he was unable
to show that the pycnidiospores (pycniospores, spermatia) play any
part in the life-history of the fungus.

In 1883 Ráthay,[3] in work summarized in this Chapter, stated:
that he had observed 135 species of insects visiting the pycnidia of
various Rust Fungi; that flies and ants feed upon the sweet fluid
exuded from the pycnidial ostioles; and that flies carry away the
pycnidiospores and deposit them elsewhere. However, like de Bary,
Ráthay was unable to show that the pycnidiospores perform any
function or are of any use to the fungi which produce them.

Finally, in 1927, the problem concerning the true significance of
the pycnidia was solved by Craigie. Craigie,[4] by means of experimental evidence, proved conclusively
that, when the pycnidial nectar containing pycnidiospores is mixed by
hand or by insects so that (+) pycnidiospores are taken to (−) pycnidia
and (−) pycnidiospores are taken to
(+) pycnidia, after two or three days
aecidia begin to develop on the under
sides of the rust pustules, thus indicating that the rust pustules have
become diploid and that nuclear
association between (+) and (−)
nuclei has taken place. In 1933,
Craigie[5] pointed out that between and
beyond the slenderly conical periphyses, there grow out from the
ostiole of a pycnidium into the drop
of nectar a number of *flexuous hyphae* with which pycnidiospores

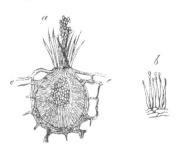

FIG. 1.—*Puccinia graminis*: a, pycnidium ("spermogonium") in
parenchyma of *Berberis vulgaris*; it has broken through the
epidermis *e e*; *b*, pycnidiosporophores bearing pycnidiospores
("sterigmata with young spermatia"). From de Bary's *Morph.
u. Phys. der Pilze*, 1866. Mag. *a*
200, *b* about 350.

[1]A. de Bary, "Neue Untersuchungen über die Uredineen, insbesondere die
Entwicklung der *Puccinia graminis* und den Zusammenhang derselben mit *Aecidium
Berberidis*. I," *Monatsber. K. Akad. d. Wiss.*, Berlin, 1865, pp. 15-49.

[2]A. de Bary, "Neue Untersuchungen über Uredineen. II," *ibid.*, 1866, pp. 205-215.

[3]E. Ráthay, "Untersuchungen über die Spermogonien der Rostpilze," *Denkschrift d. Kais. Akad. d. Wissensch.*, Wien, Bd. XLVI, 2 Abt., 1883, pp. 1-51.

[4]J. H. Craigie: "Discovery of the Function of the Pycnia of the Rust Fungi,"
Nature, Vol. CXX, 1927, pp. 765-767; "On the Occurrence of Pycnia and Aecia in
certain Rust Fungi," *Phytopathology*, Vol. XVIII, 1928, pp. 1005-1015; and "An
Experimental Investigation of Sex in the Rust Fungi," *Phytopathology*, Vol. XXI,
1931, pp. 1001-1040.

[5]J. H. Craigie, "Union of Pycniospores and Haploid Hyphae in *Puccinia
helianthi* Schw.," *Nature*, Vol. CXXXI, 1933, p. 25.

brought from a pycnidium of opposite sex unite, thereby making it possible for the nucleus of a pycnidiospore to pass through the passage made by the hyphal fusion and into the haploid mycelium contained within the pustule. It thus became clear: (1) that haploid mycelia in rust pustules of monobasidiosporal origin can be diploidised by nuclei derived from pycnidiospores of opposite sex; and (2) that the true significance of the visits of insects to the pustules lies in the fact that the insects convey (+) pycnidiospores to (−) flexuous hyphae and (−) pycnidiospores to (+) flexuous hyphae and so make it possible for flexuous hyphae and pycnidiospores of opposite sex to unite.

Craigie's discoveries are of classical importance in respect to the advance in our knowledge of the life-history of the Rust Fungi, and they have given a great stimulus to the experimental study of sex problems not only in the Uredinales but in the Ascomycetes. On this account, an attempt will now be made to depict the historical background which made them possible and to describe them more fully.

The Discovery of Pycnidia. —The fungal organs known as pycnidia (spermogonia) were discovered first of all in Lichens, then in the Rust Fungi and, finally, in various non-lichenous Ascomycetes.

The Pycnidia (Spermogonia) and Pycnidiospores (Spermatia) of the Lichens.— The pycnidia of the Lichens were seen by Dillenius[1] in 1741 and were described and discussed by Hedwig[2] in 1784. Hedwig regarded them as male organs, and this view was accepted by Tulasne[3] in 1851-1852.

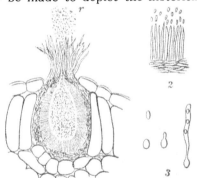

FIG. 2.—*Puccinia graminis*. No. 1, a pycnidium (Pyknide) in vertical section with periphyses at its ostiole and extruded pycnidiospores (Pyknosporen) at *r*. No. 2, a piece of the hymenium of No. 1, showing pycnidiosporophores and pycnidiospores. No. 3, germinating pycnidiospores, with three oil-drops in the longer germ-tube. ("Nach der Natur.") Magnification: No. 1, 150; No. 2, 250; No. 3, 360. From F. von Tavel's *Vergleich. Morph. d. Pilze* (1892).

[1] J. J. Dillenius, *Historia Muscorum*, Oxford, 1741. Pycnidia are represented in some of his drawings, but without comment.

[2] J. Hedwig, *Theoria generationis et fructificationis plantarum cryptogamicarum Linnei*, Petropoli, 1784, pp. 120-125.

[3] L.-R. Tulasne, "Note sur l'appareil reproducteur dans les Lichens et les Champignons," *Ann. Sci. Nat.*, Bot., Sèr. 3, T. XV, 1851, pp. 370-380; also "Mémoire pour servir à l'histoire organographique et physiologique des Lichens," *ibid.*, T. XVII, 1852, pp. 153-159 *et seq.*

The carpogonia of Lichens were discovered by Stahl[1] in 1874. In 1877[2] he described the *carpogonia* of collemaceous Lichens as consisting of a coiled *ascogonium* and of a *trichogyne* projecting above the surface of the thallus. He regarded a carpogonium as a female organ and a spermogonium as a male organ. His observations seemed to show that spermatia become attached to a trichogyne and that one of them fuses with it and thus initiates the sexual process which culminates in the production of an apothecium. Stahl's sexual theory was later supported by Baur,[3] Darbyshire,[4] Bachmann,[5] Sättler,[6] Kniep,[7] and others, but it was opposed by van Tieghem,[8] Möller,[9] Wainio,[10] Lindau,[11] the Moreaus,[12] and others. It will suffice to remark that the recent discoveries of the sexual process associated with the activity of spermatia in the Uredinales, the Pyrenomycetes, and the Discomycetes greatly strengthen the supposition that, in at least some of the Lichens, the spermatia do actually co-operate with the carpogonia and thus initiate a sexual process.

[1]E. Stahl, "Beiträge zur Entwicklungsgeschichte der Flechten (Vorl. Mitt.)," *Bot. Zeit.*, Bd. XXXII, 1874, pp. 177-180.

[2]E. Stahl, *Beiträge zur Entwicklungsgeschichte der Flechten*, Heft I, "Ueber die geschlechtliche Fortpflanzung der Collemaceen," Leipzig, 1877, pp. 1-51.

[3]E. Baur, "Zur Frage nach der Sexualität der Collemaceen," *Ber. der Deutschen Bot. Ges.*, Bd. XVI, 1898, pp. 363-367, Taf. XXIII.

[4]O. V. Darbyshire, "Ueber Apotheciumentwicklung der Flechte *Physcia pulverulenta*," *Jahrb. f. wiss. Bot.*, Bd. XXXIV, 1900, pp. 342-343.

[5]F. M. Bachmann, "A New Type of Spermogonium and Fertilization in Collema," *Annals of Botany*, Vol. XXVI, 1912, pp. 747-760.

[6]H. Sättler, "Untersuchungen und Erörterungen über die Oekologie und Phylogenie der Cladoniapodetien," *Hedwigia*, Bd. LIV, 1914, pp. 226-263, Taf. V-IX.

[7]H. Kniep, *Die Sexualität der niederen Pflanzen*, Jena, 1928.

[8]Ph. van Tieghem, *Traité de Botanique*, Paris, 1884, p. 1095.

[9]A. Möller: *Ueber die Cultur flechtenbildender Ascomyceten ohne Algen*, Münster i.W., 1887, pp. 1-52 (cited from A. L. Smith's *Lichens*, Cambridge, 1921, pp. 202-203); and "Ueber die sogenannten Spermatien der Ascomyceten," *Bot. Zeit.*, Bd. XLVI, 1888, pp. 421-425.

[10]E. A. Wainio, "Étude sur la classification naturelle et la morphologie des Lichens du Brésil," *Acta Soc. pro Fauna et Flora Fennica*, Vol. VII, Helsingforsiae, 1890, pp. XI-XII.

[11]G. Lindau, "Beiträge zur Kenntniss der Gattung Gyrophora," S. Schwendener's *Festschrift*, Berlin, 1899, pp. 19-36.

[12]F. et Mme Moreau, some ten cytological papers 1915-1932 of which here may be mentioned: (1) L'évolution nucléaire et les phénomènes de la sexualité chez les Lichens du genre Peltigera," *C.R.Ac. Sc.*, T. CLX, 1915, p. 526; (2) "Recherches sur les Lichens de la famille des Peltigéracées," *Ann. Sci. Nat.*, Bot., Sér. X, T. I, 1919, pp. 29-138; and (3) "La reproduction sexuelle chez les Lichens du genre *Collema* et la théorie de Stahl," *C.R.Ac. Sc.*, T. CLXXXII, 1926, p. 802.

The Pycnidia and Spermogonia of the Non-lichenous Discomycetes and Pyrenomycetes.—Conceptacles containing minute spore-like bodies, eventually found to be associated with the perfect stage of non-lichenous Discomycetes and Pyrenomycetes, attracted the attention of Persoon, Fries, Léveillé, and other mycologists early in the nineteenth century.

Shortly after 1850, Tulasne divided these conceptacles into *pycnidia* producing *stylospores* and *spermogonia* producing *spermatia*. The stylospores were regarded as asexual bodies which germinate and give rise to a mycelium, while the spermatia were presumed to be incapable of germination and to be male in function.

For some years after Tulasne had put forward his theory that certain spore-like bodies in the Discomycetes and Pyrenomycetes are spermatia, female organs that the supposedly male cells could fertilise were still unknown. Moreover, it was discovered in various species that spermatiiform bodies that had been regarded as spermatia do actually germinate. In 1866, therefore, de Bary, in the first edition of his text-book, regarded the significance of spermogonia and spermatia in the Discomycetes and Pyrenomycetes as uncertain.

Subsequently, support was given to the idea of the existence of true spermatia in the Discomycetes and Pyrenomycetes by Karsten's discovery of the relations of spermatia with the trichogynes of "female" organs in the Laboulbeniales (1869) and by Stahl's similar discovery in the Lichenes (1874 and 1877). At last, supposedly female organs with trichogynes were discovered in two genera of Pyrenomycetes: in Polystigma by Fisch (1882) and Frank (1883), and in Gnomonia by Frank (1886).

A strong stimulus to further investigation of the function of the so-called spermatia of Discomycetes and Pyrenomycetes was given in 1927 by Craigie's experimental proof that the pycnidiospores of the Uredinales are able to initiate the diploidisation process in haploid pustules.

The final proof that in some Discomycetes and Pyrenomycetes the "spermatia" (microconidia, phialospores) have a sexual function involving fusion with a trichogyne or other receptive structure was obtained experimentally: in the Discomycetes, by Drayton[1] in *Sclerotinia gladioli* (1932); and in the Pyrenomycetes, by B. O. Dodge[2] in

[1]F. L. Drayton, "The Sexual Function of the Microconidia in Certain Discomycetes," *Mycologia*, Vol. XXIV, 1932, pp. 345-348.

[2]B. O. Dodge, "The Non-sexual and Sexual Functions of Microconidia of Neurospora," *Bull. Torrey Bot. Club*, Vol. LIX, 1932, pp. 347-359, Plates XXIII and XXIV.

Neurospora sitophila and *N. tetrasperma* (1932), by Ames[1] in *Pleurage anserina* (1932), and by Zickler[2] in *Bombardia lunata* (1934). In 1936, Higgins[3] presented cytological evidence that in *Mycosphaerella tulipifera* "the spermatia function as male sexual elements in the development of the perithecia." Higgins shows a spermatium fused with the end of a trichogyne of a carpogonium, a nucleus of a spermatium some way down a trichogyne, and a nucleus of a spermatium associated with a nucleus of opposite sex at the base of a carpogonium. In 1939, for *Neurospora sitophila*, Bachus[4] found that the macroconidia fuse with branches of the trichogynes and that, soon after, the proto-perithecia[5] become rapidly transformed into perithecia. In 1941, he[6] further found that the "spermatia" (microconidia) of this fungus behave in the same way as the macroconidia, but even more efficiently: when microconidia were applied to the proto-perithecia, there was appreciable growth of the young fruiting bodies after 4-6 hours; whereas when macroconidia were applied, appreciable growth was not seen until after 6-8 hours.

Thus, since Craigie's discovery of the function of the pycnidio-spores of the Uredinales, a considerable amount of evidence has been accumulated in support of the view that, in at least some species of Discomycetes and Pyrenomycetes, certain spore-like bodies (spermatia, microconidia, phialospores) produced on the mycelium prior to the development of the perfect fruit-bodies have a sexual function.

The Antheridia and Spermatia (Antherozoids) of the Laboulbeniales.—The Laboulbeniales have organs which give rise to minute cells which have been called *spermatia* because they resemble the

[1]L. M. Ames: "An Hermaphroditic Self-sterile but Cross-fertile Condition in *Pleurage anserina*," *Bull. Torrey Bot. Club*, Vol. LIX, 1932, pp. 341-345; and "Hermaphroditism involving Self-sterility and Cross-fertility in the Ascomycete *Pleurage anserina*," *Mycologia*, Vol. XXVI, 1934, pp. 392-414.

[2]H. Zickler, "Genetische Untersuchungen an einem heterothallischen Askomyzeten (*Bombardia lunata* nov. spec.)," *Planta*, Bd. XXII, 1934, pp. 573-613, especially pp. 582-583.

[3]B. B. Higgins, "Morphology and Life History of some Ascomycetes with Special Reference to the Presence and Function of Spermatia. III," *American Journal of Botany*, Vol. XXIII, 1936, pp. 598-602, Text-figs. 1-12.

[4]M. P. Bachus, "The Mechanics of Conidial Fertilization in *Neurospora sitophila*," *Bull. Torrey Bot. Club*, Vol. LXVI, 1939, pp. 63-76.

[5]For the origin of the term *proto-perithecium*, *vide* A. H. R. Buller, "The Diploid Cell and the Diploidisation Process in Plants and Animals, with Special Reference to the Higher Fungi," *Botanical Review*, Vol. VII, 1941, p. 403.

[6]M. P. Bachus, "Spermatia versus Conidia as Fertilizing Agents in *Neurospora sitophila*," abstract of a paper given at a meeting of the National Academy of Sciences held at the University of Wisconsin, Oct. 13, 1941.

spermatia of the Red Sea-weeds in that they fuse with the trichogyne of a procarp of a perithecium, initiate a fertilisation process, and therefore act as male gametes; but the organ in which these spermatia are formed differs markedly in structure from the spermogonia or pycnidia of the Discomycetes and Pyrenomycetes and is generally known as an *antheridium*.[1]

The first note on the Laboulbeniales was published in 1850 by Rouget,[2] a French entomologist, and it treated of a species found on a beetle, *Brachinus crepitans*, near Dijon. Rouget could not decide whether the parasite was a plant or an animal. In 1869, Karsten[3] recognised the Laboulbeniales as fungi and perceived in them a highly developed type of sexuality including the existence of a trichogyne and its conjugation with one or more spermatia. In 1902, Thaxter[4] published the first part of his great monograph on the Laboulbeniaceae, and therein he described the male organs and the male elements as antheridia and antherozoids respectively.[5] Among his observations were the following. An antheridium consists of a single "antheridial cell" or of a group of such cells, and an antherozoid is a single naked or thin-walled cell.[6] In the aquatic genera Zodiomyces and Cerato-myces the antherozoids are *exogenous*, *i.e.* they are budded off from the anteridial branches or are produced as the result of the breaking up of slender antheridial branches into rods. In all the other genera, *e.g.* Laboulbenia and Stigmatomyces, the antherozoids are *endogenous*. In these genera the antheridia are either *simple* or *compound*.[7] A simple antheridium consists of a single cell, whereas a compound antheridium is made up of several simple antheridia which discharge their antherozoids into a common receptacle or neck. The simple antheridium is very constant in form: it has a basal somewhat inflated venter and a terminal, more slender, sub-cylindric neck, originally developed as a terminal outgrowth but, at maturity, becoming perfo-rate at its apex. The cavity of the venter is separated from the

[1]The terms spermatia and antheridia are employed by E. Gäumann in his *Vergleichende Morphologie der Pilze* (Jena, 1926, pp. 353-373) and by E. Gäumann and C. W. Dodge in their *Comparative Morphology of Fungi* (New York, 1928, pp. 364-395).

[2]A. Rouget, "Notice sur une production parasite observée sur le *Brachinus crepitans*," *Annales de la Société Entomologique de France*, T. VIII, 1850, pp. 21-58, Plate III, Figs. 1-7.

[3]H. Karsten, *Chemismus der Pflanzenzelle*, Wien, 1869, p. 78, Fig. 9.

[4]R. Thaxter, "Contribution towards a Monograph of the Laboulbeniaceae," *Memoirs of the American Academy of Arts and Sciences*, Vol. XII, 1902, pp. 187-429, Plates I-XXVI.

[5]*Ibid.*, pp. 207, 208.

[6]*Ibid.*, pp. 209-210.   [7]*Ibid.*, p. 211.

cavity of the neck by a perforated diaphragm. The contents of the venter, during its active period, pushes through the opening in the diaphragm into the cavity of the neck; and the portions thus extruded, when they have attained a certain definite size, become separated from the mass from which they were derived, assume the form of short cylindrical rods, pass upwards through the neck and, as antherozoids, make their exit from the antheridium through the terminal pore of the neck.[1] The discharge of the antherozoids from a simple antheridium is a slow process, not occurring more frequently than once every two or three hours; but from the secondary neck of a compound antheridium sometimes a dozen or more antherozoids make their exit within a few minutes after the plant has been mounted in water.[2] The antherozoids do not germinate;[3] but they readily adhere to a trichogyne, fuse with it,[4] and thus fertilise the carpogenic cell which subsequently produces the asci and the ascospores. The number of ascopores in an ascus is usually four.[5] Some genera of the Laboulbeniales are monoecious and others dioecious. "In the dioecious genera the male and female individuals are always in close proximity, their invariable association resulting from the fact that the spores always become attached to the host in pairs corresponding to those that are formed in the ascus and that, of any given spore pair, one member produces a male and the other produces a female."[6]

In the genera of Laboulbeniales with endogenous antherozoids, the antherozoids are carried to the trichogyne either by being discharged directly upon it or by floating to it passively through the water which is apt in the majority of instances to surround the individuals while their hosts are hiding in moist situations. In the genus Zodiomyces, which has exogenous antherozoids, the trichogyne always grows downwards as it develops and seems to seek the antherozoid which is almost invariably found attached to its tip; and it is only after contact with the antherozoid that the trichogyne turns upwards.[7] This seeking of the antherozoids by the trichogyne is a phenomenon which is not confined to Zodiomyces, for it has been observed: (1) in one of the Discomycetes, *Ascobolus carbonarius*,[8] where a long tri-

---

[1]*Ibid.*, pp. 211-212.   [2]*Ibid.*, p. 217.

[3]"Cultures of the antherozoids in water, continued for many days, have never shown any indication of an attempt at development." *Ibid.*, p. 217.

[4]*Ibid.*, Plate XXIII, Fig. 17. Here is shown a young perithecium of *Zodiomyces vorticellarius* with a trichogyne and conjugating antherozoid.

[5]*Ibid.*, p. 227.   [6]*Ibid.*, p. 216.   [7]*Ibid.*, p. 226.

[8]B. O. Dodge, "Methods of Culture and the Morphology of the Archicarp in Certain Species of Ascobolus," *Bull. Torrey Bot. Club*, Vol. XXXIX, 1912, pp. 173-176, Plate XII.

chogyne seeks out and coils round an antheridial conidium and (2) in the Lichen, *Collema pulposum*,[1] where the trichogyne grows through the thallus toward a group of spermatia which are borne on a hypha below the surface of the thallus, meets with a spermatium, and fuses with it.

Discovery of Pycnidia (Spermogonia) and Pycnidiospores (Spermatia) in the Uredinales.—In 1801, Persoon[2] remarked that in the spring the spots on Pear leaves, from which the aecidia of *Aecidium cancellatum*[3] subsequently grow out, are "dilute croceae et punctatae." There can be no doubt, therefore, that Persoon had noticed the pycnidia which on the upper surface of the pustules appear as dots, at first bright red and finally deep brown or black.[4]

Although Persoon was the first to notice the pycnidia of the Rust Fungi with the naked eye, it was Unger who first made an extended investigation of these organs and who must therefore be regarded as their real discoverer.

In 1833,[5] Unger described and illustrated pycnidia under the name of *Aecidiolum exanthematum*. He found them in the company of various species of Aecidium and Roestelia, and he regarded them as the fructification of a distinct fungus.

Unger's choice of the generic term *Aecidiolum* (= little Aecidium) was doubtless due to the fact that he thought that pycnidia, although much smaller than aecidia, have the same general structure. He[6] described the group of periphyses projecting around the mouth of a pycnidium as "der zerschlitzte ausgebreitete Balg," thereby indicating that he thought that a pycnidium, like an aecidium, is at first covered with a peridium that afterwards becomes slit radially into thin pieces

[1]F. M. Bachmann, "A New Type of Spermogonium and Fertilization in Collema," *Annals of Botany*, Vol. XXVI, 1912, pp. 747-760, and Plate LXIX.

[2]D. C. H. Persoon, *Synopsis Methodica Fungorum*, Gottingae, 1801, pp. 205-206.

[3]*Roestelia cancellata* (Pers.) Reb., now known as *Gymnosporangium sabinae* (Dicks.) Wint.

[4]J. E. Rebentisch, in 1804 (*Prodromus Florae Neomarchicae*, Berolini, pp. 350-351), on account of the peculiar structure of the aecidial peridium, placed *Aecidium cancellatum* in a new genus *Roestelia* (named after his friend Roestel). He cited Persoon's description of the early appearance of the pustules, but added nothing to it. De Bary (*Untersuchungen über die Brandpilze*, Berlin, 1853, p. 58) and, following him, Plowright (*A Monograph of the British Uredineae and Ustilagineae*, London, 1889, p. 16) were therefore in error when they cited Rebentisch, instead of Persoon, as being the first to notice the existence of the pycnidia of *Roestelia cancellata*.

[5]F. Unger, *Die Exantheme der Pflanzen und einige mit diesen verwandte Krankheiten*, Wien, 1833, pp. 300-301.

[6]*Ibid.*, p. 417.

and then opens out; and he referred to the pycnidiospores as "Sporidien."

Unger supposed that the pycnidia of a Rust Fungus, like other "entophytes," are produced from the diseased tissues of the host-plant as an eruptive excrescence or *exanthema*, and it was on this account that his *Aecidiolum exanthematum* received its specific name.

Unger's drawings of pycnidia,[1] hand-coloured, represent: (1) the underside of a pustule of *Aecidium bifrons* Lam., on a leaf of *Aconitum koelleanum* Reb., with numerous yellow aecidia surrounding a small group of red pycnidia; (2) a lateral view of a pycnidium on *Rhamnus catharticus*, seen from above the epidermis of a leaf, showing an upright circular fringe of long red periphyses projecting through a stoma and surrounding a number of pale concatenate pycnidiospores; and (3), on a leaf of *Ranunculus ficaria*, seen from above, a very young pycnidium beginning to grow through a stoma and a mature pycnidium with reddish periphyses stellately disposed and enclosing a dense mass of pale pycnidiospores. Unger had noticed that the conidiophores of some other leaf fungi protrude through the clefts of stomata, but he was mistaken in supposing that the periphyses of Rust pycnidia protrude in this way. In reality, a pycnidium of a Puccinia or of a Uromyces, as it opens at the surface of the host-plant, always compresses and ruptures one or more of the ordinary epidermal cells.

A Century of Observation and Speculation.—After Unger's discovery of the existence of pycnidia in the Rust Fungi in 1833 up to the publication of the results of Craigie's investigations in 1927, Rust pycnidia were re-examined from time to time by various botanists, and some new facts concerning their structure and development were gradually brought to light; but, although speculations were freely put forward as to the physiological significance of Rust pycnidia, the part played by these organs in the life-history of Rust Fungi remained a tantalising mystery.

Meyen,[2] in 1841, in his *Pflanzen-Pathologie* described the spatial relations of the pycnidia and aecidia in Rust species attacking the leaves of *Rumex acetosa*, *Rhamnus frangula*, *Ribes grossularia*, *Urtica dioica*, *Sium falcaria*, *Euphorbia cyparissias*, *Ranunculus repens*, and the Pear (attacked by *Roestelia cancellata*); and he gave an account of the development of a pycnidium. This account, admirable in itself but unaccompanied by any illustrations, included: (1) the formation

---

[1]*Ibid.*, Tab. III, Figs. 17-19.

[2]F. J. F. Meyen, *Pflanzen-Pathologie* (passed through the press by C. G. Nees v. Esenbeck after the author's death), Berlin, 1841, pp. 142-147, 149.

of a compressed, lens-shaped mass underneath the epidermis; (2) the pushing upwards of the epidermis to form a cone; (3) the pressing upwards of the outermost layer of the mesophyll (Diachym), so as to make therein a kettle-shaped (kesselförmig) impression; (4) the formation of linear pointed cells all directed toward the middle-point of the pycnidium; (5) the coming into existence in the middle of a pycnidium of a fine-grained, slimy, and reddish-yellow fluid; (6) the rupture of the epidermis; (7) the protrusion of the periphyses (nadelförmige zellen) through the opening; (8) the spreading outwards of the periphyses to form a sort of cup; (9) the passing outwards into the cup of the innumerable tiny particles (Moleküle oder Bläschen = pycnidiospores) which had been formed within the pycnidium; (10) the enclosure of these particles in a sugary sap; and (11) the fact that the particles often exhibit a molecular movement. Meyen was the first to mention slime and sugar as associated with the pycnidiospores of Rust Fungi, and he remarked that when a large number of pycnidia open at the same time "the diseased portion of the leaf is sometimes completely covered with sugary sap." Meyen does not tell us how he detected the sugar, but it was probably by tasting the pycnidial nectar with his tongue.

Meyen[1] followed Unger in supposing that Smut Fungi, Rust Fungi, *Cystopus candidus*, etc., are true entophytes which arise spontaneously between the cells of the host-plant, but he had something new to say concerning the significance of pycnidia. Having regard to the fact that, in a pustule of an Aecidium, the pycnidia are developed before the aecidia and, spatially, the pycnidia and the aecidia are closely associated, he[2] suggested that pycnidia and aecidia which occur together are respectively the male and female organs of one and the same fungus species. He referred to pycnidia as "männliche Aecidien-Pusteln" and to aecidia as "weibliche Aecidien-Pusteln," but added that he was by no means of the opinion that the pycnidia take part in any real fertilisation process.

In 1846, in the second edition of his text-book of Botany, Schleiden,[3] who, like Unger and Meyen, was an exanthematist, scornfully rejected

[1]*Ibid.*, pp. 98-150. On p. 150 he says: "The aecidia, also, are true entophytes, as may be deduced from observations on their first steps in development; and there can be no thought of an origin by reproduction from their spore-like vesicles (sporenartigen Bläschen)."

[2]*Ibid.*, pp. 142-147, 149.

[3]M. J. Schleiden, *Grundzüge der wissenschaftlichen Botanik*, ed. II, Leipzig, Bd. II, 1846, pp. 41-42.

Meyen's supposition that the pycnidia of the Rust Fungi are male organs. Says Schleiden: "As to Aecidium-anthers, an exanthema described by Unger which often appears along with aecidial eruptions, Meyen holds that a detailed investigation of this structure, as well as its time and space relations, *compels* him to regard it as a male aecidium-plant, although it can be proved by observation that there can be here no question of a real fertilisation. Anthers must indeed have become a fixed idea with Meyen when, in spite of this, he pronounces these structures to be anthers.[1] In the facts there is not only nothing which forces us to conclude that, not even an indication of the possibility that, *Aecidiolum exanthematum* Ung. (which always develops earlier than aecidia often on leaves whose *other* side forms aecidia later or not at all) stands in any other organic relation to aecidia than does *Acne punctata* to *Acne rosacea* or does one to the other of any other two human skin-diseases which often occur contemporaneously. The fanciful medical men[2] who explain a disease as an independent organism, following this analogy, have a wide field in which to look for males and females among the various pocks, pustules, and vesicles."[3]

At the present day, a century after Schleiden amused himself by writing the passage just quoted, we know, as a result of Craigie's work, that the pycnidia of the Rust Fungi do actually function in a sexual manner; and, although there are good reasons for not calling pycnidia male organs, it is clear that Meyen's views on the significance of Rust pycnidia were far nearer the truth than Schleiden's. Thus has Time brought about one of its revenges.

Tulasne,[4] in 1851, gave to certain organs produced by collemaceous Lichens the name "spermogonies," and he called the minute cells abstricted acropetally from short filaments lining the cavity of a spermogonie "spermaties"; and, in 1852, he[5] gave the Latin form of these terms as respectively *spermogonia* and *spermatia*. Having noted the striking resemblance between the pycnidia of the Rust

---

[1]It was a little unfair of Schleiden to say that Meyen had called pycnidia *anthers*. Meyen's term was "männliche Aecidienpusteln."

[2]Meyen practised medicine before becoming a botanist.

[3]We now know that a considerable number of skin diseases are actually caused by bacteria, yeasts, and fungus mycelia. There is the possibility that certain dermatophytes may cause both (+) and (−) infections.

[4]L.-R. Tulasne: "Note sur l'appareil reproducteur dans les Lichens et les Champignons," *Ann. Sci. Nat.*, Bot., Sér. 3, T. XV, 1851, pp. 372-373; also *Compt. Rend. Acad. Sci. Paris*, T. XXXII, 1851, pp. 427-430, 470-475.

[5]L.-R. Tulasne, "Mémoire pour servir à l'histoire organographique et physiologique des Lichens," *Ann. Sci. Nat.*, Bot., Sér. 3, T. XVII, 1852, p. 157.

Fungi and the spermogonia of the Lichens, he called the pycnidia "*spermogonies*."[1] He was inclined to believe that spermogonia are male organs[2] and, in 1854, he[3] referred to the pycnidiospores of the Rust Fungi as "spermaties" and gave further support to the idea that these cells are used in a fertilisation process.

De Bary,[4] in 1853, gave a detailed account of the microscopic structure of the pycnidia of the Rust Fungi, thus extending the observations of Unger and Meyen, and he illustrated the pycnidia of *Uredo suaveolens* (*Puccinia suaveolens*), *Aecidium euphorbiae* .(*Uromyces pisi*). *A. berberidis* (*Puccinia graminis*), and *A. grossulariae* (*Puccinia caricis*). He remarked on the close resemblance between the pycnidia of the Rust Fungi and the corresponding organs in species of the Lichen-genera Lichina, Urceolaria, Peltigera, and Pertusaria and adopted for the Rust Fungi the terms[5] *spermogonia*, *spermatia*, and *sterigmata*[6] which had been introduced by Tulasne for the Lichens; and he called the periphyses of the pycnidia of the Rust Fungi "*Paraphysen*,"[7] *i.e.* "threads (Fäden) which are present in an organ producing reproductive cells but which take no direct part in producing the cells."

De Bary[8] demonstrated that the pycnidia and the aecidia of a Rust pustule are produced on one and the same mycelium. He also observed that pycnidiospores are successively abstricted from the end of each pycnidiosporophore and are embedded in jelly which swells up and so enables the pycnidiospores to escape *en masse* through the mouth of the pycnidium;[9] and he supported Meyen and Tulasne

---

[1]L.-R. Tulasne, *Ann. Sci. Nat.*, 1851 (*vide supra*), p. 377.

[2]*Ibid.*, p. 374.

[3]L.-R. Tulasne, "Second Mémoire sur les Urédinées et les Ustilaginées," *Ann. Sci. Nat.*, Bot., Sér. 4, T. II, 1854, pp. 113-123.

[4]A. de Bary, *Untersuchungen über Brandpilze und die durch sie verursachten Krankheiten der Pflanzen*, Berlin, 1853, pp. 55-65.

[5]*Ibid.*, p. 62.

[6]L.-R. Tulasne, "Mémoire pour servir à l'histoire organographique et physiologique des Lichens," *Ann. Sci. Nat.*, Bot., Sér. 3, T. XVII, 1852, p. 160. Tulasne refers to the hyphae from which the "spermatia" of the Lichens are abstracted as "Ces filaments, basides ou *stérigmates* (*männliche Prophysen Bayrhoff*)" and in a footnote explains that the term *sterigma*, which means a *support*, although already used by Corda for the pedicels of the spores of the Agaricaceae, can, nevertheless, be well made use of for the pedicels of the spermatia of Lichens and, in any case, is preferable to Bayrhoff's *männliche Prophysen*. Hereafter, for the pedicels of the pycnidiospores of the Rust Fungi I shall employ the term *pycnidiosporophores*.

[7]A. de Bary, *loc. cit.*, p. 62.

[8]*Ibid.*, p. 78, Taf. III, Fig. 2, and Taf. V, Figs. 1-4.

[9]*Ibid.*, p. 60.

in their supposition that the irregular oscillations of pycnidiospores immersed in water are due to Brownian movement.[1]

The theoretical significance of pycnidia was discussed by de Bary[2] at length. Having found that pycnidia and aecidia are produced on a common mycelium he rejected the view, held by Unger and later by Schleiden,[3] that those organs are exanthemata. He admitted the possibility that the pycnidia of the Rust Fungi might be organs producing a second kind of spore differing from aecidiospores, the Uredinales thus offering an analogy with certain Algae which produce more kinds of spores than one and with certain Lichens which, as Tulasne had shown, produce not only spermatia and ascospores but also stylospores; but he was inclined to regard pycnidia as male sexual organs. He said that, in the Rust Fungi, in the absence of archegonia, the fertilisation process must differ much from that of Mosses and Ferns but that, with favourable material and more investigators sharing in the work, the problem of the relation of pycnidia to spore-producing organs, which was of the greatest botanical importance, might one day be solved by experiment.

In favour of the idea that the pycnidia of the Rust Fungi are male sexual organs, de Bary adduced the following arguments: (1) the pycnidia not only accompany aecidia but also are their forerunners; (2) the pycnidiospores are spontaneously extruded from the pycnidia before the aecidiospores are formed; (3) the pycnidiospores are poured out over the surface of the leaf from which position they might easily act upon the mycelial hyphae within the mesophyll; (4) the pycnidiospores, shortly after their extrusion, disintegrate[4]—they dissolve and become distributed in the mucilage which encloses them—from which fact the conclusion may be drawn that they do not germinate; and (5) the pycnidiospores are analogous with the spermatozoa of animals and the spermatozoids of Algae and Ferns in that they are produced in special organs, are spontaneously set free and, as shown by their taking on a bright red colour when reacting with sugar and sulphuric acid, are made up of protein compounds.

---

[1]*Ibid.*, pp. 60-62.   [2]*Ibid.*, pp. 78-82.

[3]M. J. Schleiden, *Grundzüge der Wissenschaftlichen Botanik*, ed. II, Leipzig, Bd. II, 1846, pp. 41-42.

[4]Observations on *Puccinia graminis, P. helianthi*, etc., have taught me that the pycnidiospores of the Rust Fungi do not break down and disappear shortly after their extrusion from a pycnidium but persist in good condition in the pycnidial nectar for upwards of a week. However, it appears that, under natural conditions, the vast majority of the pycnidiospores, *i.e.* all those which fail to meet with and fuse with flexuous hyphae of opposite sex, in the end die and disintegrate without ever having shown any signs of germination.

De Bary,[1] in 1884, in the second edition of his text-book on fungi, still inclined to the view that the spermogonia of the Rust Fungi are male organs producing spermatia that are incapable of germination; but he pointed out that in some Uredinales pycnidia (spermogonia) are unaccompanied by aecidia and that in other Uredinales aecidia are unaccompanied by pycnidia and he concluded that, for the time being, pycnidia must be regarded as organs of doubtful physiological significance.

Cornu,[2] in 1876, stated that the pycnidiospores of the Uredineae "peuvent émettre des sporidies secondaires" and, assuming that he had really seen Rust pycnidiospores germinating, he argued that these cells are simple conidia and not fertilising agents as Tulasne had supposed. But Cornu published no details concerning his experiments, and his statement suggests that his germinating pycnidiospores were nothing more than wild yeast-plants which, unrealised by him, had invaded his cultures.

In 1882, Ráthay[3] observed that insects are attracted to Rust pycnidia and carry away pycnidiospores with them. He therefore supported the idea that the pycnidia are male organs and he suggested that the insects facilitate a fertilisation process just as they do in many Phanerogams.

Thaxter has the distinction of being the first botanist to attempt to use the pycnidiospores of the Rust Fungi as fertilising agents. Thaxter was making cultures of various species of Gymnosporangium to find out their relation with Roesteliae. He noted that, on a host-leaf, the pycnidia are produced on the upper side of each pustule and the aecidia on the lower side; and, in 1887, he[4] described his experiment with pycnidiospores as follows: "In view of the theory that the

[1]A. de Bary, *Vergleichende Morphologie und Biologie der Pilze, Mycetozoen und Bacterien*, Leipzig, 1884, pp. 299-300.

[2]M. Cornu, "Où doit-on chercher les organs fécondateurs chez les Urédinées et Ustilaginées?" *Bull. Soc. Bot. France*, T. XXIII, 1876, pp. 120-121. Cornu here says that in a Mémoire submitted by himself and his friend M. E. Roze to the Academie des Sciences in June, 1873, he had given some details concerning the germination of the spermatia of the Uredineae. The Mémoire ("Étude de la fécondation dans la classe des Champignons," *Compt. Rend.*, T. LXXX, 1875, pp. 1464-1468) was reported upon by Brongniart and was awarded a prize, but it was never published.

[3]E. Ráthay, "Untersuchungen über die Spermogonien der Rostpilze," *Denkschr. K. Akad. Wiss. Wien*, Bd. XLVI, 1882, pp. 1-51. A more detailed account of Ráthay's observations is given in the next Section.

[4]Roland Thaxter, "On Certain Cultures of Gymnosporangium with Notes on their Rostelliae," *Proc. American Acad. Arts and Sciences*, Vol. XXII, 1887, pp. 261-262.

spermatia are sexual in function, and fertilize a female organ, the *trichogyne*, which subsequently gives rise to the aecidium, a 'fertilization' was attempted by collecting the exudations from the spermogonia in a drop of water and painting them upon the under side of different leaves in plants infected by the same species. This was carefully followed up with *Gymnosporangium globosum*, without resulting in the production of any aecidia, and in two instances where aecidia were obtained (*G. biseptatum* and *G. macropus*) no such fertilization was attempted, thus giving no definite result for or against the sexual theory." To this he added: "It should be mentioned that small flies were repeatedly observed feeding on the secretions of the spermogonia; yet as these are usually confined to the upper surface of the leaves, it is difficult to understand the agency of such insects in fertilizing a female organ borne supposedly on the under side." We now know that the proto-aecidia of the Rust Fungi do not have any trichogynes protruding from the lower surface of the host-leaf and that the flexuous hyphae with which pycnidiospores fuse project from the mouths of the pycnidia on the *upper* surface. Thaxter's ingenious experiment was therefore bound to end without throwing any light upon the function of the pycnidiospores.

In 1888 Massee,[1] from investigations made on *Puccinia poae* (on *Ranunculus ficaria*), concluded that the aecidium comes into existence as the result of the fertilization of an oogonium by an antheridium; but this was not confirmed by later workers.

In 1889 Plowright,[2] in the belief that the pycnidiospores of the Rust Fungi are conidia, employed pycnidiospores as inoculum and made a number of attempts to infect host-plants therewith; but all in vain, for his experiments gave no positive results. Plowright[3] also placed pycnidiospores in sugar solutions and thought that he saw them germinate there; but an examination of his illustrations strongly suggests that the budding which he observed was due to contamination of his cultures by wild yeasts.

Brefeld, who rejected the idea of sexuality in the Higher Fungi,[4] stated in 1891 that he had been able to observe the spermatia of the Uredineae germinating. Said he:[5] "In a culture medium, the sper-

---

[1] G. Massee, "On the Presence of Sexual Organs in Aecidium," *Annals of Botany*, Vol. II, 1888, pp. 47-51, Plate IV.

[2] C. B. Plowright, *A Monograph of the British Uredineae and Ustilagineae*, London, 1889, p. 20.

[3] *Ibid.*, pp. 14-16, Plate I, Figs. 8-16.

[4] O. Brefeld, *Untersuchungen über Pilze*, Bd. IX, Münster i. W., 1891, Vorrede, pp. IV-V, also pp. 14-20.   [5] *Ibid.*, p. 28.

matia of *Puccinia graminis* for example, like other conidia, swell up considerably and germinate. That from them no large mycelium is produced is due to the fact that, owing to their mode of life, it is scarcely possible to obtain pure spore-material and therefore the cultures are always disturbed by foreign fungi. A culture of *Uromyces pisi* gave a similar result: after lying in the culture medium for several days, they increased in size and put out a germ-tube which was at first very delicate and thin, and then became thicker and permitted one to recognise in their contents the orange-red drops characteristic of the hyphae of the Rust Fungi. Also the spermatia of *Puccinia tragopogonis* and of *P. coronata* germinated in a similar manner."

Brefeld's observations on the germination of the spermatia of the Rust Fungi as described by him in the preceding paragraph are open to grave objections. Firstly, he admits that all his cultures were contaminated. Secondly, he gave no precise details as to how he made his observations, so that there is no evidence that he watched particular spermatia and saw them putting out germ-tubes and growing from day to day. And, thirdly, he never elaborated the very brief account of his observations or published any illustrations in support of them. Moreover, subsequent investigations have not confirmed Brefeld's statement that in *Puccinia graminis* the spermatia, in a culture medium, "like other conidia, swell up considerably and germinate." Much as I have admired the work of Brefeld, so extensive and in general so accurate and so well supported with detailed and excellent illustrations, I feel that his observations on the germination of the spermatia of the Rust Fungi are unsatisfactory and unconvincing.[1]

Brefeld[2] in 1891, on the supposition that he had actually observed the pycnidiospores of several Rust species germinating, opposed the idea that the pycnidiospores of the Rust Fungi are male sexual cells and favoured the view that the pycnidiospores are conidia and that the pycnidia should be regarded not as spermogonia but as asexual reproductive organs.

If Brefeld's conclusion that the pycnidiospores of the Rust Fungi are ordinary conidia were true, then, under natural conditions in the open, we ought to find evidence that the pycnidiospores germinate on

[1]*Cf.* Chapter IV, pp. 229-233. Brefeld confined his attention to the fungi which can be grown on artificial culture media, but avoided the obligate parasites, such as Peronospora, the Erysiphaceae, and the Uredinales. It was for this reason that he added so much to our knowledge of the Smut Fungi but nothing to our knowledge of the Rust Fungi.

[2]O. Brefeld, *loc. cit.*, pp. 26-28.

leaves, etc., and so initiate the formation of new rust pustules; but﹐ during the last forty years, despite the numerous observations that have been made in the field on the life-histories of Rust Fungi by Klebahn and others, no such evidence has been forthcoming.[1]

In 1892 von Tavel, one of Brefeld's pupils, published a text-book on the comparative morphology of Fungi in which he reflected Brefeld's views and reproduced many of Brefeld's illustrations.   In his preface he remarks that sexuality in the Higher Fungi has been shown to be non-existent (ist die Sexualität als nicht bestehend erwiesen) and, in treating of the Uredineae, he gives the erroneous impression that the pycnidiospores germinate without any difficulty. He[2] says: "In culture media the pycnospores in the few cases investigated, after swelling, send out a thread-like germ-tube concerning whose further development and power of infection nothing is known."   This statement is accompanied by an original diagrammatic illustration representing the supposed germination of a pycnidiospore of *Puccinia graminis* (Fig. 2).   Von Tavel's illustration, in the absence of detailed information regarding the conditions under which it was made, can be regarded not as proof, but only as an expression of opinion, that the pycnidiospores of the Rust Fungi germinate.   In 1895, it was reproduced by Prillieux[3] and thus given a wider circulation.

In 1893, Carleton,[4] who investigated the germination of various spore-forms of the Uredineae, said: "In only one instance have I been able to grow the spermatia.   The spermatia of *Uredo Caeoma-nitens* Schwein, budded sparingly on May 31, 1893, after twenty-four hours in a dilute solution of honey, but would not germinate in water.   This is a confirmation of the results obtained by Cornu and Plowright." Carleton, like Cornu and Plowright, worked with impure cultures and the budding which he observed in his sugary solutions was in all probability due to the presence of wild yeasts.

In 1896 Sappin-Trouffy[5] stated that, after several unsuccessful attempts, he had succeeded in germinating the pycnidiospores of *Puccinia rubigo-vera*.   He placed the pycnidiospores in cell-sap which had been pressed out of the host-plant; and, 24 hours later, he fixed

[1]*Cf.* H. Klebahn, *Die wirtswechselnden Rostpilze*, Berlin, 1904, p. 40.

[2]F. von Tavel, *Vergleichende Morphologie der Pilze*, Jena, 1892, p. 127 and Fig. 58.

[3]E. Prillieux, *Maladies des plantes agricoles*, Paris, T. I, 1895, p. 218, Fig. 82, B.

[4]M. A. Carleton, "Studies in the Biology of the Uredineae. I. Notes on Germination," *The Botanical Gazette*, Vol. XVIII, 1893, p. 451.

[5]P. Sappin-Trouffy, "Recherches histologiques sur la famille des Urédinées," *Le Botaniste*, Sér. 5, 1896, pp. 105-108, Fig. 23.

them in alcohol, stained them with haematoxylin, and examined them under the microscope.   On his slides he then found what he supposed were stages in the budding of the pycnidiospores; and, in describing the structures which he had seen, he said: "This process of germination is identical with that of yeasts."   In one preparation he saw what he thought was a pycnidiospore which had produced a short germ-tube. Sappin-Trouffy took no precautions against his cultures becoming contaminated, and he studied *post-mortem* preparations only.   In the light of our present knowledge of the wide distribution of wild yeasts under natural conditions, Sappin-Trouffy's evidence that the pycnidiospores of *P. rubigo-vera* germinate is valueless.

In 1903, Carleton[1] returned to the question of the germination of Rust pycnidiospores and said: "Until recent years it was not supposed that the spermatia produced regular germ-tubes, but that the germination is always a process of budding. Dr. N. A. Cobb and the writer have shown, however, that ordinary germ-tubes are produced in the germination of the spores as in the other spore forms.   These observations make it probable that the so-called budding process simply represents instances where the germ-tubes have soon ceased their growth in length and, swelling somewhat in the middle, have given themselves the form of buds.   However, it is possible that actually budding occurs.   Among the species that have been found by the writer to be particularly good for studying the spermatia are: the rust of the Blackberry, *Aecidium oenotherae* Mont. on *Onagra biennis*, the ordinary Apple-leaf rust, and the Barberry rust.   In the first-named species, especially, the spermatia are comparatively large and are readily germinated. Spermatia, though germinating readily in water, will be found to do much better in a rather dilute sugar solution or perhaps still better in a solution of honey."   Carleton worked in the days before sterilisation methods had been generally introduced. There is no evidence that he sterilised his glass vessels, his water, or his solutions of sugar and honey, and it is certain that the nectar which he removed from pycnidial pustules must have been very often, or perhaps always, contaminated with the spores of other fungi.   Moreover, *Tuberculina persicina* is a very frequent parasite of Rust pycnidia and aecidia, and it is possible that spores of this species may have been introduced by Carleton into his cultures and have there been seen germinating.   As evidence that the spermatia of the Rust Fungi can germinate in water or in culture media, Carleton's observations, like

[1]M. A. Carleton, "Culture Methods with Uredineae," *Journal of Applied Microscopy and Laboratory Methods*, Vol. VI, 1903, pp. 2110-2111.

those of Cornu, Plowright, Brefeld, and Sappin-Trouffy, are unconvincing.[1]

Klebahn, taking advantage of the fact that *Cronartium ribicola* produces nectar containing pycnidiospores very freely, in 1889,[2] 1897,[3] and 1898[4] tried to infect healthy twigs of the White Pine (*Pinus strobus*) by applying the nectar to them, but all these experiments yielded negative results.

Repeating Thaxter's experiment, but using *Puccinia graminis* instead of *Gymnosporangium globosum*, Klebahn[5] spread pycnidiospores on the under side of pustules where aecidia were expected to appear, but apparently without any effect on the production of these organs. In 1904, he therefore concluded that "a fertilising action of the spermatia on the young aecidium-bed does not take place."

Summing up his discussion of the problem of Rust pycnidia, Klebahn[6] said: (1) there is nothing left but to look upon the spermogonia as organs which at the present time have no significance for the Rust Fungi; (2) the spermatia, formerly, may have been true conidia or have had some sexual function, but these suppositions are unsatisfactory because the spermogonia do not give the impression of being reduced organs; and (3), as a large amount of material is used up in forming spermogonia, it is difficult to accept the view that spermogonia are functionless.

In 1904 Blackman,[7] as a result of cytological observations made on *Phragmidium violaceum*, concluded that, in the spore-bed of the aecidium of this species, the nucleus of one cell migrates through a pore formed in the cell-wall into a neighbouring cell; and he interpreted the phenomenon of migration as the beginning of a sexual act, the association of two nuclei as a conjugate pair, the descendents of which are destined to fuse together in the cells of the teleutospores. There-

---

[1]For further critical remarks in respect to the question whether or not the pycnidiospores of the Rust Fungi are capable of independent germination, *vide infra*, Chapter IV.

[2]H. Klebahn, "Ueber die Formen und den Wirthswechsel der Blasenroste der Kiefern," *Ber. d. D. Bot. Gesellschaft*, 1890, p. 63.

[3]H. Klebahn, "Kulturversuche mit heteröcischen Rostpilzen," Bericht VII, *Zeitschrift f. Pflanzenkrankheiten*, Bd. IX, 1899, pp. 16-17.

[4]H. Klebahn, "Kulturversuche mit Rostpilzen," Bericht IX, *Jahrbücher f. wiss. Bot.*, Bd. XXXV, 1900, p. 694. For Klebahn's final statement that experiments with the pycnidiospores of *Cronartium ribicola* had all failed *vide* his *Die wirtswechselnden Rostpilze*, 1904, p. 387.

[5]H. Klebahn, *Die wirtswechselnden Rostpilze*, Berlin, 1904, p. 197.

[6]*Ibid.*, pp. 199-200.

[7]V. H. Blackman, "On the Fertilization, Alternation of Generations and General Cytology of the Uredineae," *Annals of Botany*, Vol. XVIII, 1904, pp. 323-373.

upon, Blackman revived the theory of the sexual nature of the pycnidia. He supposed that the pycnidiospores were once functional, like
the spermatia of the Red Seaweeds, and that they united with trichogynes which once protruded from the leaves and stems of the
host-plants; and he further supposed that the trichogynes of the Rust
Fungi disappeared in the course of evolution and that, ever since, the
pycnidiospores have not been able to perform their sexual function.
In support of the view that the pycnidiospores of the Rust Fungi are
functionless male gametes, he pointed out that pycnidiospores:
(1) have dense nuclei; (2) contain but a small amount of cytoplasm;
(3) have thin cell-walls; (4) exhibit but feeble power of vegetative
germination; and (5) are unable to bring about infection of host-plants.

Christman[1] in 1905 discovered that, in the spore-bed of the aecidia
of *Phragmidium speciosum* and of *Uromyces caladii*, the oblong cells
standing side by side in a layer fuse in pairs and so give rise to binucleate fusion cells each of which then forms a chain of aecidiospores. He
regarded the cells which fuse together as gametes and the sexual
process in the Rusts as a conjugation of sexual cells; but this view, as
he pointed out, left the function of the pycnidia and pycnidiospores
unexplained.   In 1907, he[2] questioned the supposition that the Rusts
resemble the Red Algae and that the pycnidiospores of the former
correspond to the male cells of the latter, and he inclined to the view
that Rust pycnidiospores represent asexual spores which formerly
reproduced the gametophytic generation.

McAlpine[3] in 1906, in his book on the Rusts of Australia, regarded
the pycnidiospores of the Rust Fungi as isolated structures, once
functional in some way, but now functionless.

Jaczewski[4] in 1910, following in the wake of Cornu, Plowright,
Brefeld, and Sappin-Trouffy, once again supported the idea that the
pycnidiospores of the Rust Fungi are conidia which germinate. He
said that the "stylospores" of *Puccinia graminis* germinate in water
very occasionally and that, when these spores are placed in the sweet
fluid which exudes from the pycnidia, germination—the appearance
of a small germ-tube put out from the "conidium"—can usually be
observed after 24 hours.   He also said that the same kind of germi-

[1]A. H. Christman, "Sexual Reproduction in the Rusts," *Botanical Gazette*,
Vol. XXXIX, 1905, pp. 269-274.

[2]A. H. Christman, "The Alternation of Generations and the Morphology of the
Spore Forms in the Rusts," *ibid.*, Vol. XLIV, 1907, p. 93.

[3]D. McAlpine, *The Rusts of Australia*, Melbourne, 1906, pp. 13-15.

[4]A. von Jaczewski, "Studien über das Verhalten des Schwarzrostes des Getreides
in Russland," *Zeitschrift f. Pflanzenkrankheiten*, Bd. XX, 1910, pp. 321-323.

nation takes place in artificial media such as gelatine, agar-agar, and plum-syrup, but admitted that he had not been able to follow the further development of the "stylospores" after they had germinated. In an illustration he shows eight pycnidiospores which he supposed had germinated. In five of these the "germ-tubes" are shorter than, or about the same length as, the spore and in the other three 1.5— 3 times the length of the spore. On this basis, Jaczewski suggested that the pycnidiospores of *P. graminis* are swallowed by insects and, after passing through the alimentary canal, germinate readily and infect new host-plants with the production of new pycnidia and aecidia.

Jaczewski's observations do not warrant the belief that he really saw the pycnidiospores of *Puccinia graminis* germinating. He did not observe stages in the germination of any individual pycnidiospore; he did not witness the process of growth; but, after 24 hours, on examining his cultures, he found what he thought were pycnidiospores which had already germinated and whose germ-tubes had ceased to elongate. Apparently, it never occurred to him that, among the pycnidiospores of normal size and form, there might be abnormal ones two or three times as long as the normal and of a somewhat different shape, or that the nectar removed from the rust pustules might contain foreign fungus spores which had fallen into it from the air or had been carried to it by insects. His statement that, when the pycnidiospores of *P. graminis* are placed in their own nectar, their germination can usually be observed after 24 hours is open to grave doubt; for, if the pycnidiospores in question really do germinate in their own nectar as he thought, one should be able to find plenty of germinated pycnidiospores in any nectar which has crowned the pycnidia for 3 - 10 days and this, as my own examination of such nectar in the Rust Laboratory at Winnipeg has taught me, is not so. Moreover, Hanna,[1] using modern technical methods, has failed to confirm Jaczewski's results with artificial media; for, although he kept the pycnidiospores of *P. graminis* in sucrose solution, prune-agar, etc., for nine days, they showed no signs of germination. Finally, it may be said that Jaczewski's suggestion that the pycnidiospores of *P. graminis* germinate readily after passing through the alimentary canal of an insect and then infect new host-plants is not supported by any positive evidence, while against it may be cited the fact that, as I have observed in greenhouse cultures, after flies have visited haploid rust pustules of *P. graminis* and have walked over the infected Barberry leaves, no secondary rust pustules are produced as later additions to the primary ones derived from basidiospores.

[1]W. F. Hanna, *vide infra*, Chapter IV.

Grove,[1] in 1913, in his *British Rust Fungi*, advanced eight reasons in support of the assumption that pycnidiospores are functionless male cells.

Reed and Crabill,[2] in 1915, in an account of the Cedar Rust disease, remarked that repeated attempts to germinate the pycnidiospores of *Gymnosporangium juniperi-virginianae* had failed.

Clinton and McCormick,[3] in 1924, as a result of experiments made with isolated leaves kept moist in Petri dishes, found; that they could culture many Rusts by inoculating the leaves with aecidiospores, uredospores, or basidiospores; but that, when they spread (1) pycnidiospores of *Caeoma nitens* on the leaves of *Rubus villosus*, or (2) pycnidiospores of *Gymnosporangium juniperi-virginianae* on the leaves of *Pyrus malus*, they failed to obtain any infections.   Concerning the Apple-leaf Rust they[4] remark: "Inoculations with the O stage were made on hosts known to be very susceptible but without results, which seems to indicate that the O stage is not a means of spreading the rust." Thus the negative results obtained in similar experiments made on other Rust species by Plowright[5] (1889) and Klebahn[6] (1889-1904) were confirmed.

Lindfors,[7] in 1924, on the basis of an extended cytological investigation, held that the process of the change from the uninucleate to the binucleate condition in young aecidia is to be regarded as a substitution for a formerly existing true fertilisation process in which the pycnidiospores functioned as male cells.

B. O. Dodge,[8] in 1926, at the International Congress of Plant Sciences held at Ithaca, U.S.A., in his review of the cytological evidence bearing on the sexuality of the Uredinales, declared that "no one would contend that in the Rusts known today the spermatia

[1]W. B. Grove, *The British Rust Fungi*, Cambridge (Eng.), 1927, pp. 22-25.

[2]H. S. Reed and C. H. Crabill, "The Cedar Rust Disease of Apples caused by *Gymnosporangium juniperi-virginianae* Schw.," *Virginia Agric. Experiment Station*, Bull. IX, 1915, p. 48.

[3]G. P. Clinton and Florence A. McCormick, "Rust Infection of Leaves in Petri Dishes," *Connecticut Agric. Exp. Station*, Bull. CCLX, 1924.

[4]*Ibid.*, p. 491.

[5]C. B. Plowright, *loc. cit.*

[6]H. Klebahn, *loc. cit.*

[7]T. Lindfors, "Studien über den Entwicklungsverlauf bei einigen Rostpilzen aus zytologischen und anatomischen Gesichtspunkten," *Svensk. Bot. Tidskr.*, Vol. XVIII, 1924, p. 69.

[8]B. O. Dodge, "Cytological Evidence Bearing on the Sexuality and Origin of Life Cycles in the Uredineae," presented before the International Congress of Plant Sciences, Section of Mycology, Ithaca, New York, Aug. 20, 1926; published in the *Proceedings* of the Congress, Vol. II, 1929, pp. 1751-1766.

function in fecundation," but he added that they must still be recognised as sex cells and "as such prove by their presence the existence of a sexual element in the mycelium bearing them."

Gwynne-Vaughan and Barnes,[1] in their text-book published in 1927, very definitely expressed the view that the pycnidiospores of Rust Fungi are functionless.

From the foregoing we see that, from the time of Unger's discovery of pycnidia in 1833 up to 1927, no one had succeeded in making the pycnidiospores do anything positive in respect either to the tissues of the host-plant or to the mycelium of the parasite. In the absence of new facts, the best mycologists could only make guesses concerning the function of the pycnidiospores. It was Craigie's great service to bring to a close a century of speculation by proving experimentally that the pycnidiospores of the Rust Fungi definitely function in the sexual process.

Entomological Background.—As we shall see later, it was a fly that provided the clue which led immediately to Craigie's discovery of the function of the pycnidia of the Rust Fungi. On this account, the entomological background of Craigie's work will now be presented.

The earlier workers on the Uredinales noticed that the sori producing pycnidia are brightly coloured, emit an odour, and produce nectar that is sweet to the taste; but they did not associate these facts with insect visits.

The yellow or orange colour of pycnidial sori and the darker red colour of the pycnidia could not fail to attract the attention of botanists when they first began to study Rust Fungi in the field. Thus Persoon in 1801, as noted above, remarked that the spots of infected Pear leaves are "dilute croceae et punctatae." The orange-coloured, pycnidia-producing sori of such Rusts as *Gymnosporangium globosum* on Crataegus sp., *Puccinia coronata* on *Rhamnus cathartica*, and *Puccinia graminis* on *Berberis vulgaris* can, as a matter of fact, be easily seen at a distance of several feet.

The odour of the pycnidial nectar of many Rusts is readily noticed and, particularly, in species which have a perennial mycelium and produce great numbers of pycnidia. Thus the odour of *Puccinia suaveolens* on *Cirsium arvense* (*Carduus arvensis*) and of *Puccinia minussensis* on *Lactuca pulchella* can be detected at a distance of several feet from the host-plants, and often these and other similar Rust Fungi have been recognised in the field before their host-plants

---

[1]H. C. I. Gwynne-Vaughan and B. Barnes, *The Structure and Development of the Fungi*, Cambridge (Eng.), 1927, p. 266.

have been seen. Usually the odour of pycnidial nectar appears to be pleasant, a fragrance reminiscent of flowers.

The first Rust in which the odour of the pycnidia was noticed was *Puccinia suaveolens*. In 1799, Persoon[1] described this fungus for the first time, placed it in the form-genus *Uredo* and, with the remark that for some unknown reason the Thistle leaves on which it is found emit a pleasant odour, gave it the specific name of *suaveolens*. In 1801, he[2] described the ungus again and declared that its pleasant odour is distinctive, "odorem gratum spargit quo ab omnibus differt species." The odour of *P. suaveolens* was thought by Link[3] (1825) to be disagreeable, "potius nauseosum quam suavem," while Léveillé[4] (c. 1849) likened it to the scent of Orange blossoms and de Bary[5] (1853) to the smell of the flowers of the Evening Primrose (*Oenothera biennis*). It reminded Tulasne[6] (1854) of the male catkins of Willows, and Grove[7] (1913) of the inflorescences of the Privet (*Ligustrum vulgare*). Here we have another example of a physiological fact that is well known, namely, that the impression made upon the olfactory sense of human beings by a single odoriferous substance may vary much with different individuals.

Léveillé[8] (1849) detected the scent of *Puccinia tragopogonis* (his *Aecidium tragopogi*) and said that it resembled that of *P. suaveolens*.

In 1882 Ráthay[9] mentioned the sweet scent which, in his pot cultures, had been emitted by the pycnidia of *Endophyllum euphorbiae-sylvaticae* on *Euphorbia amygdaloides*, and of *Uromyces pisi* on *Euphorbia cyparissias*.

In 1889 Klebahn[10] remarked that, in the pycnidial stage, *Melampsora larici-epitea*, sown on a Larch in a greenhouse, had given out a

---

[1]D. C. H. Persoon, *Observationes Mycologicae*, Lipsiae et Lucernae, Pars II, 1799, p. 24.

[2]D. C. H. Persoon, *Synopsis Methodica Fungorum*, Gottingae, 1801, p. 221.

[3]H. F. Link, *Caroli a Linné Species Plantarum*, ed. 4, Vol. VI, Part II, 1825, p. 19.

[4]J. H. Léveillé, *Dict. univ. d'hist. nat.* of d'Orbigny, ed. 2, undated, T. XIV, "Urédinées," p. 206. This article appeared in ed. 1 (1839-1849) which I have not seen.

[5]A. de Bary, *Untersuchungen über die Brandpilze und die durch sie verursachten Krankheiten der Pflanzen*, Berlin, 1853, p. 57, foot-note.

[6]L.-R. Tulasne, "Second Mémoire sur les Urédinées et les Ustilaginées," *Ann. Sci. Nat. Bot.*, Sér. 4, T. II, 1854, p. 118.

[7]W. B. Grove, *The British Rust Fungi* (Uredinales), Cambridge, 1913, p. 145.

[8]J. H. Léveillé, *loc. cit.*

[9]E. Ráthay, "Untersuchungen über die Spermogonien der Rostpilze," *Denkschrift d. Kais. Akad. d. Wissensch.*, Wien, Bd. XLVI, 1883 (Reprint, 1882), p. 48.

[10]H. Klebahn, "Kulturversuche mit heteröcischen Rostpilzen. VII Ber.," *Zeitschrift für Pflanzenkrankheiten*, Bd. IX, 1899, p. 88.

peculiar unpleasant sweetish odour which was so strong that one did not care to remain long in its presence.

In 1900 Klebahn[1] reported that he had inoculated two Spruce trees (*Picea excelsa*) with the basidiospores of *Thecopsora padi* and, some days later, had observed that, although no pycnidia could be found, the twigs gave out a peculiar odour that at once reminded him of pycnidia. The emission of the odour continued for about eight days, but no pycnidia made their appearance. A microscopic investigation then revealed that the mycelium of the fungus had developed freely in the intercellular spaces of the cortex of the twigs and had sent haustoria into the living cells. Klebahn,[2] in 1904, concluded that the odoriferous substance had been produced not by the pycnidia or not by them alone, but by the mycelium or by the infected tissues.

Klebahn[3] in 1904 described the odours of Rust pycnidia in general as "sweetish, usually pleasant but, when very strong, somewhat unpleasant" and said that, in the open, he had noted the scent of the peridermia on Pines and the odour of *Puccinia suaveolens* and that, indoors, he had often detected the odour of Rusts which he had had under cultivation. He[4] also remarked that a *Pinus strobus* twig which bears pycnidia of *Cronartium ribicola* covered with nectar gives out an "unpleasant sweetish odour."

Schaffnit[5] in 1909 said that Klebahn had overlooked the fact that in cereal Rusts the aecidiospores and uredospores, like the spermogonia, give out a wonderful flower-like scent resembling that of a Tea-rose; and he added that this scent can be detected especially when one has enclosed a great number of spores in a vessel and the vessel is opened the next day. Notwithstanding Schaffnit's closed-vessel observations, I am convinced that, as far as *Puccinia graminis* is concerned, *in the open* it is the pycnidia alone which give out a strong scent. In the Rust Laboratory at Winnipeg, another worker with a keen sense of smell and I, after detecting without any difficulty the strong scent given out by the haploid pycnidial pustules of *P. graminis* on some Barberry leaves,[6] immediately thereafter applied our

[1]H. Klebahn, "Kulturversuche mit Rostpilzen," *Jahrb. f. wiss. Bot.*, Bd. XXXIV, 1900, pp. 379-380.

[2]H. Klebahn, *Die Wirtswechselnden Rostpilze*, Berlin, 1904, p. 195.

[3]*Ibid.*, p. 195.

[4]*Ibid.*, p. 387.

[5]E. Schaffnit, "Biologische Beobachtungen über die Keimfähigkeit und Keimung der Uredo- und Aecidiosporen der Getreideroste," *Annales Mycologici*, Bd. VII, 1909, p. 523.

[6]For further details concerning the emission of an odour by *P. graminis* pycnidia *vide infra*, Chapter II.

noses successively to: (1) Barberry rust pustules in which the nectar had dried up and aecidia were developing; (2) well-developed uredo-spore sori on Wheat-seedling leaves; and (3) young teleutospore sori; with the result that we came to the conclusion that, under natural conditions, whereas the pycnidia of *P. graminis* are strongly scented, the aecidia, the uredospore sori, and the teleutospore sori give out no scent that is humanly appreciable. As further evidence that the uredospores of *P. graminis* and related grain Rusts do not give out any appreciable scent, it may be mentioned that such a scent has never been detected in heavily rusted grain fields nor in Rust green-houses, such as those at Winnipeg and at the University of Minnesota, where thousands of wheat plants are inoculated every year and great numbers of uredospore pustules are present on cereal seedlings confined in a small space. Whether or not insects can detect the odour of the aecidiospores, uredospores, and teleutospores of *P. graminis* growing in the open is not known; but, if they do, the stimulus given by this odour to their olfactory organs is, in all probability, far less than that given by the odour which emanates from the pycnidia. Another observation made by myself may here be mentioned. The under sides of the leaves of *Cirsium arvense*, when closely sprinkled over with the pycnidia of *Puccinia suaveolens* are at first strongly scented, but the scent disappears as the nectar dries up and the uredospore sori are developed. Here again, just as in *P. graminis*, the strong odour which the fungus emits is associated with the pycnidia and the pyc-nidia alone.

Jackson and Mains[1] in 1921, in giving an account of their dis-covery that *Puccinia triticina* has for its alternate host-plants species of Thalictrum, remarked that, as the honey-yellow pycnidia crowded together on the gall-like Thalictrum rust pustules reached maturity, they often detected "a very noticeable odour, resembling that of the Hyacinth."

Craigie[2] in 1931 stated that he had not detected any odour coming from the pycnidia of *Puccinia graminis* and of *P. helianthi* and Faull,[3] in 1934, stated that he had not detected any odour coming from the sweetish liquid excreted by the "spermogonia" of *Milesia polypodo-phila*; but, as I[4] have since discovered that the pycnidia of *Puccinia*

[1]H. S. Jackson and E. B. Mains, "Aecial Stage of the Orange Leafrust of Wheat, *Puccinia triticina* Eriks.," *Journ. Agric. Research*, Vol. XXII, 1921, p. 164.

[2]J. H. Craigie, "An Experimental Investigation of Sex in the Rust Fungi," *Phytopathology*, Vol. XXI, 1931, p. 1024.

[3]J. H. Faull, "The Biology of Milesian Rusts," *Journal of the Arnold Arboretum*, Vol. XV, 1934, p. 70.

[4]*Vide infra*, Chapter II.

graminis and of *P. helianthi* do actually emit a scent, it is possible
that a further, more careful test may reveal that an odour is produced
by the pycnidia of *Milesia polypodophila* and other Milesiae. On
the other hand, just as there are scentless flowers, so, too, there may
be scentless Rust Fungi. But, before any Rust species is definitely
set down as odourless, it should be thoroughly tested by an investigator who has a well-developed olfactory sense.

In 1882 Ráthay,[1] who worked at Klosterneuburg in Austria, made
a valuable contribution to our knowledge of the relations between
Rust Fungi and insects. One hot and sunny day in June, 1878, when
walking abroad, he[2] noticed that some ants had ascended a White-beam
tree (*Sorbus aria*) and on the leaves were eagerly feeding upon the
drops which had been exuded by the pycnidia of *Gymnosporangium
juniperinum* (*Cf. G. aurantiacum* in Fig. 3). He knew that ants like
the sugar produced by floral and extrafloral nectaries or excreted by
Aphidae, Coccidae, and Chermidae; and he therefore thought it
probable that the pycnidial drops also contain sugar. On tasting
the drops, he found that they were intensely sweet and, on testing
them with Fehling's solution, he observed that they reduced the
copper sulphate with the production of a considerable deposit of
cuprous oxide. Thereupon he concluded that the pycnidial drops of
G. juniperinum do actually contain sugar, as had been mentioned by
Meyen in 1841, and that sugar is the bait which makes the drops
attractive to ants and other insects.

In making his tests for the presence of sugar with Fehling's solution,
Ráthay[3] employed the following technique. He took 100 leaves
bearing pycnidial pustules and another 100 leaves that were healthy
and free from the Rust Fungus. Without putting the ends of the
petioles in the water, he washed the 100 diseased leaves in 80 cc.
distilled water contained in one vessel and the 100 healthy leaves in
an equal amount of distilled water contained in another vessel. He
then found that the water which had been used to wash the diseased
leaves reduced the Fehling's solution, whereas the water used to wash
the healthy leaves did not.

With Fehling's solution Ráthay[4] investigated the pycnidial nectar
of twenty-one Rust species, and he obtained: (1) copious deposits of

---

[1]E. Ráthay, "Untersuchungen über die Spermagonien der Rostpilze," *Denkschrift d. Kais. Akad. d. Wissensch.*, Wien, Bd. XLVI, 2 Abt., 1883, pp. 1-51. The
reprint is dated 1882.

[2]*Ibid.*, pp. 1-2.

[3]*Ibid.*, p. 2.

[4]*Ibid.*, pp. 5-25, summary on p. 26.

cuprous oxide from the nectar of *Aecidium* on *Euphorbia virgata*,[1] *Caeoma* on *Poterium sanguisorba* (*Phragmidium sanguisorbae*), *Endophyllum euphorbiae-sylvaticae*, *Gymnosporangium juniperinum*, *G. sabinae* (Figs. 4 and 5), *Puccinia falcariae*, *P. fusca*, *P. pimpinellae*, *P. suaveolens*, *P. tragopogi*, and *Uromyces pisi* (Fig. 6); (2) a lesser deposit from the nectar of *Aecidium magelhaenicum*, *Gymnosporangium*

FIG. 3.—*Gymnosporangium aurantiacum* (Rostrup's *G. juniperinum*) on *Pyrus aucuparia*. Left: upper side of a leaflet showing pustules bearing pycnidia. Right: under side of a leaf showing horn-shaped aecidia. Magnification, 2. From E. Rostrup's *Plantepatologi* (1902).

clavariiforme, *Puccinia coronata*, *P. graminis*, *P. rubigo-vera*, *P. sylvatica*, *P. violae*, and *Uromyces dactylis*; and (3) the least deposit of all from the nectar of *Aecidium clematidis* and *Puccinia poarum*.

Ráthay[2] found that, when applied to the tongue, the nectar tasted as follows: (1) intensely sweet, *Gymnosporangium juniperinum* and

[1]Ráthay (p. 7) remarked that he must leave it open to which species of Uromyces this Aecidium belongs. P. and H. Sydow (*Monographia Uredinarum*, Vol. II, 1910, p. 178) state that *Uromyces scutellatus* is present on *Euphorbia virgata* in Austria and this, therefore, may have been the species which Ráthay investigated.

[2]E. Ráthay, *loc. cit.*, pp. 5-25, summary on p. 6.

G. sabinae; (2) slightly sweet, *Uromyces pisi*; and (3) tasteless, *Aecidium magelhaenicum*, *Endophyllum euphorbiae-sylvaticae*, *Gymnosporangium clavariiforme*, *Phragmidium sanguisorbae*, *Puccinia coronata*, *P. falcariae*, *P. fusca*, *P. graminis*, *P. poarum*, *P. rubigo-vera*, *P. suaveolens*, and *P. tragopogi*.

FIG. 4.—*Gymnosporangium sabinae* on a twig of *Pyrus communis*. One leaf (upper left) shows on its upper side pustules with pycnidia. The nectar produced by these organs, according to Ráthay, is sweet to the taste. The two other leaves show on their under sides flask-shaped aecidia which become split to the base into laciniae, as shown six times the natural size on the right below. On the lower left is an aecidiospore as seen from without and in optical section with a magnification of 600. At the apex of the twig is an infected pear-fruit much distorted in form. From E. Rostrup's *Plantepatologi* (1902).

Ráthay also found that Fehling's solution was reduced under temperature conditions as follows: (1) at room temperature as well as when heated, *Gymnosporangium juniperinum* and *G. sabinae*; (2) slightly when cold and far more when heated, *Endophyllum euphorbiae-sylvaticae* and *Puccinia suaveolens*; and (3) only when heated, *Aecidium* on *Euphorbia virgata*, *Aecidium magelhaenicum*, *Gymnosporangium clavariiforme*, *Puccinia fusca*, *P. falcariae*, *P. sylvatica*, *P. tragopogi*, and *Uromyces pisi*.

So far as the reaction with litmus paper is concerned Ráthay[1] observed that the nectar of most of the investigated Rust species was neutral and that of a few only was weakly acid.

The nectar of *Gymnosporangium sabinae* was investigated for Ráthay by a chemist, Dr. Benjamin Haas.[2] Haas washed the nectar from the pycnidial pustules of 245 infected Pear leaves (Fig. 4), filtered the wash-water, and concentrated the solution that remained by evaporation. He then observed that the angle of rotation caused by the solution in a polarimeter was −0.2° and found that 12 cc. reduced 5 cc. of the Fehling's solution. From calculations and deductions based on these data Haas concluded that the nectar of *G. sabinae* contains a mixture of *dextrose* and *laevulose* with the sweeter laevulose prepondering in amount.

In tasting intensely sweet and in reducing Fehling's solution in the cold, the sugar in the nectar of *Gymnosporangium juniperinum* resembles that of *G. sabinae* and may therefore consist of a mixture of dextrose and laevulose; but these sugars do not appear to be present in the nectar of *Aecidium magelhaenicum, Endophyllum euphorbiae-sylvaticae, Gymnosporangium clavariiforme, Melampsora euphorbiae-dulcis, Puccinia falcariae,*

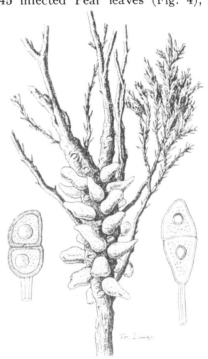

FIG. 5.—*Gymnosporangium sabinae* on *Juniperus sabina*. Irregularly conical and gelatinous masses of teleutospores project from the bark. The two kinds of teleutospores, thick-walled and thin-walled, are shown right and left of the infected stem. From E. Rostrup's *Plantepatologi* (1902).

P. fusca, P. sylvatica, P. tragopogi, and *Uromyces pisi* as the nectar of these species reduces Fehling's solution only when heated.[3] It is also certain that the nectar of *Aecidium* on *Euphorbia virgata, Endophyllum euphorbiae-sylvaticae, Puccinia falcariae, P. sylvatica, P. tragopogi,* and *Uromyces pisi* does not contain sucrose (cane-sugar)

[1]*Ibid.*, p. 26.
[2]*Ibid.*, p. 23.   [3]*Ibid.*, p. 26.

Fig. 6.—*Uromyces pisi.*, a long-cycled, heteroecious, and probably heterothallic|Rust, with pycnidia and |aecidia on the under side of the leaves of *Euphorbia cyparissias* (left) and uredospore sori and teleutospore sori on *Pisum sativum* (leaf on right). Above, a globose aecidiospore and an ovate teleutospore. The nectar excreted by the pycnidia on the Euphorbia leaves, according to Ráthay, is slightly sweet to the taste, reduces Fehling's solution, gives out an odour, and attracts insects. Flowering plants, natural size; spores magnified 400 times. From E. Rostrup's *Plantepatologi* (1902).

since, after it has been treated with yeast, it does not reduce Felhing's solution in the cold. It would appear that the nectar of *Endophyllum euphorbiae-sylvaticae* does not contain a fermentable sugar because, after long treatment with yeast, it still retains a substance which when heated reduces Fehling's solution. In the nectar of *Uromyces pisi* there is a substance that both before and after being warmed with a little hydrochloric acid to 68° - 70° C. turns the plane of the polariscope to the right.[1]

That the reducing substance present not only in *Gymnosporangium juniperinum* and *G. sabinae* but also in the other Rust species investigated is sugar is supported by the fact that insects were seen by Ráthay to visit and feed upon the nectar of twelve of these species just as they were seen to feed upon the nectar of the two species of Gymnosporangium.[2]

Ráthay[3] considered that the fact that in *Gymnosporangium juniperinum* and *G. sabinae* the nectar is sweet whereas in most of the other investigated species it is tasteless cannot be used as an argument against the supposition that the nectar of these other species contains sugar, because very dilute sugar solutions are tasteless.

In view of all the facts brought out by his biochemical investigations Ráthay[4] suggested that the sugar present in the nectar of all the Rust species which he investigated other than *Gymnosporangium juniperinum* and *G. sabinae* is *arabinose*; for this substance easily reduces Fehling's solution, does not undergo alcoholic fermentation, and turns the plane of the polariscope to the right.

Ráthay[5] informs us that he was particularly interested in finding that sugar is present in the pycnidial nectar of Rust Fungi and that it is attractive to insects because only two cases of cryptogams excreting nectar and the nectar being visited by insects had hitherto become known: (1) *Claviceps purpurea* in the Sphacelia stage which, as Kühn[6] had found, produces a yellowish, strong-smelling, sugar-containing liquid sought by beetles and flies but not by bees; and (2) the nectaries on the rachis of the large leaves of *Pteris aquilina*, which, as Francis Darwin[7] had observed, are sought by ants of the genus Myrmica.[8]

[1] *Ibid.*, p. 26.   [2] *Ibid.*, p. 27.   [3] and [4] *Ibid.*, p. 27.   [5] *Ibid.*, p. 2.

[6] J. Kühn, "Untersuchungen über die Entstehung, das künstliche Hervorrufen, und die Verhütung des Mutterkorns," Wilda, *landwirtsch. Centralbl.*, XI, Bd. II, 1863, pp. 7-13. Cited from Ráthay's paper.

[7] C. Darwin, *The Effects of Cross and Self Fertilisation in the Vegetable Kingdom*, London, 1876, p. 404.

[8] A fourth instance of a Cryptogam producing a saccharine fluid attractive to flies was found by Ráthay and B. Haas in *Phallus impudicus*. They discovered that

Ráthay[1] observed that insects not only visit Rust pycnidia and feed upon the nectar, but that they also serve to disseminate the pycnidiospores. Standing upon his window-sill there was a potted plant of *Euphorbia amygdaloides* upon whose leaves were a great many pycnidia of *Endophyllum euphorbiae-sylvaticae* crowned with drops of nectar. One day, on approaching the window where the plant was, Ráthay noticed some flies, which had been feeding upon the nectar, fly away from the leaves and alight upon the window pane. An examination of the glass showed that the flies had left their wet foot-marks there, and a microscopic investigation revealed that the foot-marks contained the pycnidiospores of the Endophyllum.

Ráthay came to the conclusion that the sugary matter in the nectar of Rust pycnidia in general is a bait to attract insects; and, working in the open, he observed the pycnidia of various Rust species being visited by no less than 135 species of insects.[2] Of these species 31 were Coleoptera, 32 Hymenoptera, 64 Diptera, and 8 Hemiptera. Ráthay remarked that insect species which visit the pycnidia in general also visit flowers.[3] To an insect, nectar, whether produced by a Phanerogam or a Rust Fungus, appears to be equally attractive.

Ráthay collected and identified the insects which he found visiting the pycnidia of eleven different Rust species. The names of these Rusts, with the number of species of insects found visiting them, are as follows:

Aecidium on *Euphorbia virgata*	38
Aecidium magelhaenicum	5
Endophyllum euphorbiae-sylvaticae	15
Gymnosporangium clavariaeforme	17
Gymnosporangium juniperinum	37
Puccinia coronata	12
Puccinia fusca	9
Puccinia graminis	11
Puccinia tragopogonis	11
Puccinia suaveolens	11
Uromyces pisi	12

---

the deliquesced gleba of this fungus is rich in reducing sugars (Über *Phallus impudicus* und einige Coprinus-Arten nach Wiesner," *Sitzungsber. d. Mathem.-Naturwiss. Klasse d. K. Akad. der Wiss. in Wien*, Bd. LXXXVII, 1883, pp. 18-44). Ráthay (*loc. cit.*, p. 2) remarked that C. Nägeli, in 1865, in a Festrede delivered to the Munich Academie ("Ueber Entstehung und Begriff der naturhistorischen Art," pp. 7-13) had declared that no Cryptogam excretes nectar.

[1] E. Ráthay, *loc. cit.*, pp. 50-51.

[2] and [3] *Ibid.*, pp. 38-44.

Ráthay gave the names of the species of insects observed visiting each of the eleven Rust species, and for *Gymnosporangium juniperinum* and *Puccinia graminis* his lists of insects were as shown in the accompanying Tables. Specimens of the insect species listed were collected in 1881: on *P. graminis*, in the last half of May; and on *Gymnosporangium juniperinum*, almost exclusively in the last half of June.

Insects observed by Ráthay visiting the Pycnidia of
Gymnosporangium juniperinum on Sorbus aria

ORDER	FAMILY	SPECIES	No.
Coleoptera	Scarabaeidae .	Phyllopertha horticola	1
	Elateridae . .	An undetermined species . . .	2
		Athous haemorrhoidalis	3
	Telephoridae .	Telephorus lividus var. dispar . .	4
		Dasytes plumbeus	5
	Cuculionidae .	Balaninus pyrrhoceras	6
	Cerambycidae .	Strangalia septempunctata . . .	7
	Chrysomelidae .	Clytra sexpunctata	8
		Halyzia vigintidus-punctata . . .	9
Hymenoptera	Formicidae . .	Camponotus lateralis	10
		Formica fusca	11
		Formica cunicularia	12
		Leptothorax Nylanderi	13
		Lasius brunneus	14
		Myrmica scabrinodis	15
	Adrenidae . .	Halictus leucozonius	16
		Adrena sp.	17
		Adrena pilipes	18
Diptera	Syrphidae . .	Chrysotoxum bicinctum	19
		Chrysotoxum elegans	20
		Pipizella virens	21
	Muscidae . .	Tachina rustica	22
		Sarcophaga pumila	23
		Sarcophaga striata	24
		Sarcophaga carnaria	25
		Lucilia caesar	26
		Hydrotaea dentipes	27
		Hydrotaea meteorica	28
		Hylemyia sp.	29
		Hylemyia cinerella	30
		Anthomyia sp.	31
		Anthomyia platura	32
		Homalomyia sociella	33
		Systata (Myennis) rivularis . . .	34
		Platystoma seminationis . . .	35
		Lauxania aenea	36
Hemiptera	Anthocoridae .	Anthocoris nemorum	37

Insects observed by Ráthay visiting the Pycnidia of
Puccinia graminis on Berberis vulgaris

ORDER	FAMILY	SPECIES	No.
Coleoptera	Telephorideae .	Telephorus lividus var. dispar . .	1
Hymenoptera	Tenthredineae .	Tenthredopsis tesselata	2
	Ichneumonidae .	Tryphon rutilator	3
	Syrphidae . .	Pipizella virens	4
	Bibionidae . .	Trineura aterrima	5
Diptera	Muscidae . .	Sarcophaga carnaria	6
		Spilogaster semicinerea	7
		Anthomyia pluvialis	8
		Anthomyia sp.	9
		Ortalis ornata	10
		Acidia heraclei	11

De Bary[1] had observed that the pycnidiospores of the Uredinales set free in a pycnidium are embedded in jelly and that the jelly swells up and so causes the pycnidiospores to be extruded through the ostiole, but he had seemed to suggest that, in the open, the exudate makes its appearance on the outside of a leaf only in rainy weather. Ráthay observed that Rusts excrete nectar from their pycnidia without the action of atmospheric moisture.

During the mid-day hours of some very hot and dry days, on some of which the maximum temperature had attained 27° C., after a week of rainless weather, Ráthay[2] observed little drops of nectar over the mouths of the pycnidia of various Rust species and he noted that these drops were being visited by insects. He also observed: (1) that, after heavy rain that must have washed the exudate away, new drops of nectar were soon formed by the pycnidia of *Puccinia suaveolens* and of *Aecidium magelhaenicum*; (2) that, a few hours after a heavy rain-storm there were fresh drops of nectar above the pycnidia of *Gymnosporangium sabinae* and that the drops tasted intensely sweet and reduced Fehling's solution, thus showing that they contained sugar; (3) that, in various Rust species, during warm and steamy weather, *i.e.* when evaporation is much reduced, the drops on the pycnidia of a pustule grow in size, then touch and flow together; and (4) that a similar formation of a compound drop also takes place

[1]A. de Bary, *Untersuchungen über die Brandpilze und die durch sie verursachten Krankheiten der Pflanzen*, Berlin, 1853, p. 60.

[2]E. Ráthay, "Untersuchungen über die Spermogonien der Rostpilze," *loc. cit.*, pp. 44-45.

when a shoot bearing pycnidial pustules is placed with its cut end in water in a damp-chamber. All these observations taken together convinced Ráthay and proved conclusively that the water in the exudate of Rust pycnidia comes from within the leaf and not from the atmosphere.

Ráthay[1] procured eighty shoots of *Euphorbia virgata* bearing pycnidial pustules and made four bouquets of them, and he then placed the cut ends of the shoots in water and left them thus for three days. During these three days, at equal intervals of time he washed the bouquets six times in a little water and after each washing plunged them momentarily in much water contained in a wooden vessel. By testing with Fehling's solution (the mixture heated) he found that: in the first four wash-waters, without exception, there was a considerable quantity of sugar; in the fifth wash-water there was but little sugar; and in the sixth and last wash-water there was no sugar whatever. Thus Ráthay proved by experiment that the pycnidia of the Rust Fungi, like the floral and extrafloral nectaries of Phanerogams, excrete sugar and water for a relatively long time.

On the basis of the results of Wilson's[2] studies of the nectaries of Flowering Plants made in Pfeffer's laboratory at Tübingen, Ráthay[3] rightly concluded that the excretion of nectar by the pycnidia of Rust Fungi is due to osmosis; and, in support of this view, he adduced the results of two experiments, (1) and (2).

(1) Ráthay[4] gathered some shoots of *Circium arvense* bearing ripe, and perhaps somewhat old, pycnidia of *Puccinia suaveolens* and washed them in a little distilled water. When warmed with Fehling's solution, the wash-water reduced the copper sulphate with the production of cuprous oxide. The shoots were then washed in much distilled water, the leaves very carefully dried with blotting-paper, the cut ends of the shoots placed in water, and the shoots and vessel containing the water enclosed in a damp-chamber. Five hours later it was observed that no fresh drops of nectar had been excreted by the pycnidia. Then, with the help of a fine brush, Ráthay deposited little drops of concentrated sugar-solution on all the pycnidia situated within one square cm. of leaf surface, and then he replaced the shoots in the damp-chamber. Three hours later, he found that the drops which he had placed on the pycnidia, and these only, had grown to a

---

[1] *Ibid.*, p. 45.

[2] W. P. Wilson, "The Cause of the Excretion of Water on the Surface of Nectaries," *Unters. a. d. bot. Inst. zu Tübingen*, Bd. I, 1881, pp. 1-22.

[3] E. Ráthay, *loc. cit.*, p. 45.

[4] *Ibid.*, p. 46.

considerable size and, later, he observed that these drops had run
together and had flowed toward the ground.

(2) The second experiment[1] was made with *Gymnosporangium
juniperinum* on *Sorbus aria*. At 1 p.m. on June 16, 1881, twigs bearing
pycnidial pustules which had excreted sweet drops of nectar were
washed in distilled water and the wash-water was found to reduce
Fehling's solution in the cold. The twigs were then thoroughly
washed with well-water, dried with blotting-paper, and placed with
their cut ends in water in a damp-chamber. Drops that were sweet
and rich in pycnidiospores were then excreted through the pycnidial
ostioles and on many of the pustules these drops intermingled and
formed a single large drop. Five additional wash-waters were obtained, like the first, from the infected twigs on June 16 at 10 p.m.,
June 17 at 7 p.m., 2 p.m., and 6 p.m., and on June 18 at 7 a.m., and
they all reduced Fehling's solution in the cold; and, after each washing
and re-washing and drying, the pycnidia excreted anew drops of
considerable size, that were sweet to the taste and also contained
pycnidiospores. However, after a seventh washing on June 18 at
7 p.m. and an eighth on June 19 at 7 a.m., the pycnidia behaved
differently; for after the seventh washing they excreted drops that
were very small but yet sweet and containing pycnidiospores, while
after the eighth washing they excreted nothing at all. The eighth
wash-water reduced Fehling's solution very slightly, and a ninth and
final wash-water left the Fehling's solution unchanged. The complete removal of the osmotically active sugar and the jelly from the
pycnidia had brought about a cessation in the excretion of water;
but Ráthay found that, after he had set some little drops of concentrated cane-sugar on the exhausted pycnidia, an excretion of water
took place once more.

The fact that in the second experiment, after the seventh washing,
the pycnidia of the Gymnosporangium excreted drops that contained
pycnidiospores goes to prove, as Ráthay[2] urged, that the pycnidiospores are produced in succession on the ends of the sterigmata
(pycnidiosporophores) for a comparatively long time.

Ráthay[3] argued that, since in *Phragmidium sanguisorbae* on
Poterium sanguisorba the pycnidia have no paraphyses (periphyses)
and yet the nectar of this fungus reduces Fehling's solution, the periphyses in the Rust Fungi in general do not, or do not exclusively,

[1] *Ibid.*, pp. 46-47.
[2] *Ibid.*, p. 47.
[3] *Ibid.*, p. 47.

excrete the nectar; and he stated that he regarded the pycnidiosporo-phores as the excretory organs.

In the field, Ráthay observed that Rust pycnidia are able to excrete nectar in diffuse daylight and even in direct sunlight; and, by means of some experiments which will now be described, he[1] proved that Rust pycnidia can excrete nectar in complete darkness. (1) In the spring he placed potted plants of *Euphorbia amydaloides* bearing the perennial mycelium of *Endophyllum euphorbiae-sylvaticae* and of *Euphorbia cyparissias* bearing the perennial mycelium of *Uromyces pisi* in a zinc dark-chamber; and he found that the etiolated shoots which grew up under these conditions excreted sweet-scented, pycnidio-spore-laden drops of nectar and that the nectar of the Endophyllum had the property of reducing Fehling's solution. (2) He cut off some twigs of *Sorbus aria* which bore ripe pycnidia of *Gymnosporangium juniperinum*, washed the twigs in water, and dried them with blotting paper. With their cut ends set in water, some of the twigs were then placed in a dark-chamber, whilst others were exposed to diffuse day-light at a north window. In the course of the next three hours the pycnidia in both sets of twigs excreted drops laden with pycnidio-spores, and the drops were found to reduce Fehling's solution in the cold.

Ráthay[2] noticed that compound drops of nectar become firmly attached to the pustules above which they are formed, and he found that, if one dips an infected leaf into water, then, on taking it out, one sees water adhering to the leaf in any quantity only over the pustules.

The visits of insects to the pycnidia of the Rust Fungi, according to Ráthay,[3] are secured by: (1) food-substances which act as a bait made up of pycnidiospores—a *nitrogenous* food-substance correspond-ing to the pollen-grains of flowers, and of slime, sugar, and water—a *non-nitrogenous* food-substance corresponding to the nectar of flowers; (2) the slow and long-continued exudation of the contents of a pyc-nidium; (3) the red colour of the periphyses in nearly all Rusts; (4) the pleasant scent emitted by rusts with a perennial mycelium; (5) the holding of a drop of nectar above a pycnidium by the peri-physes, so that in windy weather the drop remains fixed; (6) in species with a temporary mycelium, the orange colour of the pycnidial pustules and the darker periphyses serving as special guides to the presence of the bait of food-materials; (7) the strong adhesion of a

[1] *Ibid.*, pp. 47-48.
[2] *Ibid.*, p. 48.
[3] *Ibid.*, pp. 48-50.

larger compound drop of nectar to the general surface of the pustule on which it has been formed, which prevents the drop rolling off the leaf; and (8) in Rust species with a perennial mycelium, the special colour, mode of growth, and appearance of the infected host-plants.

Ráthay[1] refused to believe that the visits of insects to Rust pycnidia are of no use (zwecklos) to the fungi concerned and, in support of this attitude, he cited the fact that, as Kühn[2] had found, the visits of flies to the drops excreted by *Claviceps purpurea* in the Sphacelia stage lead to the sowing of the conidia on the flowers of grasses and to the subsequent infection of the host-plants. A recognition of the striking resemblance between the means employed by the pycnidia of the Rust Fungi and those employed by the flowers of Phanerogams for attracting and rewarding insects led Ráthay[3] to his final conclusion that "perhaps the spermogonia of the Rust Fungi are really male organs, as many mycologists for various reasons have asserted, and so insects bring about fertilisation in the Rust Fungi just as they do in so many Phanerogams."

Ráthay's views on the function of the pycnidia of the Rust Fungi and on the significance of the visits of insects to those organs were put forward in 1883. In 1927, after the passage of forty-four years, their essential correctness, as we shall see, was demonstrated by Craigie.

Plowright[4] in his *Monograph*, published in 1889, supported Ráthay's conclusion that Rust Fungi in the pycnidial stage are attractive to insects. He remarked that the shoots of host-plants which bear a perennial mycelium make themselves conspicuous by their manner of growth in that they are negatively geotropic, stand erect, usually overtop normal shoots growing near-by, and produce leaves which are pale in colour and abnormally thickened or attenuated. "The Uredineae with a short-lived mycelium, on the contrary" he continued "have their spermogonia produced upon the upper surface of the leaves on brilliantly coloured spots, which contrast more or less strikingly with the green colour of the healthy foliage. These spots are generally bright yellow or orange, often with a tinge of red. Sometimes they are white (*Aec. fabae*) or purple-red (*Aec. rumicis*). The negative geotropism is not, however, confined to the species with a perennial mycelium; I have noticed it very markedly with some

[1]*Ibid.*, p. 50.

[2]J. Kühn, "Mittheilungen aus dem physiologischen Laboratorium und der Versuchsstation des landwirtschaftlichen Institutes der Universität Halle," p. 13. Cited from Ráthay's paper.

[3]*Ibid.*, p. 51.

[4]C. B. Plowright, *A Monograph of the British Uredineae and Ustilagineae*, London, 1889, p. 13.

leaves of *Senecio jacobaea*, upon which I had produced the aecidium of *Puccinia schoeleriana*; as soon as the aecidial spots became sufficiently developed to produce spermogonia, the leaves which bore them became almost erect, while the unaffected leaves remained horizontal."

In confirmation of Ráthay's work on sugar, Plowright[1] found that the nectar of *Puccinia obscura* reduces Fehling's solution and also gives the reaction with the indigo-carmine test.

Plowright in his *Monograph* gave an excellent account of Ráthay's work. When reading it through in my student days, I became interested in the problem of the function of Rust pycnidia, and then said to myself: "The pycnidia do not seem to be degenerate; for they develop perfectly, they produce great numbers of pycnidiospores, and by means of their odour, their colour, and their sweet nectar, they are attractive to insects. De Bary's work on the life-history of *Puccinia graminis* seems to show that, at the present day at any rate, the pycnidia have no influence on the life-history. If the pycnidia are now functionless, they must once have functioned and then insect-visits must have been important to them. If pycnidia have lost their function, on account of their showing as yet no signs of degeneracy they must have lost their function in the immediate geological past."

In 1911, Lloyd and Ridgway[2] in a description of *Gymnosporangium juniperi-virginianae* as a parasite on *Juniperus virginiana* and cultivated Apple trees gave a sketch of the exterior appearance of some of the pycnidia and remarked: "These bodies secrete a large amount of nectar which oozes from their mouths and spreads out on the surrounding leaf surface. This attracts bees and other insects which devour the nectar and undoubtedly disseminate the spores." Lloyd and Ridgway are the only observers who have recorded seeing bees visiting the nectar of Rust pycnidia. A confirmation of the occurrence of such visits, which were unknown to Ráthay, is desirable. Perhaps the bees seen by Lloyd and Ridgway were accidental and not regular seekers of the Cedar-apple nectar.

Cytological Background.—The cytological background of Craigie's work will now be sketched.

In 1880, Schmitz[3] discovered paired nuclei in the mycelium and uredospores of *Coleosporium campanulae*.

[1] *Ibid.*, p. 14.

[2] F. E. Lloyd and C. S. Ridgway, "Cedar-Apples and Apples," *Alabama Department of Agriculture*, Bull. XXXIX, 1911, p. 10. Same article in *Annual Report of Alabama Dept. of Agric.*, 1911.

[3] F. K. J. Schmitz, "Untersuchungen über die Struktur des Protoplasmas und der Zellkerne der Pflanzenzellen," *Verh. des naturhistorischen Vereins der preussischen Rheinlande und Westfalens*, Jahrg. XXXVII, 1880 (*Sitzb. Niederrh. Ges. Natur. Heilk. Bonn*, 1880, p. 195).

In 1892, Rosen[1] observed in *Uromyces pisi* a single nucleus in the pycnidiosporophores and pycnidiospores and two nuclei in the aecidio-sporophores, aecidiospores, intercalary cells, and cells of the peridium. He also found two nuclei in each of the two cells of a young teleutospore of *Puccinia asarina*. He noted that, as a teleutospore ripens, the two nuclei in each pair come very close together, and he suggested that in the end they fuse.

In January, 1893, Dangeard and his pupil Sappin-Trouffy[2] con-firmed Rosen's work. As a result of investigations made on *Puccinia graminis, P. poarum, P. caricis, Uromyces pisi*, Phragmidium, Coleo-sporium, etc., they concluded that, in the Uredinales in general, the pycnidiospores contain a single nucleus and the aecidiospores (and associated cells), the uredospores, and the young teleutospores two nuclei. They supposed that, in each of the two cells of a young teleutospore of a Puccinia and in each of the three cells of a young teleutospore of Triphragmium, there is at first a single nucleus and that in each cell the nucleus divides. They therefore regarded the two nuclei of each pair in a teleutospore as sister nuclei.

In February, 1893, Dangeard and Sappin-Trouffy[3] discovered that, in *Puccinia buxi*, a young teleutospore contains two nuclei and that, when the cell-wall of the spore is becoming cutinised, the two nuclei fuse together and form a single larger nucleus. They then made similar observations on *P. graminis, P. coronata, P. menthae, Uromyces geranii, U. betae, Triphragmium ulmariae, Coleosporium euphrasiae, Melampsora farinosa*, and *Phragmidium rubi* and showed that nuclear fusion in teleutospores is general throughout the Uredi-nales. They called the fusion phenomenon *pseudo-fécondation* be-cause, as they then supposed, it had taken the place of normal sexuality and involved sister nuclei only.

In 1895, Poirault and Raciborski[4] gave an account of nuclear division in the Uredinales (aecidia of *Peridermium pini* var. *acicola* [*Coleosporium senecionis*], *Aecidium leucospermum, Uromyces pisi, Puccinia poarum*, etc.; teleutospores of *P. liliacearum* and *Coleo-sporium euphrasiae*; spermatigenous filaments of *P. liliacearum*).

[1] F. Rosen, "Beiträge zur Kenntnis der Pflanzenzellen. II. Studien über die Kerne und die Membranbildung bei Myxomyceten und Pilzen," Cohn's *Beitr. z. Biol. d. Pfl.*, Bd. VI, 1892, pp. 255-257.

[2] P. A. Dangeard and P. Sappin-Trouffy, "Recherches histologiques sur les Urédinées," *Compt. Rend. Acad. Sci. Paris*, T. CXVI, 1893, pp. 211-213.

[3] P. A. Dangeard and P. Sappin-Trouffy, "Une pseudo-fécondation chez les Urédinées," *Compt. Rend. Acad. Sci. Paris*, T. CXVI, 1893, pp. 267-269.

[4] G. Poirault and M. Raciborski, "Sur les noyaux des Urédinées," *Jour. de Bot.*, T. IX, 1895, pp. 318-332, 381-388, Text-figs. 1-19 and Plate VI.

They showed that the pairs of nuclei which are present in the aecidiospores and young teleutospores are *not sister nuclei* like the two nuclei in a pollen-grain, but have been derived from two different lines of nuclei (Fig. 7). They called the pairs of nuclei *conjugate nuclei* (noyaux conjugés) and the simultaneous division of the two nuclei of each pair *conjugate division* (division conjugée).[1] Their diagrammatic drawing illustrating conjugate nuclear division is reproduced in Fig. 8.

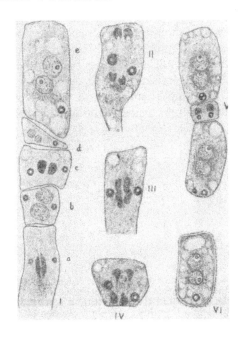

FIG. 7.—*Peridermium pini* var. *acicola* (= *Coleosporium senecionis*). Conjugate nuclear division during the formation of aecidiospores. No. I: *a*, the extremity of a sporogenous filament; beginning of the division; the chromatic masses, each of which corresponds to a nucleus, have the form of a boat with a flat bottom, right and left are nucleoli that came from the nuclei; *b*, nuclei of the rudiment of the aecidiospore before the separation of the sterile cell, the nucleoli that came from the nuclei are still seen right and left; *c*, the first stage of the division which will give the sterile cell to the lower part of the rudiment *c* of the aecidiospore; *d*, intermediary sterile cell detached from the lower part of the spore *e*; *e*, an aecidiospore with two nuclei each of which contains a nucleolus; on the right one can see a nucleolus which came from the preceding nuclear division. No. II: the extremity of a sporogenous filament shows the nuclei in anaphase; to the right and to the left are the big vacuolated nuclei; No. III: separation of the chromosomes in the sporogenous filament. No. IV: division of the nucleus in the aecidiospore (anaphase), to right and left the nucleoli. No. V: two young aecidiospores separated by a sterile cell; in the three cells one can see the extranuclear nucleoli which persist for a long time and are found in the nearly mature spore No. VI. Drawn by Poirault and Raciborski (1895).

In 1896, Sappin-Trouffy[2] gave an account of his comprehensive studies on the nuclear condition of the mycelia and fructifications of a large number of Rust species, including *Puccinia graminis*[3]; and,

[1]*Ibid.*, p. 382.

[2]P. Sappin-Trouffy, "Recherches histologiques sur la famille des Urédinées," *Le Botaniste*, T. V, 1896, pp. 59-244.

[3]*Ibid.*, pp. 92-104. The account of the nuclear condition of *Puccinia graminis*, provided with eleven admirable illustrations, is very complete.

for Rusts possessing an aecidium (eu- and -opsis forms), he presented the cycle of nuclear development as follows. The mature teleutospore is always uninucleate, and it produces one or more basidia. In each basidium the nucleus divides twice and one of the divisions is probably a reduction division. Each of the four basidiospores is uninucleate. The mycelium arising from a basidiospore contains nuclei which are arranged singly, usually one in each cell. In the aecidia, which are borne on the uninucleate mycelium, the nuclei become paired. The two nuclei in each of the basal cells which form aecidiospores divide conjugately so that all aecidiospores cut off from the basal cells are binucleate. The conjugate division of the nuclei is continued so that conjugate pairs of nuclei come to be present in: the mycelium produced by an aecidiospore; the uredo-spores if present (*cf.* Fig. 9); the mycelium produced by a uredospore (*cf.* Fig. 97, B, p. 284); and, finally, the young teleutospores. In each cell of a teleutospore the two nuclei fuse to form a single nucleus.

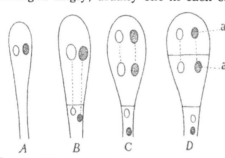

FIG. 8.—Diagram to illustrate conjugate nuclear division during the formation of the pedicel and the two cells of a teleuto-spore of a Puccinia. The first division of the two conjugate nuclei (noyaux con-jugées) in the swelling (A) of the teleuto-spore is followed by the formation of a transverse wall (B) separating the pedicel from the body, hitherto undivided. Then the two nuclei of the body divide simul-taneously in their turn (C); a wall appears separating the two groups of nuclei thus formed (D), and the bilocular teleutospore of the Puccinia is then constituted in all its essentials. *a a*, brother nuclei. Copied by the author from the drawing of Poirault and Raciborksi (1895).

Sappin-Trouffy in 1896 fully recognised that the two nuclei which fuse together in each cell of a teleutospore are not sister nuclei as he and Dangeard had thought in 1893, but are *non-sister nuclei* derived by conjugate nuclear divisions from the first pairs of nuclei formed in the young aecidium. He therefore no longer regarded the fusion of the two nuclei in a teleutospore as a "pseudo-fécondation," but as a *truly sexual process* comparable with the fusion of the two nuclei in the ascus of Ascomycetes and in the basidium of the Hymenomycetes. Indeed, Sappin-Trouffy now considered that, in the life-history of a Rust fungus, the fusion of nuclei in the teleutospore is the most im-portant stage.

Sappin-Trouffy observed that the hyphae which produce the pycnidia, the pycnidiosporophores, and the pycnidiospores are all uninucleate. However, he assumed that the pycnidiospores are conidia, thought he had seen those of *Puccinia rubigo-vera* germinating, and did not regard these cells as taking any active part in the sexual cycle.

Although impressed by the importance of nuclear fusion in the teleutospore, Sappin-Trouffy paid but little attention to the problem of the transition from the uninucleate to the binucleate phase at the

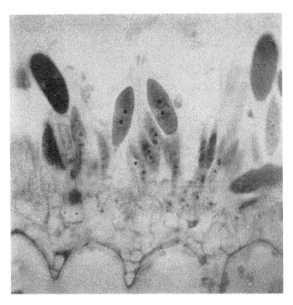

FIG. 9.—*Puccinia graminis.* A uredospore sorus on a straw of Wheat. A pair of conjugate nuclei can be seen in several of the young uredospores. Preparation made by Margaret Newton and T. Johnson, photographed by A. Savage at Winnipeg (*Phytopathology*, 1927, Fig. 4, p. 722). Magnification, about 700.

base of the young aecidium; and he contented himself with stating: (1) that from the mycelium with single nuclei binucleate cells grow up; and (2) that from these cells there are cut off a series of binucleate aecidiospore-mother-cells from which, by division, the binucleate aecidiospores are derived.

Maire[1] accepted the findings of Sappin-Trouffy; and he stated in 1900 for *Endophyllum sempervivi* and in 1902 for *Puccinia bunii* that

[1]R. Maire: "L'évolution nucléaire chez Endophyllum," *Journal de Botanique*, T. XIV, 1900, p. 85; and "Recherches cytologiques et taxonomiques sur les Basidiomycètes," *Bull. Soc. Myc. France*, T. XVIII, 1902 (mémoire with special pagination), pp. 38-39.

the cells of the young aecidium which give rise to the aecidiospore-mother-cells are at first uninucleate, but later become binucleate by a process of simple nuclear division ("association de deux noyaux-frères en synkaryons").[1] However, the idea that in the Rusts in general the first pairs of nuclei formed at the base of a young aecidium are sister nuclei was soon to be upset by Blackman.

Blackman[2] in 1904 made the important discovery that the binucleate condition of *Phragmidium violaceum* originates in the aecidial spore-bed by nuclear migration from one cell to another; and the announcement of this discovery led other workers to investigate the origin of the binucleate condition in the Rust Fungi in general.

Dangeard and Sappin-Trouffy had supposed that the fusion of the two nuclei in a teleutospore is a process of fertilisation and that the teleutospore could be regarded as an egg-cell. Blackman[3] was unable to accept this interpretation; and he held that the association of nuclei in pairs in the young aecidium by nuclear migration in *Phragmidium violaceum* is the most essential part of fertilisation, and that the long-delayed ultimate fusion of the pairs of nuclei in the teleutospores merely completes the process and prepares for chromosome reduction.

Having observed that the pycnidiospores have "a large dense nucleus, very little cytoplasm, no reserve material, and a very thin cell-wall," Blackman[4] remarked: "These characters, together with their usual association with the aecidia, their absence of function, and the peculiar, apparently reduced form of fertilisation to be observed in the aecidium of *Phragmidium violaceum*, point clearly to the view that the spermatia are male cells which formerly took part in a process of fertilisation in connexion with the aecidium, but have now become functionless." The fact that the pycnidiospores of the Uredinales are not mere vestigial structures but function at the present time was destined to be solved not by cytological methods, such as those employed by Blackman, but by means of experiment.

Christman[5] in 1905 found that, in the aecidial spore-bed of *Phragmidium speciosum*, the two nuclei of a conjugate pair are brought

[1]*Ibid.* (1902), p. 42.

[2]V. H. Blackman, "On the Fertilization, Alternation and Generations, and General Cytology of the Uredineae," *Annals of Botany*, Vol. XVIII, 1904, pp. 323-373, Plates XXI-XXIV. Preliminary announcement in *The New Phytologist*, Vol. III, 1904, pp. 24-27.

[3]*Ibid.*, pp. 364-365.

[4]*Ibid.*, p. 363.

[5]A. H. Christman, "Sexual Reproduction in the Rusts," *Botanical Gazette*, Vol. XXXIX, 1905, pp. 269-274.

together by the lateral fusion of two elongated cells forming part of a specialised palisade layer of cells directed toward the epidermis; and he suggested that the·nuclear migration observed by Blackman in *P. violaceum* might have been due to a pathological condition.

Nuclear migration during the initiation of the binucleate stage in the Uredinales has been observed not only by Blackman, but also by Blackman and Fraser (1906),[1] Welsford (1915),[2] Kursanov (1915),[3] Lindfors (1924),[4] and others; but most later workers, including Olive (1908),[5] Kursanov (1910 and 1915),[6] Fromme (1912),[7] Maire (1913),[8] Mme Moreau (1914),[9] Colley (1918),[10] Adams (1919),[11] Lindfors (1924),[12] and Dodge (1924),[13] have found that in many Rust species the conjugate condition of the nuclei in the aecidial spore-bed is due to cell-fusion.

Kursanov,[14] in 1917, came to the conclusion that the appearance of nuclear migration from one cell to another at the base of aecidia is

[1]V. H. Blackman and H. I. C. Fraser, "Further Studies of the Sexuality of the Uredineae," *Annals of Botany*, Vol. XX, 1906, pp. 35-48, Plates III and IV.

[2]E. J. Welsford, "Nuclear Migrations in *Phragmidium violaceum*," *Annals of Botany*, Vol. XXIX, 1915, pp. 293-298.

[3]L. Kursanov, "Morphological and Cytological Researches on the Uredineae," *Sci. Men. Imp. Univ. Moscou*, No. XXXVI, 1915, 288 pp. In Russian. A French summary under the title "Recherches morphologiques et cytologiques sur les Urédinées" appeared in *Bull. Soc. Nat. Moscou*, n.s. (1917), Vol. XXXI, 1922, pp. 102-114.

[4]T. Lindfors, "Studien über den Entwicklungsverlauf bei einigen Rostpilzen aus zytologischen und anatomischen Gesichtspunkten," *Svensk. Bot. Tidskr.*, Vol. XVIII, 1924, pp. 1-84.

[5]E. W. Olive, "Sexual Cell Fusions and Vegetative Nuclear Divisions in the Rusts," *Annals of Botany*, Vol. XXII, 1908, pp. 336-341, 356.

[6]L. Kursanov: "Zur Sexualität der Rostpilze," *Zeitschrift f. Bot.*, Bd. II, 1910, pp. 81-93; and L. Kursanov, 1915, *loc. cit.*

[7]F. D. Fromme, "Sexual Fusions and Spore Development of the Flax Rust," *Bull. Torrey Bot. Club*, Vol. XXXIX, 1912, pp. 113-131.

[8]R. Maire, "La biologie des Urédinales," *Prog. Rei Bot.*, Vol. IV, 1913, pp. 111-115.

[9]Mme F. Moreau, "Les phénomènes de la sexualité chez les Urédinées," *Le Botaniste*, T. XIII, 1914, pp. 145-285.

[10]R. H. Colley, "Parasitism, Morphology and Cytology of *Cronartium ribicola*," *Journal of Agric. Research*, Vol. XV, 1918, pp. 630-631.

[11]J. F. Adams, "Sexual Fusions and Development of the Sexual Organs in the Peridermiums," *Penn. State College, Agric. Experiment Station*, Bull. CLX, 1919, pp. 31-76.

[12]T. Lindfors, *loc. cit.*

[13]B. O. Dodge, "Uninucleated Aecidiospores in *Caeoma nitens* and associated Phenomena," *Journal of Agric. Research*, Vol. XXVIII, 1924, pp. 1045-1058.

[14]L. Kursanov, *loc. cit.*, pp. 104-107.

sometimes an artifact, due to fixation, and comparable with a similar phenomenon that had been observed by Miehe (1901)[1] as a result of wounds made in the leaves of Phanerogams; and he cited *Triphragmium ulmariae* as a species in which he had seen the artifact in question. Furthermore, he reminded his readers that Wahrlich (1893)[2] had shown that there is a minute open pore in the centre of each septum in the Fungi in general and in the Rust Fungi in particular, and he said that the possibility of nuclei being artificially forced through a pore from one Rust cell to another could not be left out of consideration. Nevertheless, Kursanov held: that true nuclear migration does actually take place; that he had obtained evidence thereof in *T. ulmariae* and *Puccinia suaveolens* and, perhaps sometimes, in *Trachyspora alchemillae* (*Uromyces alchemillae*); and that Blackman's findings in respect to the migration of nuclei in *Phragmidium violaceum* need no longer be doubted.

In *Puccinia graminis*, during the passage from the uninucleate to the binucleate stage at the base of the aecidial fundament, Hanna (1929)[3] observed irregular cell-fusions between two or more cells; but, in this same species, Miss Allen (1930)[4] failed to find any indications of cell-fusions whatsoever.

In *Trachyspora intrusa* and *Triphragmium ulmariae* it was found that the initiation of the binucleate condition in the young primary uredospore sorus is accomplished by both cell-fusions and nuclear migrations.

Trachyspora intrusa (*T. alchemillae*, *Uromyces alchemillae*) is an autoecious species with a systemic mycelium. The initiation of its dikaryophase in the rudiment of the primary uredospore sorus is accomplished, according to Kursanov (1917)[5] and Lindfors (1924),[6] in the following ways: (1) the complete or partial fusion of two fertile

[1]H. Miehe, "Ueber die Wanderungen des pflanzlichen Zellkernes," *Flora*, Bd. LXXXVIII, 1901, pp. 115-127. *Cf.* J. H. Sweidler, "Über traumatogene Zellsaft- und Kernübertritte bei *Moricandia arvensis* DC.," *Jahrb. f. wiss. Bot.* Bd. XLVIII, 1910, pp. 551-590; and G. Ritter, "Ueber Traumotaxis und Chemotaxis des Zellkerns," *Zeitschr. f. Bot.*, Bd. III, 1911, pp. 1-42.

[2]W. Wahrlich, "Zur Anatomie der Zellen bei Pilzen und Fadenalgen," *Scripta Botanica Horti Univers. Imp. Petropolitanae*, T. IV, 1893, pp. 101-155, Tab. II-IV.

[3]W. F. Hanna: "Nuclear Association in the Aecium of *Puccinia graminis*," *Nature*, Vol. CXXIV, 1929, p. 267; and in more detail in *Report of the Dominion Botanist for 1929*, Dept. of Agric., Ottawa, 1931, pp. 49-52.

[4]Ruth F. Allen, "A Cytological Study of Heterothallism in *Puccinia graminis*," *Journal of Agric. Research*, Vol. XL, 1930, p. 602.

[5]L. Kursanov, in Russian, 1915, and in French, 1917, *loc. cit.*, pp. 50-51.

[6]T. Lindfors, *loc. cit.*, pp. 19-23.

(palisade) cells lying side by side (K. and L.); (2) nuclear migration of the Blackman type laterally from one fertile cell to another (K.); (3) nuclear migration from a vegetative cell to a fertile cell in the same or a different hypha through a larger or smaller opening in the cell-wall (K. and L.); and (4) fusion between two vegetative cells which have no definite orientation to the surface of the leaf and often lie parallel to the epidermis (L.).

In *Triphragmium ulmariae*, Lindfors (1924)[1] observed that the initiation of the binucleate condition in the primary uredospore sorus is accomplished in three different ways: (1) cell fusion of two fertile (palisade) cells lying side by side at the same level; (2) nuclear migration of the Blackman type from a lower non-fertile cell to a higher fertile cell in the same hypha; and (3) nuclear migration from a non-fertile cell under a fertile cell laterally into a fertile cell of another hypha.

The discovery that, in one and the same species, both the Blackman and the Christman modes of initiating the dikaryophase may be employed served to cast doubt on the Blackman and Christman theories as to the nature of the fertile and non-fertile cells which co-operate in the diploidisation process.

The cytologists who worked before Craigie published his two letters to *Nature* in 1927 had no experimental knowledge of the diploidisation process in Rust Fungi and, therefore, were not in a position to interpret their findings correctly. They thought that the nuclei which come together in pairs in the young aecidiosporophores are all of local origin, whereas we now know that, in a heterothallic species such as *Puccinia graminis* or *Phragmidium speciosum*,[2] one of each pair is of local origin and the other has come from a distance (from a pycnidio-spore or from the hyphae of another mycelium). On the assumption that the two nuclei in the first conjugate pairs of nuclei formed at the base of the young aecidium come from adjacent cells, it was believed that the pycnidiospores must be functionless.

Blackman (1904),[3] as noted above, worked with the caeomoid aecidia of *Phragmidium violaceum*. He regarded the aecidiosporo-phores before their diploidisation as female cells and the buffer cells above them as degenerate trichogynes, and he supposed that the

---

[1]*Ibid.*, pp. 16-19.

[2]*Phragmidium speciosum* has been proved to be heterothallic by A. M. Brown working in the Dominion Rust Research Laboratory: "The Sexual Behaviour of Several Plant Rusts," *Canadian Journal of Research*, Section C, Vol. XVIII, 1940, pp. 18-21.

[3]V. H. Blackman, *loc. cit.*

female cells become fertilised by nuclei which migrate to them from subjacent vegetative cells. Christman (1905)[1] studied the origin of the caeomoid aecidia of *P. speciosum* and the cupulate aecidia of *Uromyces caladii*. He found that, in these species at thè base of the young aecidia, the oblong cells fuse together in pairs and so form binucleate aecidiosporophores; and he regarded the fusing cells as isogametes comparable with those of the Mucoraceae. Then Blackman and Fraser (1906)[2] observed in *Melampsora rostrupi* fusions of the fertile cells in pairs and, in *Uromyces poae*, etc., a migration of nuclei into the fertile cells from below. They regarded both of the cells which fuse together in *Melampsora rostrupi* as female cells; and they put forward the theory that, in the Rust Fungi, normal fertilisation by means of spermatia has been replaced by two different types of *reduced fertilisation*: (1) a female cell fuses (partially) with a vegetative cell, as in *Phragmidium violaceum* and *Uromyces poae*; and (2) the female cells fuse together in pairs, as in *Phragmidium speciosum* and *Melampsora rostrupi*.

Kursanov (1917),[3] in an able discussion based on cytological investigations, rejected the sexual theories of Blackman and Fraser and of Christman and defended the thesis that: (1) the cells which take part in forming the first pairs of nuclei in a young aecidium are ordinary haploid cells and not special sexual cells; and (2) the process of forming these first pairs of nuclei is entirely apogamous and one which may be defined more exactly as a *pseudomixis* (Winkler, 1908) or *pseudogamy* (Hartmann, 1909).

Lindfors (1924)[4] accepted Kursanov's explanation of the origin of paired nuclei in ordinary cupulate aecidia, but favoured Blackman's explanation for caeoma-forms and certain other Rusts, *e.g. Chrysomyxa abietis*, where the "fertile" cells are very strongly developed in a palisade layer.

In 1929, in his admirable text-book on the Uredinales, Arthur[5] remarked that it is safe to say with Harper[6] that in the Rusts "we

[1] A. H. Christman, *loc. cit.*

[2] V. H. Blackman and Helen C. I. Fraser, "Further Studies on the Sexuality of the Uredinales," *Annals of Botany*, Vol. XX, 1906, pp. 35-48, Plates III and IV.

[3] L. Kursanov, "Recherches morphologiques et cytologiques sur les Urédinées," *Bull. Soc. Nat. Moscou*, n.s. (1917), Vol. XXXI, 1922, pp. 102-114.

[4] T. Lindfors, "Studien über den Entwicklungsverlauf bei einigen Rostpilzen aus zytologischen und anatomischen Gesichtspunkten," *Svensk. Bot. Tidskr.*, Vol. XVIII, 1924, pp. 66-67.

[5] J. C. Arthur, *The Plant Rusts (Uredinales)*, New York, 1929, p. 78.

[6] R. A. Harper, "Sexual Reproduction and the Organization of the Nucleus in Certain Mildews," *Carnegie Institute of Washington*, Publication no. XXXVII, 1905, p. 87.

have sexual reproduction by vegetative fertilisation," and he added: "It is a fairly accurate statement to say that in Rusts of all sorts the sexual process begins with the fusion of the cells during or preceding the formation of the aecium, thus instituting a dikaryon, and is consummated with the maturation of the teliospore or microteliospore. The fusing cells act in the place of gametes and may be so called." This view was modified on other pages in the light of Craigie's papers, the first of which appeared while Arthur's book was in MS. or proof.

Arthur's account of sexuality in the Uredinales was based upon nearly fifty years of cytological work beginning with the observations of Schmitz and Rosen and including the prolonged and more critical investigations of Dangeard and Sappin-Trouffy, Poirault and Raciborski, Blackman and Christman, and their successors, but Craigie's experimental investigations (1927-1933) have revealed to us that in many Rusts: (1) there is a true sexual process that cannot be described as "vegetative fertilisation"; (2) the sexual process is initiated long before the formation of the first pairs of conjugate nuclei; and (3) the cells which fuse together in an aecidial rudiment can no longer be regarded as gametes.

In view of Craigie's discovery of heterothallism in the Uredinales and his proof that the pycnidiospores are functional, and in the light that has been thrown upon the diploidisation process in the Uredinales by our knowledge concerning the diploidisation process in the Hymenomycetes, the idea that, in a heterothallic Rust Fungus, the two cells which fuse together at the base of a young aecidium are (1) a female cell and a vegetative cell, or (2) two isogametes, or (3) two female cells must now be discarded. Discarded also, at least for heterothallic Rusts, must be Kursanov's supposition that the fusion process is apogamous. It is now clear that the sexual process in the Uredinales resembles very closely that of the Hymenomycetes and not that of the Mucoraceae. In later chapters[1] of this book an attempt will be made to re-interpret the observations upon which the older Rust cytologists based their views.

By discovering that the sexual process is initiated in young aecidia by nuclei derived from different cells coming together in pairs as a result of nuclear migration or of cell fusions, Blackman, Christman, and other cytologists made an important contribution to our knowledge of the life-cycle of the Uredinales and thus did their share in preparing the way for Craigie's experiments.

Experimental Background.—Craigie's experiments on the Rust Fungi were suggested by the results of experiments which had been

[1]*Vide infra*, Chapters II and VI.

made on the sexual process in certain other fungi and, in particular, the closely allied group of the Hymenomycetes. This experimental background to Craigie's work will now be examined.

Blakeslee,[1] in 1904, discovered that many of the Mucorineae are heterothallic. He recognised the existence of (+) and (−) strains and showed by means of a series of brilliant experiments that, when a (+) strain is mated with a (+) strain or a (−) with a (−), no zygospores are produced; but that, when a (+) strain is mated with a (−) strain, zygospores are produced in abundance.

In 1914, Edgerton[2] recognised in Glomerella, a genus of Pyrenomycetes, the existence of (+) and (−) strains. Those strains derived from single ascospores, when grown separately, produce perithecia with asci containing ripe ascospores, the (+) strain freely and the (−) strain sparingly and with misshapen asci. When two such strains obtained from a Glomerella on *Populus deltoides* or *Hibiscus esculentus*, etc., are mated on an agar plate, they form a ridge of perithecia along the line of contact. In single asci of these perithecia both (+) spores and (−) spores are formed. When a (+) strain or a (−) strain is mated with non-ascogenous mycelia of species of Gloeosporium or Colletotrichum or with (+) or (−) strains of a Glomerella from another host, no ridges of perithecia at the line of contact are formed. On the basis of these observations Edgerton concluded that the (+) and (−) strains of a Glomerella on *Populus deltoides* or *Hibiscus esculentus*, etc., are sexual in nature. However, such a Glomerella cannot be regarded as strictly heterothallic because the (+) and (−) strains, when grown separately, are each able to produce perithecia. It appears to be homothallic but facultatively heterothallic.

That heterothallism occurs in the Ascomycetes was first definitely proved experimentally by B. O. Dodge[3] in 1920. He found that in the Discomycete *Ascobolus magnificus*: (1) mycelia derived from single ascospores always remain completely sterile; (2) only certain pairs of monosporous mycelia give rise to ascocarps; (3) no carpogonia or antheridia are produced on monosporous mycelia; (4) functional carpogonia and antheridia are produced as a result of mating a (+) and a (−) mycelium; (5) these organs arise on different hyphae; and (6) they are produced as a result of a (+) mycelium and a (−) mycelium meeting and mutually stimulating one another.

[1]A. F. Blakeslee: "Zygospore Formation a Sexual Process," *Science*, Vol. XIX, 1904, pp. 864-866; and "Sexual Reproduction in the Mucorineae," *Proc. Amer. Acad. Arts and Science*, Vol. XL, 1905, pp. 205-319.

[2]C. W. Edgerton, "Plus and Minus Strains in the Genus Glomerella," *American Journal of Botany*, Vol. I, 1914, pp. 244-254, Text-fig. 1 and Plates XXII and XXIII.

[3]B. O. Dodge, "The Life History of *Ascobolus magnificus*," *Mycologia*, Vol. XII, 1920, pp. 115-134.

Perithecia containing asci and ascospores were obtained from *single* ascospores: of *Diaporthe perniciosa* by Miss Cayley[1] in 1923; and of *Valsa kunzei, Diaporthe albo-velata, D. binoculata, D. galericulata, Diatrype stigma*, and *Eutypella fraxinicola* by Wehmeyer[2] in 1924 and 1926. Apparently, therefore, these seven species of stromatous Pyrenomycetes are all homothallic.[3]

In 1923 Kirby,[4] in a brief abstract, stated that *Ophiobolus graminis* is heterothallic, ascocarps being produced only when mycelia derived from single ascospores are grown together in certain combinations; but, in 1926, Davis,[5] working with monosporous mycelia of the same fungus originally isolated by Kirby, found that ascocarps are developed on these mycelia in abundance. Thus, according to Davis, *O. graminis* is homothallic and not heterothallic.

In 1925 Derx[6] concluded that *Penicillium luteum* is heterothallic.

[1]Dorothy M. Cayley, "The Phenomenon of Mutual Aversion between Monospore Mycelia of the Same Fungus (*Diaporthe perniciosa* Marshall) with a Discussion of Sex Heterothallism in Fungi," *Journal of Genetics*, Vol. XIII, 1923, p. 354.

[2]L. E. Wehmeyer: "The Perfect Stage of the Valsaceae in Culture and the Hypothesis of Sexual Strains in the Group," *Michigan Acad. of Sci. Arts and Letters*, Vol. IV, 1924, pp. 395-412; "Further Cultural Life Histories of the Stromatic Sphaeriales," *American Journal of Botany*, Vol. XIII, 1926, pp. 231-247.

[3]There is the possibility that these Pyrenomycetes, while being homothallic, may be also facultatively heterothallic, so that when two mycelia of the same species meet and fuse, these mycelia may now and again exchange nuclei with the result that each pair of nuclei in a young ascus may be made up of a nucleus derived from one of the two mycelia and of another nucleus derived from the other mycelium. In this connexion it may be mentioned that Dodge mated the mycelia of two distinct strains of *Neurospora tetrasperma*, which is homothallic and normally has a (+) nucleus and a (−) nucleus in each of the four spores of each ascus, and found that some of the perithecia that were formed at the junction line were of hybrid origin as shown by genetic analysis of the constitution of dwarf uninucleate unisexual ascospores that were produced in certain abnormal asci (B. O. Dodge: "Inheritance of the Albinistic Non-conidial Characters in Interspecific Hybrids in Neurospora," *Mycologia*, Vol. XXIII, 1931, pp. 41-45; "Crossing Hermaphroditic Races of Neurospora," *Mycologia*, Vol. XXIV, 1932, pp. 7-8; and the results of other experiments communicated to the author in a personal interview).

[4]R. S. Kirby, "Heterothallism in *Ophiobolus cariceti*," an abstract, *Phytopathology*, Vol. XIII, 1923, p. 35. W. H. White, in 1939 ("The Sexuality of *Ophiobolus graminis* Sacc.," *Journ. of the Council for Sci. and Indust. Research*, Canberra, Australia, XII, 1939, 209-212) confirmed Davis's conclusion that *O. graminis* is homothallic. He cultured eight spores taken from each of three asci and found that the isolates separately or in all possible pairs all produced mature perithecia.

[5]R. J. Davis, "Studies on *Ophiobolus graminis* Sacc. and the Take-all Disease of Wheat," *Jour. Agric. Research*, Vol. XXXI, 1925, pp. 801-825.

[6]H. G. Derx, "L'hétérothallie dans le genre *Penicillium* (Note préliminaire)," *Bull. Soc. Myc. France*, T. XLI, 1925, pp. 375-381.

Several ascospores were isolated and germinated, and the mycelia
were grown in all possible combinations. The mycelia, when grown
separately, remained infertile but, when mated, ascocarps were produced in certain combinations. However, in 1932 Emmons,[1] working
with a different culture, was unable to confirm these results, for he
found that mycelia derived from single ascospores give rise to ascocarps containing asci and ascospores not only in *P. luteum*, but also
in eleven other species: *P. bacillosporum*, *P. brefeldianum*, *P. egyp-
tiacum*, *P. ehrlichii*, *P. javanicum*, *P. spiculisporum*, *P. stipitatum*,
P. vermiculatum, *P. wortmanni*, *Carpenteles asperum* Shear (= *P. glau-
cum* of Brefeld), and *Byssochlamys fulva*.

Betts[2] in 1926 proved beyond doubt that *Ascobolus carbonarius* is
heterothallic. He obtained seven mycelia from as many spores taken
from a single ascus and found that three of them were of one sex and
four of them of the opposite sex.

In 1927 Shear and Dodge[3] showed that the Red Bread-Mold Fungus
Neurospora sitophila and also *N. crassa* are heterothallic and that
N. tetrasperma is normally homothallic. *N. tetrasperma* has four
spores in each ascus and it was found that a mycelium derived from
one of these normal spores and grown by itself produces abundant
perithecia. However, Shear and Dodge observed that sometimes an
ascus of *N. tetrasperma* contains more than four spores and that then
some of the spores are under the normal size. It was found that these
dwarf spores give rise to haploid mycelia which, when grown separately,
are sterile but which, when mated, behave like the (+) and (−)
mycelia of a heterothallic species, in that in certain combinations they
fruit readily whereas in other combinations they remain sterile. Later
in 1927, Dodge[4] showed that, when first formed, the normal spores of
N. tetrasperma contain two nuclei, one (+) and the other (−), and
that the dwarf spores each contain a single nucleus, either (+) or (−).

In 1918 Mlle Bensaude,[5] following the lead given by Blakeslee,

[1]C. W. Emmons, "The Ascocarps in Species of Penicillium," *Mycologia*, Vol.
XXVII, 1935, pp. 128-150.

[2]E. M. Betts, "Heterothallism in *Ascobolus carbonarius*," *American Journal of
Botany*, Vol. XIII, 1926, pp. 427-432.

[3]C. L. Shear and B. O. Dodge, "Life Histories and Heterothallism of the Red
Bread-Mold Fungi of the *Monilia sitophila* Group," *Journ. Agric. Research*, Vol.
XXXIV, 1927, pp. 1019-1042.

[4]B. O. Dodge, "Nuclear Phenomena associated with Heterothallism and Homothallism in the Ascomycete *Neurospora*," *Journ. Agric. Research*, Vol. XXXV, 1927,
pp. 287-305.

[5]Mathilde Bensaude, *Recherches sur le cycle évolutif et la sexualité chez les Basidio-
mycètes* (These, Paris), Nemours, 1918, 156 pp.

obtained two mycelia from two basidiospores of *Coprinus lagopus* (her *C. fimetarius*) and found that: *when grown separately*, their cells were uninucleate, they were devoid of clamp-connexions, and they were persistently sterile; whereas, *when they were mated*, the mycelium resulting from the union had conjugate nuclei in its cells, it developed clamp-connexions, and it soon gave rise to normal fruit-bodies.   Thus heterothallism in the Hymenomycetes was demonstrated for the first time.   In 1919-1920, Kniep,[1] working independently, without knowledge of Mlle Bensaude's findings, showed experimentally that *Schizophyllum commune* is heterothallic; and later workers, including Mounce, Vandendries, Brunswik, Newton, Gilmore, Brodie, and Biggs, have found that, in the Hymenomycetes, while a few species are homothallic, most species are heterothallic.[2]

In 1919 Kniep[3] demonstrated heterothallism in the Ustilaginales. He observed that fusion occurs between certain sporidia of *Ustilago violacea* and that the sporidia of one sex are equal in number to the sporidia of the opposite sex.   In 1922, Bauch[4] found that, in *U. violacea*, in certain culture media, monosporidial cultures of one sex grow much better than similar cultures of the opposite sex; and in 1923, he[5] added to our knowledge of the sexual process in *U. violacea* and its variety *macrospora*.   In 1926, Kniep[6] showed that, as judged by fusions between sexually different A and B sporidia, it is possible to form hybrids between the following species of Ustilago: (1) between *U. violacea*, *U. scabiosae*, *U. cardui*, *U. utriculosa*, *U. vinosa*, *U. anomala*, and *U. tragopogonis*; and (2) between *U. longissima*, *U. grandis*,

[1]H. Kniep, "Über morphologische und physiologische Geschlechtsdifferenzierung (Untersuchungen an Basidiomyzeten)" *Verhandl. der Physikal.-med. Gesellschaft zu Würzburg.* Bd. XLVI, 1920, pp. 1-18 (Report of lecture given Nov. 27, 1919).

[2]*Vide* H. Kniep, *Die Sexualität der niederen Pflanzen*, Jena, 1928, pp. 425-427. In 1936, H. J. Brodie ("The Occurrence and Function of Oidia in the Hymenomycetes," *American Journ. of Bot.*, Vol. XXIII, 1936, p. 324) listed for the Hymenomycetes 24 heterothallic species as against 3 homothallic species.   In 1938, Rosemary Biggs, as a result of her cultural studies of the Thelephoraceae and other related lower Hymenomycetes ("Cultural Studies in the Thelephoraceae and Related Fungi," *Mycologia*, Vol. XXX, pp. 64-78), listed 28 species as heterothallic and one only as homothallic.

[3]H. Kniep, "Untersuchungen über den Antherenbrand (*Ustilago violacea* Pers.), *Zeitschr. f. Bot.*, Bd. XI, 1919, pp. 257-284.

[4]R. Bauch, "Kopulationsbedingungen und sekundäre Geschlechtsmerkmale bei *Ustilago violacea*," *Biol. Centralbl.*, Bd. XLII, 1922, pp. 9-38.

[5]R. Bauch, "Über *Ustilago longissima* und ihre Varietät *macrospora*," *Zeitschr. f. Bot.*, Bd. XV, 1923, pp. 241-279.

[6]H. Kniep, "Über Artkreuzungen bei Brandpilzen," *Zeitschr. f. Pilzkunde*, Bd. V, 1926, pp. 217-247.

U. bromivora, *U. hordei*, and *U. perennans*. He also succeeded in hybridising *U. nuda* and *U. tritici* by means of their promycelia and *U. hordei* and *U. bromivora* by means of their sporidia; but he did not succeed in hybridising an Ustilago with netted spores with one with smooth or punctate spores. Craigie's discovery of heterothallism in the Uredinales in 1927 was thus preceded by the discovery of heterothallism and even hybridism in the allied group of the Ustilaginales.

In 1927 Dickinson[1] obtained fusions between monosporidial mycelia of *Ustilago levis* and monosporidial mycelia of *U. hordei* of opposite "gender," but he did not inoculate host-plants and recover chlamydospores from them. Convincing evidence that the two Oat Smuts, *U. avenae* (Loose Smut) and *U. kolleri* (*U. levis*, Covered Smut), can be hybridised was obtained by Hanna and Popp[2] in 1930 and 1931. They inoculated one set of young Oat seedlings with a monosporidial culture of *U. avenae*; another set with a monosporidial culture of *U. kolleri* of opposite sex; and a third set with the two cultures mixed. The Oat plants inoculated with the single cultures did not become smutty, whereas the Oat plants inoculated with the mixed culture produced smutted heads of the loose type containing smut spores which were echinulate.

In 1930, in an article in *Nature*[3] and, in 1931, more fully in Volume IV of these *Researches*,[4] I described the diploidisation process in *Coprinus lagopus*. I brought forward experimental evidence which proves that, in *C. lagopus*, the diploidisation of a haploid mycelium is due to nuclei of opposite sex which (1) enter the mycelium *via* one or more cell fusions, (2) travel freely and rapidly along the leading hyphae and through bridges between the leading hyphae formed by cell-fusions, (3) divide again and again, thus increasing their numbers as required, and, finally, (4) form conjugate mates with all the nuclei originally present in the growing hyphae; and I[5] suggested that an essentially similar diploidisation process leads to the formation of conjugate nuclei in the originally uninucleate cells at the bases of the

---

[1]S. Dickinson: "Experiments on the Physiology and Genetics of the Smut Fungi. Hyphal Fusion," *Proc. Roy. Soc. London*, B, Vol. CI, 1927, pp. 126-136; "Experiments on the Physiology and Genetics of the Smut Fungi. Seedling Infection," *ibid.*, Vol. CII, pp. 174-176.

[2]W. F. Hanna and W. Popp: "Relationship of the Oat Smuts," *Nature*, Vol. CXVI, 1930, pp. 843-844; "Relationship of the Oat Smuts," Abstract, *Phytopathology*, Vol. XXI, 1931, p. 109.

[3]A. H. R. Buller, "The Biological Significance of Conjugate Nuclei in *Coprinus lagopus* and other Hymenomycetes," *Nature*, Vol. CXXVI, 1930, pp. 686-689.

[4]These *Researches*, Vol. IV, 1931, pp. 190-281.

[5]*Ibid.*, pp. 281-287.

proto-aecidia formed by the haploid mycelia of *Puccinia graminis* and *P. helianthi*. The correctness of the main features of my conception of the diploidisation process in heterothallic Rust Fungi, as set forth in 1931, became apparent in 1933 when Craigie announced his discovery that the pycnidiospores of *P. helianthi* fuse with, and deliver their contents to, flexuous hyphae of opposite sex.[1]

Discovery of Heterothallism in the Uredinales.—In 1922, in a lecture on *Sex and Social Organisation in the Hymenomycetes* at University College, London, the writer pointed out that as yet no experimental work on sex in the Rust Fungi had been undertaken and that, to find out whether or not a Rust Fungus is heterothallic, it would be necessary to pair the haploid mycelia derived from basidiospores, as had been done in the Hymenomycetes, and to observe whether or not some pairs remain haploid while others become diploid. The basidiospores would have to be sown on living host-leaves instead of on agar plates and the criterion of change from the haploid to the diploid state would be, not the production of clamp-connexions which are not developed in the Uredinales, but the production of aecidia in which conjugate nuclei are known to be present. It was argued that, if a Rust Fungus is heterothallic and bipolar, 50 per cent. of the pairs of mycelia should give rise to aecidia and the other 50 per cent. should remain sterile.

In 1926, when work directed toward the breeding of Rust-resisting wheats was initiated at the Dominion Rust Research Laboratory at Winnipeg, the writer urged the importance of elucidating the sexual phase of the life-history of *Puccinia graminis* as an aid to obtaining a fuller understanding of the Black Stem Rust disease of wheat and perhaps as a means of reducing the labour required to breed Rust-

---

[1]Certain details of my 1931 conception of the sexual process in heterothallic Rusts now seem to me to be erroneous. (1) I suggested (*loc. cit.*, p. 284) that after, let us suppose, (*a*) nuclei have arrived in (*A*) hyphae at the base of an (*A*) proto-aecidium, the (*A*) hyphae form a new series of uninucleate (*A*) cells and (*a*) cells which lie side by side and afterwards fuse with one another (or by nuclear migration co-operate) to form the binucleate aecidiosporophores. I no longer hold this view. I now believe that all the cells in an (*A*) proto-aecidium are (*A*) cells and that (*A*) cells fuse together and so make way for the (*a*) nuclei to travel and find (*A*) nuclear conjugate mates. This improvement in my 1931 conception of the sexual process in the Rust Fungi will be introduced in later Chapters of this book. (2) I said (*loc. cit.*, p. 285) that to permit of the progress, let us suppose, of an (*a*) nucleus through the hyphae of an (*A*) mycelium, the septa must break down partially or wholly; but I have since learned from Wahrlich's investigation (*vide* these *Researches*, Vol. V, 1933, pp. 89-96) that the septa of Rust Fungi are each provided with a central pore. It seems very likely that these pores are used as passage-ways when an (*a*) nucleus is passing along an (*A*) hypha from one cell to the next.

resisting wheat varieties. Thereupon, Craigie began his experimental investigations on sex in *P. graminis* and *P. helianthi*. His first series of experiments was made by sowing basidiospores (sporidia) on host-leaves and observing what happened.

In July, 1927, Craigie[1] published the results of his work on *Puccinia helianthi* which he summarised as follows:

A. Within two weeks the following happens, and usually within three weeks nothing more happens:

1. Each isolated pustule derived from a monosporidial infection usually becomes 0.6-1.2 cm. in diameter and develops pycnia which excrete nectar, but it does not give rise to any aecia (Fig. 1 [here reproduced as Fig. 10] pustule to the left). The pycnia appear about 8 days after the sowing of the sporidia.

2. In a compound pustule formed by the coalescence of two simple pustules, each simple pustule owing its origin to a monosporidial infection, when the distance between the two centres of infection is not more than about 1 mm., either: (*a*) aecia appear in the compound pustule 10-11 days after the sowing of the spores (Fig. 1, pustule to the right), or (*b*) no aecia appear.

3. When two simple pustules, each derived from a monosporidial infection, arise near to one another, coalesce, and produce aecia: the nearer they are and the sooner they coalesce, the sooner are aecia developed; while the farther apart they are and the later they coalesce, the later are aecia developed.

4. Where in compound pustules, each derived from two monosporidial infections, the centres of infection are not more than 2 mm. apart, the number of compound pustules producing aecia is about 50 per cent. of the whole. This conclusion is based on observations made upon about 175 compound pustules.

B. At the end of three weeks or more rarely less, in respect to pustules both simple and compound which hitherto have not produced aecia, the following happens:

1. A majority of the pustules (about 60 per cent.) never produce aecia, even when the pustules persist for as long as six weeks.

2. A minority of the pustules (about 40 per cent.) produce aecia of normal form and colour. In at least some of these aecia the aeciospores are uninucleate, whereas in aecia produced in a compound pustule 10-11 days after the sowing of the sporidia (*vide* A, 2, above) the aeciospores are all binucleate.

The following theoretical deductions may be drawn from the series of facts just recorded:

1. Since pycniospores[2] appear on every mycelium of monosporidial origin, it is clear that, if the pycniospores are really nothing but non-functional male gametes (spermatia), *Puccinia helianthi* is not dioecious. In other words, the monosporidial mycelia of the Sunflower Rust fungus are *not* of two kinds: (*a*) male, bearing spermatia and (*b*) female, not bearing spermatia.

2. The pycniospores are not functionless male gametes but are simply conidia corresponding to the uninucleate oidia which appear on the monosporous mycelia of such heterothallic Hymenomycetes as *Coprinus lagopus*, *C. niveus*, *Stropharia semiglobata*, and *Collybia velutipes*.

---

[1]J. H. Craigie, "Experiments on Sex in Rust Fungi," *Nature*, Vol. CXX, 1927, pp. 116-117, and Fig. 1.

[2]Spelled *pycnospores* in the original. With Dr. Craigie's consent I have changed the spelling so as to comply with current usage in North America. A.H.R.B.

3. The pycniospores produced on (+) monosporidial mycelia are (+) in their sexual nature, while pycniospores produced on (−) monosporidial mycelia are (−) in their sexual nature.

4. The sporidia are unisexual and produce unisexual mycelia. The (+) and the (−) monosporidial mycelia, and therefore the (+) and the (−) sporidia from which they originate, appear to be about equal in numbers. This suggests that segregation of the (+) and the (−) factors takes place in the promycelium during nuclear division in the same manner as it takes place in the basidium of *Coprinus Rostrupianus* and *C. radians* (= *C. domesticus*).

FIG. 10.—"Under side of a Sunflower leaf which was inoculated on its upper side with sporidia of *Puccinia helianthi*, photographed twenty-three days after inoculation. To the left, a pustule derived from a monosporidial mycelium showing absence of aecia (it had numerous pycnia on its upper side). To the right, a compound pustule formed by the coalescence of two simple pustules each derived from a monosporidial mycelium. The compound pustule has developed typical aecia. Magnified two and one-half the natural size." Photograph and description by J. H. Craigie (*Nature*, July 23, 1927).

5. When two sporidia of *opposite* sex, (+) and (−), are sown close together on a Sunflower leaf so that the pustules arising from the two infections soon coalesce, the two monosporous mycelia come into contact, fuse together, and give rise to normal binucleate aeciospores, each conjugate pair of nuclei formed in the spore-bed consisting of a (+) and of a (−) nucleus derived from a (+) and from a (−) mycelium respectively.

6. When two sporidia of the *same* sex—that is, two (+) sporidia or two (−) sporidia—are sown close together on a Sunflower leaf so that the two pustules arising from the two infections soon coalesce, the two monosporous mycelia come into contact but do not interact sexually, and therefore do not give rise to any aecia.

7. The belated aecia, which appear at the end of about three weeks on pustules of monosporidial origin or on pustules of bisporidial origin where presumably the two sporidia are of one and the same sex, probably arise without any hyphal fusions.

8. In any heterothallic Rust fungus that behaves like *Puccinia helianthi* there is a possibility of two strains of the same species being crossed by means of the union of their monosporidial mycelia within the tissues of one and the same host-plant.

A few experiments have already been made by sowing the sporidia of *Puccinia graminis* on the leaves of the Barberry. The results, so far as they have gone, appear to be similar to those already described for *Puccinia helianthi*.

Notwithstanding the production of belated aecidia in a minority of the simple and compound pustules which for a long time had remained haploid, Craigie's first series of experiments clearly indicated that *Puccinia helianthi is heterothallic*. This was not only a great advance in our knowledge of sex in the Rust Fungi but was also important as a preliminary step toward solving the problem of the function of the pycnidia.

There was the possibility that the mycelia in the exceptional pustules had spontaneously but tardily changed from the haploid to the diploid state, as do some of the haploid mycelia of certain Hymeno-mycetes, *e.g. Schizophyllum commune*,[1] *Coprinus radians*[2] and *C. ros-trupianus*;[3] and Craigie therefore concluded that the aecidia in these pustules "probably arise without any hyphal fusions." The production of these belated aecidia, as Craigie[4] subsequently explained, was doubtless due in part, if not wholly, to the mixing of pycnidial nectar by insects. At the time when the experiments with basidio-spores were being made, the function of the pycnidia was unknown and, therefore, insects were not excluded from the cultures. It happened, however, that the work was carried out during the winter months when, at Winnipeg, owing to the intense cold, there are no insects flying in the open and but very few present in a greenhouse.

In a series of later experiments with *Puccinia helianthi*, when precautions against the visits of insects to the cultures were taken by surrounding each young Sunflower plant with a screen-wire cage, Craigie[5] found that of 228 isolated haploid pustules only 11, or about

[1]Hans Kniep, "Über morphologische und physiologische Geschlechtsdifferenzie-rung," *Verhandl. der Physikal.-med. Gesellsch. zu Würzburg*, Bd. XLVI, 1920.

[2]René Vandendries, "Contribution nouvelle à l'étude de la sexualité des Basidio-mycètes," *La Cellule*, T. XXXV, 1924, pp. 129-155.

[3]Dorothy E. Newton, "The Bisexuality of Individual Strains of *Coprinus Rostrupianus*," *Annals of Botany*, Vol. XL, 1926, pp. 123-127.

[4]J. H. Craigie, "An Experimental Investigation of Sex in the Rust Fungi," *Phytopathology*, Vol. XXI, 1931, pp. 1021, 1027, and 1036.

[5]*Ibid.*, p. 1027.

5 per cent., produced aecidia. "From the time the pustules first appeared," says Craigie, "daily inspections were made of the plants in order to destroy any insects that might have gained access to them. White flies (*Aleyrodes*) and thrips (*Heliothrips*) were the only ones discovered and then only infrequently. The former were rarely found in contact with the pustules, but the latter seemed to be attracted by the nectar and fed upon it in preference to the leaf tissue. Consequently, in spite of the precautions taken, there was the possibility that thrips carried the nectar of one pustule to the pycnia of another, especially on leaves that bore two or more pustules. The assumption is supported by the fact that nine of the pustules that produced aecia occurred on leaves bearing two or more pustules, and only two on plants bearing but a single pustule. If in the experiments just described there had been an absolute exclusion of insects and a complete absence of any means by which the transfer of nectar could have taken place, it is possible that no aecia whatever would have developed in any of the pustules." To decide whether or not aecidia ever arise spontaneously on simple pustules of *P. helianthi*, it will be necessary to make a new series of experiments in which the possibility of insect visits to the cultures will be rigorously excluded.

Craigie in his first series of experiments, as we have seen (B, 2), found that "in at least some" of the aecidia produced belatedly on pustules that had long remained haploid "the aeciospores are uninucleate"; but a few months later he[1] said "Further cytological experience has convinced me that the apparent uninucleate condition of these aeciospores was due to an artefact. The young aeciospores of monosporidial origin that I have investigated more recently have all proved to be binucleate." Thus we may consider that haploid aecidia producing haploid aecidiospores are unknown in *Puccinia graminis*.

It is of interest to note, however, that haploid aecidia giving rise to uninucleate aecidiospores have actually been observed in certain other Rusts: by Mme Moreau[2] (1911) in *Endophyllum euphorbiae-sylvaticae uninucleatum*; by Kursanov[3] (1917) in a form of *Tranzschelia*

[1] J. H. Craigie, "Discovery of the Function of the Pycnia of the Rust Fungi," *Nature*, Vol. CXX, 1927, p. 767.

[2] Mme F. Moreau: "Sur l'existence d'une forme écidienne uninucléée," *Bull. Soc. Myc. France*, T. XXVII, 1911, pp. 489-493; "Les phénomènes de la sexualité chez les Urédinées," *Le Botaniste*, T. XIII, 1914, pp. 177-188; and "Note sur la variété uninucléée de l'*Endophyllum Euphorbiae* (DC) Winter," *Bull. Soc. Myc. France*, T. XXXI, 1915, pp. 68-70.

[3] L. Kursanov, "Recherches morphologiques et cytologiques sur les Urédinées," *Bull. Soc. Nat. Moscou*, n.s. (1917), Vol. XXXI, 1922, pp. 15-16.

pruni-spinosae; and by B. O. Dodge[1] (1924) in a form of *Caeoma nitens*. Kursanov[2] also observed a form of *Aecidium leucospermum* in which the aecidia contained chains of uninucleate aecidiospores with an admixture of a few chains of binucleate aecidiospores (up to 5 per cent.).

Where, in a heterothallic Rust, a self-perpetuating haploid strain producing haploid aecidia has been thrown off, we are presented in the Uredinales with a phenomenon comparable with one already known in certain heterothallic Hymenomycetes, *e.g. Coprinus lagopus*.

In this species, as Hanna[3] found, diploidised mycelia give rise to diploid fruit-bodies bearing four groups of basidiospores, (*AB*), (*ab*), (*Ab*), and (*aB*), and the haploid mycelia, when grown separately on sterilised horse dung, often give rise to haploid fruit-bodies (pale and weakly developed) which produce basidiospores of one group only, namely, the group to which belonged the parental basidiospore.

In 1931, Craigie[4] gave a fuller account of his experiments on *Puccinia graminis* and produced convincing evidence that this species, like *P. helianthi*, is heterothallic. Included in the evidence was a photograph which has been reproduced here as Fig. 11.

FIG. 11.—"Under side of a Barberry leaf showing one compound pustule of *Puccinia graminis* with aecia on the left of the midrib and three monosporidial pustules without aecia on the right of the midrib." The pocked appearance of the three haploid pustules on the right is due to the presence of proto-aecidia. "Photographed 20 days after inoculation. × 2." Photograph and description by J. H. Craigie (*Phytopathology*, November, 1931).

[1]B. O. Dodge, "Uninucleated Aecidiospores in *Caeoma nitens* and Associated Phenomena," *Journ. Agric. Research*, Vol. XXVIII, 1924, pp. 1045-1058.

[2]L. Kursanov, *loc. cit.*, pp. 17-18.

[3]W. F. Hanna, "Sexual Stability in Monosporous Mycelia of *Coprinus lagopus*," *Annals of Botany*, Vol. XLII, 1928, pp. 378-389.

[4]J. H. Craigie, "An Experimental Investigation of Sex in the Rust Fungi," *Phytopathology*, Vol. XXI, 1931, pp. 1019-1021.

In 1931, Craigie[1] also observed that, under natural conditions in the open, in *Puccinia graminis*, *P. coronata*, *P. pringsheimiana*, and *Gymnosporangium* sp. (*corniculans?*), pustules of monosporidial origin, while producing pycnidia, frequently fail to develop aecidia, especially when the pustules have been covered by a layer of cheese-cloth and the opportunity for the transference of nectar by insects has been reduced. Craigie concluded that his observations indicate that, like *Puccinia graminis* and *P. helianthi*, *P. coronata*, *P. pringsheimiana*, and *Gymnosporangium* sp. (*corniculans?*) are heterothallic.

Discovery of the Function of the Pycnidia of the Uredinales.—As soon as Craigie had conclusively proved that long-cycled Rust Fungi, *e.g. Puccinia graminis* and *P. helianthi*, are heterothallic, the way was opened for the discovery of the function of the pycnidia.

On May 17, 1927, at Craigie's request, I visited the Dominion Rust Research Laboratory to inspect his cultures prepared for further experiments on heterothallism in the Rust Fungi. On arrival I went into the greenhouse and looked around. On the benches were a number of young Sunflower plants which had been inoculated with basidiospores about ten days previously and whose leaves bore isolated and fairly large (+) or (−) haploid pustules of *Puccinia helianthi* (cf. Fig. 12). All at once I heard a fly, a solitary fly—perhaps the first to appear in the greenhouse after the winter—buzzing near me; and then I saw it settle on a Sunflower leaf (*cf.* Fig. 13, A), run up to a rust pustule (B), suck up some of the drops of nectar protruding from the pycnidia (C and D), and then fly off to another Sunflower leaf where it again visited a rust pustule and again sucked up some of the nectar. Immediately thereafter, the solution of the problem of the function of the pycnidia flashed into my mind. About five minutes later Mr. Craigie joined me. I turned my back on the fly and asked Mr. Craigie if he knew what was the function of the pycnidia. He said that they were supposed to be functionless male organs. I then asked him if he had seen flies visiting pycnidia and he said no, and so I told him about Ráthay's observations[2] on pycnidia and insects which, as a young man, I had found recorded in Plowright's book[3] and upon which, from the point of view of the possible function of pycnidia, I had long pondered. I then asked Mr. Craigie if he would like to see a fly visiting a pycnidium and he said he should. I then directed his attention to the solitary fly in the greenhouse, and we saw it visit two

[1]*Ibid.*, pp. 1028-1030, 1036.

[2]E. Ráthay, *loc. cit. Vide supra*, p. 30.

[3]C. B. Plowright, *A Monograph of the British Uredineae and Ustilagineae*, London, pp. 11-12.

rust pustules in succession, just as it had done when I was alone·
I[1,2] then said: "The solution of the problem of the function of the
pycnium is an entomological one. Copy the action of the fly. Take
(+) pycniospores to (−) pycnia and (−) pycniospores to (+) pycnia,
and it may well be that the pycniospores will germinate and bring on
the diploid phase of the mycelium, evidence of which will be given by
the development of aecia and aeciospores on the under side of each
pustule."

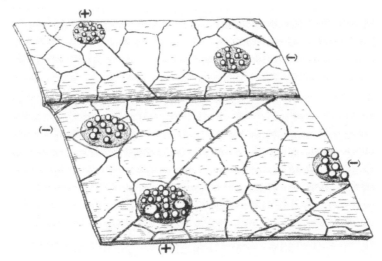

Fig. 12.—Diagram showing (+) and (−) pycnidial pustules of *Puccinia graminis*
on a leaf of *Berberis vulgaris*. The leaf is seen from above. The mycelium in
each pustule has produced a number of pycnidia and from the ostiole of each
pycnidium a drop of nectar has been excreted. Hidden from view within each
drop are numerous minute pycnidiospores and a few flexuous hyphae. Larger
than the natural size.

My suggestion was at once acted upon and, within a week, positive
results began to be obtained. During the ensuing summer, Mr.
Craigie mixed the nectar of large numbers of pycnidia of *Puccinia
helianthi* and of *P. graminis* and in November, 1927, in a letter to
Nature, he[3] described his experiments and stated his conclusions
drawn from them as follows:

[1]J. H. Craigie, "Discovery of the Function of the Pycnia of the Rust Fungi,"
Nature, Vol. CXX, 1927, p̃. 765. In citing this letter I have again changed the
spelling of the word "pycnospore" to "pycniospore."

[2]A. H. R. Buller, "The Plants of Canada, Past and Present," Presidentia
Address to the Royal Society of Canada delivered at the Winnipeg Meeting held ir
May, 1928, *Trans. Roy. Soc. Canada*, 1928, Appendix A, p. 57.

[3]J. H. Craigie, *loc. cit.*, pp. 765-767.

In two sets of experiments with *Puccinia helianthi* on Sunflower leaves, pustules of monosporidial origin, each pustule having developed numerous pycnia but no aecia, were treated as follows: in 184 pustules the pycniospore-containing nectar was mixed with the help of a scalpel, the nectar of any one pustule being mixed with nectar of several other pustules; while, as a control, in 174 pustules the nectar of each pustule was stirred up with a scalpel, but not mixed with any other nectar, the scalpel being carefully sterilised before each operation.

Five days after the experiment had begun, the condition of the pustules was as follows: of the 184 mixed pustules 176 had produced aecia, 4 no aecia, and 4 had wilted and died through leaf-injury; of the 174 unmixed pustules only 20 had produced aecia, while 154 were entirely free from aecia. Under normal conditions when the nectar is neither mixed nor stirred, a certain percentage of monosporidial pustules always produces aecia, as already recorded in my first letter. The appearance of aecia in 20 of the unmixed pustules was therefore in accordance with expectation.

From the experiments just recorded it is clear that mixing the pycniospore-containing nectar leads with rapidity and considerable certainty to the development of aecia. While the pycniospores are haploid, the aeciospores are diploid. We can therefore also say that mixing the pycnial nectar causes each pustule of monosporidial origin to change from the haploid to the diploid phase.

Experiments similar to those just described have been made with *Puccinia graminis* on Barberry leaves. In one set of experiments the pycnial nectar of 116 monosporidial pustules was mixed; while, as a control, the pycnial nectar of each of 85 monosporidial pustules was stirred up separately but not mixed with any other nectar.

Six days after the experiment had begun, the condition of the pustules was as follows: of the 116 mixed pustules 102 had produced aecia and 14 no aecia; whereas of the 85 unmixed pustules only 17 had produced aecia, while 68 were free from aecia.

In the experiments with *Puccinia graminis* just described we again have clear evidence of the function of the pycnia; for, when the nectar is mixed, aecia are rapidly formed in most of the pustules, whereas when the nectar is not mixed, most of the pustules do not develop aecia. A certain percentage of unmixed pustules always produces aecia, as in *Puccinia helianthi*.

In Fig. 1 [here reproduced as Fig. 14] is shown the under side of a Sunflower leaf. The leaf was inoculated with sporidia of *Puccinia helianthi* on July 9. Each pustule originated from a single sporidium and was therefore unisexual. On July 29 the pycnial nectar of the ten pustules on the right side of the leaf was well mixed; while, as a control, the pycnial nectar of each of the six pustules on the left side of the leaf was stirred separately but not mixed. On Aug. 3 the leaf had the appearance shown in Fig. 1, and on Aug. 4 the photograph was taken. The experiment again clearly demonstrates that the pycnia are functional, in that their pycniospore-containing nectar, when transferred from one pustule to another, brings on the diploid phase as shown by the appearance of aecia within five days of the transference.

In Fig. 2 [here reproduced as Fig. 15] is shown the under side of a Barberry leaf. The leaf was inoculated with sporidia of *Puccinia graminis* on Aug. 2. Each pustule originated from a single sporidium and was therefore unisexual. Up to Aug. 19 one pustule on the right side of the leaf had produced aecia. On that day, the pycnial nectar of all the pustules on the right side of the leaf was well mixed; while, as a control, the pycnial nectar of each of the eight pustules on the left side of the leaf was stirred separately but not mixed. On Aug. 28 the leaf was photographed and had the appearance shown in Fig. 2. The effect of mixing

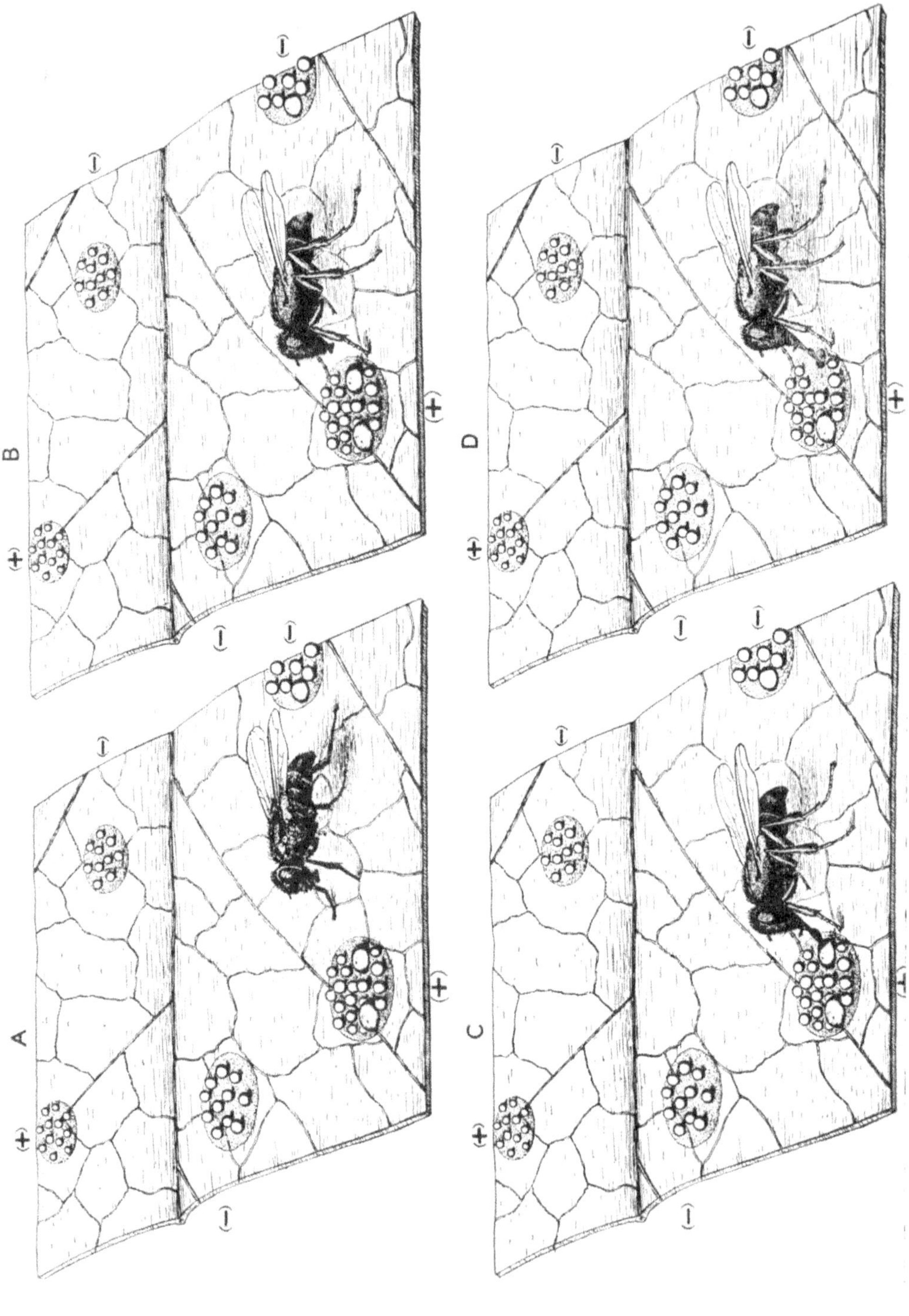
A
B
C
D

the pycnial nectar is very evident: aecia appeared on the right side of the leaf where mixing had been effected, but not on the left side where mixing had been avoided.

Proof that flies mix the pycnial nectar of separate unisexual pustules and so cause the pustules to change from the haploid to the diploid phase, as shown by the appearance of aecia, was obtained with *Puccinia helianthi* as follows:

Fifteen to twenty flies were enclosed in a large screen-wire cage with about twelve pots of Sunflower seedlings, on the foliage leaves of which there were 98 monosporidial pustules bearing pycnia but no aecia. As a control, flies were kept out of another large screen-wire cage which contained fifteen pots of Sunflower seedlings, on the foliage leaves of which there were 159 similar pustules.

Eight days after the beginning of the experiment 96 of the 98 pustules to which flies had had access had produced aecia and only 2 pustules no aecia, whereas only 5 of the 159 pustules to which flies had not had access had produced aecia.

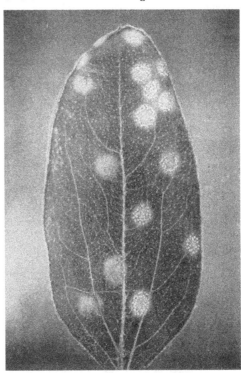

Fig. 14.—"Under side of a Sunflower leaf. × 1½." From Craigie's "Discovery of the Function of the Pycnidia of the Rust Fungi," *Nature*, Nov. 26, 1927.

Fig. 13.—Pycnidial nectar and insects. Diagram showing (+) and (−) pustules of *Puccinia graminis* (each derived from a single basidiospore and therefore haploid) on a leaf of *Berberis vulgaris*, the Common Barberry. Leaf seen from above. Each pustule has produced a number of pycnidia, and from the ostiole of each pycnidium a drop of nectar has been excreted. Hidden from view in each drop are numerous minute pycnidiospores and a few flexuous hyphae. The nectar has an orange-red appearance, emits a sweet odour, and contains sugar. A: a fly has alighted on the leaf. B: the fly has advanced to a (+) pustule. C: the fly has applied its probosis to a drop. D: the fly has sucked up the drop. When passing from pustule to pustule, flies mix (+) and (−) nectar and so convey (+) pycnidiospores to (−) flexuous hyphae and (−) pycnidiospores to (+) flexuous hyphae, thus providing the conditions for fusions between pycnidiospores and flexuous hyphae of opposite sex and the initiation of the sexual process leading to the production of the diploid aecidia on the under side of each pustule. Larger than the natural size.

It was found that in *Puccinia helianthi*, and also in *P. graminis*, nectar which had been heated to 70° C. to kill the pycniospores is not effective in inducing the production of aecia when mixed with the nectar of other pycnia on the living leaf. This indicates that it is the pycniospores which are the effective agents in inducing the formation of aecia, and not the nectar. [*Vide* Fig. 16.]

In a series of experiments with *Puccinia helianthi*, and in another series with *P. graminis*, the pycnial nectar of one monosporidial pustule was removed in a capillary tube and divided into several drops, and then the drops were applied singly to the pycnia of as many pustules as there were drops. In response to this treatment some of the pustules produced aecia and others did not, thus indicating that the pycniospores are of two kinds, (+) and (−). The full details of these experiments will be recorded elsewhere.

FIG. 15.—"Under side of a Barberry leaf. × 2." From Craigie's "Discovery of the Function of the Pycnidia of the Rust Fungi," *Nature*, Nov. 26, 1927.

It appears that, under natural conditions, there are three ways in which pustules of monosporidial origin may change from the haploid to the diploid condition: (1) by a (+) sporidium and a (−) sporidium settling on a leaf close together, so that they form pustules which coalesce in such a way that the (+) mycelium and the (−) mycelium come into contact directly; (2) by means of flies which carry (+) pycniospores from one isolated pustule to the (−) pycnia of another isolated pustule or, conversely, (−) pycniospores of one isolated pustule to the (+) pycnia of another isolated pustule; and (3) spontaneously. The cause of the spontaneous change of a certain number of the pustules of *Puccinia helianthi* and of *Puccinia graminis* from the haploid to the diploid condition is at present unknown, but the phenomenon finds its parallel among the Hymenomycetes in *Coprinus radians* investigated by Vandendries and in *C. Rostrupianus* investigated by D. E. Newton.

The pycnia attract flies and reward them for their visits in very much the same way as do flowers or the Stinkhorn Fungus. They occur chiefly on the upper side of the leaves, where they are readily accessible to insects; they are usually yellow or red in colour, by which means—and perhaps also by the refraction and reflection of light in the drops of nectar—they are made conspicuous; in some species, *e.g. Puccinia suaveolens,* and possibly in many, they emit an attractive odour; while, finally, the nectar contains sugar, and on this account is sipped by flies with avidity.

It has long been remarked that, in those rust fungi which possess them, the pycnia are the first spore-producing organs to appear. Since they play such an important part in changing the haplophase into the diplophase and in inducing the formation of aecia, their appearance on the mycelium before the aecia can now be readily understood. Pycnia precede aecia, because by pycnial action aecia are formed.

The crossing of two physiological forms of *Puccinia graminis,* etc., might be effected in the following relatively simple manner: obtain monosporidial pustules of both strains and then mix the pycnial nectar of a (+) pustule of one strain with the nectar of a (−) pustule of the other strain or, conversely, mix the nectar of a (−) pustule of one strain with the nectar of a (+) pustule of the other strain.

Fig. 16.—"Under side of a Barberry leaf showing 4 monosporidial pustules of *Puccinia graminis.* Sixteen days after inoculation, the pustules were free from aecia. At that time mixed nectar heated for 3 hours at 70° C. was applied to the upper surface of the 2 pustules on the left of the midrib, while unheated mixed nectar was applied to the upper surface of the 2 pustules on the right of the midrib. The right-hand pustules alone have developed aecia. Photographed 8 days after the nectar was applied. × 1.5." Photograph and description by J. H. Craigie (*Phytopathology,* November, 1931).

Thus, by a series of laborious and well-contrived experiments Craigie proved conclusively that the pycnidia of the Rust Fungi are functional.

Craigie, as we have seen, concluded that the sexual process in *Puccinia helianthi* and in *P. graminis* can be initiated in three different ways: (1) by the interaction of the mycelium of a (+) and of a (−) pustule within the tissues of a host-leaf; (2) by pycnidiospores being transferred from a (+) pustule to a (−) pustule, or *vice versa;* and (3) sometimes apparently spontaneously.

The first two ways of initiating the sexual process in long-cycled Rusts have been conclusively established; but the third, as Craigie[1]

[1] J. H. Craigie, "An Experimental Investigation of Sex in the Rust Fungi," *Phytopathology,* Vol. XXI, 1931, pp. 1021, 1027, 1036. Also *vide supra,* p. 70.

in 1931 himself admitted, has not. To prove beyond doubt that, occasionally, diploid aecidia are produced on haploid mycelia spontaneously would require further experimentation in which the exclusion of tiny running White-flies and Thrips, which are present in the Dominion Rust Research Laboratory often in considerable numbers, would be complete.

As promised, Craigie,[1] in 1931, gave a more detailed account of his experiments on the distribution of sex in the pycnidiospores in *Puccinia graminis* and *P. helianthi*:

> *Sex of pycniospores.* Another experiment was designed to show whether or not the pycniospores of *Puccinia graminis* are of two kinds, (+) and (−). The nectar of one monosporidial pustule of *P. graminis* was drawn off by means of a capillary tube and divided into several drops and then the drops were applied singly to the pycnia of as many pustules as there were drops. The nectar of other simple pustules was divided in like manner and distributed among other pustules. Altogether, 74 individual pustules were thus treated. Six days later, 30 of the 74 pustules had developed aecia, while 44 were free from aecia. As a control 26 similar pustules were left untreated. Aecia appeared in only one of these.
>
> Figure 13, A [here reproduced as Fig. 17] is illustrative of the results. Nectar from one pustule was applied to each one of the four pustules. Aecia arose in two of them but not in the other two. Usually, however, among pustules on the same leaf, the ratio of those with aecia to those without aecia was less evenly balanced.

Fig. 17.—"Under side of a Barberry leaf showing 4 monosporidial pustules of *Puccinia graminis*. The nectar from a monosporidial pustule on another leaf was divided into 4 small drops and added to the 4 pustules, 1 drop to a pustule, when they were 17 days old. Within 5 days aecia arose in two of them but not in the other 2. The photograph was taken 10 days after the nectar was added. × 2." Photograph and description by J. H. Craigie (*Phytopathology*, November, 1931).

[1] *Ibid.*, p. 1025.

The experiment just described was repeated with *Puccinia helianthi*. Five days after the drops of nectar were applied, 15 of the 48 pustules treated had developed aecia, whereas 33 of them were without aecia. In a control of 66 similar pustules, only 5 produced aecia within that time.

Theoretically, if each basidium bears two (+) sporidia and two (−) sporidia, the number of pustules with aecia and the number without aecia should be equal in each experiment. The ratio obtained with the pustules of *P. graminis* is possibly as near the theoretical ratio as could be expected with such a small number of pustules. With the pustules of *P. helianthi*, there is a greater divergence, yet not so great as appears in the control. The rather indifferent ratio given by the pustules of *P. helianthi* is very probably attributable, in some way or other, to the meager quantity of nectar that was being produced by pustules at the time the experiment was performed.

In 1927, Craigie[1] had observed that, in *Puccinia helianthi*, the number of compound pustules (each derived from two basidiospores which had fallen near to one another) which produce aecidia is about 50 per cent. of the whole number; and this conclusion was based on observations made upon about 175 compound pustules. His further work on the sexual reactions of the pycnidiospores of *P. graminis* and of *P. helianthi*, as recorded above, is roughly in accordance with his findings with the compound pustules of *P. helianthi*. It thus appears that in *P. graminis* and *P. helianthi*, the basidiospores, the mycelia derived from them, and the pycnidiospores produced by these mycelia are of two kinds only, (+) and (−), or (A) and (a), as in *Coprinus radians*[2] and *C. rostrupianus*,[3] and not of four kinds, (AB), (ab), (Ab), and (aB), as in *Coprinus lagopus*[4] and *Schizophyllum commune*.[5] With random pairing, where there are only two sexual groups of spores, the chances of successful sexual interaction are 50 per cent.; but, where there are four sexual groups, the chances are reduced to 25 per cent. From the practical point of view of the uredinologist making investigations on the inheritance of pathological and other characters, it is distinctly fortunate that in such Uredinales as *Puccinia graminis* and *P. helianthi* the number of sexual groups of spores is two only and not four.

[1] J. H. Craigie, "Experiments on Sex in Rust Fungi," *Nature*, Vol. CXX, 1927, p. 117. Also *vide supra*, p. 60.

[2] R. Vandendries, "Contribution nouvelle à l'étude de la sexualité des Basidiomycètes," *La Cellule*, T. XXXV, 1925, pp. 129-155.

[3] Dorothy E. Newton, "The Bisexuality of Individual Strains of *Coprinus Rostrupianus*," *Annals of Botany*, Vol. XL, 1926, pp. 105-128.

[4] W. F. Hanna, "The Problem of Sex in *Coprinus lagopus*," *Annals of Botany*, Vol. XXXIX, 1925, pp. 431-457.

[5] H. Kniep, "Über morphologische und physiologische Geschlechtsdifferenzierung," *Verhandl. Phys.-Med. Gesellschaft Würzburg*, Bd. XLVI, 1920, pp. 1-18.

In 1931, Craigie[1] published a diagram, here reproduced as Fig. 18, in which he indicated how the globules of nectar produced by pycnidia of *Puccinia graminis* enlarge and finally fuse to form a layer of nectar over the whole surface of a haploid pustule. He represented the pycnidiospores as being evenly scattered in the nectar; but, as we shall see in the next Chapter, in undisturbed drops of nectar the pycnidiospores are not evenly distributed but adhere together in a gelatinous mass (*cf.* Fig. 50, p. 153).

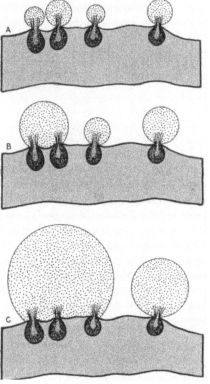

FIG. 18.—*Puccinia graminis* on a leaf of *Berberis vulgaris*. Vertical section of the leaf, to show diagrammatically how the globules of nectar excreted by the pycnidia enlarge and finally fuse. The pycnidiospores have been represented as evenly dispersed in the nectar. Drawn by J. H. Craigie (1931).

In 1929, Arthur[2] raised an objection to the validity of the evidence that Craigie had adduced to prove that, after the nectar of haploid pustules has been mixed, it is the pycnidiospores and not the nectar which initiates the sexual process. Arthur said that the heating of the pycnidial exudate to 70° might inactivate enzymes which the exudate might contain. In 1931, in reply to Arthur's criticism, Cummins[3] confirmed Craigie's conclusion by means of filtration experiments made on the Corn Rust, *Puccinia sorghi*. After applying Craigie's methods and proving that *P. sorghi* is heterothallic, Cummins filtered mixed nectar through a Berkefeld filter and thus removed the pycnidiospores from the nectar at room temperature. He then applied: (1) the filtered mixed nectar to 28 haploid pustules; and (2), as a control, unfiltered mixed nectar to 10 haploid pustules.

[1]J. H. Craigie, "An Experimental Investigation of Sex in the Rust Fungi," *Phytopathology*, Vol. XXI, 1931, p. 1017.

[2]J. C. Arthur, *The Plant Rusts*, New York, 1929, p. 243.

[3]G. B. Cummins, "Heterothallism in Corn Rust and the Effect of Filtering the Pycnial Exudate," *Phytopathology*, Vol. XXI, 1931, pp. 751-753.

During the next ten days, the 28 pustules treated with filtered nectar all remained sterile while the 10 pustules treated with unfiltered nectar all produced aecidia.

Discovery of Flexuous Hyphae in the Pycnidia of the Uredinales and of their Fusion with Pycnidiospores of Opposite Sex.—Craigie's investigations on the pycnidia of *Puccinia graminis* and *P. helianthi*, reported upon in 1927 and 1931, had shown: (1) that pycnidia develop in every pustule of monosporidial origin and all of them produce pycnidiospores; (2) that (+) pycnidia produce (+) pycnidiospores and (−) pycnidia (−) pycnidiospores; (3) that the application of pycnidiospore-containing nectar taken from a (+) pustule to the pycnidia of a (−) pustule, or taken from a (−) pustule to the pycnidia of a (+) pustule, leads to the production of aecidia in the pustule to which nectar is applied; and Cummins showed (4) that, when the pycnidiospores are removed, the nectar loses this activating property. Notwithstanding this great advance in our knowledge of pycnidiospores one question of much interest in respect to these cells still remained unanswered, namely: how do (+) pycnidiospores taken to a (−) pustule, or (−) pycnidiospores taken to a (+) pustule, initiate the diploidisation process in the mycelium in the host-leaf at the base of each rudimentary aecidium? Does a pycnidiospore taking part in the sexual process: (1) germinate and send out a germ-tube which penetrates into the host-leaf and there produces a haploid mycelium which interacts with the haploid pustular mycelium derived from a basidiospore; or does it (2) simply unite with emergent hyphae, periphyses, or other haploid fungal elements projecting above the host-leaf's surface?

It seemed unlikely that pycnidiospores should unite with the straight, conical, stiff-looking, sharply pointed periphyses which support the drops of nectar and the extruded masses of pycnidiospores; and no fusions between pycnidiospores and periphyses had ever been observed. Attention was therefore directed to the possibility of pycnidiospores uniting with hyphae which emerge from the host-leaf through stomata or between ordinary epidermal cells. After the appearance of papers by Andrus and Miss Allen, these emergent hyphae came to be known as *receptive hyphae*; but this designation is a misnomer since the so-called receptive hyphae have never been shown to receive anything and, in particular, they have never been observed to fuse with a pycnidiospore.[1]

[1]An excellent critical review of our knowledge of the so-called receptive hyphae, along with a record of original observations, is given by Dorothy Ashworth in "The Receptive Hyphae of the Rust Fungi,"*Annals of Botany*, Vol. XLIX, 1935, pp. 95-108.

In 1884 de Bary,[1] in his well-known text-book, referred to the occurrence of emergent hyphae. In discussing the mechanism of sex in the Uredineae, he said: "In young groups of aecidia, a phenomenon is observed not infrequently and without great difficulty, which seems to be in favour of the supposition that there is an archicarp duly equipped for conception; short, obtuse, hyphal branches project from some of the stomata, like the tips of the trichogyne in *Polystigma* which may be traced here and there to a young perithecium. But the chief point, the continuity of such a possible trichogyne with the supposed archicarp on the one side and on the other the distinct relation to the spermatia, has not yet been shown; there is nothing in the phenomena observed which compels us to speak of trichogynes and not simply of branches of the mycelium which may as well grow outwards from a stoma as in an inward direction into an intercellular space."

In 1896 Richards,[2] when investigating the aecidia of *Uromyces caladii* on Peltandra again saw emergent hyphae, for he says: "Occasional hyphae were seen protruding from the stomata, but they did not connect with any of the primordia and showed no evidence of specialisation."

In 1904 Klebahn[3] remarked that, when examining the aecidia of various Rusts, he had observed wefts of hyphae in open stomata and that these hyphae appeared to be swollen at the tips and often in contact with pycnidiospores; but he did not observe any fusions between these emergent hyphae and the pycnidiospores.

In 1904, in *Phragmidium violaceum*, Blackman[4] observed that, in the aecidial primordium, uninucleate cells lie side by side and each cuts off a sterile cell at its tip. Referring to such a sterile cell, Blackman said: "Its position above the fertile cell would suggest that it formerly acted as a receptive cell, pushing up between the epidermal cells as a trichogyne to which the sticky spermatia could be brought, for example, by insects. Some support is lent to such a view by the fact that occasionally cases are to be found in which the sterile cells do push up between the epidermal cells and swell out above, being merely covered by cuticle. If the development were pushed one

[1]A. de Bary, *Vergleichende Morphologie und Biologie der Pilze, Mycetozoen und Bacterien*, Leipzig, 1884, p. 300; English Translation, Oxford 1887, p. 278.

[2]H. M. Richards, "On Some Points in the Development of Aecidia," *Proc. Amer. Acad. Arts and Sci.*, Vol. XXXI, 1896, p. 257.

[3]H. Klebahn, *Die wirtswechselden Rostpilze*, Berlin, 1904, pp. 196-197.

[4]V. H. Blackman, "On the Fertilization, Alternation of Generations, and general Cytology of the Uredineae," *Annals of Botany*, Vol. XVIII, 1904, pp. 323-373.

stage further and the cuticle pierced, a very effective receptive organ would be the result."

Blackman's discovery that, in the Rust Fungi, nuclear association takes place at the base of each aecidium was soon supported by the work of other cytologists. The result of this was that for a period of twenty-three years following the publication of Blackman's paper, 1904-1927, it was generally assumed that the so-called spermatia of the Rust Fungi are no longer functional and that their original fertilising activity was taken over by cell fusions in the aecidial primordia.

As soon as Craigie, in 1927, had conclusively proved by experimental methods that the pycnidia of the Rust Fungi are functional, the old idea that the pycnidiospores are spermatia was revived, trichogynes were once more hunted for, and the emergent hyphae observed by de Bary and others again received attention.

In 1931, Andrus,[1] in *Uromyces phaseoli typica* (*U. appendiculatus*) and *U. phaseoli vignae*,[2] described hyphae as emerging at the surface of the host-leaf *via* stomata or by pushing up between the ordinary epidermal cells; and he imagined these hyphae to be the tips of *trichogynes* which he said were "much branched and highly septate organs having their terminus at the epidermis of the host-leaf where they project through the stomata or between epidermal cells and fuse with spermatia transferred by insects or by hand, or by other agencies." The trichogynes were supposed by Andrus to spring from the aecidial primordia. He considered that the two-legged cells (usually regarded as binucleate fusion cells) of the young aecidium are present *before* fertilisation, and he interpreted them morphologically as consisting of an *egg-cell* and, attached to the egg-cell, a stalk-cell and a cell which is the *basal cell of a multicellular trichogyne*. Thus the "branched trichogynes" stretch from the unfertilised "egg-cells" through the leaf-tissues in such a way that their ultimate tips protrude through the epidermis and are exposed to the outer air. The tips of the supposed trichogynes were found to stain more strongly than other parts of the mycelium, thus suggesting a "receptive spot" comparable with that in the oogonium of Vaucheria. The fact that the "trichogynes" were found protruding through the epidermis 8-10 days after the host-

[1]C. F. Andrus, "The Mechanism of Sex in *Uromyces appendiculatus* and *U. vignae*," *Journ. Agric. Research*, Vol. XLII, 1931, pp. 559-587. For remarks on another paper by Andrus supporting his first, *vide infra*, Chapter IV, in a footnote.

[2]In using the names *U. phaseoli typica* and *U. phaseoli vignae*, here as elsewhere for North American Rust species I have followed the nomenclature of J. C. Arthur (*Manual of the Rusts in United States and Canada*, Lafayette, U.S.A., 1934, pp. 296-297).

leaf had been inoculated with basidiospores, at a time when the aecidial primordia were in process of formation, was cited as fitting in with the idea that the emergent hyphae are the tips of trichogynes.

Andrus, in support of his views, failed to produce any convincing evidence that the "spermatia" do actually fuse with the tips of his "trichogynes"; and there can now be no doubt that his branched trichogynes are nothing more than ordinary parasitic intercellular mycelium and their tips merely branches of the mycelium. These branches, as de Bary[1] remarked for other Rusts, "may as well grow outwards through a stoma as inwards into an intercellular space." The idea that in the aecidial primordia of the Rust Fungi there are "egg-cells" with "trichogynes" attached has no satisfactory evidence in its favour.

Miss Allen,[2] following Andrus, examined *Puccinia triticina* for emergent hyphae. In 1932, she found that "while spermogonia are developing, certain hyphae are growing into stomatal apertures or forcing a passage way between the epidermal cells of the leaf"; and she stated that "since these hyphae reach the surface of the leaf and serve to receive the spermatial nuclei, the terms 'receptive hyphae' or 'emergent hyphae' have been applied to them." However, she failed to produce any satisfactory evidence that the "spermatia" and the "receptive hyphae" do actually fuse with one another and admitted that "the actual entry of the spermatial nuclei into the receptive hyphae was not seen."

In 1932, Miss Allen[3] gave an account of receptive hyphae in *Puccinia coronata*. These hyphae were found in rust pustules before the pycnidia are formed. With regard to the hyphae coming to the surface of a leaf between ordinary epidermal cells she says: "These hyphae seem to be unable to pierce the cuticle, but are able to effect a separation between the cuticle and the inner layers of the epidermal wall and to grow for considerable distances beneath the cuticle." Notwithstanding this, as Miss Ashworth[4] has remarked, she later considered that the spermatia scattered along the surface of the epidermis could enter at any point where the mycelium reached the surface. Said she: "The extremely small spermatial nuclei passed through the wall of the epidermal cell into the hyphae."

[1]A. de Bary, *Vergleichende Morphologie und Biologie der Pilze, Mycetozoen und Bacterien*, Leipzig, 1884, p. 300.

[2]Ruth F. Allen, "A Cytological Study of Heterothallism in *Puccinia triticina*," *Journ. Agric. Research*, Vol. XLIV, 1932, pp. 733-754.

[3]Ruth F. Allen, "A Cytological Study of Heterothallism in *Puccinia coronata*," *Journ. Agric. Research*, Vol. XLV, 1932, pp. 513-541.

[4]Dorothy Ashworth, *loc. cit.*, p. 97.

In 1931, Miss Allen,[1] in a paper on *Puccinia graminis*, did not report having seen any emergent hyphae, and she suggested that the diplophase in this Rust is initiated by the fusions of spermatia with the paraphyses of the spermogonia. In 1933, she[2] recorded that, on re-examining her material, she had found emergent hyphae protruding from the stomata of infected Barberry leaves. She concluded that, in *Puccinia*, fertilisation is chiefly effected by fusion of spermatia with spermogonial paraphyses, with fusions between spermatia and stomatal hyphae as an auxiliary process.[3]

Hanna,[4] in 1929, gave an account of his attempt to solve the problem of the means whereby nuclear association takes place in *Puccinia graminis*. After mixing the nectar of (+) and (−) pustules, he observed what seemed to be a few pycnidiospores germinating; and, on this basis, he suggested that, after nectar has been mixed, the pycnidiospores are stimulated to germinate, the germ-tubes penetrate into the host-leaf and produce mycelia, and these mycelia grow down to the hyphal wefts of the rudimentary aecidia and there fuse with cells of opposite sex. Hanna's statement that "some of the pycnidiospores have been observed to germinate" is in reality not a statement of fact, but an interpretation. It is true that he saw pycnidiospores in mixed nectar which looked to him as though they had germinated and put out short germ-tubes; but he did not see these pycnidiospores before they germinated, did not watch them putting out germ-tubes, and did not see any of the supposed germ-tubes elongating hour by hour or day by day. What he saw may have been abnormally elongated pycnidiospores or possibly pycnidiosporophores or buffer cells which had been extruded through the ostiole of one or more pycnidia along with normal pycnidiospores. In any case, he did not observe the supposed germ-tubes of his germinating pycnidiospores penetrating through the host-leaf epidermis or through a pycnidium.

Those who, during the period 1927-1933, were attempting to show that the pycnidiospores of the Rust Fungi either (1) fuse with hyphae of opposite sex emerging through a stoma or (2) germinate and send germ-tubes through the leaf to the base of the rudimentary aecidia

[1]Ruth F. Allen, "A Cytological Study of Heterothallism in *Puccinia graminis*," *Journ. Agric. Research*, Vol. XL, 1930, pp. 585-614.

[2]Ruth F. Allen, "Further Cytological Studies of Heterothallism in *Puccinia graminis*," *Journ. Agric. Research*, Vol. XLVII, 1933, pp. 1-16.

[3]For further critical remarks on emergent hyphae and upon the work of Andrus and Allen, *vide infra*, Chapter IV, including footnotes.

[4]W. F. Hanna, "Nuclear Association in the Aecium of *Puccinia graminis*," *Nature*, Vol. CXXIV, 1929, p. 267. Also *vide infra*, Chapter IV.

were on the wrong track; for, Craigie, in another letter to *Nature*, showed that the fate of the pycnidiospores in mixed nectar was far otherwise than had been supposed.

Craigie[1] in 1933 discovered that from the mouth of a pycnidium of *Puccinia helianthi* or of *P. graminis* there project not only many straight pointed hair-like cells known as periphyses or paraphyses but also a number of hyphae which had previously been overlooked and which he called *flexuous hyphae*; and in *P. helianthi*, some hours after he had mixed the pycnidiospore-bearing nectar of (+) and (−) pycnidial pustules, he observed that some of the flexuous hyphae had each fused with a pycnidiospore presumably of opposite sex; and, a little later in 1933, Pierson[2] reported that he had seen similar fusions in *Cronartium ribicola*. Craigie said:

> The discovery[3] that, in a heterothallic Rust like *Puccinia graminis* Pers. or *P. helianthi* Schw., the transfer of pycniospore-containing nectar from a mono-sporidial (haploid) pustule of one sex to a similar pustule of the opposite sex induces the development of aecia in the pustule receiving the nectar has stimulated interest in the process by which the diploidisation is effected.

> In the examination of free-hand sections of monosporidial pustules of *P. graminis* and *P. helianthi*, I have observed that usually two types of hyphae protrude through the ostiole of a pycnium: (1) the stiff tapering slightly-curved paraphyses which have been frequently figured; and (2) flexuous hyphae which show considerable variation in length, diameter, regularity of outline, and a few other features. In some of the pycnia—probably the older ones—of a pustule, these flexuous hyphae show rather profuse development; in others, only a few of them, sometimes none, are discernible. They may be shorter, but usually they are as long or longer than the paraphyses, not infrequently two or three times as long. They may branch but they rarely show septations. Some may be swollen at the tip. Occasionally a short spur or peg, of less diameter than a branch juts out at a side or tip.

> Several pycniospores in union with such hyphae have been observed in sections of haploid pustules of *P. helianthi* in which the nectar had been previously intermixed, so that presumably both (+) and (−) pycniospores were present on the surface of each pustule and in close proximity to the protruding hyphae.

> Fig. 1 [here reproduced as Fig. 19] shows such a union. The pycniospore is empty; and the hypha has lost most of its cytoplasm. Presumably the nucleus of the pycniospore has passed through the connecting tube into the hypha and proceeded down it, to associate itself in conjugate relationship with some nucleus of the mycelium.

---

[1] J. H. Craigie, "Union of Pycniospores and Haploid Hyphae in *Puccinia helianthi* Schw.," *Nature*, Vol. CXXXI, 1933, p. 25.

[2] R. K. Pierson, "Fusion of Pycniospores with Filamentous Hyphae in the Pycnium of the White Pine Blister Rust," *Nature*, Vol. CXXXI, 1933, pp. 728-729.

[3] Craigie, J. H., NATURE, **120**, 765, Nov. 26, 1927.

With regard to the short spurs or pegs on these hyphae, it is assumed that a hypha of one sex, in response to the presence in its immediate vicinity of a pycniospore of the opposite sex, sends out a short tube to establish contact with that pycniospore.

A nucleus in the act of passing from a pycniospore into a hypha has not been seen, but empty pycniospores found connected by short tubes to these hyphae furnish strong circumstantial evidence that nuclei migrate from pycniospores to these hyphae by way of fusion tubes.

Hyphae which emerge through stomata and between epidermal cells were observed by Andrus[1] in haploid pustules of *Uromyces appendiculatus* and *U. vignae,* and by Allen[2] in haploid pustules of *Puccinia triticina.* Andrus regards these

Fig. 19.—"The union of a pycniospore and a flexuous hypha of *Puccinia helianthi.* The pustule was fixed in formalin acetic alcohol 13½ hours after the intermixing of the nectar was done. × approx. 1,500." Photograph made by A. M. Brown and described by J. H. Craigie (*Nature,* Jan. 7, 1933).

hyphae as the tips of functioning trichogynes. Allen designates them "receptive hyphae." However, neither of these investigators found direct evidence of fusions between 'spermatia' (pycniospores) and trichogynes or receptive hyphae, although both assume from their studies that direct fusions do occur.

There is little doubt that, in haploid pustules of *P. helianthi* (or of *P. graminis*), the function of the protruding hyphae is to establish contact between mycelia of one sex and pycniospores of the opposite sex and thus to serve as an avenue by which the nuclei of the pycniospores reach the internal mycelia of the pustules. The type of union so far observed simulates that of oidium and hypha in the Hymenomycetes.

[1]Andrus, C. Frederic, *J. Agr. Res.,* **42,** 559-587; 1931.
[2]Allen, Ruth F., *J. Agr. Res.,* **44,** 733-754; 1932.

Hyphae emerging at stomatic clefts in rust pustules were again treated of by Miss Rice[1] in 1933 and by Miss Ashworth[2] in 1935; but these authors gave no support to the idea of Andrus and Miss Allen that "receptive hyphae" and pycnidiospores unite with one another.

In 1936, Miss Hunter[3] recorded that she had observed flexuous hyphae in the following species of Melampsoraceae:

Coleosporium helianthi,	Milesia polypodii,
Melampsora abieti-capraearum,	Milesia scolopendrii,
Milesia marginalis,	Milesia vogesiaca,
Milesia polypodophila,	Pucciniastrum americanum;

and she supported her findings with a few photo-micrographs.

Concerning the pycnidia of *Coleosporium helianthi* on the leaves of *Pinus virginiana* and *P. echinata*, Miss Hunter[4] said: "In addition to other features, attention should be called to long centrally located hyphae that extend beyond the spermatiophores and the level of the epidermis. These are 'flexuous hyphae.' Such hyphae may extend as much as 15μ beyond the epidermal level. When the spermogonia have passed full maturity and the spermatia have been discharged there is no trace left of these hyphae."

In respect to the pycnidia of *Melampsora abieti-capraearum* on the leaves of *Abies balsamea* Miss Hunter[5] remarked: "It is of interest to note that a few sections showed long hyphae extending beyond the surface of the spermogonium. They were found centrally located in spermogonia that had just matured, never in old spermogonia. Spermatia were massed around them and, unlike paraphyses, they were ephemeral. These I interpret as the 'flexuous hyphae' of Craigie."

In five species of Milesia on the leaves of Abies Miss Hunter saw structures which she regarded as flexuous hyphae; and, in her account of the pycnidia of *Milesia scolopendrii*, she[6] remarks: "Spermogonia, twenty-seven to thirty days following inoculation on *Abies alba* and from which spermatia were just beginning to discharge, contained

[1]M. A. Rice, "Reproduction in the Rusts," *Bull. Torrey Bot. Club*, Vol. XL, 1933, pp. 23-54.

[2]Dorothy Ashworth, "The Receptive Hyphae of the Rust Fungi," *Annals of Botany*, Vol. XLIX, 1935, pp. 95-108.

[3]Lillian M. Hunter, "Morphology and Ontogeny of the Spermogonia of the Melampsoraceae," *Journal of the Arnold Arboretum*, Vol. XVII, 1936, pp. 115-152, with seven Plates.

[4]*Ibid.*, p. 138.

[5]*Ibid.*, pp. 116-118.

[6]*Ibid.*, p. 130.

unusually long 'flexuous hyphae' projecting far beyond the apical 'slit.' Hand-sections of fresh material were made, then mounted in water, and finally preserved in lacto-phenol tinted lightly with acid fuchsin stain. Some of the flexuous hyphae in these sections were turgid and filled with protoplasm, while others appeared empty and flaccid." In one section Miss Hunter saw seven flexuous hyphae; and, finally, she recommended species of Milesia for further studies of sexual phenomena in Rusts because "their spermogonia are relatively large, they lack confusing paraphyses, and they produce striking 'flexuous hyphae'. "

Miss Hunter[1] also noticed that in mature pycnidia of *Pucciniastrum americanum* flexuous hyphae extend "beyond the tips of the spermatiophores through the opening of the spermogonium."

In 1940, Kamei[2] gave an account of his cultural studies of the Fern Rusts of Abies in Japan. He observed flexuous hyphae in six species of Milesia:

Milesia dryopteridis,   Milesia jezoensis,
Milesia exigua,   Milesia miyabei,
Milesia itoana,   Milesia sublevis.

Illustrations of the flexuous hyphae of three of these Milesiae are given in Kamei's Fig. 19a.

Kamei also found between the spermatiophores in the pycnidia of *Uredinopsis intermedia* and *U. ossaeformis* paraphyses-like organs that are unseptate and longer and thicker in diameter than the spermatiophores; but he expressed himself as uncertain whether or not these organs correspond to the flexuous hyphae of Craigie. It seems most probable that the organs in question are indeed flexuous hyphae and here they will be so regarded.

In 1943, L. S. Olive[3] described and illustrated flexuous hyphae in *Pucciniastrum hydrangeae* (his *Thekopsora hydrangeae*). In this species he observed: that the pycnidia on Hemlock leaves are inconspicuous and are composed of basal cells, pycnidiosporophores, and flexuous hyphae unaccompanied by paraphyses; and that the flexuous hyphae

[1]*Ibid.*, p. 122.

[2]S. Kamei, "Studies on the Cultural Experiments of the Fern Rusts in Japan," *Journal of the Faculty of Agriculture Hokkaido Imperial University*, Vol. XLVII, 1940, pp. 138-139. The flexuous hyphae of *Milesia exigua*, *M. jezoensis*, and *M. miyabei* are illustrated.

[3]Linsay S. Olive, "Morphology, Cytology, and Parasitism of *Thekopsora hydrangeae*," *Elisha Mitchell Scientific Society*, Vol. LIX, 1943, p. 47, Plate V, Figs. 2 and 9.

arise from the more central regions of the pycnidium, are readily distinguished from the pycnidiosporophores by being longer and more sinuous, and are sometimes branched. With a hand-lens he also observed the nectar containing pycnidiospores exuding from the pycnidial ostioles.

The observations of Pierson[1] (1933) on *Cronartium ribicola*, of Hunter (1936) on species of Coleosporium, Melampsora, Milesia, and Pucciniastrum, of Kamei (1940) on species of Milesia and Uredinopsis, and of Olive (1943) on *Pucciniastrum hydrangeae* indicate that flexuous hyphae are present in the pycnidia of the Melampsoraceae in general.

In 1937, Lehmann, Kummer, and Dannenmann[2] published a comprehensive treatise on the Black Stem Rust disease of cereals. In this work they presented the views of Miss Allen on the supposed fate of the pycnidiospores of *Puccinia graminis*, but they failed to make any reference to Craigie's discovery that, in *P. graminis*, just as in *P. helianthi*, flexuous hyphae protrude from the pycnidia, or to his opinion that in *P. graminis*, just as in *P. helianthi*, the flexuous hyphae and pycnidiospores of opposite sex fuse together and so initiate the sexual process.

On January 1, 1938, in a letter to *Nature*, accompanied with three illustrations, I[3] recorded that, in *Puccinia graminis*, after mixing the nectar of (+) and (−) pustules, I had observed from eighty to one hundred fusions between a flexuous hypha of one sex and a pycnidiospore of opposite sex. Thus the observations of Craigie on *P. helianthi* were extended and confirmed. In the next three Chapters of this book, my studies of the flexuous hyphae of *P. graminis* and of the union of these hyphae with pycnidiospores of opposite sex will be described in detail, and it will be shown that flexuous hyphae are present in the pycnidia of Rusts in general.

Species of Rust Fungi proved to be Heterothallic or Homothallic.— Since Craigie, in 1927, demonstrated that both *Puccinia helianthi* and *P. graminis* are heterothallic, and up to 1942, twenty-four other species have been studied experimentally. As shown in the accompanying Tables, twenty-two Rusts are now known to be heterothallic and four homothallic.

---

[1] *Vide supra*, p. 80.

[2] E. Lehmann, H. Kummer, H. Dannenmann, *Der Schwarzrost, seine Geschichte, seine Biologie, und seine Bekämpfung in Verbindung mit der Berberitzenfrage*, Berlin, 1937, 608 pp. and 87 illustrations.

[3] A. H. R. Buller, "Fusions between Flexuous Hyphae and Pycnidiospores in *Puccinia graminis*," *Nature*, Vol. CXLI, 1938, p. 33.

Heterothallic Rust Fungi

No.	SPECIES	OBSERVER	YEAR
1	Cronartium ribicola	R. K. Pierson[1] . .	1933
2	Melampsora lini	Ruth F. Allen[2] . .	1933
3	Gymnosporangium clavipes . . .	A. H. R. Buller[3] .	1940
4	Gymnosporangium globosum . . .	P. A. Miller[4] . .	1932
5	Gymnosporangium haraeanum . .	E. Kawamura[5] . .	1934
6	Gymnosporangium juniperi-virginianae	W. F. Hanna[6] . .	1929
		P. A. Miller[7] . .	1932
7	Phragmidium speciosum	A. M. Brown[8] . .	1938
8	Puccinia anomala	B. d'Oliveira[9] . .	1939
9	Puccinia bardanae	W. A. F. Hagborg[10]	1933
10	Puccinia coronata avenae	Ruth F. Allen[11]	1930
11	Puccinia graminis	J. H. Craigie[12] . .	1927
12	Puccinia helianthi		
13	Puccinia menthae	J. S. Niederhauser[13]	1942
14	Puccinia phragmitis	I. M. Lamb[14] . .	1935
15	Puccinia sorghi	G. B. Cummins[15] .	1931
		Ruth F. Allen[16] .	1934
16	Puccinia suaveolens	A. M. Brown and .	1940
		A. H. R. Buller[17] .	
17	Puccinia triticina	Ruth F. Allen[18] .	1931
18	Uromyces fabae	A. M. Brown and .	1938
		J. H. Craigie[19] . .	
19	Uromyces graminis	B. d'Oliveira[20] . .	1938
20	Uromyces phaseoli typica	C. F. Andrus[21] . .	1931
21	Uromyces phaseoli vignae		
22	Uromyces trifolii hybridi	A. M. Brown[22] . .	1939

Homothallic Rust Fungi

No.	SPECIES	OBSERVER	YEAR
1	Puccinia coronata elaeagni . . .	A. M. Brown and .	1938
		J. H. Craigie[23] . .	
2	Puccinia grindeliae	A. M. Brown[24] . .	1938
3	Puccinia malvacearum	Dorothy Ashworth[25]	1931
		H. S. Jackson[26] .	1931
		A. M. Brown[27] . .	1937
4	Puccinia xanthii	A. M. Brown[28] . .	1937

[1]R. K. Pierson, "Fusion of Pycniospores and Filamentous Hyphae in the Pycnium of the White Pine Blister Rust," *Nature*, Vol. CXXXI, 1933, pp. 728-729.

[2]Ruth F. Allen: "Heterothallism in *Puccinia graminis, P. coronata*, and *Melampsora lini*," *Phytopathology*, Vol. XXIII, 1933, p. 4; also "A Cytological Study of Heterothallism in Flax Rust," *Journ. Agric. Research*, Vol. XLIX, 1934, pp. 765-789.

[3]A. H. R. Buller, in this volume, Chapter IV.

[4]P. A. Miller, "Pathogenicity of Three Red-Cedar Rusts that occur on Apple," *Phytopathology*, Vol. XXII, 1932, pp. 735-736, 739.

[5]E. Kawamura: "On the Function of the Spermatia of *Gymnosporangium Haraeanum* Syd.," *Horticulture and Agriculture*, Vol. IX, 1934, in Japanese; also "Studies on *Gymnosporangium Haraeanum* Syd. I. Heterothallism in the Fungus," *Ann. Phytopath. Soc. Jap.*, Vol. X, pp. 84-92.

[6]W. F. Hanna, work done in 1929, results published in this volume, Chapter II.

[7]P. A. Miller, *loc. cit.*, pp. 734-735, 739.

[8]A. M. Brown: Personal communication, 1939; also "The Sexual Behaviour of Several Plant Rusts," *Canadian Journal of Research*, Vol. XVIII, 1940, pp. 18-21. Experiments made at the Dominion Rust Research Laboratory in 1938.

[9]B. d'Oliveira, "Studies on *Puccinia anomala* Rost.," *The Annals of Applied Biology*, Vol. XXVI, 1939, pp. 56-82.

[10]W. A. F. Hagborg, personal communication, 1939.

[11]Ruth F. Allen: "Heterothallism in *Puccinia coronata*," *Science*, Vol. LXXII, 1930, p. 536; also "A Cytological Study of Heterothallism in *Puccinia coronata*," *Journ. Agric. Research* Vol. XLV, 1932, pp. 513-541.

[12]J. H. Craigie, *vide* Chapter I.

The technique used in determining whether any particular Rust species is heterothallic or homothallic has in general been the same as that used by Craigie in his experiments with *Puccinia helianthi* and *P. graminis*. To indicate how this technique has been employed, some sexual experiments made on species of Phragmidium, Uromyces, and Puccinia by A. M. Brown[1] will now be briefly described.

Phragmidium speciosum on *Rosa blanda*. "Composite pycnial nectar drawn from infections of *Phragmidium speciosum* (Fries) Cooke was applied to one hundred well isolated monosporidial infections of which the nectar was not removed. Within ten days afterwards, ninety-one produced aecia. Fifty monosporidial infections, however, that served as controls remained unchanged."

Uromyces trifolii hybridi on *Trifolium hybridum*. "To twenty 18-day-old monosporidial infections of *Uromyces trifolii hybridi* (W. H. Davis) Arth. on *Trifolium hybridum* L. composite nectar was applied and twenty other infections were set aside as controls. Nineteen of the twenty infections receiving pycnial nectar produced aecia in due time, whereas the control infections remained unchanged."

The evidence just brought forward "shows that *Phragmidium speciosum* and *Uromyces trifolii hybridi* are heterothallic."

Puccinia malvacearum, *Puccinia grindeliae*,[2] and *Puccinia xanthii* (*micro*-forms without pycnidia). "One thousand and ten mono-sporidial infections of *Puccinia malvacearum* Bert. were obtained on *Althaea rosea* (L.) Cav. and *Malva rotundifolia* L., one hundred and fifty of *Puccinia grindeliae* Peck on *Grindelia squarrosa* (Pursh) Dunal,

---

[13]J. S. Niederhauser, Personal communication, 1942. The host was *Monarda didyma* and the nectar-mixing technique was employed.
[14]I. M. Lamb, "The Initiation of the Dikaryophase in *Puccinia Phragmitis* (Schum.) Körn," *Annals of Botany*, Vol. XLIX, 1935, pp. 403-438.
[15]G. B. Cummins, "Heterothallism in Corn Rust and Effect of Filtering the Pycnial Exudate," *Phytopathology*, Vol. XXI, 1931, pp. 751-753.
[16]Ruth F. Allen, "A Cytological Study of Heterothallism in *Puccinia sorghi*," *Journ. Agric. Research*, Vol. XLIX, 1934, pp. 1047-1068.
[17]A. M. Brown and A. H. R. Buller, in this volume, Chapter VIII.
[18]Ruth F. Allen, "Heterothallism in *Puccinia triticina*," *Science*, Vol. LXXIV, 1931, pp. 462-463; also "A Cytological Study of Heterothallism in *Puccinia triticina*," *Journ. Agric. Research*, Vol. XLIV, 1932, pp. 733-754.
[19]A. M. Brown and J. H. Craigie, "Studies on the Sexual Behaviour of Plant· Rusts," *Report by the Dominion Botanist for the years 1935 to 1937, inclusive*, Ottawa, 1938. Also A. M. Brown, "The Sexual Behaviour of Several Plant Rusts," *Canadian Journal of Research*, Vol. XVIII, 1940, pp. 18-21.
[20]B. d'Oliveira, "New Hosts for the Aecidial Stage of *Uromyces graminis*, (Niessl) Diet.," *Boletim da Sociedade Broteriana*, Sér. II, Vol. XII, 1938, pp. 81-89.
[21]C. F. Andrus, "The Mechanism of Sex in *Uromyces appendiculatus* and *U. vignae*," *Journ. Agric. Research*, Vol. XLII, 1931, pp. 559-587.
[22]A. M. Brown: Personal Communication, 1939; also "The Sexual Behaviour of Several Plant Rusts," *Canadian Journal of Research*, Vol. XVIII, 1940, pp. 18-21. Experiments made at the Dominion Rust Research Laboratory.
[23-28]For references *vide infra*, Chapter V. The form of *Puccinia grindeliae* investigated by Brown had no pycnidia.

---

[1]A. M. Brown: personal communication, 1939; also "The Sexual Behaviour of Several Plant Rusts," *Canadian Journal of Research*, Vol. XVIII, 1940, pp. 18-21. Experiments made in the Dominion Rust Research Laboratory, Winnipeg.

[2]The form of *Puccinia grindeliae* investigated by Brown was entirely devoid of pycnidia. *Vide infra*, Chapter IV.

and three hundred and fifty of *Puccinia xanthii* Schw. on *Xanthium commune* Britt. In addition to these infections, fifty other mono-sporidial infections were randomly selected from the three Rusts to serve as controls. All of these infections, including the controls, produced telia spontaneously before they were twelve days old. More-over, one hundred compound infections randomly selected from the three Rusts produced telia, without exception, simultaneously in both components, and within the same period of time as the monosporidial infections."

The evidence just brought forward "indicates quite clearly that *Puccinia malvacearum*, *P. grindeliae*, and *P. xanthii* are all homo-thallic."

Craigie's technique was also employed by Hagborg[1] in some experi-ments on the sexuality of *Puccinia bardanae*, a *brachy*-form, which in Canada is parasitic on the Burdock, *Arctium minus*. Hagborg's account of his experiments, hitherto unpublished, is as follows. "About one hundred and eighty artificial infections were studied. Ninety simple pustules, presumably each formed as a result of sowing a single basidiospore, remained haploid in that they did not develop any uredinia for more than twenty days, at the end of which time they were becoming necrotic. To fifty-seven other simple pustules mixed nectar was applied. From four to ten days later, fifty of these pustules developed uredinia. It was also observed that many compound pustules, formed by the coalescence of two or more simple pustules, developed uredinia. The evidence just brought forward justifies the conclusion that *P. bardanae* is heterothallic."

Miller, in his investigations on the sexuality of *Gymnosporangium globosum* and *G. juniperi-virginianae*, mixed the nectar of haploid pustules on isolated leaves floating on a sucrose solution in Petri dishes. A further account of this method and a record of some of the experi-ments made with its help will be found in Chapter III.

A technical method, supplementing the technical methods of Craigie, for finding out whether a long-cycled autoecious Rust species is heterothallic or homothallic has been devised by A. M. Brown.[2] It consists of sowing diploid (dicaryotic) uredospores or aecidiospores near a haploid pustule containing a haploid mycelium derived from a single basidiospore and then observing whether or not the diploid (dicaryotic) mycelium derived from the diploid spore diploidises the haploid mycelium, the criterion of diploidisation being the develop-

[1]W. A. F. Hagborg, personal communication, 1939. The work was carried out under the direction of Professor H. S. Jackson at the University of Toronto, in 1933.

[2]A. M. Brown, "Diploidisation of Haploid by Diploid Mycelium of *Puccinia helianthi* Schw.," *Nature*, Vol. CXXX, 1932, p. 777.

ment of the otherwise sterile proto-aecidia into aecidia. Brown obtained positive results with *Puccinia helianthi*, *Uromyces fabae*, and *U. trifolii hybridi*, thus providing evidence, additional to that supplied by the application of Craigie's methods, that these three species are heterothallic.

The details of Brown's experiments with *Puccinia helianthi*, in which he diploidised haploid mycelia with diploid mycelia will be given in Chapter VIII; but here may be cited very briefly his experiments with *Uromyces trifolii hybridi*. "Near to the peripheries of fifteen 18-day-old monosporidial infections of *Uromyces trifolii hybridi* (W. H. Davis) Arth. on *Trifolium hybridum* L. were sown uredospores, while five infections were kept as a control. All of the fifteen infections that coalesced with the uredial infections produced aecia in due time, whereas the control infections remained unchanged. This evidence, by itself, shows that *U. trifolii hybridi* is heterothallic."[1]

In undisturbed haploid pustules of any Rust having pycnidia the pycnidiospores and flexuous hyphae never fuse together. If, therefore, after mixing the nectar of haploid pustules, one observes a number of definite fusions between the pycnidiospores and the flexuous hyphae, one has clear evidence that the species concerned is hetero-thallic. Such fusions were seen by Pierson in *Cronartium ribicola* and by myself in *Gymnosporangium clavipes* (Fig. 80). On the basis of this evidence both *C. ribicola* and *G. clavipes* have been placed in the list of heterothallic species.

The twenty-two Rust species that so far have been proved to be heterothallic are included in six genera distributed in both the Mel-ampsoraceae (Cronartium, Melampsora) and the Pucciniaceae (Gymnosporangium, Phragmidium, Puccinia, Uromyces) so that we now know that heterothallism is wide-spread throughout the Uredi-nales. With Jackson[2] we may look upon heterothallism as the primitive sexual condition of the ancestors of the Rust Fungi, and it may well be that heterothallism is characteristic of all or nearly all the macrocyclic Rusts existing at the present day.

The Rust species which have been proved to be homothallic are all included in the genus Puccinia. There can be no doubt but that, in course of time, many *micro*-forms other than *P. malvacearum*, *P. grindeliae*, and *P. xanthii* will be found to be homothallic. Homo-thallism in Rust Fungi will be further treated of in Chapter V.

---

[1]A. M. Brown: personal communication, 1939; also "The Sexual Behaviour of Several Plant Rusts," *Canadian Journal of Research*, Vol. XVIII, 1940, p. 19.

[2]H. S. Jackson, "Present Evolutionary Tendencies and the Origin of Life Cycles in the Uredinales," *Memoirs Torrey Bot. Club*, Vol. XVIII, 1931, p. 100.

CHAPTER II

THE PYCNIDIA AND PROTO-AECIDIA
OF THE UREDINALES

Species Studied — Preparation of Material for Inoculation — Inoculation Method — The Formation of Pycnidia and Proto-aecidia in a Haploid Pustule — Conversion of a Proto-aecidium into an Aecidium — Nuclear Migration, Cell Fusions, Septal Pores, and the Diploidisation of a Proto-aecidium — The Structure of a Mature Pycnidium — Terminology — The Development of a Pycnidium of *Puccinia graminis* — Periphyses and their Functions — The Pycnidiosporophores and the Pycnidiospores — Horns of Pycnidiospores produced abnormally by *Puccinia graminis* — Oil-drops in Extruded Masses of Pycnidiospores — The Drop of Nectar — The Nectar is Antiseptic — The Odour of Rust Fungi including *Puccinia graminis* — Insects in their Relations with Rust Fungi — Effect of Rain on Pycnidia — The Flexuous Hyphae — Conversion of Mature Periphyses into Flexuous Hyphae.

Species Studied.—During the months of February, March, April, and May, 1937, at the Dominion Rust Research Laboratory at Winnipeg, I studied the pycnidia and pycnidial nectar of the following Rust Fungi:

> *Puccinia coronata avenae* on *Rhamnus cathartica*,
> *Puccinia graminis* . . on *Berberis vulgaris*,
> *Puccinia helianthi* . . on *Helianthus annuus*,
> *Puccinia minussensis* . on *Lactuca pulchella*.

The cultures were prepared by members of the staff of the Laboratory and were freely placed at my disposal.

Puccinia graminis, *P. coronata avenae*, and *P. helianthi* are annual species that form localized pycnidial pustules on more or less horizontally outspread dorsi-ventral leaves. In contrast therewith, *P. minussensis* (*P. hemisphaerica*) is a systemic perennial species that has its pycnidia scattered all over the leaves of its host-plant and on leaves (radical and cauline) that are upwardly inclined.

The pycnidia of *Puccinia minussensis*, as compared with those of *P. graminis*, *P. helianthi*, and *P. coronata avenae*, are more distantly separated from one another, somewhat larger, and provided with more

89

numerous periphyses; but they are pale orange-yellow instead of bright orange and are therefore less conspicuous. A Wild Lettuce leaf bearing pycnidia, when viewed from a distance of a foot or two, appears to be yellowish-green all over; but this lack of brilliant colour in the pycnidia is compensated by the infected plants being taller than the uninfected and by the pycnidia emitting a very powerful odour, an odour which is more noticeable than that given out by the pycnidia of *P. graminis, P. helianthi,* or *P. coronata avenae.*

An examination of the pycnidia of *Puccinia coronata avenae, P. graminis, P. helianthi,* and *P. minussensis* revealed that in all these species there protrude into the pycnidial nectar not only periphyses but also flexuous hyphae. After this had been established, and with a view to finding out whether or not flexuous hyphae are present in the pycnidia of Rust Fungi in general, I examined as many Rust species in the pycnidial condition as I could procure.

Late in May and early in June, 1937, whilst visiting the University of Minnesota and Cornell University (New York State), I examined the pycnidia of the following Rusts,[1] all of which were obtained in the living condition from woods, fields, or cultivated grounds:

Gymnosporangium clavipes	. on	Crataegus sp.,
Gymnoconia peckiana	. . on	Rubus sp.,
Puccinia bardanae . .	. on	*Arctium minus,*
Puccinia caricis { *grossulariata* on		{ *Ribes floridum,* { *Ribes prostratum,*
solidaginis .	on	*Solidago patula,*
urticata .	on	*Urtica dioica,*
Puccinia coronata avenae	. on	*Rhamnus cathartica,*
Puccinia graminis . .	. on	*Berberis vulgaris,*
Puccinia orbicula . .	. on	*Nabalus albus,*
Puccinia podophylli . .	. on	*Podophyllum peltatum,*
Puccinia violae . . .	. on	*Viola spp.,*
Uromyces perigynius .	. on	*Rudbeckia laciniata.*

In England, in the early summer of 1937, I examined the living pycnidia of:

Puccinia suaveolens on *Cirsium arvense,*

Puccinia poarum   on *Tussilago farfara,*

and in the spring of 1938:

Uromyces poae on *Ranunculus ficaria,*

Puccinia vincae on *Vinca major.*

[1]For North-American species I have followed the nomenclature given by J. C. Arthur in his *Manual of the Rusts in United States and Canada,* Lafayette, U.S.A., 1934.

At Winnipeg, in the spring and early summer of 1938, I examined the living pycnidia of:

Gymnosporangium juvenescens on *Amelanchier alnifolia,*
Phragmidium speciosum . . on *Rosa blanda,*
Puccinia bardanae . . . on *Arctium minus,*
Puccinia coronata avenae . . on *Rhamnus cathartica,*
Puccinia extensicola oenotherae on *Oenothera biennis,*
Puccinia rubigo-vera agropyrina on $\begin{cases} \textit{Thalictrum dioicum,} \\ \textit{Anemone quinquefolia.} \end{cases}$

The Gymnosporangium was kindly sent to me from Saskatoon by Dr. W. P. Fraser. The Rose Rust was found on a cultivated Rose-bush. *Puccinia bardanae* and *P. coronata avenae* were cultivated on pot-plants in a greenhouse. Early in May some teleutospore sori of *P. bardanae* were found on old much decayed leaves of a Burdock plant lying flat on the ground in a field, and the teleutospores were then employed for inoculating a Burdock plant in a greenhouse. The infections of *P. coronata avenae* on Rhamnus bushes were obtained by inoculating the leaves with basidiospores derived from teleutospores on Oat straw that had been kept in cold storage in the open over the winter. The Oenothera Rust was found in a garden in the city of Winnipeg. *P. rubigo-vera agropyrina* was found growing wild on *Thalictrum dioicum* by the Red River near the Rust Laboratory and on *Anemone quinquefolia* in woods at Minaki in western Ontario.

At Ottawa on July 29, 1938, I examined the living pycnidia of:
Milesia intermedia on *Abies balsamea.*

At Newton Abbot,[1] Devonshire, England, on November 6, 1938, I examined the living pycnidia of:
Puccinia smyrnii on *Smyrnium olusatrum.*

At Winnipeg in the spring of 1939, I examined the living pycnidia of:

Gymnosporangium clavariaeforme
Gymnosporangium clavipes . .
Gymnosporangium corniculans .
Gymnosporangium juvenescens . $\Big\}$ on *Amelanchier alnifolia,*
Gymnosporangium juniperi-virginianae on Apple seedlings,
Melampsorella cerastii . . . . on *Picea canadensis,*
Phragmidium potentillae . . . . on *Potentilla bipinnatifida,*
Puccinia monoica . . . . . on *Arabis retrofracta,*
Puccinia sessilis . . . . . on *Smilacina stellata.*

[1] At Newton Abbot the climate is very mild throughout the year and there, according to Dr. P. H. Gregory, the occurrence of *P. smyrnii* in the pycnidial and aecidial stages in November and other winter months is normal.

Gymnosporangium clavariaeforme and *G. clavipes* were found in the teleutospore stage on bushes of *Juniperus communis* at Minaki in western Ontario; and *G. corniculans* and *G. juvenescens* were found in a similar stage on *Juniperus horizontalis* at Pine Ridge, about ten miles north of Winnipeg. *G. juvenescens* forms conspicuous upright witches brooms and so is easily found, while *G. corniculans* gives rise to more or less spherical galls on its host's horizontal branches. All the four species pass to *Amelanchier alnifolia*.[1] I verified this fact by means of cultures. Twigs bearing well-soaked teleutospore sori of *G. clavariaeforme* and *G. clavipes* were suspended over Amelanchier bushes in a wood near the Rust Laboratory, and twigs bearing teleutospore sori of *G. corniculans* and *G. juvenescens* were suspended over pot-plants of Amelanchier in the greenhouse. All the four inoculation experiments were successful; and I thus had at my disposal the pycnidial pustules of four different species of Gymnosporangium on as many schrubs of *Amelanchier alnifolia* at one and the same time.

Cedar-apple galls bearing teleutospore sori of *Gymnosporangium juniperi-virginianae* were procured from trees of *Juniperus virginianae* growing in the United States, and this material was then used to infect Apple seedlings grown in pots.

Melampsorella cerastii (in the *Peridermium coloradense* stage) was obtained on June 11 on witches brooms of *Picea canadensis* at Victoria Beach, Lake Winnipeg. The brooms bore new leaves only. These leaves were very light-green, and each of them showed four rows of pycnidia.

Phragmidium potentillae has brilliant red aecidiospores and is often conspicuous on species of Potentilla growing on the prairies. Collections of it were made at Birds Hill. It was noticed that attached to several of the infected plants were some dead leaves of the previous year bearing teleutospore sori.

Plants of *Arabis retrofracta* bearing aecidia of *Puccinia monoica* were kindly sent to me by Dr. W. P. Fraser from Saskatoon, and they were at once planted in pots. A month later some of them produced new shoots bearing pycnidia. *P. monoica* on *Arabis retrofracta* was also found by me at Birds Hill.

Puccinia sessilis was found along the edge of a wood near Tuxedo Park, Winnipeg.

At Winnipeg in October and November, 1939, I examined the living pycnidia of:

Cronartium fusiforme on *Pinus caribaea.*

[1]*Vide* J. C. Arthur, *Manual of the Rusts in United States and Canada*, Lafayette, U.S.A. 1934. *Amelanchier alnifolia*, known as the Juneberry, is very common in Manitoba.

The *Cronartium fusiforme* material was kindly sent to me by Mr. Bailey Sleeth from the Mississippi State Nursery at Wiggins, Mississippi. It consisted of two-year-old seedlings of *Pinus caribaea* bearing *Cronartium fusiforme* cankers.

During the first four months of 1940, at Winnipeg, I examined successively the living pycnidia of:

Melampsora lini . . . . on *Linum usitatissimum,*

Puccinia coronata calamagrostidis on *Rhamnus alnifolia,*

Cronartium cerebrum . . . on *Pinus clausa,*

Puccinia sorghi . . . . on *Oxalis corniculata.*

The Melampsora material was collected in the field in November, 1939, and in January, 1940, the basidiospores were sown on young Flax plants in the greenhouse. The teleutospores of *Puccinia coronata* were overwintered in the open on the straw of *Calamagrostis canadensis.* In February, 1940, some twigs of *Rhamnus alnifolia* were cut off and placed in water in the greenhouse. Leaves grew out from the buds at the apical end of the twigs and these leaves were then inoculated with basidiospores. In March, branches of *Pinus clausa* bearing swollen cankers of *Cronartium cerebrum* with active pycnidia were sent to me from Florida by Dr. Geo. G. Hedgcock. In April, Dr. J. J. Christensen sent me from Minnesota overwintered *Puccinia sorghi* teleutospores on Corn (*Zea mays*) stems, and this material was employed to infect the leaves of *Oxalis corniculata* in the greenhouse.

Among the species of Rusts without pycnidia which I have had under observation at Winnipeg from time to time are the following microcyclic forms:

Puccinia grindeliae . on *Grindelia squarrosa,*

Puccinia malvacearum on *Malva rotundifolia,*

Puccinia rubifaciens on *Galium boreale,*

Puccinia xanthii. . on *Xanthium commune.*

Preparation of Material for Inoculation.—It is well known that the teleutospores of *Puccinia graminis* (Fig. 20) will not germinate immediately after they have been formed, but only after they have passed through a resting period some weeks or months in length. It also appears that the temperature[1] at which teleutospores are formed greatly influences their powers of subsequent germination, so that, in a greenhouse, if one wishes to produce teleutospores to be used for

---

[1]T. Johnson, "A Study of the Effect of Environmental Factors on the Variability of Physiological Forms of *Puccinia graminis tritici*, Erikss. and Henn.," *Dom. of Canada, Dept. of Agric.* Bull. CXL (New Series), 1931, pp. 47-70. The critical temperature above which teleutospore-formation results in defective germination is about 70°.

inoculation purposes, one must see that the temperature is kept down below a certain maximum.

As obtaining teleutospores of *Puccinia graminis* in a condition in which they will germinate freely is a *sine qua non* for successful inoculation of Barberry bushes, the mode in which the material for my own work was obtained and treated in the Dominion Rust Research Laboratory will here be given. I am indebted to Dr. T. Johnson for the details.

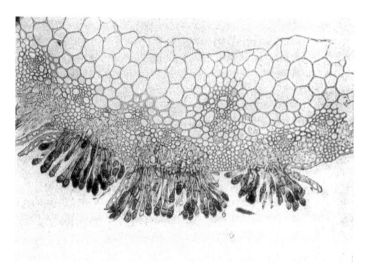

FIG. 20.—*Puccinia graminis.* Transverse section through three teleutospore sori on a straw of wheat. The teleutospores are here shown looking downwards as they do on straws used for inoculating Barberry bushes (*cf.* Fig. 21). Photographed by A. M. Brown. Magnification, about 350.

Puccinia graminis tritici. Culms of *Elymus macounii* bearing teleutospore sori, which had been frozen in the winter of 1935-1936, were collected in May, 1936. A test showed that the teleutospores germinated readily. The material was then stored in a refrigerator at a temperature of about 10° C. until used for inoculating Barberry bushes in the spring of 1937.

Puccinia graminis tritici, race 36. The uredospores were collected in the field on wheat in 1934 and used to inoculate wheat seedlings in the greenhouse. Teleutospores were formed in the low-temperature greenhouse in February and March, 1935. The culms bearing the teleutospore sori were kept frozen in ice in a refrigerator from May 21, 1935, to February 10, 1936. A test then showed that some of the teleutospores were able to germinate. The material was then allowed

to dry and, thereafter, it was stored dry in a refrigerator at a temperature of about 10° C. until used for inoculating Barberry bushes in the
spring of 1937.

Puccinia graminis avenae, race 10. The uredospores were collected
in New Brunswick on oats in 1935 and were used to inoculate oat
seedlings in the greenhouse. Teleutospores were formed in the low-
temperature greenhouse in January, 1936. The culms bearing the
teleutospore sori were kept frozen in ice in a refrigerator from February
24 to November 16, 1936. A test then showed that some of the teleuto-

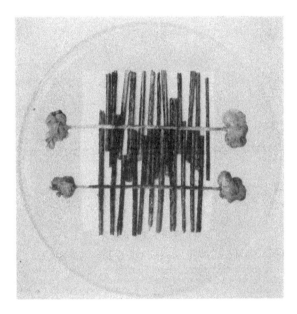

FIG. 21.—*Puccinia graminis.* Straws of wheat bearing teleutospore sori laid on wet
blotting paper in the lid of a Petri dish. Straws held down by plasticine.
Preparation to be used for inoculating a Barberry bush with basidiospores (*cf.*
Fig. 23). Photographed by J. H. Craigie (*Phytopathology*, 1931).

spores were able to germinate. The material was allowed to dry and,
thereafter, it was stored dry in a refrigerator at a temperature of about
10° C. until used for inoculating Barberry bushes in the spring of 1937.

The procedure for obtaining teleutospores of *Puccinia helianthi*
in a condition in which they will germinate is much simpler than in the
case of *P. graminis*. Sunflower leaves bearing teleutospore sori of
P. helianthi were collected by Mr. A. M. Brown in the open in the
autumn of 1936 and were stored by him dry in a drawer in his desk at
ordinary room temperature. In the spring of 1937, a leaf bearing

teleutospore sori was soaked in water for 24 hours, then dried, and then soaked again; and this was repeated two or three times. Then tests showed that the teleutospores were able to germinate. The leaf was then dried and kept at room temperature until it was required for inoculating Sunflower plants.

Brown found that teleutospores of *Puccinia helianthi* formed on wild Sunflower plants in the open in September and collected in

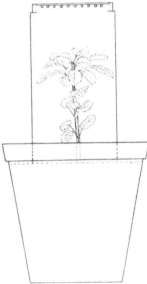

FIG. 22.—*Puccinia graminis.* Arrangement used by J. H. Craigie for inoculating a Barberry bush. Straws bearing teleutospore sori are adhering to wet blotting paper in the Petri-dish cover above (*Phytopathology*, 1931).

September or October, *i.e.* after the spores had been cooled nightly in the autumn, could be induced to germinate early in December. The resting period for the teleutospores of *P. helianthi* therefore appears to be a relatively short one.

Inoculation Method.—The method employed to inoculate the leaves of Barberry bushes with *Puccinia graminis* and of Sunflower plants with *P. helianthi* was in essentials the same as that used by Craigie[1] (Figs. 21 and 22).

For inoculations with *Puccinia graminis* the procedure was as follows. Pieces of rusted culms of wild grasses or cereals which bore teleutospores that would readily germinate were removed from the refrigerator and soaked in water for about fifteen minutes; and then

[1]J. H. Craigie, "An Experimental Investigation of Sex in the Rust Fungi," *Phytopathology*, Vol. XXI, 1931, pp. 1013-1014.

they were cut into suitable lengths and laid on wet blotting paper lining the lid of a Petri dish (*cf.* Fig. 21). A potted Barberry bush, which had been kept small by pruning and which had one or two twigs developing new leaves, was then prepared for inoculation. Its leaves were breathed upon and then sprayed from a distance of a few

Fig. 23.—Arrangement for inoculating a small potted Barberry bush with basidiospores of *Puccinia graminis*. The bush is surrounded by a lamp-shade which is capped above by a Petri-dish cover containing wheat straws bearing teleutospore sori.

feet with water from a fine atomiser. The tiny droplets settled evenly over each leaf surface. Two or three glass lamp-shades were then placed around the Barberry bush and one above the other in a series; but, with a very low bush, a single lamp-shade sufficed (Fig. 23). The Petri-dish cover, with the culms looking downwards (*cf.* Fig. 22), was then set over the uppermost lamp-shade (Fig. 23); and, finally, the whole combination was placed in a water-sealed damp-chamber[1] in

[1]This damp-chamber was cylindrical. It consisted of a low metal tray, which was partly filled with water, and of a tall metal cylinder which had a glass top and was open below. The pot was placed in the middle of the tray and then the cylinder, which was less in diameter than the tray, was placed over the tray with its mouth immersed in the water.

the low-temperature greenhouse. During the next day or two the teleutospores germinated, and the basidiospores fell upon the young leaves and stems of the host-plant, where they produced germ-tubes and caused infections. The Barberry bush was then removed from the damp-chamber and set in a screen-covered cage on a greenhouse bench. The infections became visible on about the fifth day after inoculation.

For inoculating Sunflower plants with *Puccinia helianthi*, the procedure was as follows. A leaf, or part of a leaf, bearing teleuto-spores that would readily germinate, was soaked in water for a few minutes and then placed on wet blotting paper contained in a Petri-dish cover. This cover, with the blotting paper and leaf adhering to it, was then inverted and placed over the top of a series of glass lamp-shades enclosing a sprayed Sunflower plant which was developing its first foliage leaves; and then the whole was treated in the same manner as that already described in connexion with the inoculation of Barberry bushes. Just as with *P. graminis* on Barberry leaves, the infections of *P. helianthi* on Sunflower leaves became visible on about the fifth day after inoculation. On the eighth day after inoculation, the pycnidia had broken through the epidermis and were crowned with little drops of pycnidiospore-bearing nectar.

The Formation of Pycnidia and Proto-aecidia in a Haploid Pustule.—When a basidiospore (sporidium) of a long-cycled Rust, *e.g. Puccinia graminis* or *P. helianthi*, has germinated on the leaf of a host-plant, the mycelium derived from the germ-tube develops in the intercellular spaces of the mesophyll, sends haustoria into the mesophyll cells, and causes the formation of a rust pustule. The pustule takes on a more or less discoid form, grows radially, becomes a few mm. wide, and then ceases to extend. To the naked eye a fully developed pustule appears as a leaf-spot.

After basidiospores of *Puccinia graminis* have been well sprinkled over a bush of *Berberis vulgaris* growing in a pot in a greenhouse, one finds that the rust pustules which develop are all formed on the light-green leaves of the young shoots and that the dark-green leaves of shoots that were developed in an earlier period of growth remain entirely free from infection. It is evident that, in respect to *Puccinia graminis*, the Barberry has what may be called *mature-leaf resistance*.[1]

[1]This resistance was noticed in 1920 by I. E. Melhus, L. W. Durrell, and R. S. Kirby ("Relation of the Barberry to Stem Rust in Iowa," *Iowa Agric. Expt. Sta. Research Bull.*, No. LVII, pp. 283-325) who, having regard to the fact that infection is accomplished by penetration of the tissues and not by entry through the stomata, suggested that the marked difference in the susceptibility of young and old leaves

The pustules formed on the leaves of a young vertical shoot of a Barberry bush vary considerably in size, and one of the factors concerned in this variation is the state of development of the leaves. The older leaves on the lower part of the shoot, that were fully developed at the time of inoculation, have the smallest pustules, 1-2 mm. wide, while the youngest leaves on the upper part of the shoot, that were still only partially grown at the time of inoculation, have the largest pustules. Two of these large pustules were found to measure 10 X 7 mm. and 8 X 7 mm. respectively. It thus appears that, other things being equal, the younger the Barberry leaf at the time of inoculation, the wider are the pustules which subsequently develop.

The width of a pustule which has originated from a single basidiospore also varies with the general condition of growth of the host-plant. Thus at Winnipeg, in a greenhouse, during the short dark days of winter, when snow and ice are on the roof, other things being equal, the pustules of *Puccinia helianthi* are only 1-3 mm. wide, whereas in the summer they are usually 3-6 mm. wide.

The mycelium of *Puccinia graminis* stimulates the tissues of the Barberry leaf and thus causes hypertrophy. The epidermal cells, palisade cells, and spongy parenchymatous cells do not undergo cell-division but they all increase considerably in *volume*, with the result that the part of a leaf forming a pustule becomes from two to four times as thick as it would have been had it not been attacked by the parasite (Fig. 40, p. 130). Owing to the lateral increase in the leaf-tissues, that part of a leaf constituting a well-developed pustule bulges, usually upwards, but sometimes downwards. The upward-bulging of the pustules makes the pustules more conspicuous than they would otherwise be.

The amount of hypertrophy of the host-leaf caused by the mycelium derived from a basidiospore, as might be expected, varies with

---

may be due to the thickness of the cuticle and outer epidermal wall. In 1927, L. W. Melander and J. H. Craigie ("Nature of Resistance of Berberis spp. to *Puccinia graminis*," *Phytopathology*, Vol. XVII, pp. 95-114), in their study of the comparative resistance of different species of Barberry to *Puccinia graminis*, measured the thickness and resistance to puncture of the outer epidermal walls of leaves of different ages, and they found: (1) that the species whose leaves have an epidermis that is very resistant to puncture are usually resistant to rust; and (2) that the leaves of susceptible species become practically immune with age. In *Berberis vulgaris* they noticed a marked resistance in leaves only ten days old.

Whether or not the mature-leaf resistance of *Berberis vulgaris* to *Puccinia graminis* can be broken down by wounding the leaves as P. R. Miller ("Pathogenicity of Three Red-Cedar Rusts," *Phytopathology*, Vol. XXII, 1932, p. 733) found for the leaves of susceptible varieties of Apple inoculated with basidiospores of *Gymnosporangium juniperi-virginianae* awaits determination by experiment.

different Rust species.   In Sunflower leaves attacked by *Puccinia helianthi* it is not appreciable.

The mycelium derived from a basidiospore is haploid, and it gives rise to a number of haploid organs which are here referred to as *pycnidia*[1] and also to a number of slightly larger, rudimentary, haploid structures which precede the formation of aecidia and which I have named *proto-aecidia*[2] (Fig. 24).

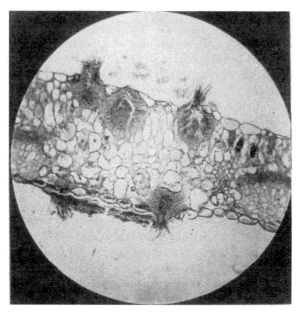

FIG. 24.—*Puccinia graminis* on *Berberis vulgaris*.   Microtome section through middle part of a haploid pustule, 57 hours after the nectar had been mixed, showing: above, three pycnidia; and below, two pycnidia and two proto-aecidia (on extreme right and left, *cf*. Fig. 26).   Section prepared and photographed by W. F. Hanna.   Magnification, 100.

A pycnidium (pycnium, spermogonium) of *Puccinia graminis* is a flask-shaped organ which produces a large number of pycnidiospores (pycniospores, spermatia) that are extruded from its ostiole in a drop of nectar.

The pycnidia in any pustule are formed between the enlarged palisade cells and just below the upper epidermis, or between the spongy parenchymatous cells and just above the lower epidermis.

[1]See footnote on p. 3.

[2]A. H. R. Buller, "Fusions between Flexuous Hyphae and Pycnidiospores in *Puccinia graminis*," *Nature*, Vol. CXLI, 1938, pp. 33-34.

They come into existence in radial succession, so that the first-formed are situated at the centre of the pustule and the last-formed near the pustule's periphery.

In *Puccinia graminis*, after the basidiospores have been desposited on a Barberry leaf, under favourable conditions for growth the first signs of the formation of pustules appear on the fifth day. At that time the pustules can be seen as minute pale-green flecks entirely covered by the epidermis. On the sixth day, the pustules are larger and yellower and the first-formed pycnidia are bursting through the epidermis; and, on the seventh day, the pycnidia in the centre of the pustules are each covered with a drop of orange-yellow nectar. On the eighth day, the drops of nectar on adjacent pycnidia in the centre of each pustule have grown in size, are more orange in colour, and are beginning to coalesce. For about another week the pustules gradually increase in width, develop more pycnidia, extrude more nectar from the pycnidia ostioles, and become redder. Finally, the pustules cease to extend laterally and to produce more pycnidia, and then all that one can see with the naked eye on their upper surface is one or a few orange-coloured drops of nectar. The nectar itself is colourless, and its apparent redness is due chiefly to the carotinoid pigment of the pycnidiospores suspended within it.

In *Puccinia graminis*, when a haploid pustule is very young, the first-formed pycnidia are produced on both the upper and the lower surfaces of the Barberry leaf and in about equal numbers; but, as the pustule extends laterally, whereas more and more pycnidia make their appearance on the upper side of the pustule, on the lower side the production of pycnidia ceases. This cessation in the production of pycnidia on the lower side of the leaf synchronises with the beginning of the development of the rudiments of the aecidia. After the sexual process has been initiated and the aecidia have begun to develop aecidiospores, no more pycnidia are produced on either side of the leaf.

In *Puccinia graminis*, in very large haploid pustules formed on leaves that were only partly grown at the time of inoculation, it sometimes happens that on the *under* side of the leaf, many days after the formation of the first set of pycnidia in the middle of the pustule, a new set of pycnidia is formed in a ring around the pustule's periphery. These belated pycnidia, which are widely separated from the first-formed pycnidia, increase the chances that the pustules concerned may at last be visited by insects.

In *Puccinia helianthi*, as a rule, all the pycnidia are situated on the upper side of the Sunflower leaf and there are none on the under side. However, in exceptional pustules, while the great majority of the pycnidia are on the upper side, a few (1-6) may be found on the under side. On the upper side of a large pustule, the pycnidia tend to be produced in a series of concentric circles (Fig. 25).

The production of most of the pycnidia on the upper side of a leaf rather than on the lower is of considerable advantage to the fungus; for it serves to place the pycnidia in a position where these organs can be easily seen by, and visited by, flying insects.

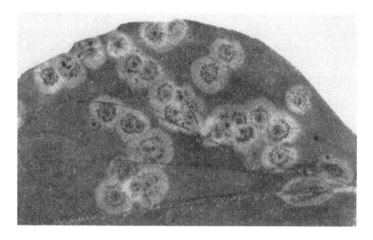

FIG. 25.—Upper side of a leaf of *Helianthus annuus* bearing 15-day-old pustules of *Puccinia helianthi* which are now past maturity and have their nectar dried-up. Each pustule contains mycelium derived from a single basidiospore and has produced numerous orange *pycnidia* which here appear as black dots. Many of the pustules have met, fused together, and have diploidised one another, with the result that, where (+) and (−) pustules have interacted, aecidia have been formed. These are on the under side of the leaf and therefore not visible here. Culture and photograph made by A. M. Brown at the Dominion Rust Research Laboratory. Magnification, 1.5.

The number of pycnidia on the upper side of a pustule of *Puccinia graminis* or *P. helianthi* varies with the size of the pustule. In a very small pustule there may be only half a dozen or even less, whereas in a large pustule there may be from thirty to one hundred.

The pycnidia in a pustule are scattered fairly evenly and there are usually upwards of ten in each square mm. of upper pustular surface. The number of pycnidia per square mm. on the upper side of a large pustule was found to be: for *Puccinia helianthi*, about 14; for *P. graminis* and *P. coronata avenae*, about 20.

In each haploid pustule of *Puccinia graminis*, on about the eighth day after inoculation of the Barberry leaf with basidiospores and about a day after the first-formed pycnidia have become covered with drops of nectar, preparations are being made for the production of aecidia; for, at this time, one can observe that there are being formed at fairly regular intervals in the middle layer of the leaf wefts of hyphae which, after the sexual process has been initiated, are destined to form the aecidial spore-beds. These wefts of hyphae, when fully developed, are rounded in outline, convex above, and concave below, and they

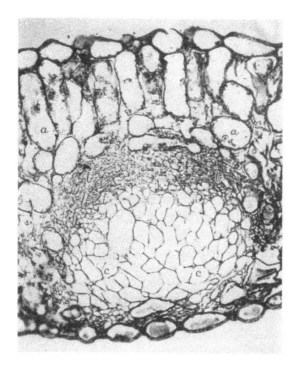

Fig. 26.—A proto-aecidium of *Puccinia graminis* in a leaf of *Berberis vulgaris*. Part of pustule shown in transverse section: *a a*, swollen palisade and spongy mesophyll cells; *b b c c*, the rounded proto-aecidium; *b b*, a concavo-convex mass of small basal cells; *c c*, pseudoparenchyma. The flexuous hyphae of the pycnidia of the pustule, 57 hours previously, were supplied with mixed nectar containing (+) and (−) pycnidiospores, with the result that the proto-aecidium is now becoming diploidised. The nuclei in the cells at the base of *b b* have become enlarged (*vide* Fig. 27). Section and photograph made by W. F. Hanna. Magnification, 280.

therefore resemble in form a concave-convex lens with the concavity directed downwards (Fig. 26, *b b*). The cells in each weft are packed closely together and are filled with protoplasm. Below a weft, the cells of the hyphae in the intercellular spaces swell up, displace the cells of the spongy parenchyma, press against one another, and form a more or less globular mass of comparatively large, isodiametric, colourless,

watery-looking, highly vacuolate pseudoparenchymatous fungal cells
which occupy the space between the concavity of the weft and the
lower epidermis (Figs. 26 and 27, *c c*). The concave-convex weft of

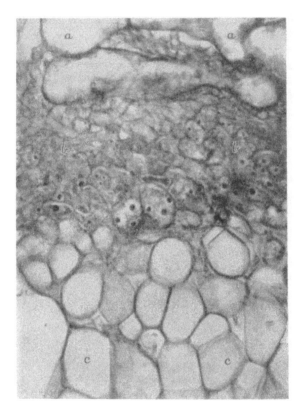

Fig. 27.—The upper central part of the proto-aecidium of *Puccinia graminis* shown
   in Fig. 26, now much enlarged: *a a*, spongy mesophyll cells; *b b*, dense basal cells
   of the proto-aecidium, 57 hours after mixing the nectar of the associated pycnidia;
   c c, pseudoparenchyma of the proto-aecidium. The lower cells of the basal
   cell-layer are becoming diploidised, as shown by the great enlargement of their
   nuclei and the association of their nuclei in pairs or in greater numbers. Section
   and photograph made by W. F. Hanna. Magnification, 900.

hyphae above and the mass of pseudoparenchymatous cells below
form a more or less globular organ. This is the organ which I have
called a *proto-aecidium*.

If that part of a proto-aecidium which abuts on the lower epidermis
of the host-leaf is considered to be the *apex* of the organ, then the weft
of hyphae above the pseudoparenchyma lies at the *base* of the organ,
and the small densely protoplasmic cells of which the hyphal weft is

composed may be called *basal cells*. Thus in a proto-aecidium we can distinguish two kinds of cells: the *basal cells* and the *pseudo-parenchymatous cells*. From the physiological point of view: the basal cells are *potentially fertile* in that, if they become diploidised,

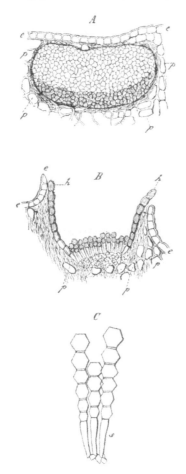

FIG. 28.—A, *Puccinia tragopogonis*: a proto-aecidium (shown looking upwards instead of downwards), described by de Bary as an early stage in the development of an aecidium, the shaded part indicating the region where the "hymenium" (palisade of aecidiosporophores) would be formed later; *e*, epidermis of host-leaf; *p*, parenchyma. B, a mature aecidium of *Uromyces sp.* on *Trifolium repens* (shown looking upwards): *e*, epidermis of host-leaf; *p*, parenchyma; *h*, peridium; in the open cup are aecidiosporophores and a few aecidiospores; most of the aecidiospores have been shot away. C, *Puccinia graminis*: outline of three aecidiosporophores ("basidia") springing from the base of an aecidium; each of the aecidiosporophores is continued apically into a chain of aecidiospores. From A. de Bary's *Morph. u. Phys. der Pilze*, 1866, p. 185. Magnification: A, 100; B, 90; C, 390.

they are able to produce aecidiospores; while the pseudoparenchymatous cells are *sterile* in that under no conditions do they ever produce
aecidiospores.

The function of the sterile pseudoparenchymatous cells is a
mechanical one: during their formation they displace the cellulose-
bounded spongy-parenchymatous cells lying below the basal cells and
substitute for the host-cells a very thin-walled delicate pseudoparenchyma which, in its turn, can be readily displaced by the chains
of aecidiospores when these are pushing their way downwards toward
the lower epidermis (Fig. 28). The pseudoparenchymatous cells are
displacement cells in a double sense, for they displace host-cells and,
thereafter, are themselves displaced by the chains of aecidiospores.

The function of the densely protoplasmic basal cells is the formation
of aecidiosporophores and chains of aecidiospores, but the performance
of this function is dependent on the basal cells becoming diploidised.
The basal cells, as already remarked, are potentially fertile but not
unconditionally fertile.

From the cytological investigations of Blackman, Christman,
Colley, Dodge, and others[1] it is known that, in Phragmidium, Cronartium, Gymnoconia, and other Rust genera which have broad flat
aecidia, the basal or fertile cells of the proto-aecidia which are destined
to produce the aecidiosporophores and chains of aecidiospores are
arranged in a more or less definite palisade layer in which the oblong
cells are all directed toward the exterior surface of the epidermis or
bark (*cf.* Fig. 35, p. 116); but no such arrangement can be seen in the
proto-aecidium of *Puccinia graminis*. In this species the basal cells
are packed closely together, but in no regular order.

The organs consisting of densely protoplasmic basal cells and of
pseudoparenchymatous cells, which have just been described, were
first brought to the notice of the botanical world by de Bary[2] in 1866
in the first edition of his Text-book. He regarded them as an early
stage in the development of aecidia, and, in an illustration of *Puccinia
trogapogonis*, he showed one of the organs with its two kinds of cells
clearly distinguished (Fig. 28, A). In 1884, in the second edition of
his Text-book, he[3] reproduced Sach's illustration of a very young
aecidium of *P. graminis* and referred to it as an aecidial primordium

---

[1]For references to the papers of these authors, *vide* Chapter I, pp. 43 and 53.

[2]A. de Bary, *Morphologie und Physiologie der Pilze, Flechten und Myxomyceten*
(Bd. II, Abt. I of W. Hofmeister's *Handbuch der physiologischen Botanik*) Leipzig,
1866, Fig. 74, p. 185.

[3]A. de Bary, *Vergleichende Morphologie und Biologie der Pilze, Mycetozoen und
Bacterien*, Leipzig, 1884, Fig. 124, p. 297; English Translation, 1887, Fig. 124, p. 275.

(Fig. 35, A, p. 116). Olive,[1] in 1911, remarked that sometimes the primordia do not develop into aecidia but remain sterile. Aecidial primordia were studied by Fromme (1914),[2] by Kursanov (1917)[3] and, subsequently by other cytologists. Fromme called the cells of the hyphal weft *fertile cells* and the pseudoparenchymatous cells *sterile cells*, and he rightly recognised that the sterile cells which form the pseudo-parenchyma of the primordium of an aecidium are homologous with the so-called *buffer cells* of a caeoma. Craigie[4] observed the primordia in 1928. In 1929, Hanna[5] referred to the primordia in *P. graminis* as being "evidently haploid rudiments of aecial cups waiting to be stimulated into further developmental activity" and, in the same year, Miss Allen[6] said that they resemble aecia but "consist of haploid mycelium only." In 1931, Craigie,[7] in describing a photograph (Fig. 33, p. 114) showing the under side of a Barberry leaf 23 days after inoculation with basidiospores, said: "On the left-hand side of the mid-rib appears a compound pustule bearing aecia. On the right-hand side are three simple pustules in which no aecia developed. The pocked appearance of these three pustules is due to the formation of wefts of haploid hyphae which develop in simple pustules just beneath the epidermis and simulate a young aecium." Bessey,[8] in 1935, in his *Text-book of Mycology* says that the structure here under discussion "may be called an *aecial primordium*."

As the more or less globular fungal organs which have just been described (Figs. 24, 26, and 28, A) are the precursors of aecidia and therefore of much importance and as, nevertheless, they had not hitherto received any name consisting of a single word and suitable

[1]E. W. Olive, "Origin of Heteroecism in the Rusts," *Phytopathology*, Vol. I, 1911, p. 143.

[2]F. D. Fromme, "The Morphology and Cytology of the Aecidium Cup," *Botanical Gazette*, Vol. LVIII, 1914, pp. 1-35.

[3]L. Kursanov, "Recherches morphologiques et cytologiques sur les Urédinées," *Bull. Soc. Nat. Moscou*, (1917), T. XXXI, 1922, pp. 1-38. On p. 1, in Fig. 1, he showed a transverse section of a *primordium écidien* of *Puccinia graminis* developing into an aecidium.

[4]J. H. Craigie, "On the Occurrence of Pycnia and Aecia in certain Rust Fungi," *Phytopathology*, Vol. XVIII, 1928, pp. 1005-1015.

[5]W. F. Hanna, "Nuclear Association in the Aecium of *Puccinia graminis*" *Nature*, Vol. CXXIV, 1929, p. 267.

[6]Ruth F. Allen, "Concerning Heterothallism in *Puccinia graminis*," *Science*, Vol. LXX, 1929, pp. 308-309.

[7]J. H. Craigie, "An Experimental Investigation of Sex in the Rust Fungi," *Phytopathology*, Vol. XXI, 1931, pp. 1020-1021.

[8]E. A. Bessey, *A Text-book of Mycology*, Philadelphia, 1935, p. 259.

for international usage, in January, 1938, I[1] proposed that they should
be called *proto-aecidia*. A mature proto-aecidium bears the same relation to an aecidium as a mature ovule does to a seed: an ovule is a

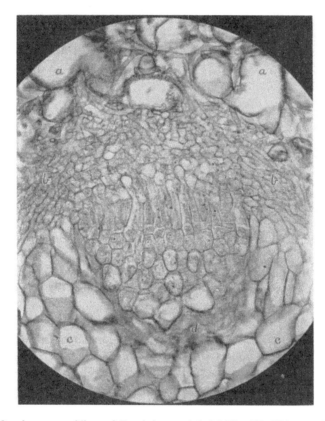

Fig. 29.—A proto-aecidium of *Puccinia graminis* (*cf.* Fig. 26), 82 hours after mixing
the nectar of the associated pycnidia: *a a*, swollen spongy mesophyll cells with
intercellular mycelium; *b b*, dense basal cells of the proto-aecidium; *c c*, pseudoparenchyma of the proto-aecidium. The proto-aecidium is now becoming
converted into an aecidium, as shown by: (1) the down-growth of the lowest
basal cells in the centre of the basal-cell layer *b b* so as to form aecidiosporophores;
(2) the production of short chains of aecidiospores; and (3) the association of
nuclei in pairs in the aecidiosphorophores and aecidiospores. The growth of the
chains of aecidiospores downwards is accompanied by the gradual disorganisation
and destruction of the pseudoparenchyma, as may be seen at *d*. Section and
photograph made by W. F. Hanna. Magnification, 760.

potential seed and a proto-aecidium is a potential aecidium. A proto-
aecidium, like an ovule, awaits the initiation of the sexual process
before it can undergo any further development.

[1]A. H. R. Buller, "Fusions between Flexuous Hyphae and Pycnidiospores in
Puccinia graminis," *Nature*. Vol. CXLI, 1938, pp. 33-34.

Those who, following Arthur,[1] prefer to call an aecidium an *aecium* should call a proto-aecidium a *proto-aecium*.

Aecidia, as Arthur and his co-workers[2] have remarked, may be cupulate, cornute, operculate, naked, stylosporic, or hyphoid. The non-cupulate, flattened, non-peridial and therefore *naked* aecidia which are characteristic for the genera Phragmidium, Melampsora, and Gymnoconia are known as *caeomata*. The primordium of a caeoma may be conveniently designated a *proto-caeoma*.

A proto-aecidium of *Puccinia graminis*, as we have seen, consists of two parts: (1) the weft of hyphae in the middle layer of the leaf and (2) the subjacent mass of pseudoparenchymatous cells. These two parts have two entirely different functions. After the sexual process has been initiated (Fig. 27), the weft of hyphae becomes converted into a spore-bed from which are produced chains of aecidiospores (Fig. 29) which become enclosed by a thick-walled peridium (Figs. 30 and 34, pp. 110 and 115). The spore-bed, the chains of aecidiospores, and the peridium together constitute the diploid organ which we call an *aecidium*. When this organ is young it occupies the middle and lower-middle layer of the host-leaf and is separated from the exterior of the leaf by the mass of pseudoparenchymatous cells and the lower epidermis. The young aecidium grows downwards through the leaf and, as it has a dissolving action and consists of a solid mass of aecidiospore-chains tightly packed within a thick-walled peridium, it has no difficulty in displacing the thin-walled pseudoparenchymatous cells and in breaking through the epidermis. The function of the pseudoparenchymatous cells appears to be merely that of preparing a path along which the growing aecidium can readily make its way out of the leaf. It is only after an aecidium has forced its way out of a leaf that the peridium opens at its apex, splits into valves, and exposes the chains of aecidiospores to the air.

Proto-aecidia are somewhat larger organs than pycnidia. In *Puccinia graminis* the diameters of pycnidia and proto-aecidia were found to be about 0.1 mm. and 0.17 mm. respectively.

Fully developed proto-aecidia of *Puccinia graminis* and of *P. helianthi* can readily be seen by looking at the under surface of the pustules with the naked eye or with the aid of a hand-lens. When a

---

[1] J. C. Arthur: "Terminology of the Spore-structures in the Uredinales," *Botanical Gazette*, Vol. XXXIX, 1905, pp. 219-222; J. C. Arthur and F. D. Kern, "The Problem of Terminology in the Rusts," *Mycologia*, Vol. XVIII, 1926, pp. 90-93; and J. C. Arthur in collaboration with F. D. Kern, C. R. Orton, F. D. Fromme, H. S. Jackson, E. B. Mains, and G. R. Bisby, *The Plant Rusts*, New York, 1929, pp. 9-10.

[2] *Ibid.* (*The Plant Rusts*), pp. 11-12.

Barberry leaf 11-14 days after inoculation is placed between the un-clouded sun and one's eyes, it is possible to see the proto-aecidia in the form of translucent spots which are reminiscent of the appearance of the oil-glands in the leaves of *Hypericum perforatum*.

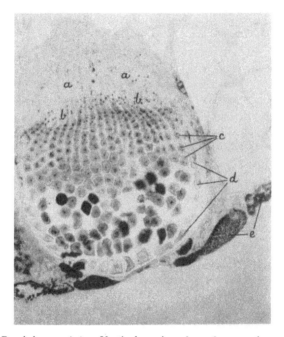

FIG. 30.—*Puccinia graminis*. Vertical section through a nearly mature aecidium about to break through the host epidermis. The pseudoparenchymatous cells of the proto-aecidium have now disappeared and their place has been taken by chains of aecidiospores surrounded by a peridium. A conjugate pair of nuclei can be seen in many of the aecidiospores: *a a*, basal cells, now largely drained of their food-contents; *b b*, aecidioporophores; *c*, binucleate aecidiospores; *d*, peridium, not yet split at the apex; and *e*, lower epidermis of Barberry leaf. Photographed by A. M. Brown. Magnification, about 800.

Often, on the under side of a haploid pustule of *Puccinia graminis*, a *black dot* can be seen in the middle of each proto-aecidium (Figs. 31 and 33). The blackness of the dot is due not to a shadow caused by a depression of the epidermis, but to the fact that where a proto-aecidium abuts on the lower epidermis more light enters the leaf than comes out again. The mass of pseudoparenchymatous cells of a proto-aecidium is watery and very transparent, whereas the hyphal weft above it is dense and relatively dark and opaque. Rays of light readily pass through the lower epidermis and the mass of pseudoparenchymatous cells above it; but, on coming to the weft of hyphae beyond, they are largely absorbed. Thus the central part of a proto-aecidium is black for the same reason that the pupil of an eye is black or that the windows of a house are dark when seen in daylight from outside.

By boiling portions of Barberry leaves bearing pycnidial pustules of *Puccinia graminis* in lacto-phenol and then staining the preparations with cotton blue, it is possible to look through the transparent leaf-pieces with the low power of the microscope and to observe all the pycnidia on both sides of the leaf and the positions of all the proto-

Fig. 31.—*Puccinia graminis* on *Berberis vulgaris*. Under side of a Barberry leaf showing an old, undiploidised, haploid pustule in which the proto-aecidia, more than 50 in number, can be seen as black circular dots. Photographed by J. H. Craigie. About five times the natural size.

aecidia. Such a preparation of a pustule of medium size, eight days old, and excentrically developed, has been drawn and is represented in Fig. 32. Here there are 42 pycnidia (shaded) on the upper side of the pustule, 7 pycnidia (unshaded) on the lower side, and 16 proto-aecidia (dotted) in the lower half of the pustule on the right-hand side.

With the help of the lacto-phenol cotton-blue technique I have observed that the proto-aecidia of *Puccinia graminis* are in course of formation on the eighth day after the inoculation of the Barberry leaf and that well-developed proto-aecidia are present in the leaf on the tenth day. It therefore appears that the proto-aecidia come into existence immediately after the first-formed pycnidia have opened out on the surface of the leaf. Thus they are ready to develop into aecidia as soon as the sexual process has been initiated as a result of the first visits of insects to the pustules.

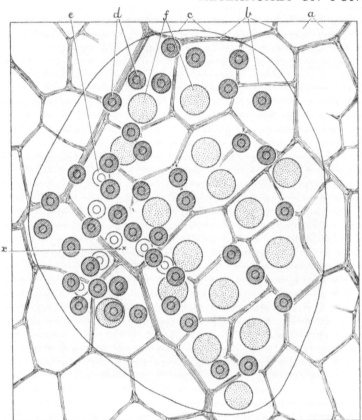

Fig. 32.—*Puccinia graminis* on *Berberis vulgaris*. Distribution of the haploid organs, *pycnidia* and *proto-aecidia*, in a leaf pustule. Pustule, derived from a single basidiospore, eight days old, seen from above after being boiled in lacto-phenol and stained with cotton-blue. The basidiospore germinated at or near to the position marked by the cross at *x*, and the development of the pustule was therefore excentric. *a*, the transparent leaf substance; *b*, leaf veins; *c*, the periphery of the pustule; *d*, pycnidia, 42, on the upper side of the leaf; *e*, pycnidia, 7 only, on the lower side of the leaf; and *f*, proto-aecidia, 16, in the middle and lower half of the leaf, in the mesophyll between the veins. Magnification, 47.

In one 14-day-old pustule of *Puccinia graminis*, which had been prepared by the lacto-phenol cotton-blue technique, it was observed that there were 88 pycnidia scattered over the upper surface of the pustule, 10 pycnidia close together in the centre of the lower surface of the pustule, and 60 proto-aecidia scattered through the middle and lower layer of the pustule. It was also observed that the 10 pycnidia on the lower surface of the leaf had *no proto-aecidia between them* but were surrounded by these organs (*cf.* Fig. 32). The development of the proto-aecidia had evidently checked any further development of pycnidia on the under surface of the leaf. All the ultimate veinlets of the leaf, occupying the middle layer of the leaf, were clearly in view, and it was observed that the *proto-aecidia had all been formed between the veinlets* so that, in the middle layer of the leaf, there was an alternation of veinlets and *proto-aecidia* (*cf*. Fig. 32). On the other hand, many of the pycnidia, which are sub-epidermal organs, stood either above or below the position of a veinlet. It would appear that, in a well-grown pustule, the total number of pycnidia somewhat exceeds the number of proto-aecidia which are ultimately formed.

The aecidia on the under side of a well-developed pustule of *Puccinia graminis* derived from a single basidiospore are in general set closely together, side by side, with just enough room left between them for each to open terminally without interfering with its neighbours. However, in the middle of the group of aecidia there may often be seen a *small unfilled space* where aecidia are absent, and two such spaces are shown in the compound pustule represented in Fig. 33. The bare space in a simple pustule corresponds to the position of the small group of pycnidia which were formed in the middle of the under side of the pustule and between which no proto-aecidia were developed.

In *Puccinia graminis*, as we have seen, on the under side of a pustule on a Barberry leaf, as soon as the proto-aecidia begin to develop no more pycnidia are formed. This fact helps us to understand why it is that more pycnidia are formed on the upper side of each pustule than on the lower side. But we may still ask: why are proto-aecidia and the aecidia into which they develop formed on the *under* side of the leaf? Hanna[1] sought to answer this question. He inoculated the lower surface of some Barberry leaves with basidiospores of *P. graminis* and *kept the leaves upside down*. Nevertheless, the fungus developed in respect to each leaf in the usual manner, *i.e.* in every pustule most of the pycnidia were formed below the upper epidermis and therefore looked toward the soil, while the aecidia broke through the lower epidermis and looked toward the sky. Hanna therefore concluded that the position of the pycnidia and aecidia in a Barberry leaf is decided not by the stimulus of gravity but by stimuli derived from different leaf-regions.

In *Puccinia graminis*, *P. helianthi*, and similar Pucciniae, the pycnidia in a single pustule are at such distances apart that the nectar-drops on the pycnidia are independent of one another when they are

[1]W. F. Hanna, "Nuclear Association in the Aecium of *Puccinia graminis*," in *Report of the Dominion Botanist for 1929*, Dept. of Agric., Ottawa, 1931, p. 50.

Fig. 33.—*Puccinia graminis* on *Berberis vulgaris*. The leaf was inoculated on its upper side with basidiospores and now, 20 days later, its under side is shown. On the right-hand side of the leaf are three simple, haploid, (+) or (−) pustules each of which originated from a single (+) or (−) basidiospore. In these undi-ploidised pustules the proto-aecidia appear as black dots. On the left-hand side of the leaf is a compound pustule derived from two spores, one (+) and the other (−), which fell on the leaf very near to one an-other. The two simple pustules derived from the two spores fused with one another and diploidised one another, with the result that the proto-aecidia in each of the two components of the compound pustule became converted into aecidia. The aecidia have opened and are discharging their spores. In the centre of each of the two simple pustules which fused to form the compound pustule is seen a dark spot free from aecidia. In each of these dark spots were several pycnidia but no proto-aecidia and so, when diploidisation took place, no aecidia were formed there. Photograph made by J. H. Craigie. Magnification, 5 times the natural size.

still very small, but are bound to meet and melt together as they grow in size. In the end, under favourable conditions of moisture, each pustule becomes covered over with a *compound drop* formed by the coalescence of several or many simple drops. One can look down upon a compound drop on a pustule of *P. graminis*, *P. helianthi*, etc., and observe the pycnidia with their masses of pycnidiospores well covered by the drop and scattered at intervals beneath its surface.

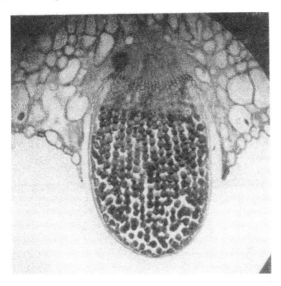

FIG. 34.—Microtome section of an aecidium of *Puccinia graminis* on a leaf of a Barberry bush in a greenhouse culture. The aecidium is closed because the humidity of the greenhouse air was not sufficient for it to open. It is surrounded by a thick-walled peridium and contains chains of ripe aecidiospores. Owing to shrinkage on fixing, the leaf and the sides of the aecidium are no longer in contact, and the spores which, when living, filled the whole space within the aecidial chamber and were packed tightly together are now much shrunken. Photographed by A. M. Brown. Magnification, about 400.

Conversion of a Proto-aecidium into an Aecidium.—Figs. 26 and 27 (pp. 103 and 104) show a proto-aecidium in which the diploidisation process is taking place and which is beginning to be transformed into an aecidium. Certain of the lowermost basal cells abutting on the pseudoparenchymatous cells have already become enlarged and their nuclei have increased in size and in number (so that there is more than one in a cell).

Fig. 29 (p. 108) shows a further stage in the conversion of a proto-aecidium into an aecidium. Already, in the lower-middle part of the mass of basal cells, a considerable number of basal cells have become binucleate and have grown downwards so as to form a palisade layer of aecidiosporophores, and these structures have produced chains of two or three young aecidiospores. Thus the basal cells have become fertile. It will be seen in Fig. 29 that wherever the advancing chains of aecidiospores are in contact with the pseudoparenchymatous cells, these cells are becoming disintegrated and thus are making room for the aecidiospores: the *displacement cells* which during their development displaced certain spongy parenchymatous cells are themselves now becoming displaced.

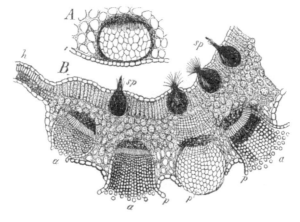

FIG. 35.—*Puccinia graminis* in a leaf of *Berberis vulgaris*. A, an aecidial primordium or proto-aecidium becoming converted into an aecidium. In this structure, called by de Bary a young aecidium, one can see the proto-aecidial pseudo-parenchyma and above it a palisade of young aecidiosporophores which are about to form chains of aecidiospores. B, vertical section through a pustule showing pycnidia *sp* above and aecidia *a a* below. At *h* is shown an uninfected part of the leaf by comparison with which it is seen that the infected or pustular part is much thickened. Of the four aecidia, No. 1 (on the left) is opening, Nos. 2 and 4 have just opened and are violently discharging their spores, while No. 3, still unopened, is shown enclosed by its peridium *p*. Drawn by Julius Sachs (*Lehrbuch der Botanik*, ed. II, 1870, Fig. 170), reproduced by de Bary (*Verg. Morph. u. Phys. d. Pilze*, ed. II, 1874). Magnified.

Fig. 30 (p. 110) shows a nearly mature aecidium. The chains of aecidiospores have become much elongated and have completely displaced and replaced the pseudoparenchyma, and the mass of aecidiospores is now enveloped by a thick-walled peridium. The binucleate condition of many of the aecidiospores is apparent. The aecidium as a whole is still developing: the aecidiosporophores are still cutting off aecidiospores, the peridium is being added to, and the aecidiospores and peridium are beginning to press against the lower epidermis. The development of the mass of aecidiospores and of the peridium has been accompanied by a flow of protoplasm from the mycelium and basal cells into the new structures; and now the mass of aecidiospores (Fig. 30, *c c*) is densely protoplasmic, whereas the mass of basal cells (*a a*) is very largely exhausted.

As the aecidiosporophores in such an aecidium as that represented in Fig. 30 continue to add new aecidiospores to the chains of aecidio-spores already in existence, the aecidium gradually elongates; and soon it ruptures the lower epidermis and exposes the apical part of the peridium to the air (Fig. 34). Under moist atmospheric conditions, an aecidium that has burst through the lower epidermis of the host-leaf dehisces: the peridium, at its apex, splits into valves which bend

Fig. 36.—*Puccinia graminis* on *Berberis vulgaris*. Long, closed, finger-like aecidia produced in a greenhouse at the Dominion Rust Research Laboratory, Winnipeg, under dry atmospheric conditions. Photographed by A. M. Brown. Magnification, about 1.5.

outwards (Fig. 35). The ends of the chains of aecidiospores thus become exposed to the air. The spores are then shot violently away from each chain in basipetal succession.[1] While the older aecidiospores are being discharged, new ones may still be in course of formation. Finally, when all the spores have been shot away, the aecidial cup is left empty of contents (*cf*. Fig. 28, B, p. 105).

Fig. 34 shows an aecidium which, under dry atmospheric conditions in a greenhouse at Winnipeg, has elongated considerably but has not opened. Under such dry conditions, the aecidia of *Puccinia graminis* often become about 2 mm. long without dehiscing, and then in form they resemble fingers (Fig. 36). In these finger-like aecidia the chains of spores are very long indeed. If a Barberry plant bearing finger-like aecidia is put into a chamber containing air saturated with water vapour, the aecidia quickly dehisce and begin to discharge their spores.

If a finger-like aecidium resembling those shown in Fig. 36 is picked off a pustule with the help of forceps and is then placed in a drop of

[1]These *Researches*, Vol. III, 1924, pp. 552-559.

water, the mass of spores rapidly swells up with the result that the peridium bursts and the chains of spores are set free (Fig. 37). Often the peridium bursts with so great a force that chains of aecidiospores are shot out of the drop of water to a distance of several millimeters beyond the drop's boundary.[1]

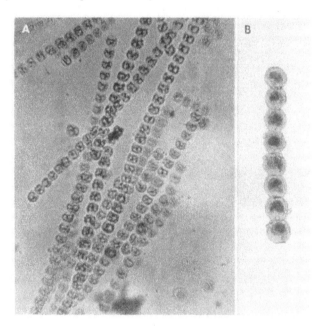

FIG. 37.—*Puccinia graminis.* A: chains of aecidiospores set free from a long, unde-hisced aecidium (*cf.* Fig. 36) after the aecidium had been placed in water on a glass slide and had burst open. B: a part of a single chain. Photographed by M. Newton and T. Johnson. Magnification: A, 175; B, 355.

The peridium surrounding an aecidium of *Puccinia graminis* is a layer of cells of peculiar form with the outer walls much thickened, and it is formed around the aecidium whilst this organ is growing through the pseudoparenchyma of the proto-aecidium and before it has forced its way through the lower epidermis of the Barberry leaf (*cf.* Figs. 29 and 30).

In 1896, Richards[2] showed that, in a Puccinia, the peridium of an

[1]Margaret Newton and T. Johnson, "Specialization and Hybridization of Wheat Stem Rust, *Puccinia graminis tritici* in Canada," *Dom. of Canada Depart. of Agric.,* Bull. No. CLX, New Series, 1932, p. 38. Supplemented by personal communication with the authors, 1939.

[2]H. M. Richards, "On Some Points in the Development of Aecidia," *Proceedings of the American Academy,* Vol. XXXI, 1896, pp. 255-269, with one Plate.

aecidium does not grow up over the spore-chains from below and meet at the centre, but has a two-fold origin: the *central arch* of the peridium, made up of what may be called *cap-cells*, is formed from the uppermost cells of the more central spore-chains; whilst the *lateral walls* of the peridium are formed (as de Bary[1] had already observed) from vertical chains of cells developed from a ring of basal cells corresponding to aecidiosporophores. It is admitted that the cells making up a peridium are metamorphosed aecidiospores and aecidiosporophores.

In 1914, the observations of Richards were confirmed by Fromme and Kursanov. Fromme[2] concluded that, in species of Puccinia and Uromyces: "The first peridial cells are produced at the centre of the arch and the peridium enlarges from this point centrifugally until the bases of the lateral walls are reached. Its subsequent enlargement is by the basipetal growth and sterilization of the peripheral spore chains". Kursanov[3] observed the dual mode of development of the peridium in various species of Puccinia and Uromyces and, in particular, in *Puccinia graminis* (Fig. 38, A and B).

In the very young aecidium of *Puccinia graminis* shown in Fig. 29 (p. 108) one can observe that the terminal cells of the central spore-chains have already become enlarged and are evidently developing into peridial cap-cells, and one can also perceive that the lateral walls of the peridium have not yet begun to develop. The aecidium appears to be adding new aecidiosporophores on its sides, and it further appears that the sporophores and spore-chains which, ultimately, will become metamorphosed into peridial cells that will form the lateral walls of the peridium have not yet come into existence.

The *intercalary cells* in the ordinary spore-chains of *Puccinia graminis* and other similar Rust Fungi are cut off basally from the aecidiospore mother-cells with the result that in any spore-chain the aecidiospores and intercalary cells alternate with one another. A few intercalary cells can be distinguished in Fig. 29 and others are clearly shown in Fig. 38, A. As the spore-chains mature, the intercalary cells degenerate and disappear and so function as *disjunctor cells*, *i.e.* cells which aid in the ultimate separation of the aecidiospores when these are forcibly discharged, one by one, from the aecidium.

[1]A. de Bary, *Morphologie und Physiologie der Pilze, Flechten und Mycetozoen.* Leipzig, 1866, p. 186. English Translation of Second Edition, Oxford, 1887, pp. 275-276.

[2]F. D. Fromme, "The Morphology and Cytology of the Aecidium Cup," *The Botanical Gazette*, Vol. LVIII, 1914, p. 33.

[3]L. Kursanov, "Über die Peridienentwicklung im Aecidium," *Ber. d. Deutschen Bot. Gesellschaft*, Jahrg. XIV, 1914, pp. 317-327, Taf. VI.

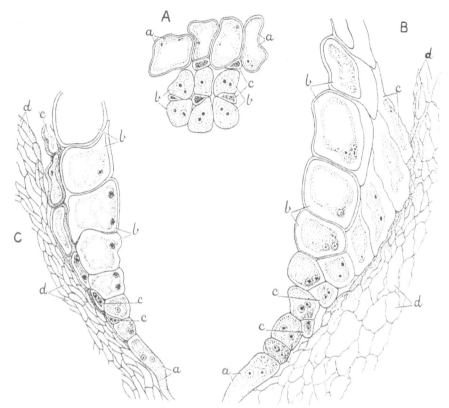

F<small>IG</small>. 38.—Development of the peridium in aecidia. A, *Puccinia graminis*: *a a*, peridial cap-cells terminating chains of aecidiospores; *b b*, aecidiospores; *c*, intercalary cells. The cap-cells and the aecidiospores have had intercalary cells cut off below them. B, *P, graminis*: longitudinal section of a basal cell, *a*, and lateral peridial wall-cells, *b b*. The cells, *b b*, have had cut off from them obliquely below and toward the exterior of the aecidium a series of intercalary cells, *c c*, which are becoming gelatinous. The jelly comes to lie between the peridium and hyphal cells, *d d*, surrounding the peridium. C, *Gymnosporangium juniperinum* (Kursanov's *G. tremelloides*), a section similar to that at B: *a*, a basal cell; *b b*, peridial cells; *c c*, intercalary cells becoming gelatinous; *d d*, hyphal cells on the exterior of the aecidium. The peridial cells in A, B, and C are all morphologically equivalent to aecidiospores. Figures, copied and lettered by the author, from Kursanov's *Über die Peridienentwicklung im Aecidium* (1914). Magnification: A, 600; B, 900; C, 300.

Kursanov[1] has shown that, in *Puccinia graminis*, intercalary cells are produced in the spore-chains that become metamorphosed into cells of the lateral wall of the peridium (Fig. 38, B). These intercalary cells are cut off from the aecidiospore mother-cells at an outer corner of the mother-cells so that they come to lie not between successive

[1]*Ibid.*, pp. 318–320.

mother-cells but *on the outside of the peridium* (Fig. 38, B, *c*) and it is therefore impossible for them to function as disjunctor cells. The cells which make up the lateral wall of the peridium are exclusively the cells corresponding to aecidiospores. The intercalary cells of the peridial chains at first form a layer of cells just outside of, and surrounding, the peridium. This cell-layer, however, as Kursanov observed, soon becomes slimy and breaks down (Fig. 38, B, *c*). Thus the wall of the peridium, at maturity, comes to be separated from the surrounding fungal and host elements by a layer of slime. Kursanov[1] has suggested that the slime is a mechanical aid to the aecidium at the time when this organ is growing outwards through the surrounding sterile hyphae, and he has compared it to the slime produced by certain cells in a root-pocket during the development of a lateral root of one of the Higher Plants.

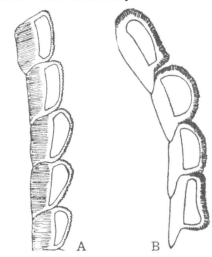

Fig. 39.—A, *Puccinia graminis*. B, *P. helianthi*. Side view of peridial cells. The walls toward the outside of the aecidium are thicker than those toward the inside. From Fischer's *Die Uredineen der Schweiz* (1904).

The peridial cells of an aecidium of *Puccinia graminis* and of *P. helianthi*, when mature, adhere tightly together so as to form a continuous membrane; and they all have a thickened outer wall and a much thinner inner wall (Fig. 39, B). Moreover, they are imbricate and, by means of their outer thicker walls, are hinged together[2] (Fig. 39). These arrangements are concerned with the mechanism for opening the aecidium when this has attained maturity and is ready to discharge its spores. When the air surrounding a mature but unopened aecidium changes in respect to its moisture contents from a relatively dry condition to a relatively very moist or saturated condition, the thinner inner wall of each peridial cell expands in area more than does the thicker outer wall, with the result that the apical arch of the peridial membrane becomes split radially into valves which turn out-

[1] *Ibid.*, p. 323.
[2] E. Fischer, *Die Uredineen der Schweiz*. Bern. 1904, Fig. 190, p. 244.

wards and become revolute, thus opening the aecidium in preparation for spore-discharge.

After the peridium of an aecidium of *Puccinia graminis* has once dehisced and its valves have become reverted, the aecidium remains open under both wet and dry conditions of the atmosphere.

The violent discharge of the aecidiospores from an aecidium of *Puccinia graminis* and similar Rust Fungi has already been treated of in Volume III.[1] The discharge takes place only in a very moist atmosphere and when the spore-chains are fully turgid. The aecidio-spore-chains fit very tightly together and press against one another with such force that the individual aecidiospores become flattened against their neighbours and hexagonal in optical cross-section. The more or less cylindrical lateral wall of the peridium is a very strong and continuous sheet of cells and, when a mature aecidium is about to discharge or is actually discharging its spores, the spore-chains as a whole are pressing against the lateral wall of the peridium and distending it whilst the peridium is pressing with an equal force against the spore-chain and compressing them. The strength of these antagonistic forces, which doubtless is equal to several atmospheres pressure, may be judged from the already recorded fact that, when an unopened finger-like aecidium resembling those shown in Fig. 36 is removed from a leaf-pustule and placed in a drop of water, the peridium very soon bursts along its sides and the spore-chains are violently discharged (Fig. 37). Normally on a leaf, the aecidiospores in the spore-chains of an aecidium are shot away in succession to a distance of several millimetres. The discharge of the spores in this way is dependent on the whole mass of spore chains being in a high state of turgidity and upon their pressing strongly against one another and against the spores about to be discharged. There can be but little doubt that, in keeping up the pressure necessary for spore-discharge, the counter-force of the lateral wall of the peridium is an important factor.

The number of aecidiospores present on three Barberry leaves heavily infected with *Puccinia graminis* was calculated to be 3,700,000, 5,400,000 and 8,000,000 respectively.[2]

The aecidiospores of *Puccinia graminis*, in damp still air under laboratory conditions, are shot up individually to a height of 3-4 mm. above the mouths of the aecidia, and to a maximum height of about 7-8 mm. Sometimes clusters of aecidiospores composed of many

---

[1]These *Researches*, Vol. III, 1924, pp. 552-559.
[2]*Ibid.*

cells (60-150), which may be called *aecidiospore-bombs*, are shot from an aecidium to a height of 7-8 mm. or to a horizontal distance of about 1 cm.[1]

Owing to adhesion due to moisture, one turgid aecidiospore could not separate from surrounding turgid aecidiospores without violence. By violent discharge, due to turgor pressure and the sudden rounding off of flat adhering cell-walls, the individual aecidiospores are projected into the air below the host-leaf whence they may be readily carried off by the wind.

Between the aecidiospores of *Puccinia graminis* there are tiny *pore-plugs* resembling those described by Dodge[2] in *Gymnosporangium ellisii* and *Puccinia podophylli*. They can be easily detected in aecidio-spore deposits mounted in water as little rounded lumps of matter clinging to the sides of aecidiospores or lying free (Fig. 37, B, p. 118 and see Fig. 34 in Ingold's *Spore discharge in land plants*, 1939). It has been suggested by Dodge that pore-plugs are of mechanical assistance in the discharge of aecidiospores.[3]

The peridium, as the hard external covering of an aecidium, not only protects the aecidiospore-chains from rupture but also affords a good pressure-surface for an aecidium at the time when this organ as a whole is forcing its way through the epidermis of its host.

Under natural conditions in the open, where the air is often very moist, the aecidia of *Puccinia graminis* dehisce as soon as they have burst through the lower epidermis of a Barberry leaf (Fig. 35, p. 116); but, under dry atmospheric conditions such as obtain in an experimental greenhouse during the winter at Winnipeg, the aecidia, as we have seen, after breaking through the host-epidermis, remain closed and continue to elongate until they become several times as long as they are broad (Figs. 34 and 36, pp. 115 and 117). Under dry atmospheric conditions, the peridium as a continuous layer of cells with a very thick outer cell-wall (Figs. 34 and 39, pp. 115 and 121) serves to protect the chains of aecidiospores within the aecidium from an excessive loss of water; whilst, on the advent of moist weather, when the chains of aecidiospores become fully turgid and ready for spore-discharge, the peridium soon bursts open at its apex and so allows the aecidiospores to be freely discharged into the air.

[1] *Ibid.*

[2] B. O. Dodge: "Aecidiospore Discharge as related to the Character of the Spore Wall," *Journ. Agric. Research*, Vol. XXVII, 1924, pp. 749-756; and "Expulsion of Aecidiospores by the Mayapple Rust, *Puccinia podophylli* Schw.," *Journ. Agric. Research*, Vol. XXVIII, 1924, pp. 923-926.

[3] For a more complete account of pore-plugs *vide infra*, Chapter XIII.

It thus appears that the peridium of an aecidium functions in three ways: (1) as a mechanical protection for the spore-chains and as an effective pressure-surface for an aecidium as a whole when this organ is forcing its way through the host-epidermis; (2) as a protection for the spore-chains against excessive loss of water, during dry weather and before the aecidium has opened; and (3) as a means for providing a counter-pressure for the mass of turgid spore-chains, after the aecidium has opened and during violent spore-discharge.

Nuclear Migration, Cell Fusions, Septal Pores, and the Diploidisation of a Proto-aecidium.—A proto-aecidium is borne on a uninucleate mycelium and is composed of uninucleate cells. A proto-aecidium, therefore, is a haploid organ.

An aecidium is made up of binucleate cells, for one finds conjugate nuclei in the aecidiosporophores, the aecidiospores, the intercalary cells, and the cells making up the peridium (Fig. 30, p. 110). An aecidium, therefore, is a diploid organ.

In 1896, Sappin-Trouffy[1] recorded and illustrated the fact that, in *Puccinia graminis*, from a mycelium with single nuclei binucleate cells grow up and produce chains of binucleate aecidiospores; but he did not discuss the origin of the binucleate condition.

In 1917 Kursanov[2] observed that in *Puccinia graminis*, when a proto-aecidium (his aecidial primordium) begins to develop into an aecidium, its basal cells, originally uninucleate, become binucleate; but, owing to the lack of sufficiently young material, he could not attempt to solve the transformation problem.

In *Puccinia graminis*, as shown by the observations of Kursanov, Miss Allen, Hanna, and others, the basal cells of a proto-aecidium do not form a palisade layer with the hyphae directed toward the host-epidermis as in *Phragmidium speciosum*, *Melampsora lini*, and *Cronartium ribicola*, but rather an irregular mycelial weft. This weft in cross-section has the appearance of a small-celled pseudoparenchyma and, as such, it appears in the drawings of Miss Allen[3] and also in Hanna's photograph reproduced in Fig. 29 (p. 108).

---

[1]P. Sappin-Trouffy, Recherches histologiques sur la famille des Urédinées," *Le Botaniste*, T. V. 1896, pp. 92-104.

[2]L. Kursanov, "Morphological and Cytological Researches on the Uredineae," *Sci. Mem. Imp. Univ. Moscou*, No. XXXVI, 1915, 288 pp. In Russian. A French translation under the title "Recherches morphologiques et cytologiques sur les Uredinees" appeared in *Bull. Soc. Nat. Moscou*, n.s.(1917), Vol. XXXI, 1922. In this *vide* pp. 6-8 and Text-fig. 1; also Plate 1, Figs. 1-7.

[3]Ruth F. Allen, "A Cytological Study of Heterothallism in *Puccinia graminis*," *Journ. Agric. Research*, Vol. XL, 1930, Plate IX, A, B, and C.

In 1929 Hanna[1] concluded from his observations that, in *Puccinia graminis*, when a proto-aecidium is becoming diploidised, the nuclei in the basal weft of hyphae become enlarged and "neighbouring hyphae then fuse in pairs in a manner similar to that described by Christman for *Phragmidium speciosum*, and two nuclei become associated in each fusion cell."

On the other hand, in 1930, Miss Allen,[2] in her account of an investigation on the heterothallism of *Puccinia graminis*, recorded that, when the proto-aedicia in her material were becoming diploidised, she had failed to observe any "2-legged" cells (such as had been seen by others in *Phragmidium speciosum*, *Melampsora lini*, *Cronartium ribicola*, etc.) or any other definite evidence of cell fusions.

It is desirable that the question to what extent, if any, hyphal fusions take place in the basal parenchyma of *Puccinia graminis* during the diploidisation process should be further investigated. However, for a discussion of the way in which diploidisation is effected in the proto-aecidia of heterothallic Rusts in general and of *P. graminis* in particular, it will be assumed that, so far as fusions are concerned, *P. graminis* resembles other heterothallic Rusts in which the formation of basal-cell fusions is definitely known.

Craigie's investigations, as recorded in Chapter I, have established that: (1) the haploid pustules of *Puccinia graminis* are of two kinds, (+) and (−); (2) a (+) proto-aecidium in a (+) pustule becomes converted into an aecidium only after its basal cells have been diploidised by (−) nuclei which have been derived from one or more pycnidiospores or mycelial hyphae of a (−) mycelium; and, similarly, (3) a (−) proto-aecidium in a (−) pustule becomes converted into an aecidium only after its basal cells have been diploidised by (+) nuclei which have been derived from one or more pycnidiospores or mycelial hyphae of a (+) mycelium. These conclusions have been strengthened by the discovery that, in *P. graminis*, by using the nectar-mixing technique, it is possible to hybridise such varieties as *P. graminis tritici* and *P. graminis avenae* and to cross races of *P. graminis tritici* differing in pathogenicity and spore colour.[3]

Now let us attempt to visualise what takes place in a proto-aecidium of a heterothallic Rust, such as *Puccinia graminis*, *Phrag-*

---

[1]W. F. Hanna, "Nuclear Association in the Aecium of *Puccinia graminis*," *Nature*, Vol. CXXIV, 1929, p. 267; also with the same title, another paper, with four photomicrographs, in *Report of the Dominion Botanist* for 1929, Dept. of Agric., Ottawa, 1931, p. 50.

[2]Allen, *op. cit.*, p. 603.

[3]For the literature *vide* the Introduction to Chapter IV.

midium speciosum, Melampsora lini, or *Cronartium ribicola,* during the last stages of the diploidisation process. Obviously, two factors are involved: (1) hyphal fusions, and (2) the migration of nuclei. During the sexual process, the nuclei of Rust Fungi, just like those of the Hymenomycetes, can move rapidly along a hypha from cell to cell and this migration is of prime importance in the diploidisation process. Hyphal fusions may not be necessary for the process; but, where they occur, they allow of the nuclei taking short cuts to their destination.

From the morphological and experimental data brought forward above, in respect to heterothallic Rusts such as *Puccinia graminis, Phragmidium speciosum, Melampsora lini,* and *Cronartium ribicola* in which cell-fusions are formed in the basal cells of the proto-aecidium during the transition from the haplophase to the diplophase, we can make two important deductions: (1) when a (+) proto-aecidium is being diploidised, the fusion of the (+) basal cells in pairs or in greater numbers is accomplished in response to the stimulus of invading (−) nuclei which have come from a distance through the (+) mycelium on which the (+) proto-aecidium originated, and the fusions provide passages which are used by the invading (−) nuclei in finding (+) nuclear mates; and, similarly, (2) when a (−) proto-aecidium is being diploidised, the fusion of the (−) basal cells in pairs or in greater numbers is accomplished in response to the stimulus of invading (+) nuclei which have come from a distance through the (−) mycelium on which the (−) proto-aecidium originated, and the fusions provide passages which are used by the invading (+) nuclei in finding (−) nuclear mates.

In 1893, Wahrlich,[1] in an important paper published in both Russian and German, demonstrated that septal pores and proto-plasmic bridges from cell to cell through the pores are characteristic of the Ascomycetes and Basidiomycetes in general and, in particular, so far as we are now concerned, he recorded finding a small central open pore in the septa of the mycelium of *Puccinia graminis* and *P. fusca* and in the septa of the mycelium and chains of unripe aecidio-spores of *P. caricis.* In the illustrations in his Plates he shows a central pore threaded by protoplasm: in a hyphal septum of *P. graminis;*[2] and in each of the two septa in a chain of three very young aecidio-spores of *P. caricis.*

[1]W. Wahrlich, "Zur Anatomie der Zelle bei Pilzen und Fadenalgen," *Scripta Botanica Horti Universitatis Imperialis Petropolitanae,* T. IV, 1893, pp. 101-155, Tab. II-IV. For an extended abstract of this paper in so far as it applies to Fungi and a reproduction of many of the Figures *vide* these *Researches,* Vol. V, 1933, pp. 89-96.   [2]*Ibid.,* Tab. III, Fig. 26.

In view of the fact that there is a central pore in each septum of a Rust mycelium, it seems probable that, in *Puccinia graminis* and in heterothallic Rusts in general, the nuclei which travel through a haploid mycelium to a proto-aecidium and eventually diploidise that organ's basal cells pass along the mycelial hyphae from cell to cell *via* the septal pores.

If we accept (1) Wahrlich's observations on septal pores and also (2) the findings of cytologists on fusions between the basal cells of proto-aecidia, it seems probable that, in such a Rust as *Puccinia graminis* or *Phragmidium speciosum*, when, for example, (−) nuclei are invading a (+) proto-aecidium, the (−) nuclei pass through the mass of (+) basal cells along the hyphae *via* the septal pores and laterally from one row of cells to another *via* newly-made breaches in the double walls (cell-fusions). In the case under consideration, the business of the invading (−) nuclei is to gain access to the (+) nuclei and with them to form conjugate pairs of nuclei in cells destined to become aecidio-sporophores. For the (−) nuclei to be successful in their efforts it is necessary for them: (1) to undergo nuclear division so as to provide a sufficient number of conjugate mates for the (+) nuclei; and (2) to move freely through single rows of (+) cells and from one (+) row of cells to another (+) row.

The suggestion, already put forward, that, in a heterothallic Rust, the nuclei which travel to a proto-aecidium and diploidise its basal cells pass along from cell to cell *via* the septal pores may not find ready acceptance. It may be objected that: (1) Rust cytologists since Wahrlich have not seen the pores; (2) the nuclei that travel from cell to cell are much thicker than the pores are wide; and (3) no positive evidence that the nuclei do pass through the pores has been presented. To these objections a reply will now be attempted.

(1) To demonstrate the existence of a pore in each septum Wahrlich used special technical methods.[1] These methods have not been employed by any Rust cytologist since his time.

(2) The nuclei are not only thicker than the septal pores are wide, but also thicker than the necks of the sterigmata of the basidia. However, since it is known that, in Rust Fungi, as well as in Hymeno-mycetes, the nuclei in a basidium-body pass upwards into the four spores *via* the channels of the extremely narrow sterigmatic necks, from the mechanical point of view there seems to be no good reason why Rust nuclei should not pass along hyphae and down into a proto-aecidium *via* the pores of the septa.[2]

[1]For a description of Wahrlich's technique and the results that I have obtained with it *vide* these *Researches*, Vol. V, 1933, pp. 89-91, 98-100.   [2]*Cf.* Vol. V, p. 167.

Cytoplasm has been seen moving rapidly through successive septal pores of one of the Hymenomycetes and of certain Pyrenomycetes and Discomycetes; and in *Fimetaria fimicola* large vacuoles have been seen slipping through septal pores with the greatest apparent ease.[1] If cytoplasm can pass through a septal pore, why should not a nucleus do likewise? It is known that, during its passage through a narrow sterigmatic neck, a nucleus of a Rust or of a Hymenomycete becomes greatly drawn out and deformed but, after entering a spore, soon becomes rounded once again. A similar deformation would permit of a Rust nucleus passing through a septal pore.

(3) The pores in the septa of the Ascomycetes are similar to those in the Hymenomycetes and Rust Fungi; and, in an account of a special investigation on the migration of nuclei during the sexual process in the Pyrenomycete *Gelasinospora tetrasperma*, Dowding and Buller[2] have adduced strong experimental and histological evidence in support of the conclusion that, in *G. tetrasperma*, the migration of a (+) nucleus along a (−) hypha, or of a (−) nucleus along a (+) hypha, from cell to cell, for a distance of several centimeters at a rapid speed (20-25 mm. in 5 hours, *i.e.* 4-5 mm. per hour), does actually take place *via* the septal pores and without the septa being broken down or otherwise altered. If nuclear migration takes place in hyphae from cell to cell *via* the septal pores in the Ascomycetes, there seems to be no good reason why the same mode of migration should not also take place in the Rust Fungi.

Certain illustrations published by Rust cytologists suggest that stages in nuclear migration through septal pores have actually been seen. As examples one may cite Figs. 3, 4, 5 and 6 on Plate III in Blackman and Fraser's[3] account of the 'fertilisation' process in the proto-aecidium of *Puccinia poarum*. In each of these Figures a nucleus is seen about half-way through a very narrow hole in the middle of a septum and the Figures are described as "examples of nuclear migrations into fertile cells." Blackman and Fraser do not discuss the origin of the holes in the septa; but, having regard to the position and size of the holes, we are justified in interpreting the holes as ordinary septal pores.

[1]*Ibid.*, pp. 103-122.

[2]Eleanor Silver Dowding and A. H. R. Buller, "Nuclear Migration in Gelasinospora," *Mycologia*, Vol. XXXII, 1940, pp. 471-488, with 6 Text-figs.

[3]V. H. Blackman and H. C. I. Fraser, "Further Studies on the Sexuality of the Uredineae," *Annals of Botany*, Vol. XX, 1906, pp. 35-48. For a reproduction of three of the Figures showing nuclear migration *vide* H. C. I. Gwynne-Vaughan and B. Barnes, *The Structure and Development of the Fungi*, Ed. 2, Cambridge, 1937, Fig. 261, p. 313.

The Structure of a Mature Pycnidium.—A general description of a mature and undisturbed pycnidium, such as is applicable to the pycnidia of *Puccinia graminis*, *P. helianthi*, *P. coronata avenae*, etc., will now be given.

A pycnidium (pycnium) is an oval or somewhat pear-shaped structure containing a small *cavity* which opens by means of an *ostiole* (Fig. 40). The ostiole is situated above the top of a little conical mound that is made up of raised epidermal cells and pycnidial hyphae. We may call this mound the *epipycnidial papilla* (*j*). In the substance of the pycnidium two layers can be distinguished: an outer *wall* and an inner *hymenium*. The *wall* of the pycnidium (*k*) is made up of compacted hyphae; and it is of uneven thickness, for it merges locally into masses of hyphae filling up the intercellular spaces in the host-tissue. The *hymenium* covers the wall-layer and lines the pycnidial cavity; and it is composed of three kinds of elements: *pycnidiosporophores*, *periphyses*, and *flexuous hyphae*.

The *pycnidiosporophores* (Fig. 40, *l*) are numerous, short, straight, slenderly conical cells that arise from the wall of the pycnidium. They are slightly convergent and packed closely together, side by side, so that collectively they constitute a curved palisade layer with all the elements pointing toward the ostiole. From the free ends of the pycnidiosporophores are constricted off a series of minute, oval, orange-yellow, gelatinous, uninucleate *pycnidiospores*. These (*m*) accumulate by thousands in the cavity of the pycnidium and, as they increase in number and their outer walls swell up, they are gradually extruded through the ostiole.

The *periphyses* (Fig. 40, *o*) arise in the hymenium in a zone around the pycnidiosporophores and just below the ostiole. They are about 60-70, stiff-looking, straight or very slightly curved, non-septate, orange-yellow, tapering cells, definitely limited in respect to their growth in length. They pass through the ostiole and extend about 0.1 mm. beyond this opening as slenderly conical divergent hairs.

At the ostiole, the periphyses are encircled by a band made up of cuticle and other remains of the epidermal cell or cells which they ruptured at the time of their emergence into the outer air. The ostiolar band just described may be called the *cuticular collar* (Fig. 78, *c*).

The periphyses are enclosed and hidden from view by a *drop of nectar* (Fig. 40, *n*) which has been excreted through the ostiole. In this drop, also, localised between and just above and around the periphyses, is a *slimy orange-yellow mass of pycnidiospores* (*p*) which has been extruded through the ostiole. The drop of nectar extends beyond the

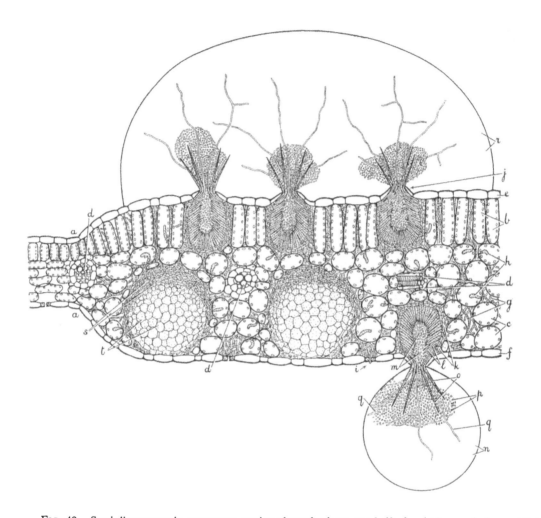

FIG. 40.—Semi-diagrammatic transverse section through about one-half of a (+) or a (−) pycnidial rust pustule formed by the mycelium of *Puccinia graminis* in a leaf of *Berberis vulgaris*, to show the hypertrophied tissues of the leaf and the haploid mycelium and haploid organs (pycnidia and proto-aecidia) of the fungus. To the left of *a a*, part of the leaf not invaded by mycelium and therefore in its normal state. To the right of *a a*, about one-half of a pustule which originated from the growth and stimulation of a mycelium derived from a single basidiospore: *b*, elongated palisade cells; *c*, swollen spongy mesophyll cells; *d d d*, vascular bundles; *e*, upper epidermis; *f*, lower epidermis; *g*, intercellular mycelium; *h*, haustoria; *i*, a stoma through the cleft of which a hypha very slightly protrudes; *j*, an epipycnidial papilla; *k*, a pycnidial wall; *l*, pycnidiosporophores; *m*, the cavity of a pycnidium, containing pycnidiospores with swollen gelatinous walls; *n*, a drop of nectar crowning the pycnidial papilla; *o*, periphyses, red and pointed; *p*, a mass of gelatinous pycnidiospores that have been extruded through the pycnidial ostiole; *q q*, flexuous hyphae, fully grown in length; on the upper side of the pustule are three pycnidia covered by a compound drop of nectar, *r*, formed by the coalescence of three simple drops, and from the ostiole of each of these pycnidia there project a mass of pycnidiospores, pointed periphyses, and cylindrical flexuous hyphae of which some are branched; in the lower half of the pustule in the mesophyll between the vascular bundles are two proto-aecidia, each of which consists of a concavo-convex mass of small cells, *s*, densely filled with protoplasm, and of a rounded mass of larger pseudoparenchymatous cells, *t*, which contain very little protoplasm and are very transparent. After a proto-aecidium has been diploidised, the cells at *s* give rise to the aecidiosporophores and aecidiospores, while those at *t*, as the aecidiospores are pushed downwards in compact chains, are dissolved and completely disappear.

pointed ends of the periphyses and the undisturbed mass of pycnidio-spores as a clear and colourless liquid, free from solid particles and as transparent as a drop of dew.

The *flexuous hyphae* (Fig. 40, *q*) are situated in the hymenium in a zone around the pycnidiosporophores and just beneath the ostiole, and they originate between the periphyses or (as represented in Fig. 40) just below the periphyses. They are much elongated, cylindrical, usually unbranched, more or less wavy, thin-walled, unicellular hyphae, about 3-20 in number. They extend through the ostiole, pass more or less radially outwards from the ostiole between the periphyses, push their way beyond the pointed periphyses and the static mass of pycnidiospores, and so come to have their ends freely exposed in the clear part of the drop of nectar, there to await the visit of an insect and the coming of pycnidiospores of opposite sex. They elongate slowly in the nectar and their length therefore varies with their age.

After the initiation of the sexual process, the periphyses are per-sistent, but the flexuous hyphae soon become gelatinous and disappear.

Terminology.—Now that the structure of a typical pycnidium of a Puccinia has been described and illustrated, we may turn for a moment to a discussion of terminology.

The *spermogonium* of the older uredinologists (the *pycnidium* of this volume) was so called because it was supposed to be a male organ that produced *spermatia*. As a result of Craigie's discoveries, we now know that a "spermogonium" not only produces thousands of "sper-matia," but also has growing out from its hymenium a number of *flexuous hyphae* which are able to receive "spermatial" nuclei and conduct them into the mycelium leading to the proto-aecidia. The flexuous hyphae are certainly not male structures and therefore "spermogonium" is unsatisfactory as the name of the organ in which they are developed.

A *flexuous hypha* is a *receptive hypha* in that, after fusing with a pycnidiospore of opposite sex, it receives the nucleus of the pycnidio-spore and conducts it down to the mycelium below. It may therefore be asked: is not a flexuous hypha in reality a female structure that may be called a *trichogyne?* The answer is in the negative. A flexuous hypha is an element produced in the hymenium of a pycnidium and it is not directly connected with one or more proto-aecidia, as may be seen from an inspection of Fig. 40. Between a flexuous hypha and the basal cells of the nearest proto-aecidium there is a vegetative mycelium whose hyphae have haustoria inserted into the host-cells. Therefore, there is no justification for calling a flexuous hypha a trichogyne.

A flexuous hypha is to be regarded as a specialized hypha that grows out from the mycelium into the nectar on the exterior of the host and that functions by making contact with a pycnidiospore of opposite sex, fusing with it, and receiving its nucleus. A flexuous hypha, therefore, is comparable with one of the leading hyphae of a haploid mycelium of such an agaric as *Coprinus lagopus* when, in the course of its growth, it meets with and fuses with the germ-tube or young mycelium produced by an oidium of opposite sex.

A *proto-aecidium* is not to be regarded as a female organ, but only as a rudimentary haploid organ destined, after certain of its basal cells have been diploidised, to develop into an *aecidium* producing chains of *aecidiospores*.

In such a Rust as *Puccinia graminis* or *P. helianthi*, as we have seen, there are two kinds of mycelium, (+) and (−); and these are *of opposite sex*, for sexual reactions take place not between two (+) or two (−) mycelia, but only between a (+) mycelium and a (−) mycelium.

All the nuclei in a (+) mycelium are (+), and all the nuclei in a (−) mycelium are (−). As indicated by Craigie's experiments, any of the nuclei of a (+) mycelium or of a (−) mycelium, whether situated in the vegetative hyphae or in the pycnidiospores are capabel of effecting diploidisation in the basal cells of the proto-aecidia of a mycelium of opposite sex. On this account, one cannot admit that the nuclei of the pycnidiospores are "male" whereas those of the mycelium on which the pycnidiospores are borne are "not male." It seems best, therefore, to regard the *pycnidiospores* not as male cells, not as spermatia, but merely as cells which differ from the cells of the mycelium in that they are *cellulae disseminulae*, *i.e.* cells that can readily be transported by a suitable agency to mycelia of opposite sex.

Arthur[1] has called the "spermogonium" of the Uredinales a *pycnium*, and in this connection he remarked: "Pycnium is derived from the Greek πυκνὸς, dense or compact. The term pycnidium, derived from the same root, has been used by some writers in this connexion, but it seems best to confine its use to the similar but probably not homologous structure in ascomycetous fungi."

Pycnium is a brief and distinctive term, and there is no objection to using it for a "spermogonium" of a Rust Fungus; but for convenience in the historical discussion given in Chapter I, for comparisons in general, and to avoid increasing the number of mycological terms unduly, I have preferred to employ the older term *pycnidium*. It may

[1]J. C. Arthur, *et al.*, *The Plant Rusts* (*Uredinales*), New York, 1929, p. 9. **See** footnote on p. 3.

well be that both the pycnium of the Uredinales and the pycnidium of
the Ascomycetes were in origin merely organs that produced minute
spores capable of germination and that the loss of the power of germination, where that has occurred, was a secondary development
associated with a greater specialisation of the sexual process.

As we saw in Chapter I, it was de Bary[1] who introduced the term
paraphyses for the stiff-looking tapering hairs that arise in a pycnidium
of a Rust Fungus and protrude through the ostiole, and this term has
been freely used by other uredinologists, *e.g.* by Arthur and his
colleagues.[2]

The paraphyses ($\pi\alpha\rho\dot{\alpha}$ = by the side of, $\phi\acute{v}\sigma\iota s$ = a growth) of the
Hymenomycetes and Discomycetes are situated *at the side of* and
between the spore-bearing organs (basidia or asci), whereas the so-called
paraphyses of a Puccinia do not lie in between the pycnidiosporophores
but form a fringe around those organs just within the pycnidial
ostiole (Fig. 40 and 43, pp. 130 and 140). Hence the "paraphyses" of
a Puccinia are in reality *periphyses* ($\pi\epsilon\rho\acute{\iota}$ = around, $\phi\acute{v}\sigma\iota s$ = a growth).
On this account, in this work, I have called the conical hairs that pro-
trude from the ostiole in a Puccinia *periphyses* instead of paraphyses.

The term *paraphyses* in the Rust Fungi has for a long time been
correctly used in connexion with uredo-sori and teleuto-sori where it
designated true paraphyses, *i.e.* cylindrical or capitate structures occurring around and between and by the side of the uredospores or
teleutospores. For this reason alone it would be advisable not to use
the term paraphyses for the hair-like structures situated around the
ostioles of the pycnidia of Puccinia.

A further justification for calling the ostiolar hairs of a Puccinia
periphyses is that the term has already been applied in the Pyrenomycetes to somewhat analogous structures, namely, the short hairs
which, in many species, cover the interior surface of the neck of the
perithecium and are directed upwards in the neck-canal toward the
ostiole.

In Gymnosporangium the tapering bristle-like hairs passing
through the pycnidial ostiole resemble in appearance those of Puccinia.
Nevertheless, they arise not as in Puccinia as a fringe around the
inside of the ostiole but, separately, in various places among and
between the pycnidiosporophores (*vide* Fig. 41, p. 137). In Gymnosporangium, therefore, the ostiolar hairs are truly *paraphyses* and not
periphyses.

[1] *Vide supra*, p. 15.

[2] J. C. Arthur *et al.*, *The Plant Rusts (Uredinales)*, New York, 1929, p. 126, Fig. 11,
c, p. 8.

Since neither the term *periphyses* nor the term *paraphyses* is strictly applicable to the tapering pointed hairs of the pycnidia of *all* the Rust genera that possess them, another more general term is required that is not based on the exact place of origin of the hairs and that could be employed for the Uredinales in general. The hair-like organs in question may be called *ostiolar trichomes* or *ostiolar bristles*.

In some genera of the Rust Fungi, *e.g.* Puccinia and Gymnosporangium, ostiolar trichomes are present, while in other genera, *e.g.* Phragmidium and Melampsora (*vide infra*), they are absent. In taxonomic descriptions of Rust genera, it should be clearly indicated for each genus whether or not ostiolar trichomes are present. The structures in question in descriptions of this kind might be referred to merely as *trichomes*. This would serve to distinguish them from the *flexuous hyphae* which (1) look like hyphae rather than like trichomes, (2) are present in all active Rust pycnidia, and (3) differ from all the various forms of trichomes in Flowering Plants in that they have a sexual function. It would also obviate the difficulties that arise from the use of the terms *periphyses* and *paraphyses*, the correct employment of which involves histological knowledge.

The Development of a Pycnidium of Puccinia graminis.—The steps leading to the formation of a mature pycnidium of *Puccinia graminis* have been elucidated by Miss Allen,[1] and in what follows an endeavour will be made to summarize her account of them.

The germ-tube of a basidiospore of *Puccinia graminis*, after piercing the outer wall of an epidermal cell, enters the cell and there forms a *primary hypha* of 4-6 cells. Then each of these cells sends out a branch which grows down into the intercellular spaces of the mesophyll and there develops into *haploid mycelium*. The primary hypha degenerates rapidly and disappears.

Pycnidial development begins on the fourth day after inoculation. Hyphae grow up between palisade cells and form a *mat* between the epidermal and palisade layers. Then hyphal branches are formed which converge at a common point under the epidermis. At this stage of development the young pycnidium has a bi-convex form. Large upright hyphae packed side by side in the centre of the pycnidium grow in length, press against the epidermis, and lift the epidermis up, thus forming a little cone. Miss Allen has called the upright hyphae *buffer cells*, and the conical protuberance which they form is my *epipycnidial papilla*.[2]

---

[1]Ruth F. Allen: "A Cytological Study of Heterothallism in *Puccinia graminis*," *Journ. Agric. Research*, Vol. XL, 1930, pp. 585-614; and "Further Cytological Studies of Heterothallism in *Puccinia graminis*," *Journ. Agric. Research*, Vol. XLVII, 1933, pp. 1-16. [2]*Vide supra*, p. 129.

The pycnidium soon becomes organised into an *outer wall* from which slender-tipped *pycnidiosporophores* grow into the central cavity. The cavity appears to be formed by the growth in size of the pycnidium brought about by an increase in the number of wall hyphae and pycnidiosporophores. After the cavity has been formed, the buffer cells no longer form a compact mass, for by this time they have been separated by pycnidiosporophores which have squeezed in between them; but they are still long, coarse, blunt-tipped cells with vacuolated cytoplasm. The buffer cells gradually deteriorate, die, and disappear. They produce no spores, and it would appear that they "serve only the transient purpose of helping to lift the epidermis above the young growing pycnium."

From around the upper edge of the wall of the pycnidium *periphyses* (paraphyses) grow toward one another at a point, meet, turn upwards, and pierce the epidermis. Outside the epidermis they continue their growth and form a brush of tapering stiff-looking hyphae. Each periphysis contains 1, 2, or 3 nuclei.

A young pycnidiosporophore, present in the pycnidium before the ostiole is opened, is thick and often curved, and it tapers to a rather blunt tip. It contains dense cytoplasm and a single nucleus. As it grows, its nucleus divides, and sometimes this is followed by the formation of a septum near its base and the consequent cutting off of a short stalk-cell. The terminal cell continues to grow in length. The pycnidiosporophores are crowded together, and the details of their shape, length, and thickness vary with the available space. Each pycnidiosporophore grows in length at its free end and there constricts off in a row a series of *pycnidiospores*, each of which contains a single nucleus. The pycnidiospores are formed in great numbers and the pycnidial cavity becomes full of them. The first spores to emerge from the pycnidium escape shortly after the ostiole has opened.

If a pustule remains in the haploid condition, the pycnidia remain active as long as the fungus lives, and they continue to grow and broaden out and to form fresh periphyses, and their drops of exudate (pycnidiospores plus a liquid) are maintained. A day or two after the nectar has been mixed and the sexual process has been initiated, whether this takes place early or late, the formation of pycnidiospores stops, the exudate dries, and the pycnidia die.

Miss Allen's account of the development of a pycnidium of *Puccinia graminis* will now be supplemented by my own observations on the development of the flexuous hyphae.

The flexuous hyphae grow from the upper part of the wall of the pycnidium, through the ostiole, and into the drop of nectar. As a pycnidium becomes older, the number of its flexuous hyphae usually increases. The first-formed flexuous hyphae develop at about the same time as the periphyses. A flexuous hypha becomes cylindrical, its apical growing-point remains blunt, as it elongates it becomes wavy or flexuous, and it gradually grows within the drop of nectar until it stretches out far beyond the tips of the pointed periphyses. While some flexuous hyphae remain simple, others become branched.

The Periphyses and their Functions.—In the genus Puccinia, the average number of periphyses per pycnidium doubtless varies from species to species. It is not easy to observe all the periphyses in a pycnidium and, therefore, counts for individual pycnidia are usually not perfectly accurate. Such counts yielded: for *Puccinia graminis*, 57, 59, 65, 71, and 78; for *P. helianthi*, 59, 63, 65, 67, 71, and 93; and for *P. minussensis* as many as 119, 123, 131, 139, and 140. We thus see that *P. minussensis* has about twice as many periphyses per pycnidium as either *P. graminis* or *P. helianthi*.

One of the methods which was employed for observing the periphyses and flexuous hyphae in a pycnidium is as follows. Cut a surface section of a pustule from a leaf, mount in water under a cover-glass supported by pieces of another cover-glass, boil for a few moments, add a drop of cotton-blue at the edge of the cover-glass, lift the cover-glass temporarily to allow the dye to run over the top of the section, then observe with the microscope. The periphyses and flexuous hyphae are then seen to be deep blue in colour and standing out in contrast with the orange-yellow colour of the main mass of the pycnidium forming the background. In the case of *Puccinia minussensis*, which has the largest pycnidia, the results given by this method were especially good: the fringe of over 100 blue periphyses around each orange ostiole made the pycnidia look like strangely-coloured miniature sea-anemones.

The periphyses of *Puccinia graminis* originate as special outgrowth from pycnidial wall-cells situated just below the ostiole and within the little mound that I have called the epipycnidial papilla[1] (Fig. 40, p. 130).

The pycnidia of *Gymnosporangium juniper-virginianae* and other species of Gymnosporangium are relatively large and more or less globose, and their ostiolar trichomes differ from those of *Puccinia graminis* and other species of Puccinia in that they arise not close together and just below the mouth of a pycnidium but isolated from one another at various places among the pycnidiosporophores, so that before they reach the ostiole they are obliged to pass through the pycnidial cavity. This was first shown in 1911 by Lloyd and Ridg-

[1] *Vide supra*, p. 129.

way[1] in an illustration of a vertical section of a pycnidium of *Gymnosporangium juniperi-virginianae* (Fig. 41). I, myself, have observed the scattered arrangement of the paraphyses not only in *G. juniperi-virginianae* but also in *G. clavariiforme*. In both of these species the more or less cylindrical pedicel of a paraphysis situated inside a pycnidium and arising at its base is often about equal in length to the conical shaft of the same paraphysis projecting outwards beyond the ostiole in the pycnidial nectar.

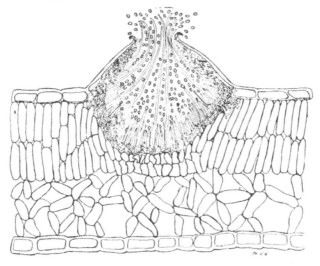

FIG. 41.—*Gymnosporangium juniperi-virginianae*. A pycnidium on the upper side of an Apple leaf, showing pycnidiosporophores, pycnidiospores, and long pointed ostiolar trichomes. The trichomes arise in the hymenium at various places among the pycnidiosporophores and are therefore *paraphyses*. The protuberant parts of the paraphyses should have been represented as much longer and as slenderly conical in form (*cf.* Figs. 80, A, and 71). The flexuous hyphae were overlooked. Drawn by F. E. Lloyd and C. S. Ridgway. From their *Bulletin* on "Cedar-Apples and Apples," Alabama, 1911. Highly magnified.

From the point of view of position of origin of the ostiolar trichomes in the pycnidial hymenium, as already intimated in the Section on *Terminology* it is clear that whereas the ostiolar trichomes of Puccinia are *periphyses*, those of Gymnosporangium are *paraphyses*.

In a very young pycnidium of *Puccinia graminis*, the periphyses,

[1]F. E. Lloyd and C. S. Ridgway, "Cedar-Apples and Apples," *Alabama Department of Agriculture*, Bull. XXXIX, 1911, Fig. 7, p. 11. Same article in *Annual Report of Alabama Dept. of Agric.*, 1911.

in the form of somewhat thick, cylindrical, bluntly-ended hyphae, grow radially toward a single point under the epidermis.   Here, collectively forming a solid cone and supported by the underlying mass of pycnidial hyphae, they press against the epidermis, push it upwards, crush and break through one or more of its cells, and thus form the ostiole through which the pycnidial exudate is destined to be extruded.   The first function of the periphyses, therefore, is a mechanical one.

After breaking through the epidermis of the host-plant, the periphyses of a pycnidium continue to elongate at their apex.   As they grow in length they become more and more pointed, so that, when they have attained their full development, their ends have less than a quarter of the thickness of their bases.   It is often easy, in surface sections of a pustule stained with cotton-blue, to observe the contrast between the thick blunt ends of periphyses that have just made their way through the epidermis and the finely pointed ends of mature periphyses close by.

Whilst the periphyses are growing to their full length outside the ostiole, they can be seen (in side view with a lens or the microscope) projecting away from the ostiole and inclined toward one another so as to form an acutely pointed cone (Fig. 42, A).   This cone, which surmounts the epipycnidial papilla, reminds one of a Red Indian tent or tepee.

The tent-like closed-up mass of periphyses which has just been described retains its conical form until the periphyses are caused to separate by the exudation of the pycnidial nectar.   As the nectar, bearing pycnidiospores, is excreted and the periphysal cone opens, one can see that the nectar from the first extends to the ends of the periphyses (Fig. 42, B).

When excretion of the pycnidial drop has begun, the drop steadily grows in size (Fig. 42, C and D), and soon it becomes at least twice as high as the projecting periphyses.   At this stage in development, the drop is seated on the top of the epipycnidial cone.   Here it touches the cuticular collar of broken epidermal cells surrounding the ostiole, but is otherwise entirely free from the epidermis (D).   At the same time it wets and envelops all the periphyses which, now being free to move, diverge from the axis of the drop so that they become spaced like the hairs of a paint-brush in water or so that they form a wide circular fringe projecting around the ostiole (D).

The periphyses are easily wetted by the exuding drop.   They are therefore not pushed outwards on the surface of the drop by surface tension.   In side views of drops seen under the microscope one can observe that the periphyses are actually immersed.

In a vertical section taken through the middle of a mature pycnidium it is seen that the periphyses have their bases well inside the pycnidium and that they form the roof of the pycnidial chamber (Figs. 40 and 43). Each periphysis (Fig. 63, A, p. 185), therefore, has (1) a *basal portion* or *pedicel* inside the pycnidium under the epidermal cells of the epipycnidial core, which stretches from its point of origin in the wall of the pycnidium to the ostiole, and (2) a much longer *apical portion* or *shaft* which extends freely into the drop of nectar and is bent somewhat away from the drop's axis. The basal portion of a periphysis, hidden below the epipycnidial papilla, is cylindrical. As a periphysis passes out through the ostiole it becomes much thickened, and then it tapers evenly to a fine point. The length of a periphysis

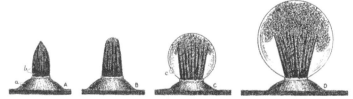

FIG. 42.—*Puccinia graminis.* Stages in the opening of a pycnidium at the surface of a leaf of *Berberis vulgaris.* A: *a,* an epipycnidial papilla with its ostiole surrounded by a cuticular collar; through the ostiole protrude the periphyses in the form of a cone. B: through the ostiole pycnidiospores and nectar are being extruded and the periphyses are beginning to move apart. C: the periphyses have now separated from one another, and a drop of nectar is visible. D: the drop has increased in size and in its upper half there is a gelatinous mass of pycnidiospores held in place by the periphyses. Magnification, about 120.

of *Puccinia graminis,* from its thickest part at the mouth of the ostiole to its pointed tip, is about 0.1 mm. or about equal to the height of the pycnidial chamber measured from its base in the mesophyll of the leaf to its ostiole.

A periphysis of *Puccinia graminis* (Fig. 63, A, p. 185) was found to have widths as follows: inside a pycnidium, 2.0-2.5μ; at its thickest part just outside an ostiole, 4.6μ; and at its pointed end, about 1μ.

A mature periphysis, as seen projecting from the ostiole of a pycnidium, by its straightness, tapering form, and apparent stiffness reminds one of those epidermal hairs of Flowering Plants which are known as *bristles.* However, while bristle-hairs are often multicellular and usually have thickened cell-walls, periphyses are always unicellular and have thin walls. Even at their pointed ends periphyses are thin-walled.

The name *periphyses* sufficiently suggests cell-organs which are produced around the ostiole of a pycnidium and to that extent is

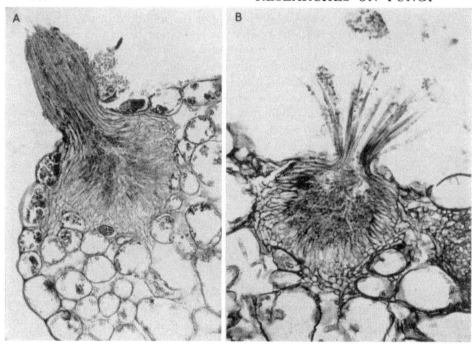

Fig. 43.—Pycnidia of *Puccinia graminis* on leaves of *Berberis vulgaris*, as seen in microtome sections. A: a pycnidium in which the nectar had dried and in which the periphyses and the pycnidiospores had all become stuck together. B: another pycnidium in which the periphyses diverge from the pycnidial ostiole as they do when living and functioning. One can also observe: some of the mycelium in the intercellular spaces between the host-cells; the dense weft of hyphae making up the wall of the pycnidium; the pycnidiosporophores lining the pycnidial cavity and directed to its centre; the mass of gelatinous pycnidio-spores filling the cavity; other masses of pycnidiospores among and beyond the periphyses; and, projecting from the ostiole on its extreme right, a truncated cylindrical structure which appears to be the lower half of a flexuous hypha (*cf.* Fig. 40). In thin microtome sections, like the one shown here, a flexuous hypha, even if present, cannot usually be recognised as such because it has been cut into several pieces. Sections prepared and photographed by W. F. Hanna. Magnification, about 350.

appropriate for international usage. However, by analogy with epidermal hairs of like form, we might well call periphyses *epipycnidial bristles*. Such a designation would fit in well with the term *pycnidial brush* which shortly will be introduced.

The periphyses, as we have seen, function mechanically when they are young in that they pierce a way through the epidermis of the host-plant, thus providing the pycnidium with its ostiole. But we may still enquire whether or not the periphyses, when fully grown and immersed in the drop of nectar, have other functions to perform.

The periphyses of *Puccinia graminis*, *P. helianthi*, etc., are filled with protoplasm and in this protoplasm are numerous tiny oil droplets containing a carotinoid pigment. As one can readily perceive by examination with a hand-lens, the periphyses impart a good red colour to the base of every individual pycnidial drop of nectar. The pycnidiospores and the flexuous hyphae, as well as the hyphae within the epipycnidial papillae, also contain a carotinoid pigment, so that, in an older pustule, the orange-red colour of the compound drop or drops of nectar is due in part to the colour of the pycnidiospores, the periphyses, and the flexuous hyphae and in part to the red background of the epipycnidial papillae. It may now be asked: is the orange-red colour of the periphyses and other fungus elements of a pycnidial pustule of any ecological significance?

It may be supposed that the red colour of the pycnidial drops is a factor in making the drops attractive to flies. There is as yet no conclusive experimental proof of this idea, but it may be urged in its favour that the pitchers of Pitcher Plants (Sarracenia, Nepenthes) and the leaves of the Sundew (Drosera) which attract and catch flies are red, and that certain Gastromycetes, namely *Mutinus caninus* and *Clathrus cancellatus*, which attract flies and have their spores dispersed by these insects, have red glebae. If we assume that the red colour of the drops of nectar of *Puccinia graminis* and similar Rusts does actually serve to attract flies to the pycnidial pustules, we must then admit that the carotinoid pigment of the periphyses is a contributory factor in making the attraction effective.

In Flowering Plants carotin has more than one function. In a green leaf it may well be indispensable for photosynthesis, while in the petals of a yellow flower it doubtless assists in attracting insects. So, too, in Fungi, carotin may have various uses. In Pilobolus carotin accumulates at the base of the subsporangial swelling just at that spot where sunlight, after being refracted through the upper part of the swelling, is most concentrated; and, as I have suggested elsewhere,[1] the pigment probably plays some important role in the phenomenon of heliotropism. In the Uredinales, it is mainly in pycnidial pustules, if at all, that the carotinoid pigment is useful in attracting insects. Its presence in aecidiospores, uredospores, teleutospores, and the mycelium generally suggests that it is of importance in some unknown

---

[1]These *Researches*, Vol. VI, 1934, pp. 71 and 99. *Cf.* E. Bünning, "Phototropismus und Carotinoide. II. Das Carotin der Reizaufnahmezonen von Pilobolus, Phycomyces, and Avena," *Planta*, Bd. XXVII, 1938, pp. 148-158. Bünning regards the red pigment in *Pilobolus kleinii* as B-carotin, like 90 per cent. of the carotin in carrots.

way for the general metabolism of Rust Fungi. There is also the possibility that it may be a waste product.

In transverse sections of living pycnidial pustules of *Puccinia graminis* it was noticed with the help of the microscope that the periphyses of the pycnidia on the upper well-lighted side of a Barberry leaf are brighter in colour than those of the pycnidia on the under and darker side of the same leaf. Thus light appears to promote the accumulation of carotinoid pigment in periphyses.

The conclusion that light increases the redness of periphyses is in accord with certain observations made by Newton and Johnson[1] on the colour of uredospores. These workers have observed that, on a Wheat leaf, the uredospores on the upper well-lighted side are much redder than those on the lower less well-lighted side; and they also noticed that this difference in colour in respect to light was especially pronounced in a "white" race of *Puccinia graminis tritici* which contained but a very small amount of carotinoid pigment, for the uredo-spores of this race were deep-buff on the upper side of a leaf and almost pure white on the lower side. It would appear that, in the Rust Fungi, just as in Flowering Plants,[2] light is a factor which is of considerable importance for the production of the carotinoid pigment.

A periphysis, as we have seen, is a stiff, straight, tapering, sharply pointed hypha. The sixty or seventy or more periphyses that stick outwards from the ostiole of a pycnidium in a drop of nectar, taken collectively, form what may be called a *pycnidial brush*. Through this brush, from its base to its exterior, slides the gelatinous mass of pycnidiospores as it is exuded through the ostiole; and, when exudation ceases, the mass of pycnidiospores extends from the ostiole upwards and outwards in such a way that the pycnidial brush holds it firmly in place.

Let us now suppose that a fly, after visiting a (+) pycnidium and getting its trunk smeared with slimy (+) pycnidiospores, visits a (−)

[1]M. Newton, H. Johannson, and T. Johnson, "A Study of the Carotinoid Pigments of Uredospores of Wheat Stem Rust and Four of its Color Variants," *Phytopathology*, Vol. XXV, 1935, p. 30. These authors found that spectroscopic determinations of the naphtha extracts of uredospores of four colour types (orange, normal, antique brown, and white) "showed for each type a higher pigment content in spores produced under normal than under subnormal light." They also found that "spectral distribution curves indicated that the pigments were carotinoid in nature and in most instances consisted chiefly of carotene."

[2]W. A. Beck and R. Redman, "Seasonal Variations in the Production of Plant Pigments," *Plant Physiology*, Vol. XV, 1940, pp. 81-84. These authors show by means of graphs how, in Dutch Sweet Clover, the amounts of carotene, xanthophyll, and chlorophyll vary with the seasonal changes in light and temperature.

pycnidium. What will happen? The fly applies its trunk roughly and firmly to the nectar, in a few seconds drinks much or most of the sweet juice up, and then departs. So much can be seen with the naked eye. Since experiments made by Craigie[1] on *Puccinia helianthi* have demonstrated that flies do actually bring about an exchange of pycnidiospores between pycnidia, in the particular case now under consideration we are justified in imagining that: (1) the fly's trunk, in addition to removing clear nectar, dislodges part or the whole of the slimy mass of (−) pycnidiospores previously held between the periphyses; and that (2), at the same time, owing to the fact that the fly's trunk is pushed down upon and rubbed against the bristles of the pycnidial brush and against the flexuous hyphae, some of the (+) pycnidiospores which have been brought from the (+) pycnidium will be brushed off the trunk by the periphyses and will be left behind on the outside of the (−) pycnidium in contact with the flexuous hyphae. When this has been accomplished, the way has been prepared for (+) pycnidiospores and neighbouring (−) flexuous hyphae to unite and thus form channels through which the nuclei of the pycnidiospores may pass. From the facts and suppositions which have just been presented, we may conclude that, whilst flies are visiting the drops of nectar, the pycnidial brush is of considerable mechanical importance.

Thus it appears that the periphyses of a pycnidium have several functions: (1) they form a cone which is used for penetrating the epidermis of the host-plant, (2) they impart their red colour to the drop of nectar and thus, in association with the red pycnidiospores, etc., are probably a factor in increasing the chances that the drops will attract the attention of insects; (3) they hold the mass of extruded pycnidiospores up above the surface of the leaf in a drop of nectar and so make the pycnidiospores readily accessible to flies; and (4) they form a brush which assists in the exchange of (+) and (−) pycnidiospores when insects are feeding on the nectar.

The Pycnidiosporophores and the Pycnidiospores.—In each pycnidium, directed toward the small pycnidial cavity and at the same time toward the ostiole, are some hundreds of unicellular pycnidiosporophores (Fig. 40, p. 130). These are short tapering reddish hyphae, relatively thick as they spring from the wall of the pycnidium and very thin at their free ends. The pycnidial cavity from the first is full of fluid and in this the pycnidiospores are produced. Each pycnidiosporophore abstricts from its free end, one by one, a series of minute, more or less oval, reddish pycnidiospores. During spore-production,

[1]J. H. Craigie, "An Experimental Investigation of Sex in the Rust Fungi," *Phytopathology*, Vol. XXI, 1931, p. 1023.

nuclear division takes place in each pycnidiosporophore, with the result that every pycnidiospore is provided with a single nucleus (*cf*. Chap. IX). As soon as a pycnidiospore has been produced it separates from its pycnidiosporophore and comes to lie with its fellows in the pycnidial chamber. The pycnidial chamber soon becomes filled with pycnidiospores and then these spores, ever increasing in number, press outwards through the ostiole. After passing through this aperture, along with the nectar with which they are mixed, they force their way up into the as yet unexpanded tent-like cone of periphyses. As more and more of the exudate is extruded, the cone of periphyses opens out at its apex and the individual periphyses become separated from one another like the hairs of a paint-brush dipped in water (Fig. 42, p. 139).

Owing to the fact that the pycnidial nectar contains sugar and has a high osmotic pressure, the drop of nectar which exudes with the pycnidiospores from an ostiole draws water from the leaf and grows for some time, so that it comes to extend far beyond the tips of the periphyses. It is a noteworthy fact that the pycnidiospores of a pycnidium, unless disturbed by an insect or by some other means, *do not spread freely throughout such an enlarged drop, but hang together in a mass* which is held in one place by the periphyses. The mass of pycnidiospores extends from the ostiole through the spaces between the periphyses and usually to some distance beyond, or to one side of, the pycnidial brush (Fig. 40, p. 130); but the pycnidiospores do not float freely as individual granules and, in the absence of Brownian movement, they do not become evenly dispersed in the drop (*cf*. Fig. 50, p. 153).

The pycnidiospores are cut off from their pycnidiosporophores in succession during several successive days; and, as more and more of them are extruded from the outside, the mass which they form grows in size, occupies more space in the drop of nectar, and tends to break up into smaller masses.

In a pustule of *Puccinia graminis* or *P. helianthi* the simple drops of nectar grow in size to such an extent that often they touch one another and fuse together. At the moment when two simple drops unite, the two masses of pycnidiospores are disturbed and changed in form, but the union does not cause the pycnidiospores to become evenly scattered throughout the compound drop.

The reason why the pycnidiospores adhere together is that they have thick gelatinous walls. Associated with each mass of pycnidiospores is a free gelatinous substance which appears to be a matrix in which the pycnidiospores are immersed.

If one examines a group of living pycnidiospores adhering to a cover-glass and lying in one plane just beneath its surface, one can see that the individual pycnidiospores are well separated from one another by their rather thick gelatinous cell-walls. Where two or more pycnidiospores are in contact, their gelatinous walls become merged into a single wall with no discernible limits at the surfaces of contact (Fig. 44, A).

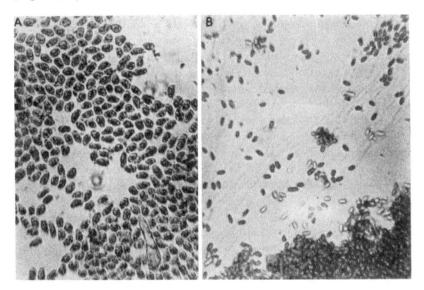

FIG. 44.—Photomicrographs of pycnidiospores: A, *Puccinia helianthi*; B, *P. graminis*; equally magnified. The nectar bearing the pycnidiospores was removed to a glass slide, allowed to dry, then stained with cotton-blue, stirred with a needle, and covered with a cover-glass. It will be seen that: (1) in A the pycnidiospores are separated by their colourless gelatinous walls and among the pycnidiospores is a conical cell (a pycnidiosporophore which was extruded from the pycnidium with the spores); (2) in B a gelatinous trail shows the direction in which the pycnidiospores were pushed by the needle over the surface of the glass slide; and (3) the pycnidiospores of *P. helianthi* are much larger than those of *P. graminis*. Magnification, 540.

To stain the gelatinous substance connected with the pycnidiospores of (say) *Puccinia helianthi*, one touches a clean glass slide to the upper side of a pycnidial pustule of a Sunflower leaf and then adds a little water, some cotton-blue, and a cover-glass. Under the microscope one can then see that there is a bluish mucilaginous covering around each individual pycnidiospore and that the more fluid mucilaginous substance is revealed as a bluish cloud adhering to the cover-glass.

The extrusion of the pycnidiospores of a pycnidium is due to the gelatinisation and swelling of the cell-walls of the individual pycnidiospores and to the consequent increase in bulk of the pycnidiospore mass. This mode of extrusion of spores from conceptacles is well known to be used by many Fungi Imperfecti.

On several occasions I have seen pycnidiospores rapidly and forcibly extruded from a pycnidium. (1) A Barberry leaf bearing a pycnidial pustule of *Puccinia graminis* was folded so that part of the pustule, including some of the pycnidia, could be seen in *side view* under the microscope. A cover-glass was set over the preparation and then water was added with a pipette so that the pycnidia were immersed. Shortly thereafter I saw four parallel and adhering rows of pycnidiospores pushed steadily away from the pycnidium and beyond the other pycnidiospores to a distance equal to the length of the periphyses. (2) In another preparation I saw a single row of pycnidiospores pushed out from the pycnidium. (3) In yet another preparation I saw two rods of pycnidiospores oozing out of a pycnidium simultaneously. One of the rods (Fig. 45, *g*) contained a number of rows of pycnidiospores. It continued to elongate slowly for about five minutes, became twisted and, in the end, attained a length of about 0.5 mm. The other rod (Fig. 45, *e*) was made up of a single row of pycnidiospores set with their long axes transversely to the long axis of the rod. It soon ceased to ooze out from the pycnidium and attained a length of not more than about 0.1 mm.

Whilst making the observations just described, I gained the impression: that the water supplied in making the preparations had caused the gelatinous walls of pycnidiospores still contained within the pycnidial cavities to swell up; and that the pressure thus created in a cavity had forced some of the pycnidiospores outwards through a narrow channel made in the relatively solid main mass of pycnidiospores already held between the periphyses. It is not to be supposed that the rows of pycnidiospores existed as such in the pycnidium, but rather that they were formed at the time the pycnidiospores were being forced radially outwards from the pycnidium.

Pycnidiospores, on account of their mucilaginous walls, adhere not only to one another, but also to the surface of cover-glasses, etc.; and, therefore, we may suppose that they readily become adherent to the proboscides and feet of flies at the time when these insects are visiting pycnidial pustules. This adherence of masses of pycnidiospores to the insects must, under natural conditions, much increase the chances that, between (+) and (−) pycnidia, pycnidiospores will be freely exchanged.

The pycnidiospores, like the pycnidiosporophores from which they are abstricted and like the periphyses, contain a carotinoid pigment and, in the mass, they appear to be orange-yellow. When young pycnidia of *Puccinia graminis*, each bearing a drop of nectar, are so placed under the low power of the microscope that they can be seen in air in side view with direct sunlight impinging upon them, one can observe the individual pycnidiospores in each drop appearing as brilliant orange-coloured particles. There can be no doubt, therefore,

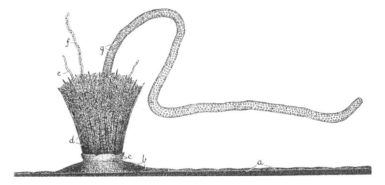

Fig. 45.—*Puccinia graminis*. Extrusion of two gelatinous spore-horns after a thick transverse section of a pustule had been immersed in water: *a*, surface of leaf of *Berberis vulgaris*; *b*, epipycnidial papilla; *c*, cuticular collar; *d*, periphyses and pycnidiospores; *e*, a flexuous hypha; *f*, a gelatinous spore-horn with one row of pycnidiospores; *g*, a similar spore-horn with several rows of spores. Both *f* and *g* were seen oozing out of the pycnidium. Magnification, 240.

that much of the red coloration of a pycnidial drop is due to the carotinoid pigment contained within the pycnidiospores. If the red colour of a pycnidial drop is actually a factor in causing the drops to be visited by insects, this must be due in large measure to the colour of the pycnidiospores themselves.

The dried-up pycnidiospores of *Puccinia graminis* that were present in the gelatinous horns about to be described became turgid as soon as they were given access to water. Moreover, when water is added to a drop of nectar that has been kept dry for some time on a glass slide, the pycnidiospores quickly swell up and resume their turgidity. From these observations we may conclude that pycnidiospores can withstand temporary desiccation.

In a preparation of fresh nectar on a glass slide, when the nectar is replaced by water, the osmotic pressure of the medium surrounding the pycnidiospores left clinging to the slide and cover-glass is greatly and suddenly reduced, but the pycnidiospores remain turgid and do

not appear to suffer any injury. Also, Newton and Johnson[1] have
observed that diluting mixed nectar of *Puccinia graminis* with distilled
water does not prevent the pycnidiospores from initiating the sexual
process when the diluted nectar is applied to haploid pustules. These
observations teach us that pycnidiospores are not appreciably injured
by rapid and considerable changes in the osmotic pressure of the
surrounding liquid and they allow us to draw the conclusion that,
under natural conditions in the open, when the nectar is diluted by
rain, the pycnidiospores suffer no harm.

**Horns of Pycnidiospores produced abnormally by Puccinia
graminis.**—A Barberry bush had been inoculated with basidiospores

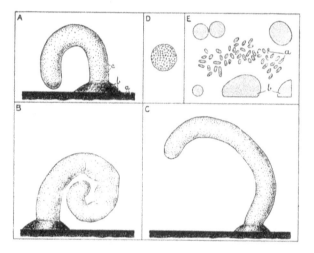

Fig. 46.—*Puccinia graminis.* Extrusion of pycnidiospores from pycnidia which
    failed to excrete any nectar. A, B, and C, lateral view of three pycnidia pro-
    jecting from a leaf of *Berberis vulgaris*, each with a pale-orange, cylindrical,
    gelatinous mass of pycnidiospores projecting from its ostiole: *a*, the leaf; *b*, the
    pycnidial papilla; and *c*, the pycnidiospores. D, a transverse section of one of
    the cylinders represented diagrammatically, showing the pycnidiospores im-
    mersed in jelly. E: *a*, pycnidiospores, and *b*, orange oil-drops, seen after one of
    the cylinders had been placed in water. Magnification: A-D, 120; E, 520.

from a wild strain of *Puccinia graminis tritici* on March 1, 1937, and
on March 12 several leaves belonging to the upper six internodes of
the main shoot exhibited well-developed pycnidial pustules. All the
pustules, except one, bore drops of nectar and appeared to be quite
normal. In the exceptional pustule which was near the apex of a leaf,

[1]Margaret Newton and T. Johnson, "Specialization and Hybridization of Wheat
Stem Rust, *Puccinia graminis tritici*, in Canada," *Dominion of Canada, Depart. of
Agric.*, Bull. CLX, 1932, p. 40.

as could be seen with the naked eye, some of the pycnidia bore tiny drops of nectar while the others were nectarless and had little orange-yellow hornlike structures protruding from their ostioles (Fig. 46).

A piece of the leaf bearing the abnormal pustule was cut away from the Barberry bush and the pustule was then examined with the low power of the microscope first in *face view* as it lay on a slide on the stage of the microscope and then in *lateral view* as it was held attached by a film of water to a slide set up vertically. Subsequently some of

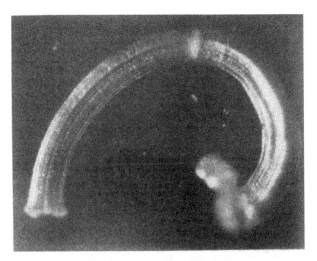

Fig. 47.—*Xanthomonas translucens* f. sp. *undulosa* (an organism causing Black Chaff disease of Wheat). Gelatinous horn of bacteria exuded from a peduncle of Acme wheat after an artificial inoculation. Grown at 25° C. Observed and photographed (dark-field illumination) by W. A. F. Hagborg at the Dominion Rust Research Laboratory. Magnification, about 100.

the horns were removed from the pustule, placed on a slide, and examined first in the dry condition and then as they absorbed water passed to them beneath a cover-glass.

The horns were found to be solid, gelatinous, bluntly-ending rods, 0.3-0.5 mm. in length and about 0.05 mm. in diameter, either twisted irregularly or bent backwards toward the leaf, and made up of numerous pycnidiospores cohering together by their gelatinous walls. On being supplied with water, each horn swelled up and quickly disintegrated into its component parts.

It would seem that the abnormal pycnidia had produced the usual number of pycnidiospores, but had failed to excrete the sugar which is necessary for withdrawing water osmotically from the leaf, with the result that, in the absence of the liquid nectar, the pycnidiospore-mass was not able to spread out in a drop of nectar in the normal manner but was constrained to form a relatively dry gelatinous column.

Another instance of a pycnidial pustule of *Puccinia graminis* producing gelatinous horns made up of pycnidiospores was observed by me on the already-mentioned Barberry bush on March 17.

It is well known that in Cytospora, Phyllosticta, Phoma, Melanconium, and many other Fungi Imperfecti the conidia are extruded from the pycnidia in the form of twisted or winding gelatinous threads or horns; and the little yellow horns of pycnidiospores produced by the abnormal pycnidia of *Puccinia graminis* reminded me, at first view, of the much longer golden-yellow horns of conidia which I have

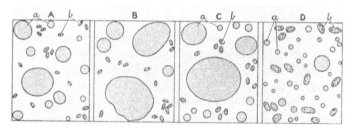

FIG. 48.—Oil-drops, *a a*, and pycnidiospores, *b b*, in the pycnidial nectar of the Rust Fungi. A, *Puccinia graminis*, 12-day-old culture; drop of nectar obtained from a pustule by contact with a glass slide. B, *P. graminis*, oil-drops and spores from one of the gelatinous tendrils shown in Fig. 45. C, *P. minussensis* (*P. hemisphaerica*), largest oil-drop 28μ long. D, *P. helianthi*, oil-drops numerous and mostly smaller than the spores. Magnification, 520.

so often seen hanging from the ostioles of the pycnidia of *Cytospora chrysosperma* and beautifully displayed against the dark bark of dead or dying Black Poplar trees.

The horns of pycnidiospores of *Puccinia graminis* also resembled a tiny gelatinous horn of bacteria which was formed by *Xanthomonas translucens* f. sp. *undulosa*, the pathogen which causes the Black Chaff disease of wheat. This horn (Fig. 47), which was observed at the Dominion Rust Research Laboratory by W. A. F. Hagborg, was exuded from a peduncle of Acme wheat after the peduncle had been artificially inoculated.

The jelly making up the outer part of the walls of Rust pycnidiospores absorbs water so freely that, when it has access to pure water, it usually flows and disappears from view leaving the pycnidiospores surrounded merely by a very thin, inner, more resistant, cell-wall membrane. When one of the gelatinous horns of *Puccinia graminis* described above was supplied with water, not only did the constituent pycnidiospores come into view, but also drops of oil (Fig. 46, E) resembling those about to be described as being normally present in pycnidial nectar.

Oil-drops in Extruded Masses of Pycnidiospores.—In *Puccinia graminis*, *P. minussensis*, and *P. helianthi*, the pycnidiospores which are extruded through the ostiole of each pycnidium are accompanied by *drops of oil*. To prove this, all that it is necessary to do is to take a clean glass slide and to touch one side of the slide very gently to a drop of nectar on a host-leaf and then to examine the drop under the microscope. If to the preparation a cover-glass and a little cotton-blue are added, one can then see the pale-orange drops of oil standing out in contrast with the blue pycnidiospores. Sudan III stains the drops bright orange-red.

In *Puccinia graminis* and *P. minussensis* there are scores of oil-drops in each large drop of nectar and the largest oil-drops are about 10-15μ in diameter (Fig. 48, A-B and C). The largest drop observed among the pycnidiospores of *P. minussensis* was oval in form and measured 17 × 12μ. In *P. helianthi* (Fig. 48, D) the oil-drops are very small, usually

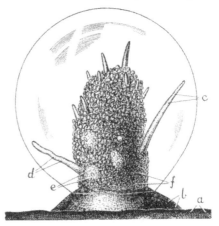

Fig. 49.—*Puccinia graminis.* Lateral view of a young pycnidium showing oil-drops enclosed between the pycnidiospores; pycnidial elements drawn with the *camera lucida*, the drop of nectar added semi-diagrammatically: *a*, a Barberry leaf; *b*, the pycnidial papilla; *c*, a periphysis; *d*, a very young flexuous hypha; *e*, an oil-drop; and *f*, pycnidiospores. Magnification, 500.

even smaller than the pycnidiospores, but they are numerous. Most of them are not more than about 2-3μ in diameter; but, here and there among the little drops, there may be a large drop with a diameter equal to about 10μ.

When hand-cut water-mounted sections of pustules containing pycnidia of *Puccinia graminis* are examined with the microscope, one can often see fairly large pale-orange drops of oil held among the periphyses and extruded mass of pycnidiospores a little way above the pycnidial ostioles (Fig. 49) and, occasionally, one can see one or more large oil-drops inside a pycnidial cavity. Moreover, drops of oil were found to be present in the abnormal gelatinous horns of pycnidiospores that have already been described (Fig. 46, E). These observations indicate that the oil-drops in pycnidial nectar are produced inside the cavities of the pycnidia and are extruded through the ostioles along with the pycnidiospores.

If the oil-drops in pycnidial nectar are extruded from the pycnidia, a further question arises: how do the oil-drops get into a pycnidial cavity? All the fungal cells of a pycnidium contain minute droplets of oil bearing a carotinoid pigment, and we may therefore suppose that the oil-drops in a pycnidial cavity come into existence by the breaking down of certain cells. The buffer cells which push up the epidermis above a developing pycnidium are short stout hyphae and, according to Miss Allen,[1] with the further development of the pycnidium they break down and disappear. Perhaps the little oil-droplets thus liberated run together and form larger droplets which are eventually extruded with the pycnidiospores.

It is known that certain tropical Orchids, *e.g. Maxillaria rufescens*,[2] offer as a bait to insect visitors, instead of nectar, a mat of thin-walled food-hairs containing protein particles and numerous tiny oil-drops; and it may therefore be that the presence of oil-drops in the nectar of the Rust Fungi is a factor in making the nectar more palatable and attractive to flies.

As we shall see, the nectar of *Puccinia graminis* and of many other Rust Fungi is strongly scented, but where the odoriferous substance is located is at present unknown. Probably the nectar itself is scented and, if it is, the odoriferous substance, at least in part, may be dissolved in the free oil-drops which are mixed with the pycnidiospores.

The Drop of Nectar.—The drop of nectar excreted from the ostiole of a pycnidium is of primary importance in securing the visits of insects to pycnidia and that exchange of (+) and (−) pycnidiospores which leads very soon to the production of dicaryotic aecidiospores and, ultimately, to nuclear fusions in the teleutospores and the formation of zygotes. As a factor in making the sexual process possible, nectar plays the same role in the Rust Fungi as it does in such nectar-excreting flowers as those of the Violet, the Honeysuckle, the Larkspur, and the Sunflower.

The nectar excreted by a pycnidium is a clear, colourless, mucilaginous, saccharine fluid, resembling in appearance the nectar of flowers (Figs. 49, 50, and 55, pp. 151, 153, and 163).

[1]Ruth F. Allen, "A Cytological Study of Heterothallism in *Puccinia graminis*," *Journ. Agric. Research*, Vol. XL, 1930, p. 594.
[2]L. Kny, *Botanische Wandtafeln mit erläuterndem Text*, Berlin, XII Abt., 1909, pp. 498-500, Taf. CXI.

It is true that a drop of nectar covering one or more pycnidia of *Puccinia graminis*, *P. helianthi*, etc., appears to be reddish; but this is due solely to the optical effect of the carotinoid pigment contained in the enclosed periphyses, flexuous hyphae, and pycnidiospores and to the background provided by the reddish epipycnidial papilla. In *P. helianthi*, *P. minussensis* (Fig. 50), and *P. graminis*, if one examines

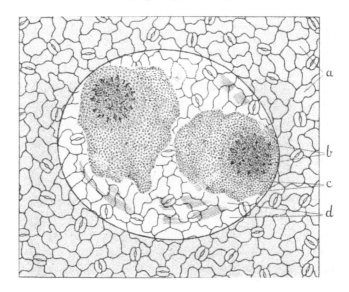

FIG. 50.—*Puccinia minussensis* (*P. hemisphaerica*) on *Lactuca pulchella.* Upper surface of a host-leaf showing two pycnidia covered by a common drop of nectar: *a*, the epidermis; *b*, the pointed ends of the periphyses of one of the pycnidia; *c*, a gelatinous mass of pycnidiospores which has been extruded through the ostiole of the pycnidium and is immersed in the nectar; and *d*, a compound drop of nectar formed by the coalescence of two simple drops of nectar which were excreted through the ostioles of the two pycnidia. Magnification, 112.

an undisturbed young compound drop in face view with a lens or the low power of a microscope, one can often see quite clearly that the part of the drop which lies between the pycnidia or extends beyond them and which has not been invaded by masses of pycnidiospores is as clear and colourless as water, so that one can look down through it and through the epidermis to the green palisade cells below.

The degree of the apparent redness of drops of nectar varies with different species. Thus the drops of nectar in the very strongly scented *Puccinia minussensis* are not nearly as red as those of *P. graminis*. In *P. graminis*, in young pycnidial pustules that have been active for 3-7 days and have produced great numbers of pycnidiospores, the compound drops of nectar in the centre of each pustule, and the little simple drops surrounding them, all have a brilliant bright-orange colour that stands out in striking contrast with the purplish-green of the host-leaf. The larger compound drops of nectar in *P. graminis* are often 1-2 mm. in diameter and hemispherical in form. Such drops can easily be seen at a distance of several feet from the bushes and, on account of their colour, to insects they may appear just as attractive as tiny orange-coloured flowers.

It was noticed that some large drops of nectar of *Puccinia graminis* that had been in existence for upwards of three weeks had lost their red colour and resembled drops of milk. It was found that this change

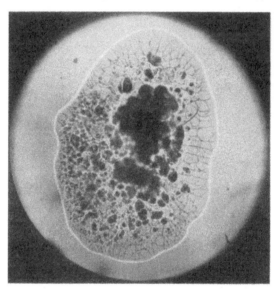

FIG. 51.—*Puccinia graminis*. A dried-up drop of nectar on a glass slide. The pycnidiospores, which were orange-red, are adhering together in dark gelatinous masses, and the clear part of the drop is cracked. The cracks indicate that the nectar of *P. graminis* is gelatinous. Magnification, about 25.

in colour was due, not to an invasion of the drops by yeast cells or bacteria, but to a loss of carotinoid pigment from the pycnidiospores.

Freshly excreted drops of nectar are not only reddish, but they also affect the eye by their *glitter*. Light is reflected from their surface and refracted through them so that their chance of attracting the attention of insects is much enhanced thereby. Particularly in direct sunlight do the drops shine like little drops of dew or with a golden glow. A pycnidial pustule covered with a hemispherical compound nectar-drop is reminiscent of the head of a tentacle of Drosera covered with its globule of slimy secretion: in both fungus and flowering plant the redness and glitter exhibited by fluid drops combine in drawing insects to the excreting organs.

A drop of nectar consists chiefly of *water*. In addition to the periphyses, flexuous hyphae, and many thousands of extruded pycnidiospores, it encloses *drops of oil* and such *dust particles* as may fall upon it from the air during the days or weeks of its existence; and, mixed with it or dissolved in it, are: *mucilage, sugars,* an *odoriferous substance* and, doubtless, several as yet unknown other substances.

A study of drops of nectar in course of becoming desiccated revealed the fact that the drops contain a considerable amount of mucilage. Drops of nectar of *Puccinia graminis, P. helianthi,* and *P. minussensis*

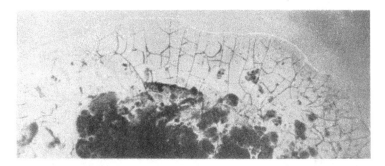

Fig. 52.—*Puccinia graminis.* An enlarged portion of Fig. 888. The cracks in the transparent, pycnidiospore-free part of the drop now come more clearly into view. Magnification, about 50.

were removed from pustules by touching the drops with glass slides (*cf.* Fig. 51). On examining the drops under the microscope it was seen that they were colourless and that they contained irregularly distributed larger and smaller masses of orange-yellow pycnidiospores mixed with minute oil globules. As the drops dried up, they became reduced to hummocky mounds, their free surface became irregularly folded, and finally their substance became cracked (Figs. 51, 52, and 53). The drops, when nearly dry, looked like, and had the physical consistency of, drops of gum. Pressure with a needle indented them. And when they were quite dry, the cracks which had formed in them as the water evaporated resembled the cracks which I have seen in dried-up drops which had been collected in a liquid condition from the slimy pilei of *Panaeolus solidipes.*

Some drops of nectar of *Puccinia graminis* were smeared on to white paper and, after they had dried, in their yellow colour and shiny appearance they resembled the dried gum on the flaps of envelopes.

It was also found that the nectar of *P. graminis* can be used as an adhesive for sticking two pieces of paper together.

The mucilage of pycnidial nectar originates within the cavity of the pycnidium, and it is probably derived from the swelling of the outer part of the mucilaginous walls of the pycnidiospores. It causes the nectar to adhere more readily to the proboscides and feet of flies; and it assists in the extrusion of the pycnidiospores from the pycnidium.

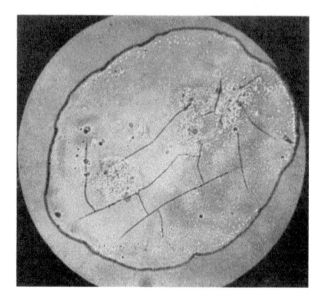

FIG. 53.—*Puccinia minussensis.* A tiny drop of nectar, removed from a pustule by touching with a slide, that has dried up. The cracks reveal that the nectar is gelatinous. Magnification, about 50.

In *Puccinia graminis*, as drops of nectar on haploid leaf-pustules become older and older, they dry up. The larger compound drops, when half solidified, have a dull appearance and their consistency is that of a drying drop of gum; but later on, as they dry up completely, they become reduced to a thin crust which adheres to the epidermis. Each simple drop which crowns a single pycnidium shrinks up into a little solid bead-like body consisting of mucilage and pycnidiospores and held in place by the periphyses. When such a little solid body is removed from a pycnidium, placed on a slide, and supplied with water one can observe that the mucilage surrounding the mass of pycnidio spores swells up. On pressing the cover-glass, one can see that the spores escape into the water and separate from one another while the mucilage is left behind in the form of a loose transparent skin.

By Craigie and myself the excertion produced by Rust pycnidia has been called *nectar*,[1] and this term was chosen because evidence has accumulated which supports the view that the excretion contains sugar. This evidence will now be reviewed.

Ráthay[2] observed that: (1) the nectar of Rust Fungi is eagerly sought by ants and flies; (2) it reduces Fehling's solution; (3) in certain species, *e.g. Gymnosporangium juniperinum* and *G. sabinae*, it is intensely sweet to the taste; and (4) the nectar of *G. sabinae*, as shown by appropriate tests, contains both dextrose and laevulose.

The nectar of species of Cronartium is produced in such abundance that it can be readily applied to the tongue; and that it is sweet to the taste has been noticed in both *Cronartium quercuum* and *C. ribicola*.

Cronartium quercuum forms large excrescences on the trunks and branches of certain Pines in Japan. According to Shirai:[3] "The spermogonia of this fungus are formed in the month of January in the intercellular spaces between the corky bark and the cortical parenchyma as flat continuous layers, when large drops of viscous fluid of sweet taste loaded with an immense number of spermatia flow out from the fissures in the cracked bark. These viscid drops of sweet taste are known by the name of *Matsumitsu (Pine-honey)* and are eaten by boys and girls when they happen to find them."

In respect to the "spermogonia" of *Cronartium ribicola* Klebahn[4] remarks: "As yellow patches, 2-3 mm. wide, they shine through the bark and empty the spermatia in drops of sap tasting distinctly sweet."

Ráthay,[5] as noted on pp. 31 and 32, found the nectar of *Gymnosporangium juniperinum* and *G. sabinae* intensely sweet, that of *Uromyces pisi* only slightly sweet, and that of a dozen other species tasteless. The lack of sweetness in the nectar of these last species may have been due to the small amount of sugar in the nectar or to the sugar being a compound different from that in the nectar of *Gymnosporangium juniperinum* and *G. sabinae*.

[1]J. H. Craigie, "Discovery of the Function of the Pycnia of the Rust Fungi," *Nature*, Vol. CXX, 1927, pp. 765-767.

[2]E. Ráthay, "Untersuchungen über die Spermogonien der Rostpilze," *Denkschrift d. Kais. Akad. d. Wissensch*, Wien, Bd. XLVI, 2 Abt., 1883, pp. 1-51.

[3]M. Shirai, "On the Genetic Connection between *Peridermium giganteum* (Mayr) Tubeuf and *Cronartium quercuum* (Cooke) Miyabe," *Bot. Mag.*, Tokyo, Vol. XIII, 1899, p. 76.

[4]H. Klebahn, *Die wirtswechselnden Rostpilze*, Berlin, 1904, p. 387.

[5]E. Ráthay, *loc. cit.*, p. 26.

Craigie,[1] having come upon a Barberry bush in the open infected with *Puccinia graminis* whose pustules were producing an abundance of nectar, tasted the drops and found then distinctly sweetish. He also heated some nectar of *P. graminis* with Fehling's solution and, like Ráthay, observed that the nectar reduced the copper sulphate with a moderate production of cuprous oxide. On the campus of Cornell University I found a Barberry bush with a few large haploid pustules of *P. graminis* covered with nectar, and I requested a student of Botany who was with me to taste the nectar and tell me how he reacted to it. He licked the nectar off a leaf and said that it tasted sweet.[2]

Faull[3] observed that, when the weather is humid, the liquid excreted from the "spermogonia" of *Milesia polypodophila* spreads over the under sides of the needles of *Abies balsamea* and, at times, fairly drips from them. On examining the liquid, Faull found it "somewhat sticky and sweetish."

Some twigs of *Amelanchier alnifolia* on whose leaves there were a few haploid pustules of *Gymnosporangium juvenescens* were kept for a few days in water under a bell-jar. A large drop of nectar covering one of the pustules was then transferred to a glass slide and allowed to dry. After it had dried up it was seen to contain abundant crystals and also mucilage (Fig. 54). Since, according to Ráthay, the nectar of *G. juniperinum* and *G. sabinae* react readily to Fehling's solution and are distinctly sweet to the taste, it seems likely that the crystals in dried-up nectar of *G. juvenescens* are composed in the main, if not entirely, of sugar.

Since the nectar of some Rusts definitely contains sugar, it seems not unlikely that sugar is present in the nectar of all other Rusts, even in those in which the nectar has not yet been found to be sweet. In support of this view it may be added that, as Ráthay observed, insects eagerly suck up the nectar of *Puccinia graminis* and other Rusts which appeared to him tasteless, and the nectars of these Rusts all reduce Fehling's solution.[4]

It is a remarkable fact that the numerous simple drops excreted

[1] J. H. Craigie, "An Experimental Investigation of Sex in the Rust Fungi," *Phytopathology*, Vol. XXI, 1931, p. 1024; and a personal communication.

[2] I licked the nectar off one pustule, but it seemed to me to be tasteless. Thus my reaction to the nectar of *Puccinia graminis* was like that of Klebahn. Like Klebahn, too, I have not been able to detect any sweetness in the nectar of *P. suaveolens*.

[3] J. H. Faull, "The Biology of Milesian Rusts," *Journal of the Arnold Arboretum*, Vol. XV, 1934, pp. 69-70.

[4] *Vide supra*, p. 157.

from a pustule or the compound drop or drops formed by their coalescence, when once they have come into existence, maintain themselves not only under moist weather conditions, but even on hot dry days when host and parasite are fully exposed to the direct rays of the sun.

A simple experiment serves to illustrate the fact that, under the same atmospheric conditions, a drop of pycnidial nectar and a drop of water behave very differently. Some little drops of pure water,

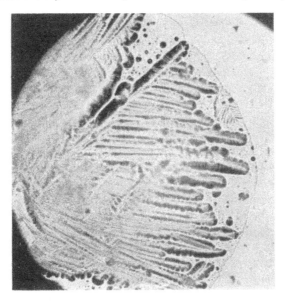

Fig. 54.—Drop of nectar of *Gymnosporangium juvenescens* which was removed to a glass slide and allowed to dry up. In it one can see mucilage and crystals. Magnification, about 45.

1-2 mm. in diameter, were placed on a Sunflower leaf which bore pycnidial pustules of *Puccinia helianthi*. The drops of water dried up within about half an hour; whereas, at the end of an hour, the volume of the pycnidial drops showed no apparent alteration.

The drops of nectar, after once being formed, are able to maintain themselves for an indefinite time under relatively dry atmospheric conditions because they contain sugar and therefore have a considerable osmotic pressure. As a drop of nectar exposed to unsaturated air loses water by evaporation, a compensating amount of water is drawn into the drop from the mycelium and, ultimately, from the leaf-cells by osmosis.

Using the Barger method,[1] Hanna[2] determined the osmotic pressure of the pycnidial nectar of *Puccinia graminis* grown on Barberry bushes under greenhouse conditions in January and February. Sucrose solutions of known osmotic pressure and nectar were placed in the capillary tubes in the form of alternating drops. A Barberry plant had been inoculated with basidiospores on January 25, 1929. On February 14, the osmotic pressure of the nectar was found to be between 12 and 24 atmospheres and, on February 18, approximately 20 atmospheres. These experiments therefore indicate that the osmotic pressure of the nectar of *P. graminis* under greenhouse conditions in the winter at Winnipeg is greater than 12 atmospheres, less than 24 atmospheres, and probably about 20 atmospheres.

A drop of nectar which under dry conditions of the atmosphere is small increases greatly in size when the atmosphere becomes very humid. Moreover, the osmotic pressure of a drop depends on the amount of sugar (and perhaps other osmotic substances) dissolved within it. Therefore there can be no doubt that, under natural conditions, the osmotic pressure of the nectar of *Puccinia graminis*, *P. helianthi*, etc., must be subject to considerable variation.

Thatcher[3] has shown that, in general, fungus parasites have a higher osmotic pressure than their hosts and that, in particular, the Rust fungus, *Uromyces fabae*, has a higher osmotic pressure than the mesophyll cells of one of its host-plants, *Pisum sativum*. By plasmolytic methods it was determined that the osmotic pressure in the haustoria of *Uromyces fabae* is 18-25 atmospheres and that the osmotic pressure in the host-cells is 7-10 atmospheres. The osmotic pressure of the haustoria of *U. fabae* is equal to, or somewhat greater than, that determined by Hanna for the pycnidial nectar of *Puccinia graminis*. As the periphyses, flexuous hyphae, and pycnidiospores contained within a well-developed drop of nectar of *P. graminis* are not plasmolysed but turgid, we must suppose that the osmotic pressure of these cells is greater than that of the nectar and therefore sometimes greater than 20 atmospheres. The osmotic pressure of the chlorenchymatous cells of the leaves of *Berberis vulgaris* has apparently not been measured, but it is probably considerably less than 20 atmospheres. Doubtless, in every Rust fungus having a pycnidial stage, the osmotic

[1]For a description of this method as it was used for determining the osmotic pressure of the cell-sap of the sporangiophores of Pilobolus, *vide* these *Researches*, Vol. VI, 1934, pp. 140-144.

[2]W. F. Hanna, personal communication. Experiments made at the Dominion Rust Research Laboratory, Winnipeg.

[3]F. S. Thatcher, *Amer. J. Bot.*, XXVI, 1939, pp. 449-458 and *Canad. J. Res.*, Sect. C, XX, 1942, pp. 283-311.

pressure of the periphyses, flexuous hyphae, and pycnidiospores is greater than that of the nectar and still greater than that of the cells of the host-leaf.

Where the sugar contained within a drop of nectar comes from is a mystery. Ultimately, no doubt, it is derived from the mesophyll cells of the host-leaf. Since the wall of the pycnidium is made up of compacted hyphae and the drop of nectar crowns the mouth of the pycnidium, it may be supposed that the fungus does not force the host to excrete the sugar, but rather excretes the sugar itself. Presuming that the fungus excretes the nectar, it still remains to be determined whether the periphyses or pycnidiosporophores are the active hyphae in the excretory process.

In Cronartium, Gymnoconia, Phragmidium, Milesia, and some other genera periphyses are absent. Therefore, in the species of these genera, periphyses cannot be concerned with nectar excretion. All things considered, it seems most probable that, in *Puccinia graminis* and in Rust Fungi generally, the sugar contained in the pycnidial nectar is excreted by the cells lining or forming the basal layer of the pycnidial cavity, *i.e.* by the pycnidiosporophores.

A further problem concerned with the saccharine matter in pycnidial nectar is the mechanism by which the sugar is excreted. Is the sugar excreted through the cell-wall or is it produced on the outside of the excretory cells by means of some chemical change in the substance of the cell-wall? To this question, at present, there is no obvious answer.

In a pycnidial pustule, at first, there are as many drops of nectar as there are mature pycnidia. Under adverse conditions of light and moisture, the drops may remain so small that they do not fuse with one another. Under more favourable conditions, however, the drops, after once having made their appearance, soon grow so large that many of them come to touch one another and so to melt together. In this way there may be formed several compound drops, or one compound drop in the centre of a pustule surrounded by a number of simple drops, or a single very large compound drop covering all the pycnidia (*cf.* Fig. 55).

The amount of liquid excreted by pycnidial pustules appears to be dependent (1) on the condition of the host-plant and (2) on the amount of moisture in the atmosphere.

Under greenhouse conditions at Winnipeg, it has been observed that far more nectar is produced by pycnidial pustules of *Puccinia graminis*, *P. helianthi*, etc., during spring and early summer than in winter. This difference is largely due to the condition of the host-plant which depends on sunlight for its photosynthetic activity. In winter the individual pycnidial pustules are smaller than in spring and summer, and it also appears that the individual pycnidia excrete nectar less freely in winter than in spring and summer.

A Barberry bush bearing pycnidial pustules of *Puccinia graminis* was growing in a greenhouse under rather dry atmospheric conditions and the drops of nectar covering the pycnidia were quite small. Some of the pustule-bearing leaves were then removed from the bush and floated on water in a closed Petri dish. After thus being surrounded with air saturated with water vapour, the drops grew rapidly in size and after two days had become very large. This simple experiment proves what might be expected on physical grounds, namely, that the size of a drop of nectar on a leaf pustule varies with the amount of moisture in the atmosphere. Other things being equal, the more humid the atmosphere the larger the drop, and the less humid the atmosphere the smaller the drop. The maximum size of any drop of nectar is determined in part by the amount of sugar and perhaps other osmotic substances that are dissolved in the drop and in part by the rate of evaporation from the drop's surface.

Under greenhouse conditions at Winnipeg, not only do Barberry bushes, young Sunflower plants, and other host-plants flourish less in winter than in summer; but, also, owing to the extraordinary dryness of the winter weather, the air inside the greenhouse is much dryer in winter than in summer, so dry indeed that the aecidia of *Puccinia graminis*, when mature, are unable to open and discharge their spores but instead form finger-like processes (Figs. 34 and 36, pp. 115 and 117). The physiological condition of the host-plant and the physical state of the atmosphere therefore combine in causing such fungi as *P. graminis* and *P. helianthi* to produce, under greenhouse conditions at Winnipeg, a much smaller volume of nectar in winter than in spring and summer.

The nectar of *Puccinia graminis*, *P. helianthi*, etc., is so liquid that it can easily be sucked up into, and expelled from, a fine glass pipette; and the pipette method was actually employed by Craigie in mixing the nectar of (+) and (−) pycnidia when making his classical discovery of the function of pycnidia. With a needle applied to a pycnidial drop, one cannot draw out the drop into a glutinous string and the nectar, apart from the gelatinous masses of pycnidiospores which it encloses is not then as viscous as the drops which crown the heads of the tentacles of the Sundew, *Drosera rotundifolia*.

If, with the help of a pipette, one tries to place drops of water on a leaf of a Barberry bush, one finds that the drops roll off the leaf and, in spite of one's best efforts, will not stay there. On the other hand, if one touches the surface of a Barberry leaf with a drop of nectar, the nectar immediately and with ease adheres to the epidermis in streaky drops and masses. It is evident that the surface tension of the nectar

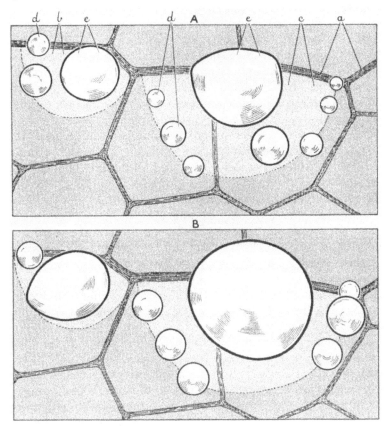

FIG. 55.—*Puccinia graminis.* Increase in size and fusion of drops of nectar. A, upper surface of a piece of leaf of *Berberis vulgaris* bearing two small haploid pustules of *P. graminis* each derived from a single basidiospore: *a*, veins of the leaf; *b*, one pustule; *c*, the other pustule; *d d*, simple drops of nectar each protruding above the ostiole of a single pycnidium; *e e*, compound drops of nectar formed by the fusion of 2-4 simple drops. At this stage of the development of the drops, the leaf was floated on water in a Petri dish, the cover of the dish was put on, and then the drops were drawn with the help of a *camera lucida.* B, the same piece of leaf as in A, but two hours later. All the drops have increased in size and the two largest in each pustule have touched one another and have coalesced. Magnification, 48.

with its sugary and mucilaginous contents is far less than that of water. The low surface tension of drops of nectar is advantageous for the working of the pycnidial mechanism in that: (1) it enables the drops to cling more readily to the surface of the leaf on which they are produced; and (2) it is a factor in causing the nectar to adhere to the proboscides and legs of visiting insects.

A drop of nectar produced by a pycnidium, when it is very small, surrounds the periphyses and mass of pycnidiospores and, as a tiny globule, crowns the top of the epipycnidial papilla; but, as it grows in size, its base spreads down the exterior of the papilla on to the general surface of the host-epidermis where the drop may come to meet and fuse with neighbouring drops which also have descended to the epidermis. In a compound drop, a large area of the epidermis is wetted. The wetting of the epidermis by drops of nectar takes place even where the epidermis is waxy, as in *Berberis vulgaris*, and it is due in large measure to the low surface tension of the nectar itself.

A Barberry leaf bearing pustules of *Puccinia graminis* was floated on water in a closed Petri dish and, during the next two hours, with the low power of the microscope the growth of the little spherical drops crowning individual pycnidia and the fusion of neighbouring drops were observed (Fig. 55). At the time when two drops fused with one another it was seen that the drops did not form a sphere instantaneously as they would have done had they been drops of water, but that they united relatively slowly, so that one could observe a dumbbell stage and watch the compound drop gradually becoming oval in form. The slowness of this change in form was doubtless due to the fact that the drops contained much mucilage and, in consequence, had a low surface tension.

The Nectar is Antiseptic.—Drops of nectar produced by pycnidia appear to be antiseptic; for, even after they have been exposed on the upper side of Barberry and Sunflower leaves, etc., for many days or for several weeks, they remain free from bacteria and usually free from moulds. If these organisms were not prevented from developing in the nectar, the drops, after their first appearance, would soon lose their sugar and undergo such chemical changes as would make them no longer attractive to insects.

It has yet to be determined whether the antiseptic properties of pycnidial nectar are due merely to the concentration of the nectar or to the presence in small quantity of some special chemical substance.

The attack of *Claviceps purpurea* on ears of Rye leads to the excretion of drops of honey-dew containing reducing sugars and great numbers of conidia. The concentration of the solution is very high, for it is equal to a 2.33 molar solution of cane sugar.[1]   Kirchhoff

---

[1] H. Kirchhoff, "Beiträge zur Biologie und Physiologie des Mutterkorns," *Centralblatt f. Bakteriol., Parasitenk., und Infektionskr.*, Zweite Abt., Bd. LXXVII, 1929, pp. 310-369, especially p. 313.

observed that the conidia do not germinate in undiluted honey-dew but will germinate readily in honey-dew that has been sufficiently diluted. When the concentration was reduced to one-tenth the conidia were found to germinate in 10-12 hours.

In the Rust Fungi the pycnidiospores do not produce germ tubes either in concentrated or in diluted nectar or in water; but it may well be that, just as the high concentration of the honey-dew of *Claviceps purpurea* prevents the immersed conidia from germinating, so the high concentration of the nectar of the Rust Fungi prevents the growth of mould fungi, yeasts, and bacteria.

A detailed investigation of the nectar of the Rust Fungi in relation to foreign fungi and bacteria is desirable.

The Odour of Rust Fungi including Puccinia graminis.—The sense of smell varies much with different individual human beings, but mycologists are agreed that, in the pycnidial stage, many Rust Fungi emit a strong odour. The odour was noticed by the older mycologists, *e.g.* Persoon[1] (1801), Léveillé[2] (*c.* 1849), Tulasne[3] (1854), and Ráthay[4] (1882), and Ráthay recognized that the odours, resembling as they do the perfume of flowers, are attractive to insects.

Among the Uredinales in the pycnidial stage with odours that I myself have detected are: (1) *Puccinia suaveolens* on *Cirsium arvense;* (2) *Puccinia monoica* on *Arabis retrofracta;* (3) *Puccinia minussensis* on *Lactuca pulchella;* (4) *Melampsorella cerastii* on *Picea canadensis;* (5) *Puccinia graminis* on *Berberis vulgaris;* (6) *Puccinia helianthi* on *Helianthus annuus;* and (7) *Puccinia coronata* on *Rhamnus cathartica.* The first four species have systemic mycelium and are perennial on their hosts.

(1) *Puccinia suaveolens.* In England, when walking in pastures, I have often seen clumps of the Creeping Thistle in which the under side of every leaf was covered with the pycnidia of *Puccinia suaveolens.* The pycnidial nectar of such thistles gives out so strong an odour that one can detect the presence of the Rust Fungus before one comes up to the thistles and at a distance of fully ten feet from them. The odour of the pycnidial nectar seems sickly sweet, like that of honey; and, for this reason, the fungus was given the specific name of *suaveolens.*

[1]C. H. Persoon, *Synopsis Methodica Fungorum*, Gottingae, 1801, p. 221.

[2]J. H. Léveillé, *Dict. univ. d'hist. nat. d'Orbigny*, ed. 2, undated, T. XIV, Urédinées, p. 206. This article appeared in ed. 1 (1839-1849) which I have not seen.

[3]L. R. Tulasne, "Second Mémoire sur les Urédinées et les Ustilaginées," *Ann. Sci. Nat. Bot.*, Sér 4, T. II, 1854, 1854, p. 118.

[4]E. Ráthay, "Untersuchungen über die Spermogonien der Rostpilze," *Denkschr. d. K. Akad. d. Wissensch.*, Wien, Bd. XLVI, 2 Abt., 1883, pp. 48-49. The Reprint is dated 1882.

(2) *Puccinia monoica.* Near Saskatoon in western Canada a Rock Cress, *Arabis retrofracta*, is abundant, and sometimes many plants growing near to one another are infected with *Puccinia monoica*. The mycelium is systemic, and the individual plants become so badly parasitised that, as a rule, they fail to flower. The pycnidia are formed on both sides of each radical and cauline leaf. Just as with *P. suaveolens*, the pycnidial nectar gives out a heavy honeylike odour that can be smelled at a distance of several feet from a group of infected plants. The bank of the Saskatchewan river at Saskatoon is very steep and high. Professor W. P. Fraser, who first showed me the fungus in the field, has assured me that, under favourable conditions in the spring, when large numbers of rusted Arabis plants have been present at the bottom of the bank, he has detected the odour of *P. monoica* at the top of the bank, one hundred feet or more distant from the nectar on the infected leaves.

(3) *Puccinia minussensis* (*P. hemisphaerica*). This Rust grows in western Canada on the perennial wild Lettuce, *Lactuca pulchella*, and it commonly occurs near Winnipeg. Its mycelium hibernates in the rhizomes of its host and, in the spring, invades the young leaves and there forms numerous pycnidia. Single infected plants which were being grown in pots in the greenhouse of the Dominion Rust Research Laboratory were found to be strongly scented. I asked several people to smell infected plants and, without exception, they all detected the honey-sweet odour at once.

(4) *Melampsorella cerastii.* This Rust, in the stage formerly known as *Peridermium coloradense*, causes the formation of large "witches brooms" on certain Coniferae. I have observed it on the White Spruce, *Picea canadensis*, at Victoria Beach on Lake Winnipeg. In the spring, all the numerous leaves on the twigs making up a broom bear four rows of pycnidia (Figs. 65 and 66, pp. 196 and 197). The pycnidial nectar gives out so strong a scent that, under favourable atmospheric conditions, the presence of a broom on a Spruce tree can be detected at a distance of 20-40 feet.

(5) *Puccinia graminis.* In the greenhouse at the Dominion Rust Research Laboratory there was a Barberry bush with several of the leaves heavily infected with *Puccinia graminis*. On one leaf there were 46 larger and smaller pustules, and from every pustule the pycnidia had excreted an abundance of nectar which stood out above the leaf-epidermis as bright-orange hemispherical drops. The un-infected lower leaves of the Barberry bush appeared to be scentless; but, whenever I brought my nostrils close to the pycnidial drops on the more heavily infected leaves, I could detect with ease and certainty that the nectar was emitting a sweet and honeylike odour.

There were two lady clerks in the laboratory who knew nothing about the pycnidia of *Puccinia graminis*. I asked them to smell first the uninfected leaves of the Barberry bush and immediately thereafter the infected leaves, and to tell me their impressions. They were taken into the laboratory separately and no information was given them as to what to expect. The first lady said that the infected leaves had an odour that was "sweet" and "like Petunias," while the second said that the infected leaves had a distinct odour that was "sweet," "like the scent in a florist's shop," and "like the scent of Easter Lilies." Three men engaged in making investigations in the laboratory also agreed that the infected leaves emitted an odour not given out by the uninfected leaves, and one of them said the odour reminded him of the aroma of apples.

About a year after making the observations just recorded, I had in the greenhouse at Winnipeg a heavily infected Barberry bush. The young shoots had been inoculated on May 11 and, sixteen days later, the leaves and stems bore several hundred little pustules covered with drops of nectar. The scent given out by the fungus was very strong, and I found that Mr. A. A. Krohn, an assistant in the greenhouse, on closing his eyes and approaching the bush, was able to detect the pleasant odour of the pycnidia when his nose was still about eight inches distant from the infected leaves.

Mr. Krohn and I, after smelling the pycnidial nectar on the Barberry bush, as just described, found that the aecidia-bearing pustules, the uredospore sori, and the teleutospore sori of *Puccinia graminis* are scentless. It therefore appears that the scent given out by *P. graminis* is limited to the pycnidial stage, *i.e.* to just that stage of development at which alone the fungus can be benefitted by insect visits. Scented flowers emit their odour before pollination has been effected and they soon cease to emit it after the ovules have been fertilised and have begun to develop into seeds. The economy of scent production in *P. graminis* thus resembles that in an Angiosperm.

One humid morning early in June, 1935, at Macdonald College (Quebec), Dr. H. J. Brodie,[1] a former pupil, was examining a very large Barberry bush that was heavily infected with *Puccinia graminis*. He noted the large drops of nectar covering the leaf-pustules and, on smelling them, he detected that they were giving off a "honey-like odour." It is therefore clear that the sweet scent of the pycnidial exudate of *P. graminis* can be appreciated even when the infected Barberry bush is growing under natural conditions in the open.

[1] H. J. Brodie, personal communication, Sept., 1940.

(6) *Puccinia helianthi.* On smelling heavily infected young Sun-flower leaves, I was at first inclined to believe that the nectar of *Puccinia helianthi* is odourless; but, after persisting in my attempts, I thought I could sometimes detect the emission of a very faint scent. Not satisfied with my own observations, I asked two workers in the Rust Research Laboratory to smell the leaves for me. The first, one of the lady clerks who had detected the odour of *P. graminis*, said that the pustules of *P. helianthi* gave out a distinct odour, like that of *P. graminis* but weaker; while the second, a man with a keen sense of smell and a non-smoker, was quite certain that the pustules of *P. helianthi* give out an odour which, he said, was like that of Lilac flowers (*Syringa vulgaris*).

(7) *Puccinia coronata.* A small *Rhamnus cathartica* bush was in-oculated in the greenhouse with basidiospores of *Puccinia coronata.* On some of the leaves several large pycnidial pustules were formed and upon them nectar duly appeared. The scent given out by the nectar was easily detected by myself, Mr. B. Peturson, and a visitor to the laboratory, Professor C. W. Lowe. The scent seemed sweetish, but not as pleasant as that given off by *P. graminis*.

It has already been remarked that *Puccinia graminis* emits its sweet odour only in the pycnidial stage, and that its aecidia, uredo-spore sori, and teleutospore sori are scentless. The same appears to be true for the Rust Fungi in general. Rust Fungi therefore resemble Monocotyledons and Dicotyledons in which, as a rule, sweet odours are given out by the flowers alone and not by the leaves, stems, or roots. There can be no doubt that the odour emitted by the pycnidial nectar of such Rust Fungi as *P. suaveolens* and *P. graminis*, like the odour emitted by flowers, is attractive to insects, and that it assists these animals in finding pycnidial pustules. Probably, as soon as a fly has visited one or two pycnidial pustules it has learnt to associate the scent, general appearance, and colour of a pustule with the presence of sugary food. If so, such an association would assist the fly in searching for and finding new pustules, and scent would be a factor in causing the insect to pass from one pustule to another.

Tilletia tritici, T. laevis, and *T. indica*[1] are three pathogens which cause the Stinking Smut Disease of Wheat. The odour given out by their dark chlamydospores *en masse* is very unpleasant; it is due to

---

[1]M. Mitra: "A New Bunt on Wheat in India," *Ann. Appl. Biol.,* Vol. XVIII, 1931, pp. 178-179; also "Stinking Smut (Bunt) of Wheat with special reference to *Tilletia indica* Mitra," *Indian Journ. Agric. Sci.,* Vol. V, 1935, pp. 51-74 (especially p. 54). In his first communication he said that the spore-mass had no smell of de-caying fish, but in his second he recorded that he had detected this odour.

trimethylamine,[1] and various observers have said that it makes them think of rotten fish. It is of interest to note that the odour, if any, of the Ustilaginales which do not depend on insects for their welfare is repulsive, whereas the odour of the Uredinales for which insect visits are indispensable is sweet and reminiscent of flowers.

Insects in their Relations with Rust Fungi.—Taking into consideration what has been said in the preceding Sections of this Chapter, we may conclude that the pycnidial pustules of the Uredinales are as well organised for attracting insects and causing them to interchange (+) and (−) pycnidiospores between pycnidia as are flowers for attracting insects and causing them to bring about cross-pollination. Summarising the factors of pycnidial organisation that favour the visits of insects it may be said that: (1) the pycnidia are mostly situated on the *upper* surface of leaves where they can readily be seen by flying insects and where they are easily accessible to insects which settle near; (2) the pycnidial drops are *reddish* and they *glitter*, so that they stand out conspicuously against their green background and are likely to attract the attention of insects' eyes; (3) the drops in a number of species are strongly *scented*, so that they appeal to the olfactory sense of insects; and (4) the drops contain *sugar, oil-drops,* and *pycnidiospores* which, as food substances, reward insects for the time and energy expended in passing from pustule to pustule.

There can be no doubt that many species of insects do actually visit Rust pycnidia and feed upon the nectar. Ráthay,[2] as we have seen, observed 135 species of insects visiting the pycnidia of various Rusts; and, in a greenhouse where Rusts are under cultivation, one of the difficulties of the experimenter is in keeping insects away. Craigie[3] had trouble with Aleyrodes and Hemithrips, and Miss Allen[4] found her haploid pustules of *Puccinia coronata* much interfered with not by insects which flew about but by those which crawled. She said: "The inoculated plants were covered with tarlatan cages, but it was discovered that minute yellow insects not excluded by the cages were going from one infection to another, feeding on the spermogonial

[1]W. F. Hanna, H. B. Vickery, and G. W. Pucher, "The Isolation of Trimethylamine from spores of *Tilletia laevis*, the Stinking Smut of Wheat," *Journ. Biol. Chem.*, Vol. XCVII, 1932, pp. 351-358. Hanna ("The Odor of Bunt Spores," *Phytopathology*, Vol. XXII, 1932, pp. 978-979) has observed that trimethylamine is present in some strains of *Tilletia tritici* and absent in others.

[2]E. Ráthay, *vide* Chapter I, pp. 30-42.

[3]J. H. Craigie, "An Experimental Investigation of Sex in the Rust Fungi," *Phytopathology*, Vol. XXI, 1931, p. 1027.

[4]Ruth F. Allen, "A Cytological Study of Heterothallism in *Puccinia coronata*," *Journ. Agric. Research*, Vol. XLV, 1932, pp. 521-522.

exudate and incidently transferring spermatia from one infection to another. As a result, out of 248 single infections, 37 had developed open aecidia by the twentieth day after inoculation, 71 per cent by the thirtieth day, and 83 per cent a week or so later. The effectiveness of insect transfer at least, is amply demonstrated."

Craigie,[1] working with *Puccinia helianthi*, as already recorded, put fifteen to twenty flies in a wire-screen cage containing about twelve pots of Sunflower seedlings on the foliage leaves of which there were 98 monosporidial pustules bearing pycnidia but no aecidia; and, as a control, flies were kept out of another large screen-wire cage which contained fifteen pots of Sunflower seedlings on the foliage leaves of which there were 159 monosporidial pustules. Eight days later, he observed that 96 of the 98 pustules to which flies had had access had produced aecidia, whereas only 5 of the 159 pustules to which the flies had not had access had produced aecidia. Here we have a definite clear-cut experimental proof that, in a Rust Fungus, flies are most efficient agents in mixing the nectar of haploid pustules and so assisting in the initiation of the sexual process.

Another experimental proof that in a Rust Fungus insects effectively mix the nectar of (+) and (−) pustules and so initiate the sexual process was given by R. Kamamura, in 1941, as a reult of his work on *Gymnosporangium haraeanum*. This fungus produces pycnidia and aecidia on the Japanese Pear and teleutospore sori on Juniper (*Juniperus chinensis* and *J. chinensis* var. *procumbens*). By means of the nectar-mixing technique Kamamura[2] first proved that the *G. haraeanum* is heterothallic. Subsequently he[3] made the following observations. The nectar excreted from the pycnidial pustules on pear leaves containing a reducing sugar, is sweetish to the taste, but has no distinct odour. Under field conditions, when the pustules on pear leaves were covered in by cheese-cloth, the formation of aecidia was reduced to one-half as compared with those infections that were not covered. It was proved experimentally that the flies *Eristalis cerealis*, *Musca domestica*, and *Calliphora erythrocephala*, and the ant *Iridiomyrmex itoi* mix the nectar of separate monobasidiosporic

[1] J. H. Craigie, "Discovery of the Function of the Pycnidia of the Rust Fungi," *Nature*, Vol. CXX, 1927, pp. 765-767.

[2] E. Kamamura: "On the Function of the Spermatia of *Gymnosporangium Haraeanum* Syd.," *Horticulture and Agriculture*, Vol. IX, 1934; and "Studies on *Gymnosporangium Haraeanum* Syd. I. Heterothallism in the Fungus," *Ann. Phytopath. Soc. Jap.*, Vol. X, 1940, pp. 84-92.

[3] E. Kamamura, "Studies on *Gymnosporangium Haraeanum* Syd. II. The Rôle played by Insects in the Transfer of Spermatia in the Fungus," *Ann. Phytopath. Soc. Jap.*, Vol. X, 1941, pp. 297-303.

pustules and thus cause the production of the aecidial stage of the fungus.

In the fields about the city of Hukuoka, Kamamura[1] observed the following species of insects visiting the pycnidia of *Gymnosporangium haraeanum* and mixing the nectar of (+) and (−) pustules:

Diptera

Musca domestica,   Eristalis cerealis,
Sarcophaga melanura,   Syrphus corolla,
Calliphora erythrocephala,   Syrphus ribesii,
Ophyra calcogaster,   Syrphus balteatus.
Chorophila cinerella,

Hymenoptera

Formica fusca japonica,   Lasius fuliginosus,
Para trechina flavipes,   Camponotus herculaneus japonicus,
Iridiomyrmex itoi,   Bassius laetalorius,
Pristomyrmex japonicus,   Arge nigrinodosa.

Coleoptera

Rhaphidopalpa femoralis,   Anthrenus verbaci.

As, if we except the *Micro*-forms without pycnidia, *e.g. Puccinia malvacearum*, most species of Uredinales appear to be heterothallic and as, in heterothallic species, simple pycnidial pustules, when isolated from other similar pustules, do not produce aecidiospores unless they have been visited by insects and have had brought to them pycnidio-spores of opposite sex, it is clear that insects are almost as important for the existence of most Rust Fungi as are bees for the existence of Lady Slipper Orchids and many other Flowering Plants. It seems probable that, if insects were all exterminated or if they ceased entirely to visit pycnidial pustules, many of the heterothallic species of Rust fungi would die out in the course of a very few generations.

Effect of Rain on Pycnidia.—-Most or all of the pycnidia of *Puccinia graminis, P. coronata, P. helianthi*, etc., are on the upper side of the host-leaves so that the drops of nectar look upwards to the sky. With the advent of rain these drops must soon be washed away and the exuded pycnidiospores and sugar thus be removed from the pustules. But, as Ráthay[2] has proved by field observations made on *P. suaveo-lens, Aecidium magelhaenicum*, and *Gymnosporangium sabinae*, a few

[1] *Ibid.*   [2] *Vide supra*, p. 38.

hours after heavy rain the pycnidia form fresh drops of nectar so that the conditions which favour the visits of insects are quickly restored. After the cessation of heavy rain, new drops of nectar excreted from the pycnidia of *G. sabinae* were found by Ráthay to be sweet to the taste; and he also proved by repeated washings of the pycnidia and by tests of the wash-water that the pycnidia of *G. juniperinum* are able to excrete sugar, water, and pycnidiospores for several successive days. I too have observed that, in *Puccinia graminis*, the growth in size of the pycnidial drops and the extrusion of pycnidiospores continues for some days. When all these facts are taken into consideration, it seems probable that, for Rusts in general, a rain-storm disturbs the pycnidia for a short time only and without destroying their power of recuperation.

The pycnidia of the Rust Fungi are specialised for functioning in fine weather and with the help of insects. When it rains, the sweet and sweet-smelling nectar is washed away and insects hide themselves.

While in fine weather insects are of the greatest importance as agents for mixing the nectar of isolated (+) and (−) pustules, in wet weather, when there are many haploid pustules on a host-plant, the splashing of rain-drops may well result in some of the isolated (+) pustules receiving (−) pycnidiospores and some of the isolated (−) pustules (+) pycnidiospores. Since Newton and Johnson[1] have observed that, in *Puccinia graminis*, diluting mixed nectar with distilled water does not injure the diploidising power of the pycnidiospores when subsequently the mixed nectar is applied to haploid pustules, it seems likely that, if and when pycnidiospores are brought to flexuous hyphae of opposite sex by a shower of rain, fusions between these cell-elements will follow and thus the sexual process will be initiated.

Dr. J. J. Christensen of the University of Minnesota has gained the impression that rain actually does assist the sexual process in *Puccinia graminis*. He[2] says: "Insects have frequently been observed to visit Barberry bushes infected with *Puccinia graminis* and most certainly play an important rôle in the transference of pycnial nectar. In Minnesota, however, the splashing and washing of rain and the mechanical movement of leaves and twigs against infected plant parts, particularly during spring rains, appear to be even more important agents than insects in the mixing of the nectar. This is especially true

[1]Margaret Newton and T. Johnson, "Specialization and Hybridization of Wheat Stem Rust, *Puccinia graminis tritici*, in Canada," *Dominion of Canada, Dept. of Agric.*, Bull. CLX, 1932, p. 40.

[2]J. J. Christensen, *in litt.*

when Barberry bushes are heavily infected, which is not uncommon in Minnesota. On one occasion it was calculated that there were on a single Barberry bush more than 200,000 aecial infections, the majority of which were in about the same stage of development, regardless of the part of the plant on which they occurred. As there were more than 150 bushes in this particular planting, the mixing of pycnial nectar from such an enormous number of pycnia early in the spring could perhaps be best explained by the action of rain and the mechanical movement of plant parts."

Dr. Christensen's communication was not accompanied by precise data on weather conditions and the visits of insects to the Barberry bushes during the critical days preceding the diploidisation of the haploid pustules, and it is therefore impossible for me to form an opinion whether or not his general impression that rain was the chief agent in mixing the nectar was well based. Owing to the density of the infections it is probable that the leaves on the Barberry bushes bore a large number of compound pustules (*cf.* Fig. 25, p. 102) containing both (+) and (−) components. All such compound pustules would have become diploidised by interchange of nuclei between the mycelia in the leaf tissues without the intervention of any outside agency such as insects or rain splashings. It may be, too, that certain days prior to the diploidisation of the isolated pustules were much more favourable to insect visits than Dr. Christensen supposed. Nevertheless, since it is conceivable that rain may cause (+) pycnidiospores to be splashed or washed to (−) pycnidia and (−) pycnidiospores to be splashed or washed to (+) pycnidia, it is desirable that the problem of rain as a possible agent in facilitating the sexual process in the Rust Fungi should be investigated experimentally.

The Flexuous Hyphae.—The flexuous hyphae which in such Rusts as *Puccinia graminis* grow out through the ostiole of a pycnidium and extend into the drop of nectar beyond the tips of the periphyses and beyond the mass or masses of pycnidiospores have already been briefly described.[1] They will now be treated in more detail.

To observe the presence of flexuous hyphae in pycnidia of species of Puccinia, Uromyces, etc., various methods may be employed.

Method I. Craigie, when discovering flexuous hyphae in *Puccinia graminis* and *P. helianthi*, cut transverse sections of a piece of a host-leaf including a rust pustule with a hand-razor; and this method I have used very frequently, especially when examining the pycnidia of species investigated for the first time. After the somewhat thick sections have been cut (with water on the razor-blade), they are

[1] *Vide supra*, p. 131.

mounted in water under a cover-glass and are examined first of all with the low power of the microscope. If flexuous hyphae are present in any of the pycnidia, they will then be seen projecting into the water beyond the tips of the periphyses (Fig. 56, also *cf.* Fig. 40, p. 130). This method has the advantage of being very simple and it permits of the flexuous hyphae being examined whilst still living and fully turgid.

Method II. A simple and rapid method which avoids section cutting and has been successfully employed by myself for examining pycnidia and their flexuous hyphae *in face view* (*cf.* Figs. 59 and 60, pp. 177 and 179) involves the following technique.

FIG. 56.—*Puccinia graminis.* Three pycnidia showing flexuous hyphae. About five weeks after inoculation. Free-hand section, material unstained and living. Photographed by A. M. Brown. Magnification, about 100.

(1) Cut away from a living host-plant a small piece of leaf including one or two pustules containing pycnidia covered with simple or compound drops of nectar; (2) place the piece of leaf with the pycnidia looking upwards on a drop of water on a glass slide; (3) set bits of a broken cover-glass on each side of the piece of leaf; (4) cover the whole with a cover-glass; (5) with a pipette fill the space beneath the cover-glass with water; (6) boil the drop over a flame for a few moments; (7) with a pipette add to one side of the drop and cover-glass a spot or two of lacto-phenol cotton-blue; (8) raise the cover-glass on one side for about 15 seconds so as to allow the cotton-blue to pass freely over the upper surface of the piece of leaf to the pycnidia; (9) lower the cover-glass to its original position; and (10), using a sufficiently intense source of illumination, examine the preparation with the microscope. The boiling serves to rid the piece of leaf of most of the air contained within the intercellular spaces and so to make the preparation more translucent; and it also serves to disperse the outer part of the mass of pycnidiospores held in place by the periphyses. The cotton-blue is unable to make its way through the epidermis of the leaf, and it therefore stains the periphyses, the flexuous hyphae, and the loose pycnidiospores without staining the leaf-cells below them. Moreover, the dye does not penetrate downwards through the ostiole so that the pycnidiosporophores which line the cavity of the pycnidium

and the unextruded pycnidiospores retain their original orange-yellow colour. It is also to be noted that the cotton-blue does not easily diffuse into the interior of a gelatinous mass of pycnidiospores, so that the dense mass of pycnidiospores still held among the bases of the periphyses, as well as the bases of the periphyses themselves, usually remain unstained.

FIG. 57.—A pycnidium of *Puccinia minussensis* (*P. hemisphaerica*) on *Lactuca pulchella*. A surface section of a leaf bearing pycnidia was boiled in water on a slide for a few seconds, stained with cotton-blue, covered with a cover-glass which pressed down upon the pycnidium and so brought its outer elements into one plane, and then photographed: *a*, the middle part of the pycnidium; *b*, periphyses ending in a point; *c c c*, four flexuous hyphae. To three of the flexuous hyphae pycnidiospores are clinging by means of their gelatinous walls. Magnification, about 500.

In a preparation made in the manner just described (Fig. 57) the periphyses, flexuous hyphae, and loose pycnidiospores are all stained deep-blue, so that they stand out in contrast with the orange-yellow of the other parts of the pycnidium contained within the leaf below them.

The time required for making a whole-pustule cotton-blue-stained preparation, from the moment of removing a piece of leaf from a host-plant to the moment of observing the periphyses and flexuous hyphae is less than five minutes. There is no reason why the employment of this method of studying pycnidia should not become a standard exercise for students in botanical laboratories.

Method III. A third method, which I have employed for observing periphyses and flexuous hyphae in the living condition, is as follows: (1) cut away from a host-plant a piece of leaf containing a large pustule and of sufficient size to permit of it being folded; (2) fold the leaf across the middle of the pustule, so that some of the pycnidia are directed outwards at the fold; (3) place the folded piece of leaf on a slide; (4) add a cover-glass and press this down sufficiently; (5) using a pipette, wash the pycnidia (now projecting horizontally outwards from the folded leaf-edge) with water, so as to clear away some of the pycnidiospores; and (7) examine with the microscope. In such a

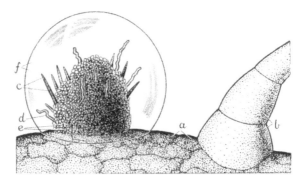

Fig. 58.—One of several pycnidia in a haploid pustule of *Puccinia helianthi* on a leaf of *Helianthus annuus*, nine days after the leaf was inoculated with basidiospores and about 24 hours after the drop of nectar began to be excreted: *a*, the leaf which was folded to permit of the pycnidium being seen in side view; *b*, a multi-cellular hair; *c*, a periphysis, pointed and fully grown; *d*, one of five flexuous hyphae, cylindrical and still elongating; *e*, a mass of gelatinous pycnidiospores which has been extruded through the ostiole of the pycnidium; and *f*, a drop of nectar, still increasing in volume. Magnification, 227.

preparation, which can be made in less than five minutes, one sees the brush of periphyses and the flexuous hyphae in side view (Fig. 58). A little practice and skill is required in folding the piece of leaf across the pustule, so as to bring pycnidia into view, but apart from this, the technique of making side-view preparations of living periphyses and flexuous hyphae presents no difficulty.

Such side-view preparations of living periphyses and flexuous hyphae as have just been described can be readily stained with cotton-blue or they can first be boiled and then stained.

Flexuous hyphae differ from periphyses in the following ways: (1) they are *flexuous* instead of being straight or slightly curved; (2) they are *cylindrical* and have *bluntly-rounded ends* instead of being slenderly-conical and tapering to a point; (3) they grow in length for some time and *become longer than the periphyses;* (4) occasionally they are *branched,* whereas periphyses are never branched; (5) they are *less red* than periphyses; (6) they *fuse with pycnidiospores* of opposite sex, whereas periphyses do not; and (7) they are relatively *ephemeral,* so that in old pycnidia of Puccinia, etc., one can find periphyses but no flexuous hyphae.

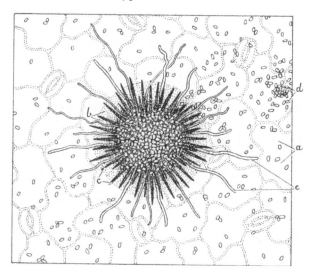

Fig. 59.—*Puccinia helianthi* on *Helianthus annuus.* Surface view of a pycnidium on the upper side of a host-leaf, 16 days after the leaf had been sprinkled with basidiospores, as revealed in a preparation of a piece of leaf boiled in water and mounted in water containing cotton-blue: *a* the epidermis; *b,* the orange-red wall and chamber of the pycnidium below the epidermis; *c,* the straight, pointed, orange-red periphyses supporting a mass of mucilaginous pycnidiospores which have been extruded through the pycnidial ostiole; *d,* pycnidiospores which were scattered over the epidermis in the making of the preparation; *e,* flexuous hyphae of which there are twenty. Magnification, 255.

We have seen that, in *Puccinia graminis,* the periphyses originate inside a pycnidium and from inner wall-cells in the upper part of the roof of the pycnidium just inside and below the ostiole. The flexuous hyphae have a similar origin and they arise just below the lowest periphyses as shown in Fig. 40 (p. 130) or, possibly, among the periphyses. Their short interior basal parts are not connected either directly or indirectly with any special fungal cells that could be regarded as oogonia, and their only connexion with the basal cells of the proto-aecidia is by means of a vegetative mycelium which is present between, and sends haustoria into, the chlorenchymatous cells of the host-leaf (Fig. 40).

In Gymnosporangium juniperi-virginianae and *G. clavariiforme* the flexuous hyphae in a pycnidium arise, as the periphyses are known to do, from hymenial cells surrounded by pycnidiosporophores and scattered about the sides and base of the pycnidium. In *Gymnoconia peckiana* (Fig. 69, p. 200) and *Phragmidium speciosum* (Fig. 74, p. 209) where periphyses are absent and the pycnidiosporophores form a more or less flat hymenial layer, the flexuous hyphae arise in this layer singly and in various places among the pycnidiosporophores.

In *Puccinia graminis* and *P. helianthi*, in a single mature pycnidium, the number of flexuous hyphae varies from about three to about twenty, and it is a rule that the flexuous hyphae are always far fewer than the periphyses.

In a developing pycnidium, the periphyses come through the epidermis first, but the flexuous hyphae follow hard after them. Both in *Puccinia graminis* and in *P. helianthi*, in pycnidia whose drops of nectar were not more than one-day old,[1] I have observed flexuous hyphae, still quite short, coming out among the periphyses. It therefore appears that, as soon as young pycnidia become crowned with nectar and therefore attractive to insects, they have flexuous hyphae ready to fuse with such pycnidiospores of opposite sex as visiting insects may bring to them.

Under moist conditions, as we have seen, the simple drops of nectar on a pycnidial pustule melt together and become compound. Into such enlarged drops the flexuous hyphae gradually extend, so that in older pustules these hyphae are often considerably longer than in younger ones. In this connexion the reader should compare: for *Puccinia helianthi*, the 9-day-old pycnidium shown in Fig. 58 with the 16-day-old pycnidium shown in Fig. 59; and for *P. graminis* the 9-day-old pycnidia shown in Fig. 60 (p. 179) with the much older pycnidia represented semi-diagrammatically in Fig. 40 (p. 130).

In old pustules of *Puccinia helianthi* that have developed under greenhouse conditions in February at Winnipeg the flexuous hyphae sometimes elongate to such an extent that they not only come up to the surface of the pycnidial drop but grow beyond it so that their *tips protrude into the air*. A preparation which showed the ends of flexuous hyphae of *P. helianthi* projecting out of a compound drop of nectar into the air was made and manipulated as follows: (1) A piece of Sunflower leaf including an old pustule with five pycnidia covered with a single compound drop of nectar was cut away from a leaf, placed on a slide (without water or cover-glass) and viewed with the

[1]The cultures, from the time of inoculating the host-plants with basidiospores to examination, were only eight days old.

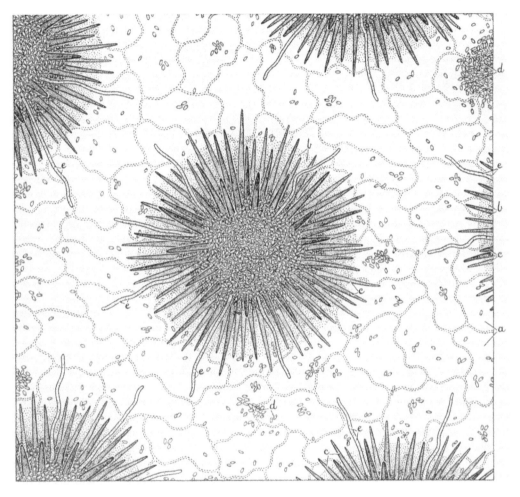

FIG. 60.—*Puccinia graminis* on *Berberis vulgaris*. Surface view of pycnidia on the
upper side of, and in the middle of, a pustule, nine days after inoculation of the
leaf with a basidiospore. The pustule was boiled in water, stained with cotton-
blue, and mounted in water: *a*, epidermal cells of leaf; *b b*, orange-yellow bodies
of pycnidia below the epidermis; *c c*, periphyses, fully grown, supporting and
surrounding a mass of pycnidiospores; many of the pycnidiospores were washed
away from the pycnidia during the making of the preparation, so that now the
periphyses are more freely exposed to view than they were originally; *d d*,
pycnidiospores scattered over the epidermis; *e e*, flexuous hyphae, of which in
the central pycnidium there are six, all young and still growing in length.
Magnification, 453.

low power of the microscope. It could then be seen that a few flexuous hyphae from each of the pycnidia had grown out of the drop into the air: they had the appearance of thin white filaments. (2) The same piece of leaf was stuck (with water as adhesive) to the side of a vertical glass slide, and the compound drop was observed laterally with the low power of the microscope. Again, and still more clearly than before, the ends of several flexuous hyphae, a few coming from each pycnidium, could be seen projecting into the air a little way beyond the surface of the compound drop.

Flexuous hyphae in *Puccinia graminis* and *P. helianthi*, when fully grown, are often twice as long as the periphyses and sometimes longer (Fig. 61, B-E, I). Indeed, in a compound drop that covers many pycnidia, the flexuous hyphae are sometimes so long that they stretch from one pycnidium to another (Fig. 61, A). It is conceivable that, where a (+) pustule and a (−) pustule have originated very close to one another and, when very young, have joined to form a compound pustule, a (+) flexuous hypha from a (+) pycnidium may sometimes unite with a (−) flexuous hypha from a (−) pycnidium and that, *via* the fusion canal, (+) and (−) nuclei may be exchanged. Brown's[1] experiments with 288 compound pustules of *P. helianthi*, in which by means of a wall-like barrier[2] he prevented the spontaneous intermixing of the nectar of the two component simple pustules and in which, nevertheless, 110 of the compound pustules produced aecidia, proved conclusively that, as Craigie[3] had supposed, in a compound pustule formed by the union of a simple (+) and a simple (−) pustule, (+) and (−) hyphae fuse together and initiate the diploidisation process *within* the host-leaf; but, hitherto, the theoretical possibility that the diploidisation process may be occasionally initiated by hypha-to-hypha fusions *outside* the host-leaf has not been envisaged.

Against the supposition that, by means of fusions between (+) and (−) flexuous hyphae *outside* the host-leaf, the diploidisation process may be initiated, may be set the consideration that where two mycelia, one (+) and the other (−), arise very close to one another, the ordinary (+) hyphae and (−) hyphae meet *inside* the host-leaf several days before the pycnidia have developed to the point which would permit of contacts between the (+) and (−) flexuous hyphae.

[1]A. M. Brown, "A Study of Coalescing Haploid Pustules in *Puccinia helianthi*," *Phytopathology*, Vol. XXV, 1935, pp. 1085-1090.

[2]The barrier was made with a preparation used for touching up photographic negatives (Eastman's Opaque No. 1). This substance adhered readily to the leaf, dried quickly, and caused no perceptible injury. *Vide infra*, Chapter V.

[3]J. H. Craigie, "Experiments on Sex in the Rust Fungi," *Nature*, Vol. CXX, 1927, pp. 116-117.

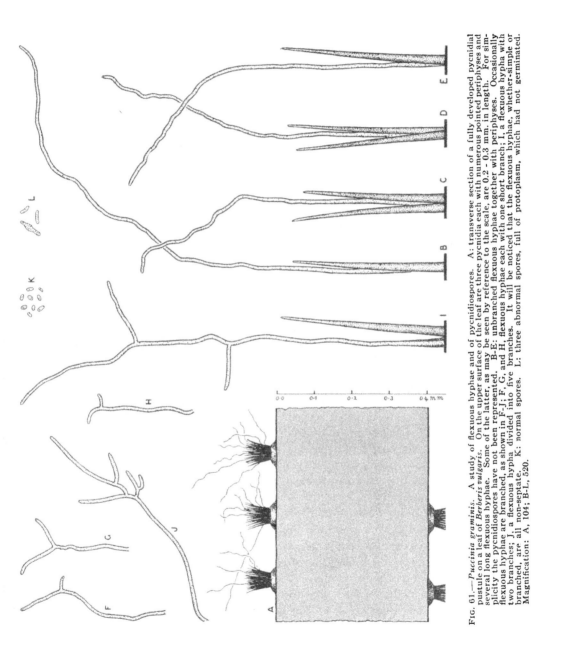

Fig. 61.—*Puccinia graminis*. A study of flexuous hyphae and of pycnidiospores. A: transverse section of a fully developed pycnidial pustule on a leaf of *Berberis vulgaris*. On the upper surface of the leaf are three pycnidia each with numerous pointed periphyses and several long flexuous hyphae. Some of the latter, as may be seen by reference to the scale, are 0.2 - 0.3 mm. in length. For simplicity the pycnidiospores have not been represented. B-E: unbranched flexuous hyphae together with periphyses. Occasionally flexuous hyphae are branched, as shown in F-J; F, G, and H, flexuous hypha each with one short branch; I, a flexuous hypha with two branches; J, a flexuous hypha divided into five branches. It will be noticed that the flexuous hyphae, whether simple or branched, are all non-septate. K: normal spores. L: three abnormal spores, full of protoplasm, which had not germinated. Magnification: A, 104; B-L, 520.

It therefore seems that, even where a (+) mycelium and a (−) mycelium arise very close together, the diploidisation process would be initiated rather by a union of ordinary vegetative hyphae inside the host-leaf than by a union of flexuous hyphae outside the leaf.

In *Puccinia graminis*, in young pycnidia which have borne drops for not more than a day or two, I have observed that many of the flexuous hyphae, while cylindrical and relatively thin where they stretch out between the outer halves of the periphyses and beyond the pycnidiospores into the drop of nectar, have *straight, conical, and decidedly thick bases* which extend from the ostiole of the pycnidium to about half way up the periphyses. One such flexuous hypha is shown at *e* on the left side of the central pycnidium of Fig. 60 and another at I in Fig. 61. It appears that many flexuous hyphae, when they are very young and about one-third the length of a periphysis, must resemble a one-third-grown periphysis in being equally straight, thick, tapering, and bluntly-ended, and that, with further growth, whereas a periphysis continues to grow in a straight line, continues to taper conically, and eventually becomes sharply pointed, a flexuous hypha suddenly changes its mode of growth, so that it curves about as it elongates, ceases to taper, becomes cylindrical in form, and retains its blunt apex indefinitely. Summarising this discussion, it may be said that the flexuous hyphae and periphyses of *P. graminis*, although very different in appearance when mature, are often indistinguishable from one another in their first stages of development.

In *Puccinia graminis*, the base of a flexuous hypha, like that of a periphysis, may be 4-6 μ in diameter, but the distal cylindrical part that extends outwards between the tips of the periphyses into the drop of nectar is usually not more than 1.5-3.0 μ in diameter. As already remarked, the length of a flexuous hypha varies with age, being shorter in very young pycnidia and longer in older ones. The maximum length attained by the flexuous hyphae of *Puccinia graminis* and *P. helianthi* is about one-third of a millimetre (Fig. 61, A).

In an older pustule of *P. graminis* or *P. helianthi*, it may be observed that, while many or most of the flexuous hyphae are simple, others have one or sometimes two or three or even more branches (Fig. 61, F-I). A branch often comes away from the parent hypha at a right angle, and sometimes it bends backwards instead of forwards.

The flexuous hyphae, like the periphyses, but unlike the hyphae attacking the mesophyll of the host-leaf, are *non-septate*[1] (Fig. 61, B-I). In the mycelia of Discomycetes, Pyrenomycetes, Hymeno-

[1] J. H. Craigie ("Union of Pycniospores and Haploid Hyphae in *Puccinia helianthi* Schw.," *Nature*, Vol. CXXXI, 1933, p. 25) said that the flexuous hyphae of *P. graminis* and *P. helianthi* "may branch but they rarely show septations." The

mycetes, and Gasteromycetes, septa are numerous and, in many Hymenomycetes, new branches usually originate at a short distance below a septum.[1] The branched flexuous hyphae of the Rust Fungi appear to differ from the mycelial hyphae of the Rust Fungi and of the Higher Fungi in general in that they are branched without being septate. The absence of septa from a flexuous hypha, whether branched or unbranched, leaves the nucleus of a pycnidiospore, when once it has entered a flexuous hypha, a free passage down the hypha to the mycelium within the leaf.

The protoplasm of a flexuous hypha contains tiny oil-drops laden with a carotinoid pigment, and it is usually decidedly vacuolate (Fig. 62, A, a-d).

The cell-wall of a flexuous hypha, like that of a periphysis, is thin and colourless. To it the gelatinous pycnidiospores readily adhere (Fig. 62, A, c, d, e). After a flexuous hypha has lost its contents and has died, its outer wall appears to become very slimy, and it often happens that such a hypha becomes so clustered about by pycnidio-spores that it can scarcely be seen (Fig. 62, A, f and g).

In the mycelia of Discomycetes, Pyrenomycetes, Hymenomycetes, Gastromycetes, and Rust Fungi, hyphal fusions between neighbouring hyphae are of frequent occurrence, and it is a rule that a mycelium of any species of the Higher Fungi, as it grows older, becomes converted into a three-dimensional network.[2] On the other hand, the flexuous hyphae in a (+) or in a (−) Rust pustule, notwithstanding the fact that they are often branched and often cross one another, never unite with one another. The only fusions that they have been observed to form are with pycnidiospores of opposite sex.

The flexuous hyphae in a (+) or in a (−) pustule of a Rust fungus differ from the hyphae that are present in the mesophyll of the host-leaf in that: (1) they do not attack the host-plant; (2) they are produced at the expense of the mycelium within the host-plant; (3) they do not fuse with one another; (4) they are non-septate; and (5) they form unions with pycnidiospores of opposite sex.

Flexuous hyphae, owing to their length, their large surface area, their position in the drop of nectar, and their non-septation appear to be very well adapted for coming into contact with and fusing with pycnidiospores, and for conducting the nuclei of pycnidiospores into

---

septations which Craigie saw were probably thin discs of cytoplasm separating vacuoles. I have never seen a single wall-septum in any of the numerous flexuous hyphae which I have examined.

[1]*Vide* these *Researches*, Vol. IV, 1931, Fig. 118, p. 202; and Vol. V, 1933, Fig. 83, p. 164.

[2]*Vide* these *Researches*, Vol. IV, pp. 152-180; also Vol. V, pp. 1-74.

the host-leaf in the first stage of their journey to the hyphal wefts of the proto-aecidia.

There are no female sexual organs in the haploid mycelium of a Rust fungus and, in *Puccinia graminis*, as I have observed, the flexuous hyphae grow out from the walls of the first-formed pycnidia into the drops of nectar before the formation of the first proto-aecidia. There is therefore no justification for calling flexuous hyphae trichogynes.

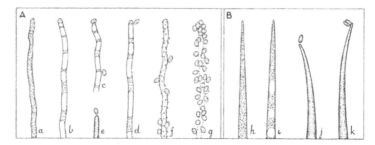

FIG. 62.—*Puccinia graminis.* A: exhaustion and disappearance of flexuous hyphae; *a*, a young flexuous hypha full of protoplasm; *b*, an old flexuous hypha, highly vacuolated; *c* and *d*, similar flexuous hyphae showing pycnidiospores sticking by their gelatinous walls to the end or side; *e*, an old flexuous hypha with a pycnidiospore sticking to its apex; *f*, a flexuous hypha which has died; it has lost its protoplasm and attached to its outer wall are small particles (possibly oil-drops) and pycnidiospores; *g*, an old dead flexuous hypha which has lost its contents; its wall, to which many pycnidiospores are attached, has become very gelatinous and is now rapidly becoming disintegrated. B, a study of periphyses: *h*, a young periphysis full of orange protoplasm; *i*, an old periphysis with vacuoles; *j* and *k*, two periphyses, seen from without, with pycnidiospores sticking to their tips. Magnification, about 520.

Flexuous hyphae are merely slightly specialised extensions of the vegetative mycelium, which grow beyond the surface of the host into a drop of fluid where fusions can take place between themselves and pycnidiospores of opposite sex.

While flexuous hyphae are of great importance for the sexual process in the Rust Fungi, they are not indispensable; for the sexual process can be initiated not only by the fusion of flexuous hyphae and pycnidiospores but also by the union of a (+) mycelium and a (−) mycelium, within a leaf.[1] In a haploid mycelium of a Rust Fungus

---

[1]This was indicated by Craigie's experiments (J. H. Craigie, "Experiments on Sex in Rust Fungi," *Nature*, Vol. CXX, 1927, pp. 116-117) and was conclusively proved by Brown in experiments made on *Puccinia helianthi* in which the spontaneous mixing of the nectar in compound pustules was prevented (A. M. Brown, "A Study of Coalescing Haploid Pustules in *Puccinia helianthi*," *Phytopathology*, Vol. XXV, 1935, pp. 1085-1090).

what is required for initiating the sexual process resulting in the formation of aecidia is a nucleus of opposite sex and not necessarily a nucleus derived from a pycnidiospore.

Conversion of Mature Periphyses into Flexuous Hyphae.—In older pycnidial pustules of *Puccinia graminis*, fully grown periphyses which have tapered to a point sometimes renew their growth and become converted into flexuous hyphae. The end of a periphysis which is about to grow in length first swells up and becomes knoblike,[1] and then the swelling becomes extended as a flexuous hypha. Periphyses in various stages of conversion into flexuous hyphae are shown in Fig. 63, C-I.

Very occasionally a mature periphysis, on renewing its growth, instead of developing into a flexuous hypha, produces another periphysis. Thus one periphysis comes to bear another periphysis on its tip. This curious abnormality has been observed in *Puccinia graminis* (Fig. 63, K) and in *Gymnosporangium juniperi-virginianae*.

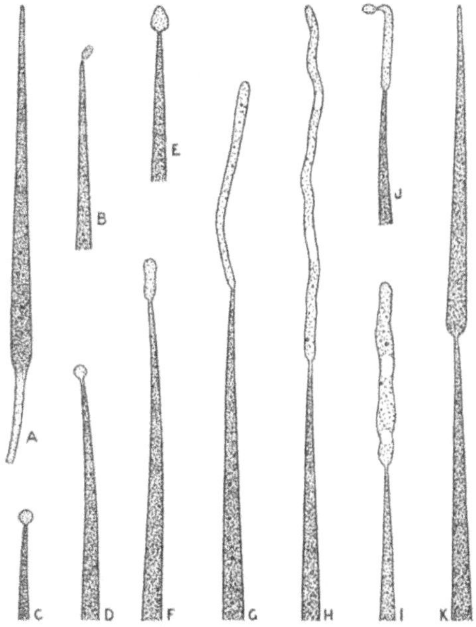

In some pustules of *Puccinia graminis*, owing to conversion of mature periphyses into flexuous hyphae as the pustules become older, the number of flexuous hyphae per pycnidium is considerably increased.

FIG. 63.—*Puccinia graminis*. Rejuvenescence of periphyses. A, a normal periphysis with its narrow base (inside the pycnidium), its conical shaft, and its pointed apex. B, the end of a periphysis, to which a pycnidiospore is sticking. C, D, and E, three periphyses the ends of which have become knob-like. F, G, H, and I, four periphyses which have formed flexuous hyphae at their apical ends. J, another flexuous hypha formed from a periphysis; this hypha, after the nectar of (+) and (−) pustules had been mixed, fused, as shown, with a pycnidiospore. K, a periphysis which by rejuvenescence produced a second periphysis. Magnification, about 520.

[1]Periphyses that are "bulbous at their tips" were observed in *Puccinia phragmitis* by I. M. Lamb ("The Initiation of the Karyophase in *Puccinia phragmitis*," *Annals of Botany*, Vol. XLIX, 1935, p. 5, and Fig. 8), but he made no remark on their origin.

While, as we have seen, mature periphyses may become converted into flexuous hyphae, flexuous hyphae never become converted into periphyses.

The increase in the number of flexuous hyphae through transformation of mature periphyses must increase the chances of a union being formed between a flexuous hypha and a pycnidiospore of opposite sex (Fig. 63, J).

The exact conditions under which pointed periphyses sometimes renew their growth and develop into flexuous hyphae have not been ascertained.

CHAPTER III

FLEXUOUS HYPHAE IN THE UREDINALES IN GENERAL

Introduction — Previous Observations — New Observations — Cronartium — *Melampsora lini* — *Melampsorella cerastii* — Milesia — *Gymnoconia peckiana* — Gymnosporangium — Phragmidium — *Tranzschelia pruni-spinosae* — Puccinia and Uromyces — Ashworth's Observations on *Melampsoridium betulinum* and *Melampsora larici-capraearum* — Conclusion.

Introduction.—From previous publications and from facts recorded in this Volume it is known that flexuous hyphae fuse with pycnidiospores of opposite sex in *Puccinia helianthi*,[1] *P. graminis*,[2] *P. coronata avenae*,[3] *Cronartium ribicola*,[4] and *Gymnosporangium clavipes*; and it is clear that, in the Uredinales, flexuous hyphae play a very important part in initiating the sexual process.

We may ask: are flexuous hyphae present in the pycnidia of Rust Fungi in general? An attempt to answer this question will now be made.

Previous Observations.—Hitherto, as recorded in Chapter I, flexuous hyphae have been observed in twenty species of Rusts. The names of these species are shown in the Table on page 188.

New Observations.—With a view to extending our knowledge of the occurrence of flexuous hyphae in the Uredinales in general I have examined as many Rust Fungi as possible in the pycnidial stage. *Gymnosporangium corniculans*, *G. juniperi-virginianae*, *G. juvenescens*, *Melampsora lini*, *Puccinia bardanae*, *P. coronata avenae*, *P. coronata calamagrostis*, *P. graminis*, *P. helianthi*, and *P. minussensis* were

---

[1]J. H. Craigie, "Union of Pycnidiospores and Haploid Hyphae in *Puccinia helianthi* Schw.," *Nature*, Vol. CXXXI, 1933, p. 25.

[2]A. H. R. Buller, "Fusions between Flexuous Hyphae and Pycnidiospores in *Puccinia graminis*," *Nature*, Vol. CXLI, 1938, p. 33.

[3]This Volume, Chapter IV.

[4]R. K. Pierson, "Fusion of Pycnidiospores with Filamentous Hyphae in the Pycnium of the White Pine Blister Rust," *Nature*, Vol. CXXXI, 1933, pp. 728-729.

grown in the greenhouse of the Dominion Rust Research Laboratory; *Gymnosporangium clavariiforme* and *G. clavipes* were grown on Amelanchier shrubs in a wood near the greenhouse; and about twenty-five other species, whose names are given in the subjoined Table, were collected in Canada, the United States of America, or England in gardens, woods, or fields. As soon as possible after wild material had

Rust Fungi in which Flexuous Hyphae have been observed by Previous Wórkers

SPECIES	OBSERVER	YEAR
Puccinia graminis	J. H. Craigie . . .	1933
Puccinia helianthi		
Cronartium ribicola	R. K. Pierson . . .	1933
Coleosporium helianthi		
Melampsora abieti-capraearum . . .		
Milesia marginalis		
Milesia polypodophila	Lillian M. Hunter . .	1936
Milesia polypodii		
Milesia scolopendrii		
Milesia vogesiaca		
Pucciniastrum americanum . . .		
Milesia dryopteridis		
Milesia exigua		
Milesia itôana		
Milesia jezoensis	S. Kamei	1940
Milesia miyabei		
Milesia sublevis		
Uredinopsis intermedia		
Uredinopsis ossaeformis		
Pucciniastrum hydrangeae	L. S. Olive	1943

been collected, it was taken to a laboratory. Here the basal ends of the herbaceous shoots or twigs were set in a beaker half-filled with water, and then the whole was covered with a bell-jar. Under these conditions the drops of nectar increased in size. With most of the species, sections of the living rust pustules were then made with a hand-razor, mounted in water, and examined with the microscope.

I have observed flexuous hyphae projecting from the ostioles or at the surface of pycnidia in thirty-one species of Uredinales.[1] The names of all these species are given in the accompanying list.

So far as North American species are concerned the nomenclature followed, except for *Cronartium cerebrum* Hedgcock and Long, *C. fusi-*

[1] A. H. R. Buller, "The Flexuous Hyphae of *Puccinia graminis* and Other Uredinales." An abstract. *Phytopathology*, Vol. XXXI, 1941, p. 4.

forme (A. and K.) Hedgcock and Hunt,[1] and *Puccinia suaveolens*,[2] has been that of J. C. Arthur's *Manual of the Rusts of United States and Canada* (U.S.A., 1934). The host species are given at the beginning of Chapter II.

Rust Fungi in which Flexuous Hyphae have been observed by the Author

Cronartium cerebrum
Cronartium fusiforme
Melampsora lini
Melampsorella cerastii
Gymnoconia peckiana
Gymnosporangium clavariiforme
Gymnosporangium clavipes
Gymnosporangium juniperi-virginianae
Gymnosporangium juvenescens
Phragmidium potentillae
Phragmidium speciosum
Puccinia bardanae

Puccinia caricis { grossulariata / solidaginis / urticata

Puccinia coronata { avenae / calamagrostidis

Puccinia extensicola oenotherae

Puccinia graminis { avenae / tritici
Puccinia helianthi
Puccinia minussensis
Puccinia monoica
Puccinia orbicula
Puccinia poarum
Puccinia podophylli
Puccinia rubigo-vera agropyrina
Puccinia sessilis
Puccinia smyrnii
Puccinia sorghi
Puccinia suaveolens
Puccinia vincae
Puccinia violae
Tranzschelia pruni-spinosae
Uromyces poae

Cronartium cerebrum, C. fusiforme, Melampsora lini, and *Melampsorella cerastii* belong to the supposedly lower family Melampsoraceae, while all the other species in the list are included in the Pucciniaceae.

In the pycnidia of all the genera of the Melampsoraceae and Pucciniaceae so far investigated *pycnidiospores* and *flexuous hyphae*, *i.e.* the two elements that are essential for initiating the sexual process by pycnidial agency, are always present.

In the pycnidia of the Melampsoraceae, slenderly conical, pointed, *ostiolar trichomes*, comparable with the *periphyses* of *Puccinia graminis* or the *paraphyses* of *Gymnosporangium juniperi-virginianae* appear to be entirely lacking, for none was seen by Miss Hunter in her species of Coleosporium, Melampsora, Milesia, and Pucciniastrum and it is certain from my own observations that none is present in *Cronartium cerebrum, C. fusiforme, Melampsora lini* and *Melampsorella cerastii*.

---

[1]Arthur has included these two species of Cronartium in *Cronartium quercuum* (Berk.) Miyabe, but Dr. Hedgcock, who has recently made a comparative study of Cronartium species, has informed me that *C. quercuum* of Japan does not occur in North America.

[2]For a justification for using the name *Puccinia suaveolens* rather than *P. obtegens* of Arthur, Grove, and other taxonomists *vide infra*, Chapter X.

In the Pucciniaceae, on the other hand, as has been observed by various workers and by myself, ostiolar trichomes are present in some genera, *e.g.* Gymnosporangium, Puccinia, and Uromyces, and absent in other genera, *e.g.* Gymnoconia, Phragmidium, and Tranzschelia. What is involved is the greater or lesser degree of specialisation in the organisation of the pycnidia. Ostiolar trichomes are accessory to, and not indispensible for, the initiation of the sexual process in pycnidia and, in this respect, they resemble the petals of the Angiosperms which are accessory to, and not indispensible for, the process of pollination.

Cronartium.—In 1933, Pierson[1] recorded that flexuous hyphae (his *filamentous hyphae*) are present in the pycnidia of *Cronartium ribicola* and that, after the nectar of different pustules has been mixed, fusions take place between the pycnidiospores of one sex and flexuous hyphae of opposite sex. Pierson's observations indicate that *C. ribicola* is heterothallic.

An examination of the living pycnidia of *Cronartium fusiforme* and *C. cerebrum* enabled me to observe that these species are provided with flexuous hyphae resembling those described by Pierson for *C. ribicola.*

According to Dr. Hedgcock,[2] who is preparing a monograph on the genus Cronartium, *Cronartium fusiforme* and *C. cerebrum* are distinct species and not to be confused with *C. quercuum* which occurs in Japan but not in North America.[3]

(1) *Cronartium fusiforme.* Through the kindness of Mr. R. W. Davidson, in October, 1939, the stems of some two-year-old seedlings of *Pinus caribaea* that bore cankers caused by *Cronartium fusiforme* were sent to me by Mr. Bailey Sleeth from the Mississippi State Nursery in the southern part of the United States of America. The seedlings had become infected when they were nine months old and the first-formed pycnidia were now just beginning to exude nectar. As soon as the material came to hand in Winnipeg, the bases of the infected stems were set in water under a bell-jar in the laboratory. Under these conditions the pycnidia became active and soon excreted slimy drops of nectar which hung down on the sides of the cankers.

[1]R. K. Pierson, "Fusion of Pycniospores with Filamentous Hyphae in the Pycnium of the White Pine Blister Rust," *Nature*, Vol. CXXXI, 1933, pp. 728-729.

[2]G. G. Hedgcock, personal information given at Washington, D.C., October, 1939.

[3]J. C. Arthur in his *Manual of the Rusts in United States and Canada* cites *C. quercuum* as a North American species and includes therein both *C. fusiforme* and *C. cerebrum.*

Hand-sections cut with a razor transversely through pycnidia and then mounted in water revealed that the pycnidia of *Cronartium fusiforme* are formed under a thin layer of bark and consist of a broad flat layer of pycnidiosporophores, a large mass of pycnidiospores, and a number of flexuous hyphae, without the accompaniment of any periphyses or paraphyses. It became evident that, with the development of the dense layer of pycnidiosporophores and the production by them of more and more pycnidiospores, the pressure on the thin overlying layer of bark increases until the bark is ruptured, and that the rupture permits of the escape of the nectar and results in exposing the pycnidiospores and the flexuous hyphae to the visits of insects. The exposed pycnidia, including the masses of pycnidiospores, are bright orange-red.

The pycnidiosporophores of *Cronartium fusiforme* are slenderly conical and are densely compacted, and each of them at its apex gives rise by abstriction to pear-shaped pycnidiospores (*cf.* Fig. 101, p. 130).

A glass slide was touched to a drop of nectar hanging from a pycnidium and the nectar was then examined under the microscope. In it were found great numbers of isolated more or less pear-shaped pycnidiospores, a few chains of pycnidiospores, and some malformed pycnidiospores. The existence of chains of pycnidiospores in which disjunction of the individual spores for some reason or other failed to take place teaches us that each individual pycnidiosporophore normally forms and sets free a considerable number of pycnidiospores.

When one considers how vast a number of pycnidiospores are abstricted from the tens of thousands of pycnidiosporophores in a single pycnidium of *Cronartium fusiforme* it is not surprising that some of the pycnidiospores should be malformed. Some of the malformed spores look, at first sight, as if they might have germinated; but, as they are unswollen, full of protoplasm, and always of very limited length, any one with experience in watching the germination of conidia of fungi in general can realize that they have not really put out germtubes. In the nectar, too, were a few pycnidiosporophores, but how they got there is not clear.

The flexuous hyphae of *Cronartium fusiforme* arise among the pycnidiosporophores at scattered intervals and grow out beyond the mass of pycnidiospores covering the pycnidiosporophores into the slimy nectar. They are parallel-sided, slightly flexuous, bluntly-ending, unbranched, and non-septate structures, 0.05-0.2 mm. long and about 2μ wide, thus not differing in any essential point from the flexuous hyphae of other Uredinales.

(2) *Cronartium cerebrum.* Branches of *Pinus clausa* bearing large

cankers caused by *Cronartium cerebrum* were kindly sent to me by Dr. Geo. G. Hedgcock early in March, 1940, from Ormond in Florida. On arrival in Winnipeg it could be seen that the pycnidia in the cankers had already excreted drops of orange-red nectar. The lower ends of the branches were placed in water and the specimens were covered with a bell-jar.

Two days later drops of orange nectar could be seen exuding from some of the pycnidia. Transverse sections of these pycnidia were made and then it was observed that the pycnidia resembled those of *Cronartium fusiforme* in their general structure and in possessing long, parallel-sided, slightly flexuous, bluntly-ending, unbranched non-septate flexuous hyphae.

Now that flexuous hyphae have been observed in three species of Cronartium, *Cronartium ribicola*, *C. fusiforme*, and *C. cerebrum*, there can be but little doubt that flexuous hyphae are present in species of Cronartium in general. Since we have evidence that in *C. ribicola* the flexuous hyphae are actually employed for the initiation of the sexual process, we may infer that the flexuous hyphae of Cronartium species in general are similarly employed. So far as is known all species of Cronartium produce large and active pycnidia. This fact alone suggests that the genus is heterothallic.

Melampsora lini.—*Melampsora lini*, well known as a parasite of Flax, is an autoecious, macrocylic, heterothallic Rust. That it produces pycnidia, aecidia, uredospore sori, and teleutospore sori on one and the same host was established culturally by Arthur[1] in 1907, and that it is heterothallic was determined experimentally by Miss Allen[2] in 1934. *M. lini* confines its attacks to Linaceae and, in North America, it has been recorded on thirteen Linum species.[3] In western Canada (Saskatchewan and Manitoba) it is commonly present on *Linum usitatissimum* and it has also been found on *L. lewisii* and *L. rigidum*.[4]

In 1912, Fromme,[5] on the basis of a cytological investigation, stated that the pycnidia of the Flax Rust are "without ostiolar fila-

[1]J. C. Arthur, "Cultures of Uredineae in 1906," *Journal of Mycology*, Vol. XIII, 1907, p. 201.

[2]Ruth F. Allen, "A Cytological Study of Heterothallism in Flax Rust," *Journ. Agric. Research*, Vol. XLIX, 1934, pp. 771-772.

[3]J. C. Arthur, *Manual of the Rusts in United States and Canada*, Lafayette, U.S.A., 1934, p. 57.

[4]G. R. Bisby, *The Fungi of Manitoba and Saskatchewan*, National Research Council, Ottawa, 1938, p. 63.

[5]F. D. Fromme, "Sexual Fusions and Spore Development of the Flax Rust," *Bull. Torrey Bot. Club*, Vol. XXXIX, 1912, p. 118.

ments." By "ostiolar filaments" he doubtless meant paraphyses or periphyses (slenderly conical structures tapering to a point); and, with this interpretation, his statement is correct.

In 1934, Miss Allen[1] studied the structure and the development of the pycnidia of *Melampsora lini* and discovered that each pycnidium has "paraphyses" projecting from its ostiole; but, as will appear in what follows, her "paraphyses" were in reality unrecognized flexuous hyphae.

At Winnipeg, in November, 1939, some dead Flax stems bearing teleutospore sori were collected in the field and brought into the laboratory. Thereafter, at intervals of about a week and for some six weeks, the stems were alternately wetted and dried. Late in December basidiospores began to be discharged and, by the end of that month, a single infection on a Flax seedling was obtained. On January 16, 1940, basidiospores were sown on several Flax seedlings and, a fortnight later, the leaves and stems of the inoculated plants bore rust pustules in which were a number of tiny inconspicuous orange-coloured pycnidia. The pycnidia were covered with drops of nectar and with the microscope it could be seen: (1) that from each pycnidium there were projecting numerous still living and turgid flexuous hyphae; and (2) that the flexuous hyphae were unaccompanied by periphyses.

A transverse section through a leaf revealed that the pycnidia are amphigenous and subepidermal, as described by Arthur, and that each pycnidium is flask-shaped and bulges outwards from the leaf so as to form a little epipycnidial papilla.

A small piece of Flax leaf including a rust pustule was placed on a slide, boiled for a few moments in lacto-phenol cotton-blue, and covered with a cover-glass. The flexuous hyphae, stained deep blue, could then be clearly seen against the orange background of each epipycnidial papilla (Fig. 64).

The flexuous hyphae of the pycnidia of *Melampsora lini* are slender, parallel-sided, bluntly-ending, flexuous structures, 0.7-0.15 mm. long and about 2μ wide. Many of them are unbranched but some have one or two or more branches and then they are reminiscent of the antlers of deer. The number of flexuous hyphae per pycnidium varies from about 12 to about 50.

Miss Allen[2] observed that at the apex of each young pycnidium of *Melampsora lini* there is a stoma and that through the stomatic cleft flexuous hyphae grow outwards and nectar and pycnidiospores are

---

[1]Ruth F. Allen, *loc. cit.*, pp. 769-771.   [2]Ruth F. Allen, *loc. cit.*, p. 770.

excreted.   On examining cotton-blue surface preparations I was able
to confirm Miss Allen's finding; for, in several pycnidia, I could clearly
see that the epidermis at the apex of the epipycnidial papilla was
unbroken and that the flexuous hyphae were threaded through a
central stomatic cleft (Fig. 64).   According to Miss Allen, as pycnidia
enlarge, the pressure which they exert may cause the epidermis to
rupture and, when that happens, a number of flexuous hyphae grow
outwards through the aperture.   *M. lini* is the only species of the
Uredinales in which I have seen a stomatic cleft used as an ostiole.

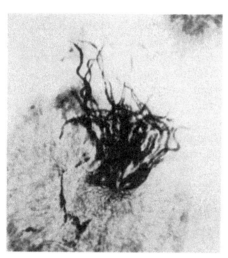

It is now known that flexuous hyphae unaccompanied by periphyses occur in the pycnidia of *Melampsora lini*, *M. abieti-capraearum*,[1] and *M. larici-capraearum*[2] and we can therefore draw the conclusion that flexuous hyphae are characteristic of species of Melampsora in general.   In systematic works the pycnidia of Melampsora might be thus described: "Pycnidia subepidermal or subcuticular; flexuous hyphae numerous, slender, occasionally branched; ostiolar trichomes absent."

FIG. 64.—*Melampsora lini*. Surface view of a pycnidium on a Flax leaf, showing numerous flexuous hyphae (unaccompanied by pointed periphyses) protruding through a stomatic cleft. Hyphae deeply stained with cotton-blue. Magnification, about 500.

Melampsorella cerastii.—
Melampsorella cerastii, in Europe and North America, causes the formation of large and conspicuous witches' brooms on Spruce and Fir trees.   On the leaves of these brooms, which are pale in colour, the fungus passes through its pycnidial and aecidial stages.

In my student days at Munich (1899-1901), Robert Hartig gave
me one of the brooms that had been formed on *Abies pectinata*, the
Silver Fir.   At that time the fungus was known only as *Aecidium*
(or *Peridermium*) *elatinum*.   Shortly thereafter, Fischer determined

[1]Lillian M. Hunter, "Morphology and Ontogeny of the Spermogonia of the
Melampsoraceae," *Journal of the Arnold Arboretum*, Vol. XVII, 1936, pp. 116-118,
142.
[2]Dorothy Ashworth, *vide infra* in this Chapter.

the alternate hosts: he[1] was successful in 1901 in sowing aecidiospores from *Abies pectinata* on *Stellaria nemorosa* and in 1902 in making the reverse culture.

Arthur and Kern[2] studied American material, describing the witches'-broom Rust on Picea as *Peridermium coloradense* and the corresponding Rust on Abies as *P. elatinum*. Successful cultures with American material were made in 1911 by Arthur,[3] who sowed aecidiospores from *Abies lasiocarpa* on *Cerastium oreophilum*, and in 1917 by Weir and Hubert,[4] who sowed aecidiospores from *Picea engelmanni* on *Stellaria longifolia* and *S. borealis*.

Arthur[5] in 1934 recognised in North America the single species *Melampsorella cerastii* (Pers.) Schroet. having its pycnidial and aecidial stage on Picea and Abies and its uredospore and teleutospore stages on Cerastium and Stellaria; but Pady,[6] after carefully comparing witches' brooms, pycnidia, aecidia, and geographical distribution patterns, has recently come to the conclusion that, in reality, *M. cerastii* consists of two distinct species, one on Picea and the other on Abies. Pady's conclusion still remains to be verified by cultural observations and, up to the present, he has not given the two species distinctive names.

As Pady[7] has pointed out, the witches' brooms on *Picea* are very large (up to 6 feet in diameter), reddish-orange in colour, devoid of a gall, having a diffuse irregular type of growth, with the branches radiating in all directions and some pendant, and with the host-leaves not greatly altered (Fig. 65); whereas, on *Abies*, the, brooms are smaller (up to 1-2 feet in diameter), yellowish, with a gall, having a very compact somewhat oval type of growth, with the branches all pointing upwards, and with the host-leaves greatly shortened and thickened.

[1]Ed. Fischer, *Zeitsch. f. Pflanzenkr.*, Bd. XI, 1901, pp. 321-343, 1902 and Vol. XII, 1902, pp. 193-202.

[2]J. C. Arthur and F. D. Kern, "North American Species of Peridermium," *Bull. Torrey Bot. Club*, Vol. XXXIII, 1906, pp. 403-438.

[3]J. C. Arthur, *Mycologia*, Vol. IV, 1912, p. 58.

[4]J. R. Weir and E. E. Hubert, "Notes on Forest Tree Rusts," *Phytopathology*, Vol. VIII, 1918, pp. 114-118.

[5]J. C. Arthur, *Manual of the Rusts in United States and Canada*, Lafayette, U.S.A., 1934, pp. 20-21.

[6]S. M. Pady: "Preliminary Observations on the Aecial Hosts of Melampsorella," *Trans. Kansas Acad. of Science*, Vol. XLIII, 1940, pp. 147-143; "Further Notes on Witches' Brooms and Substomatal Pycnia of Melampsorella," *Trans. Kansas Acad. of Science*, Vol. XLIV, 1941, pp. 190-201; and "Distribution Patterns in Melampsorella in the National Forests and Parks of the Western States," *Mycologia*, Vol. XXXIV, pp. 606-627. [And see *Mycologia*, XXXVIII, 1946, pp. 477-499.]

[7]S. M. Pady, 1940, *loc. cit.*, p. 149.

Pady[1] observed that the pycnidia on Abies are subcuticular and flattened, whereas those on Picea are subepidermal (not subcuticular as Arthur[2] had supposed), substomatal, and spherical. These dif-

Fig. 65.—Twigs from a "witches broom" on *Picea canadensis* caused by *Melampsorella cerastii* (*Peridermium coloradense* stage). Gathered at Victoria Beach, Lake Winnipeg, June 11, 1939. The fungus causes defoliation, so that the infected twigs bear young leaves only. There were pycnidia on every leaf. Photographed by A. M. Brown. Natural size.

ferences, together with others concerned with the size, colour, markings, and date of maturity of the aecidiospores and with the manner of growth of the witches' brooms, served to convince Pady that the Melampsorella on Abies is distinct from that on Picea.

[1]S. M. Pady, 1940 and 1942, *loc. cit.*
[2]J. C. Arthur, *loc. cit.*, p. 21.

In Manitoba and Saskatchewan, *Melampsorella cerastii* has been recorded by Bisby[1] and his colleagues on *Picea canadensis* and *Picea mariana* and on *Cerastium arvense*.   I, myself, have noted the witches' brooms on *Picea canadensis* on a number of trees near Victoria Beach

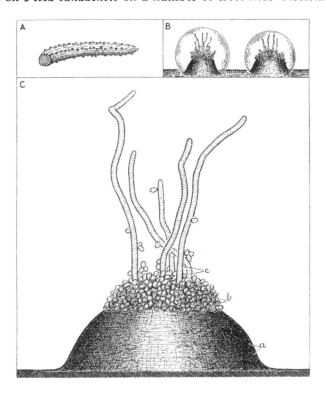

FIG. 66.—*Melampsorella cerastii* (*Peridermium coloradense* stage) on *Picea canadensis*. A, a Spruce needle with four rows of pycnidia, each pycnidium crowned with a drop of nectar.   B, two pycnidia projecting from a needle as seen with the low power of the microscope.   Drops added diagrammatically.   Flexuous hyphae are projecting into the drops.   C, a pycnidium as seen after a Spruce needle had been mounted in water: *a*, the epipycnidial papilla; *b*, a mass of pycnidiospores which cover over and hide the pycnidial ostiole; projecting beyond the ostiole and the pycnidiospores are five flexuous hyphae. Periphyses are absent. Magnification: A, 4; B, 140; C, 700.

on Lake Winnipeg.   In the early summer, when the numerous pycnidia on the leaves are excreting nectar, the brooms have so strong an odour that they can be detected 10-20 feet away by scent alone.   So far, in Manitoba and Saskatchewan, the fungus has not been recorded on the only species of Abies growing in this region, namely, *Abies balsamea*,

[1]G. R. Bisby, *loc. cit.*, p. 63.

the Balsam Fir. However, in the forest on the east shore of Lake Winnipeg, I saw a few Balsam-Fir trees with witches' brooms which may well have been caused by *Melampsorella cerastii*. Unfortunately, these brooms were inaccessible.

Twigs bearing leaves infected with *Melampsorella cerastii* were gathered by me from two large witches' brooms on White Spruce, *Picea canadensis*, at Victoria Beach, Lake Winnipeg, on June 11, 1939. The fungus causes local defoliation of its host so that, in the spring, the brooms bear new leaves only (Fig. 65). All the broom leaves are infected. Each leaf or needle is four-sided and has four rows of stomata. The pycnidia on a leaf are produced in four rows corresponding in position with the four rows of stomata (Fig. 66, A). Each pycnidium, when active, has a reddish appearance owing to the fact that its ostiole is covered by a little drop of nectar enclosing a mass of orange-yellow pycnidiospores.

FIG. 67.—*Melampsorella cerastii*. Nine long flexuous hyphae protruding from a pycnidium of a Spruce needle mounted in water. Mag., about 250.

Some of the Spruce needles infected with *Melampsorella cerastii* were cut away from the twigs, placed whole in water on a slide, covered with a cover-glass, and examined with the microscope. Even with low-power magnification it could then be seen that the pycnidia of *M. cerastii* are each provided with from one to nine flexuous hyphae and that these hyphae project from the ostioles and extend far beyond the masses of pycnidiospores (Fig. 66, C, and Fig. 67).

The genus Melampsorella resembles Melampsora, Milesia, and other genera of the Melampsoraceae in that its pycnidia have no periphyses or paraphyses; and the absence of these trichomes from the pycnidia of *Melampsorella cerastii* is a factor which greatly facilitates the detection of flexuous hyphae in that species.

Milesia.—On July 28, 1938, near Ottawa, I collected leaves of *Abies balsamea* bearing pycnidia and aecidia of *Milesia intermedia*. The aecidia—as in other Milesiae characteristically white—were cylindrical and had already ruptured at their apices. With the help of a binocular microscope the tiny mouths of the pycnidia on the under sides of the leaves could be seen, but nothing protruded from them.

In the genus Milesia ostiolar trichomes are lacking,[1] so that if flexuous hyphae had been present in the specimens of *M. intermedia* under investigation they would have been quickly detected. Since flexuous hyphae have actually been observed in eleven other species of Milesia (in five by Miss Hunter[2] and in six by Kamei,[3] see p. 83), it is most probable that they are present also in *M. intermedia*. Doubtless, my material had been gathered several weeks too late, long after the diploidisation process had been initiated and the flexuous hyphae, if once present, had gelatinised and disappeared.

Gymnoconia peckiana. — Kursanov,[4] in 1910, described the development, mature structure, and nuclear condition of a pycnidium of *Gymnoconia peckiana* as seen in microtome sections (Fig. 68); but he failed to observe the flexuous hyphae, and he left unmentioned certain other pycnidial characteristics which are ecologically significant so far as insect visits are concerned, namely, scent, colour, and the excretion of nectar. My own observations were made on living material cut with a hand-razor.

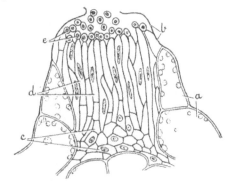

FIG. 68.—A pycnidium of *Gymnoconia peckiana*: *a*, epidermis; *b*, cuticle; *c* mycelium *d* pycnidiosporophores, and *e* pycnidiospores, all uninucleate. From Kursanov's *Zur Sexualität der Rostpilze* (1910), copied and labelled by A. H. R. Buller.

In *Gymnoconia peckiana*,[5] which is systemic and attacks Rubus species, the pycnidia are peculiar in that they are seated on short cylindrical multicellular pedicels made up of epidermal and sub-

[1]J. C. Arthur, *Manual of the Rusts in United States and Canada*, Lafayette, U.S.A., 1934. On p. 6, he says: "Pycnia without paraphyses."

[2]Lillian M. Hunter, "Morphology and Ontogeny of the Spermogonia of the Melampsoraceae," *Journal of the Arnold Arboretum*, Vol. XVI, 1936, pp. 115-152, with seven Plates.

[3]S. Kamei, "Studies on the Cultural Experiments of the Fern Rusts in Japan," *Journal of the Faculty of Agriculture, Hokkaido Imperial University*, Vol. XLVII, 1940, pp. 138-139.

[4]L. Kursanov, "Zur Sexualität der Rostpilze," *Zeitschrift f. Bot.*, Bd. II, 1910, pp. 83-85, Taf. I, Figs. 3-6.

[5]For identifying this species I am indebted to Miss Ruth Remsberg of Cornell University. She found that the aecidiospores produced long germ-tubes as happens in the long-cycled *Gymnoconia peckiana*, and not basidia, as happens in the short-cycled *Kunkelia nitens* with which *G. peckiana* might be confounded. Both of these species cause Orange Rust of Brambles.

epidermal host-cells (Figs. 68 and 69). They are scattered over both sides of the infected leaves and, at first glance, they look like little brown warts. Under moist conditions, each pycnidium is crowned by a drop of nectar (Fig. 69). The nectar is scented. The pycnidio-sporophores are red and they are arranged parallel to one another (Figs. 68 and 69), as in a Phragmidium. There are no periphyses or paraphyses. The exterior of each pycnidium is partly enclosed by a dome-shaped extension of the epidermal cuticle, and the apical ostiole

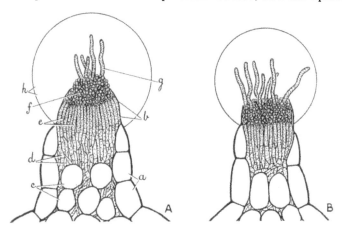

FIG. 69.—*Gymnoconia peckiana* on a Rubus species. A and B, two pycnidia at the ends of papillae formed by the host tissues. Drawn free-hand and shown in optical section: *a*, epidermis; *b*, cuticle which has been stretched and broken and now forms the pycnidial ostiole; *c*, internal host cells represented diagram-matically; *d*, mycelium; *e*, pycnidiosporophores; *f*, a gelatinous mass of pycnidio-spores; *g*, flexuous hyphae; *h*, a drop of nectar. Magnification, undetermined, over 250.

is a broad opening in the cuticle. The pycnidiospores form a gelatinous mass in the inside of, and on the exterior of, the pycnidium. Beyond the mass of pycnidiospores and extending into the drop of nectar are several, short but fairly stout, colourless, flexuous hyphae. In one pycnidium I saw five flexuous hyphae and in another seven (Fig. 69).

Gymnosporangium.—In the species of Gymnosporangium with roestelioid aecidia, the development and lacerate dehiscence of the aecidia and the powdery dispersal of the aecidiospores will be treated in Chapter XI. Here we shall consider the structure and function of the pycnidia of Gymnosporangium and, in particular, we shall enquire whether or not these organs are provided with flexuous hyphae.

With the help of the nectar-mixing technique it has been shown that *Gymnosporangium clavipes*, *G. globosum*, *G. haraeanum*, and

G. juniperi-virginianae are all heterothallic.[1] This in itself indicates that, in Gymnosporangium species in general, flexuous hyphae are present in the pycnidia and that these hyphae fuse with pycnidiospores of opposite sex, just as they do in *Puccinia graminis* and *P. helianthi*.

Here it will be shown that flexuous hyphae are actually present in *Gymnosporangium clavariiforme*, *G. clavipes*, *G. juniperi-virginianae*, and *G. juvenescens*.

The pycnidia of Gymnosporangium resemble those of Puccinia and Uromyces and differ from those of Gymnoconia and the Melampsoraceae in that, in addition to flexuous hyphae, they are provided with numerous straight, stiff, tapering, pointed ostiolar trichomes.

Whereas in Puccinia the ostiolar trichomes and flexuous hyphae arise on the wall of the pycnidium close to the ostiole, in Gymnosporangium the trichomes, and doubtless also the flexuous hyphae, arise at points scattered over the wall of the pycnidium (Fig. 41, p. 137).

(1) *Gymnosporangium clavipes and G. clavariiforme.*—At Cornell University (at Ithaca, New York), in the first days of June 1937, I examined some very young pycnidial pustules of a Gymnosporangium abundantly present on the leaves of a Hawthorn (Crataegus sp.). The ostiolar trichomes or paraphyses were just breaking through, or had just broken through, the host epidermis, and no flexuous hyphae could be seen among them. Unfortunately, I could not prolong my visit to examine older material. About ten days later, Dr. H. H. Whetzel kindly sent to me in London, England, some older material. The leaf pustules appeared to have made no further progress in development and no structures were found in the pycnidia that could be definitely regarded as flexuous hyphae. On July 21, Dr. Whetzel sent to me further material including Hawthorn leaves and fruits. The numerous pustules on the leaves were all sterile, whereas the fruits all bore aecidia. These aecidia were numerous, cylindrical and 1.5-3 mm. high, and their peridia were lacerate and very white. On inspection they were at once identified by Dr. B. O. Dodge and Dr. P. R. Miller as belonging to *Gymnosporangium clavipes* (*G. germinale*), and a later examination made by myself, which included a microscopic study of the shape, size, and wall-markings of the peridial cells,[2] con-

---

[1]For a list of heterothallic Rusts with citations from the literature *vide supra*, p. 85.

[2]Illustrations of the peridial cells of *Gymnosporangium clavipes* (*G. germinale*) and allied species have been given by F. D. Kern ("A Biologic and Taxonomic Study of the Genus Gymnosporangium," *Bulletin of the New York Bot. Garden*, Vol. VII, 1909-1911, issued separately, 1911) and D. E. Bliss ("The Pathogenicity and Seasonal Development of Gymnosporangium in Iowa," *Iowa Agric. Experiment Station*, Research Bulletin No. CLXVI, 1933).

vinced me that this identification was correct. Possibly, if I had
examined the pycnidia on the fruits, instead of those on the leaves, I
should have found flexuous hyphae in them. My failure to find flexuous hyphae in a Gymnosporangium at the first attempt stimulated me
to make further investigations.

In the spring of 1939, I visited Minaki in western Ontario and
there, on the shore of the Winnipeg River, near Holst Point, found
bushes of *Juniperus communis* of which some were infected with
Gymnosporangium clavipes and others with *G. clavariiforme* (Fig. 70).
Twigs bearing the gelatinous teleutospore sori of both these species
were taken to Winnipeg and, as already recorded,[1] they were used to

FIG. 70.—Teleutospore sori of *Gymnosporangium clavariiforme* projecting from a
branch of *Juniperus communis*. Nat. size.

inoculate Juneberry bushes (*Amelanchier alnifolia*) in a wood. The
pustules which appeared in large numbers on the inoculated leaves
were examined, and flexuous hyphae were found in the pycnidia of
both *Gymnosporangium clavipes* (Fig. 80, p. 239) and *G. clavariiforme*.
Subsequently, as will be recorded in another Chapter, the *G. clavipes*
material was used for the study of fusions formed between the flexuous
hyphae and pycnidiospores of opposite sex.

At Victoria Beach, Lake Winnipeg, where bushes of *Juniperus
communis* bearing the teleutospore sori of *Gymnosporangium clavipes*
grow close to and intermingled with bushes of *Amelanchier alnifolia*,
haploid pustules of *Gymnosporangium clavipes* are formed on the
leaves, twigs, and fruits of *Amelanchier alnifolia*. Later in the year,
however, one finds the roestelioid aecidia on the twigs and fruits of
the host, but not on the leaves. Why the pustules should remain
sterile on Amelanchier leaves, and apparently also on Crataegus
leaves, and yet be fertile on the fruits and twigs is an unsolved problem.

[1]*Vide supra*, p. 92.

(2) *Gymnosporangium juvenescens and G. corniculans*. In the spring of 1938, some living and some pickled leaves of *Amelanchier alnifolia* bearing *Gymnosporangium juvenescens* were kindly sent to me at Winnipeg by Dr. W. P. Fraser from Saskatoon. Nectar was present above the living pustules, but a prolonged investigation of the material, made with hand-sections and many lacto-phenol cotton-blue surface preparations, failed to reveal any structures that I could definitely regard as flexuous hyphae. The ostiolar trichomes (paraphyses) of *G. juvenescens* are of the usual type, straight, stiff-looking, and tapering to a point. A few shorter, straight, slightly tapering hyphae with blunt ends were seen. These elements may have been young flexuous hyphae, but they could also be regarded as partially developed trichomes still growing to a point.

My first attempts to find flexuous hyphae in *Gymnosporangium juvenescens* having failed, nothing remained but to obtain better material and to make further investigations.

In the spring of 1939, I visited Pine Ridge, near Winnipeg, and there, on *Juniperus horizontalis*, found the teleutospore stages of *Gymnosporangium juvenescens* and *G. corniculans*. *G. juvenescens* causes the formation of conspicuous upright witches brooms and *G. corniculans* globose galls from which in wet weather cylindrical-acuminate teleutospore sori are put forth. Both these species pass to *Amelanchier alnifolia*, and on this shrub near to patches of the Creeping Juniper I found young rust pustules. Some Amelanchier twigs with infected leaves were taken to Winnipeg and placed in water under a bell-jar, whereupon the pustules excreted an abundance of nectar. A few days after the material had been collected I examined the pustules in hand-sections and in surface preparations and found that flexuous hyphae were present in the pycnidia; but to which of the two species of Gymnosporangium the pycnidia belonged could not be decided.

Pine Ridge is upwards of ten miles to the north of Winnipeg and about seventeen miles north of the Dominion Rust Laboratory, and many acres of it are more or less covered with *Juniperus horizontalis* bearing both *Gymnosporangium juvenescens* and *G. corniculans*. Every spring scattered rust pustules are found on Amelanchier bushes around Winnipeg and at the Rust Laboratory, and it seems most likely that the source of the inoculum is located at Pine Ridge. I procured some infected Amelanchier twigs from bushes growing near the laboratory, each with a leaf bearing a single pustule. The twigs were placed in water under a bell-jar and left for two or three days. At the end of this time, each pustule was covered over by a large drop of nectar. On examining these pustules in cross-section and in surface preparations made with lacto-phenol and cotton-blue, in many of the pycnidia flexuous hyphae were discovered (Fig. 71). These hyphae were stout, straighter than those of *Puccinia graminis*, but they projected beyond the masses of pycnidiospores and into the drop of nectar in the usual way.

Twigs of *Juniperus horizontalis* of which some bore teleutospore sori of *Gymnosporangium juvenescens* and others teleutospore sori of *G. corniculans* were gathered at Pine Ridge, taken to the Rust Labora-

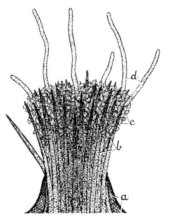

FIG. 71.—Gymnosporangium sp. (*G. juvenescens*?) on *Amelanchier alnifolia*. Upper part of a pycnidium: *a*, epidermis of host leaf; *b*, periphyses; *c*, pycnidiospores; *d*, flexuous hyphae. Mag., 600.

tory, and used to inoculate pot-plants of *Amelanchier alnifolia*. Owing to the poor growth of the host-plants, the pustules which appeared on the leaves were very small. Those of *Gymnosporangium juvenescens* were examined by means of surface preparations made with lacto-phenol and cotton-blue, and it was then found that flexuous hyphae were present in the pycnidia. The pycnidia of *G. corniculans* dried up soon after they appeared and could not therefore be advantageously investigated.

(3) *Gymnosporangium juniperi-virginianae*.—Hanna,[1] in 1929, procured some galls of *Gymnosporangium juniperi-virginianae* from the United States and with them, at Winnipeg, he infected the leaves of Apple seedlings. He observed that simple pustules, each derived from a single basidiospore, did not form aecidia, but that aecidia appeared on a number of compound pustules formed by the union of two or more mycelia. He thus obtained evidence that *G. juniperi-virginianae* is heterothallic.[2]

Miller,[3] in 1932, employing the nectar-mixing technique, also found that *Gymnosporangium juniperi-virginianae* is heterothallic. He carried out experiments on: (1) apple leaves on trees naturally infected in an orchard, (2) leaves of apple seedlings that had been atomised

[1] W. F. Hanna, personal communication.

[2] Of 32 simple pustules only one produced aecidia. Of 12 compound pustules 9 produced aecidia.

[3] Paul R. Miller, "Pathogenicity of Three Red-Cedar Rusts that Occur on Apple," *Phytopathology*, Vol. XXII, 1932, pp. 734-735.

with a suspension of basidiospores in a greenhouse, and (3) isolated
apple leaves inoculated with single basidiospores and kept floating on
6 per cent. sucrose solution in closed Petri dishes.[1] Even with the
leaves of orchard trees exposed to the visits of insects it was found
that, after the nectar had been mixed, the percentage of lesions with
aecidia was increased from 56 (in controls) to 90; and with methods
(2) and (3) the results obtained conclusively proved that *G. juniperi-
virginianae* is heterothallic.

Miller thus describes his experiments with *Gymnosporangium
juniperi-virginianae* and potted Apple seedlings: "A very dilute spore
suspension of basidiospores was atomized on the leaves of potted
seedlings, May 15. On certain leaves only a single lesion resulted.
These leaves bearing single lesions were bagged separately and, at the
proper time, which was about June 8, the pycnial exudate on 12 of
those isolated pustules was mixed. All of these lesions had produced
aecia bearing aecidiospores by July 30. On 7 isolated lesions the exudate was not mixed and no aecia were formed." He thus describes
one of his experiments on isolated Apple leaves in Petri dishes: "Drops
of a very dilute spore suspension were put on slides and examined
under a microscope. When a drop was found containing less than
5 basidiospores, an attempt was made to withdraw one spore by the
aid of a bulb pipette attached to the mechanical stage. The spores
remaining in the drop were counted and, if one less spore remained, it
was assumed that one spore was in the pipette. The contents of the
pipette were released on a leaf and by this method of inoculation a
lesion was obtained on each of 8 leaves floated on sucrose solution.
These were considered as monosporidial lesions. The pycnial exudate

---

[1]Miller (*loc. cit.*, p. 732) remarks: "It was found early in the work that the time
of day that the leaves were removed from the tree was of great importance. Leaves
removed from the tree in the late afternoon remained viable much longer than those
taken in the morning. This may be accounted for by the probability that leaves
taken in the late afternoon contain much starch, while the leaves taken in the
morning are low in starch." It was found (p. 739) that apple leaves removed from
the tree at the end of the day remained "alive on a 6 per cent. sucrose solution in
Petri dishes for two months which was sufficient time for aecidia to develop."
Miller also states that in his later works: "The leaf petioles were inserted
through holes in thin layers of cork that were floated in the nutrient solution. The
leaves were first washed in running water. The nutrient solution was changed every
5 days and dead portions of the leaves that were noticed were cut off and removed."
In susceptibility tests for different varieties of Apple: "The leaves were inoculated
in the dishes by atomising with a basidiospore suspension." Dr. Miller informed
me in conversation that the only part of an isolated leaf to touch the sucrose solution
in a Petri dish was the petiole.

of 4 of these was mixed and each of these lesions produced aecia, while
no aecia were formed by the other 4 lesions where the exudate was
not touched."

Hanna and Miller did not look for flexuous hyphae in the pycnidia
of *Gymnosporangium juniperi-virginianae* because, when they made
their cultures, flexuous hyphae in the Rust Fungi had not yet been
discovered.   To find out whether or not *G. juniperi-virginianae* has
flexuous hyphae, and thus to attempt to fill up a gap in our knowledge
of this well-known species, in the spring of 1939 I procured some
Cedar-apple galls from the United States, used them for infecting the
leaves of Delicious Apple seedlings, and then examined the pycnidial
pustules in the usual way.   On some of the more resistant seedlings
the fungus grew badly, producing pycnidia, a few pycnidiospores, and
little or no nectar; but, on other more susceptible seedlings, development was normal and, under moist conditions, abundant nectar was
excreted.   Hand-sections of some pustules which were about a month
old and which had been covered with nectar for about a week, when
mounted in water and examined whilst still alive, revealed somewhat
pale but quite typical flexuous hyphae stretching out far beyond the
very red and sharply pointed periphyses; and some of these preparations were used to demonstrate the existence of flexuous hyphae in
the genus Gymnosporangium to members of the laboratory staff.

(4) *Gymnosporangium globosum.*—Miller,[1] making use of the nectar-
mixing technique, has proved that *Gymnosporangium globosum* on
Hawthorn leaves is heterothallic.   He made two sets of experiments,
one with 50 leaves and another with 200 leaves.   The details of the
experiment with 200 leaves are as follows.   Two hundred leaves, each
having an immature single pycnidial lesion caused by *G. globosum*,
were taken from Hawthorn trees on June 4, 1931, and were floated in
Petri dishes on a 6 per cent. sucrose solution.   At the time of exudation, the pycnidial exudate was transferred from one leaf to another
on 100 leaves, so that each lesion received the exudate from several
others.   The pycnidial exudate on the other 100 leaves was not
touched.   The result was as follows: "Of the 100 leaves on which the
exudate was mixed, 92 per cent. had produced aecia on August 15,
whereas only 3 per cent. of the remaining leaves bearing the unmixed
exudate bore aecia."   It may here be suggested that the three lesions
which bore aecidia in the control leaves may have developed aecidia
either (1) because the lesions were not simple in origin but arose from
two basidiospores, one (+) and the other (−), which happened to

[1]Paul R. Miller, "Pathogenicity of Three Red-Cedar Rusts that Occur on
Apple," *Phytopathology*, Vol. XXII, 1932, p. 735.

fall on the leaf very close to one another or (2) because the lesions had been visited by nectar-mixing insects before the leaves were gathered.

Gymnosporangium golbosum does not occur in Manitoba and I have not yet had the opportunity of examining the pycnidia in the living state. However, Miller's demonstration that in this species nectar-mixing leads to the initiation of the sexual process strongly suggests that *G. globosum*, just like *G. juniperi-virginianae*, is provided with flexuous hyphae.

The results of my study of the pycnidia of *Gymnosporangium juvenescens*, *G. clavariiforme*, *G. clavipes*, and *G. juniperi-virginianae* provide evidence sufficient to enable us to conclude that flexuous hyphae are just as regularly present in the pycnidia of species of Gymnosporangium as they are in species of Puccinia.

Phragmidium.—The genus Phragmidium is autoecious and confined to the Rosaceae. Its teleutospores are freely spaced in the teleutospore sori and, unlike those of Puccinia, Uromyces, and Gymnosporangium, are usually multicellular. In England, the species of Phragmidium that attack Brambles and Rose-bushes are often encountered by the field-mycologist and, as long ago as 1665, Robert Hooke[1] in his famous *Micrographia* described and illustrated some Phragmidium teleutospores which he regarded as "a plant growing in the blighted or yellow specks of Damask-rose leaves, Bramble leaves and some other kind of leaves." Most of the species of Phragmidium are *eu*-forms having pycnidia, aecidia, uredospore sori, and teleutospore sori. Throughout the

[1]Robert Hooke, *Micrographia*, 1665, pp. 121-127, Fig. 2, Schema XII.

Fig. 72.—*Phragmidium speciosum* on twigs of *Rosa blanda*, showing teleutospore sori. Collected at Winnipeg, Dec. 7, 1939, and photographed by A. M. Brown. Natural size.

genus the aecidia are caeomoid in form, the pycnidia are subcuticular, and pycnidial periphyses are absent.

(1) *Phragmidium speciosum.* This species is injurious to wild and cultivated Rose-bushes and, in the autumn and winter, the black irregular teleutospore sori on the stems have the appearance shown in Fig. 72. *Phragmidium speciosum* is remarkable as a Phragmidium in that (1) it has no uredospores, (2) its aecidiospores do not infect Rose leaves,[1] and (3) the teleutospore sori are formed not as a result of aecidiospore infection but on the mycelium in the old aecidiospore pustules.[2] The propagation of the fungus on Rose bushes appears to be due solely to the natural sowing of the basidiospores produced in the spring when the teleutospores germinate and, in the light of our present knowledge, the aecidiospores would seem to be nothing more than vestigial structures, once of use to the ancestors of *P. speciosum*, but now performing no function whatsoever. *P. speciosum* is heterothallic.[3]

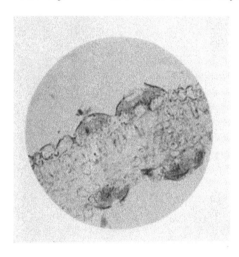

Fig. 73.—Subcuticular pycnidia of *Phragmidium speciosum* on upper and lower sides of a leaf of *Rosa blanda*. Photographed by A. M. Brown. Mag., 150.

Mr. A. M. Brown informed me that, in 1936, he cut hand-sections of living pustules of *Phragmidium speciosum* on *Rosa blanda* and observed that flexuous hyphae were present in several of the pycnidia: the hyphae were rather stout, they projected beyond the pycnidiospores, and there were from three to five in each pycnidium. Brown also made some microtome sections of his material (Fig. 73) and in some of them parts of the flexuous hyphae could still be recognised. He kindly placed these slides at my disposal and an examination of them permitted me to confirm his finding. Subsequently, in the first week of July, 1938, Dr. H. J. Brodie brought me a few living pycnidial pustules of *P. speciosum* and, on examining them in hand-sections, I saw flexuous hyphae projecting from the pycnidia (Fig. 74). In

[1]A. M. Brown, "The Sexual Behaviour of Several Plant Rusts," *Canadian Journal of Research*, Section C, Vol. XVIII, 1940, pp. 20-21.
    [2]*Ibid.*   [3]*Ibid.*

P. speciosum, the flexuous hyphae are not numerous; but, as they are relatively thick, project well beyond the gelatinous masses of pycnidiospores, and are unaccompanied by ostiolar trichomes (paraphyses or periphyses), they can be readily observed.

(2) *Phragmidium potentillae.* In May, 1939, I collected *Phragmidium potentillae* on *Potentilla bipinnatifida* (the Plains Cinquefoil) and, on examining hand-sections of the rust pustules, observed in the pycnidia a number of flexuous hyphae resembling those of *Phragmidium speciosum*.

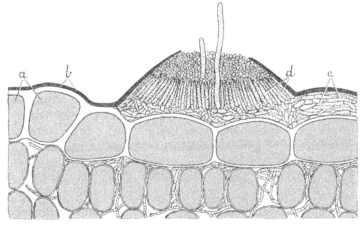

Fig. 74.—*Phragmidium speciosum* on *Rosa blanda*. Part of a vertical section through a leaf showing a pycnidium with two flexuous hypha: *a*, the epidermis; *b*, the cuticle represented as if stained; *c*, mycelium between the cuticle and the epidermal cells; *d*, a palisade layer of pycnidiosporophores above which is a mass of pycnidiospores. In the centre of the pycnidium are two flexuous hyphae. Periphyses are absent. The drop of nectar which covered the pycnidium has not been represented. Magnification, 600.

The finding of flexuous hyphae in *Phragmidium speciosum* and *P. potentillae* indicates that flexuous hyphae are present in the pycnidia of Phragmidium species in general. That flexuous hyphae were not discovered by Blackman[1] in *P. violaceum* in 1904 and by Christman[2] in *P. speciosum* in 1905 was doubtless due to the nature of the technique used in making the preparations. This technique, although admirable for studying cell-sections and nuclei, was unsuited for revealing hyphal organs projecting from the pycnidial ostioles and extending in three directions of space.

[1]V. H. Blackman, "On the Fertilization, Alternation of Generations, and General Cytology of the Uredineae," *Annals of Botany*, Vol. XVIII, 1904, pp. 323-373.

[2]A. H. Christman, "Sexual Reproduction in the Rusts," *Botanical Gazette*, Vol. XXXIX, 1905, pp. 269-274.

Tranzschelia pruni-spinosae.—*Tranzschelia pruni-spinosae* forms pycnidia and aecidia on certain Ranunculaceae and uredospores and teleutospores on certain Amygdalaceae. In December, 1942, at Cornell University, Mr. J. S. Niederhauser kindly showed me some cross-sections of pycnial pustules of *T. pruni-spinosae* borne on leaves of *Hepatica acutiloba*, in which, as he supposed, flexuous hyphae could be clearly seen. The sections had been cut by hand from living leaves and then stained with cotton-blue. On examining the sections, it was at once apparent that a pycnidium of *Tranzschelia pruni-spinosae* resembles that of *Phragmidium speciosum* in being subcuticular, in lacking periphyses and paraphyses, and in having flexuous hyphae projecting among the pycnidiospores outwards to some distance from the pycnidial ostiole. The number of flexuous hyphae seen projecting from the ostiole of a single pycnidium varied from three to eight.

Puccinia and Uromyces.—In the pycnidia of nineteen species of Puccinia and Uromyces whose names are given in the list on page 189 I have observed flexuous hyphae clearly distinct from the slenderly conical pointed ostiolar trichomes (periphyses). Illustrations of both these kinds of cell-organs are shown: for *Puccinia graminis* in Figs. 40 (p. 130) and 60 (p. 179); for *P. helianthi* in Figs. 58 (p. 176), 75 (p. 218), and 78 (p. 236); for *P. minussensis* in Fig. 57 (p. 175); and for *P. coronata avenae* in Fig. 79 (p. 237).

By reference to the list on page 189 it will be seen that flexuous hyphae have been found in three varieties of *Puccinia caricis*, two varieties of *P. graminis*, and two varieties of *P. coronata*.

Puccinia coronata elaeagni is an exceptional variety of *P. coronata* in that its pycnidia are much reduced in size, do not excrete nectar, and develop neither pycnidiospores nor flexuous hyphae. These vestigial pycnidia will be treated of more fully in a later Chapter.

My observations on the nineteen species of Puccinia and Uromyces listed indicate conclusively that flexuous hyphae are present in the pycnidia of Puccinia and Uromyces species in general.

In studying the pycnidia of certain species of Puccinia and Uromyces, Lamb and Miss Allen failed to find flexuous hyphae and Savile was unable to distinguish clearly between flexuous hyphae and periphyses. These three workers all employed the microtome method in making their preparations, and their negative results may well have been due to this fact.

The difference in the effectiveness of hand-sections of living material and microtome sections of dead material in respect to observing flexuous hyphae may be illustrated by reference to *Puccinia coronata avenae.* In 1932 Miss Allen,[1] with the help of microtome sections, made a minute study of the haploid pustules of this fungus; but, although she saw periphyses in the pycnidia, she failed to observe the flexuous hyphae. In 1937 I, too, investigated the pycnidia of *P. coronata avenae*, but with hand-sections; and, in these living preparations, I had no difficulty in finding the structures that Miss Allen had overlooked (Fig. 79, p. 237).

Lamb[2] in 1935, using the nectar-mixing technique, found that *Puccinia phragmitis* is heterothallic; but, on examining the pycnidia, although he observed numerous pointed periphyses, he failed to discover any flexuous hyphae. Since in *P. graminis, P. helianthi, P. coronata avenae,* and *Gymnosporangium clavipes* it is known that the pycnidiospores fuse with flexuous hyphae of opposite sex and not with the pointed periphyses, it seems probable that Lamb's failure to find flexuous hyphae in *Puccinia phragmitis* was due to his use of the microtome.

When examining a Rust Fungus to find out whether or not it has flexuous hyphae, it is important that the pycnidial pustules should be neither too young nor too old. In a very young pycnidium, one may see ostiolar trichomes (periphyses or paraphyses) and not flexuous hyphae because the flexuous hyphae, as yet, have not had time to grow out beyond the trichomes; and, in a very old pycnidium, still persisting after aecidia have been formed, one may see trichomes and not flexuous hyphae because the flexuous hyphae—unlike the trichomes—have become gelatinous and have disappeared. In searching for flexuous hyphae one should always choose pustules in which the nectar is liquid and well developed and which show no signs of aecidia. It is haploid and not diploidised pustules which should be examined.

An example of a failure to find flexuous hyphae in a species of Uromyces on account of the advanced condition of the material may here be cited. Late in May, 1937, at the University of Minnesota, all the available pustules of *Uromyces perigynius* on *Rudbeckia laciniata* had aecidia forming or already formed on their under sides. In hand-sections of the pustules periphyses were seen, but no flexuous hyphae. On account of the pustules all having been diploidised, the flexuous hyphae, if once present in the pycnidia, had gelatinised and disappeared.

[1]Ruth F. Allen, "A Cytological Study of Heterothallism in *Puccinia coronata*," *Journ. Agric. Research*, Vol. XLV, 1932, pp. 513-541.

[2]I. M. Lamb, "The Initiation of the Dikaryophase in *Puccinia Phragmitis* (Schum.) Körn," *Annals of Botany*, Vol. XLIX, 1935, pp. 403-438.

In 1939, Savile,[1] in giving an account of nuclear structure and nuclear behaviour in species of the Uredinales stated that in *Puccinia sorghi*, *Uromyces fabae*, *U. hyperici*, and *U. lespedezae-procumbentis*, there are *ostiolar filaments* projecting from the mouths of the pycnidia, but maintained that these filaments cannot be divided into two distinct categories, slenderly conical pointed periphyses and longer cylindrical bluntly-ending flexuous hyphae. Said he: "In all the species studied, the writer found that, if all the hyphae in a central section were drawn, there was nearly always a complete gradation from the typical paraphysis (Fig. 70) at the edge to the blunt and often branched hypha toward the centre. It is not possible to divide them into the two clearly separate types shown by Buller[2] (1938) in his sketch of the pycnia of *Puccinia graminis*." As a result of finding easily distinguishable ostiolar trichomes and flexuous hyphae in *Puccinia sorghi* (one of Savile's species) and in twenty-two other species of Rust Fungi included in the genera Puccinia, Uromyces, and Gymnosporangium, I cannot help but feel that Savile's failure to find that flexuous hyphae and ostiolar trichomes are distinct morphological structures and not intergrading ones was due to his faulty technique. My own investigations were made with thick hand-sections of living material and his with dead material, fixed, stained, and cut into thin sections with the help of a microtome. The microtome method is most unsuitable for making out the difference between periphyses and flexuous hyphae because: (1) both periphyses and flexuous hyphae, when being pickled, lose their turgidity and often become altered in form; (2) the flexuous hyphae, when fully grown, are relatively long; and (3) the knife cuts up the flexuous hyphae into small pieces, so that long intact flexuous hyphae are but rarely, if ever, seen in the preparations. Moreover, Savile does not tell us whether he sliced up very young pycnidia in which the flexuous hyphae are quite short and still developing, or older pycnidia in which the hyphae have grown to their full length, or still older pycnidia in which the flexuous hyphae have disappeared. It seems probable that, when the pycnidia of the three species of Uromyces investigated by Savile come to be further studied in the living condition and by the methods described in this Volume, just as with *Puccinia sorghi* the two categories of ostiolar filaments will be found to be readily distinguishable.

[1]D. B. O. Savile, "Nuclear Structure and Behavior in Species of the Uredinales," *American Journal of Botany*, Vol. XXVI, 1939, p. 598.

[2]A. H. R. Buller, "Fusions between Flexuous Hyphae and Pycnidiospores in *Puccinia graminis*," *Nature*, Vol. CXLI, 1938, pp. 33-34.

Miss Ashworth's Observations on Melampsoridium betulinum and Melampsora larici-capraearum.—Shortly after Craigie announced his discovery of flexuous hyphae in *Puccinia helianthi* and *P. graminis*, the late Dorothy Ashworth looked for and observed flexuous hyphae in two Rusts that occur in England and have their pycnidial stage on *Larix europaea*, the Larch. An account of this work, hitherto unpublished, will now be given.[1]

(1) *Melampsoridium betulinum.* The pycnidia are subcuticular, and it was found convenient to study them in epidermal strips torn away from the infected Larch leaves. After removal, the strips were fixed in 70 per cent. alcohol and mounted in cotton-blue dissolved in glycerine or lactic acid.

In the pycnidia of material twelve to thirteen days old long bristle-like hyphae, corresponding to Craigie's flexuous hyphae, were frequently found protruding far beyond the pycnidiosporophores. These hyphae were non-septate and uninucleate and, usually, they could be traced to the matrix of tissue below the pycnidium. Their number in a pycnidium ranged from seven to thirty-four.

Examination of fixed and critically stained material supported the observations just recorded. The bristle-like hyphae (flexuous hyphae) make their appearance at the time the aecidial primordia are beginning to develop in the substomatal spaces; but they are fugitive, and they were not seen in material older than fourteen days. In their early stages the hyphae appear to be slightly stouter than the pycnidiosporophores, and they can also be distinguished from those structures by their non-tapering tip. At the top of a mature pycnidium the tuft of flexuous hyphae can be seen with the naked eye. The pycnidiospores are formed in large numbers before the flexuous hyphae appear, and the hyphae grow up through the spore-film. The hyphae can be distinguished from the pointed periphyses of a Puccinia by (1) their occurrence at the centre of the pycnidium rather than at the periphery and (2) by their being fugitive.

(2) *Melampsora larici-capraearum.* A sharp watch was kept on material of suitable age for the occurrence of hyphae in pycnidia comparable with the bristle-like flexuous hyphae seen in *Melampsoridium betulinum*, and such hyphae were found in material that was from nine to thirteen days old. These hyphae are more delicate in form

[1]Miss Ashworth, in 1938, very kindly communicated to me the results of her work embodied in a MS., and from that MS. the account given above has been drawn up. For her terms spermogonium, spermatia, spermatiophores, and paraphyses I have substituted the terms pycnidium, pycnidiospores, pycnidiosporophores and periphyses.

than those of *M. betulinum*; but, as in that species, they are non-septate and they protrude a considerable distance beyond the tips of the pycnidiosporophores. In younger pycnidia the hyphae are rather short, the cytoplasmic cell-contents are dense, and the number of the hyphae is about five or six; but, in older pycnidia, the hyphae are much longer, the cell-contents are less dense, and the number of the hyphae is much increased.

It thus appears that the flexuous hyphae of *Melampsoridium betulinum* and *Melampsora larici-capraearum* observed by Miss Ashworth resemble the flexuous hyphae observed by Miss Hunter and by myself in other Melampsoraceae.

Conclusion.—From the observations that have been made by Craigie, Pierson, Miss Hunter, Kamei, Olive, Miss Ashworth, and myself flexuous hyphae are now known to be present in fifty-one species of Rust Fungi distributed in fourteen genera. Eight of these genera belong to the supposedly more primitive family, Melampsoraceae, and six to the supposedly more advanced family, Pucciniaceae. The names of the genera are given in the accompanying Table.

Genera of Rust Fungi and Number of Species in which Flexuous Hyphae have been observed

FAMILY	GENUS	SPECIES
Melampsoraceae	Coleosporium	1
	Cronartium	3
	Melampsora	3
	Melampsorella	1
	Melampsoridium	1
	Milesia	11
	Pucciniastrum	2
	Uredinopsis	2
Pucciniaceae	Gymnoconia	1
	Gymnosporangium	4
	Phragmidium	2
	Puccinia	18
	Tranzschelia	1
	Uromyces	1
Total		51

From the data at our disposal, as set forth above, we may draw the conclusion that, in the Uredinales in general, flexuous hyphae are normal constituents of all mature and active pycnidia.

CHAPTER IV

THE UNION OF PYCNIDIOSPORES AND FLEXUOUS
HYPHAE IN PUCCINIA GRAMINIS
AND THREE OTHER RUSTS

1. *Puccinia graminis.* Introduction — Methods — Fusions between Pycnidiospores and Flexuous Hyphae described and discussed — Time elapsing between Mixing the Nectar and the Appearance of Fusions — Pycnidiospores and Pointed Periphyses — Pycnidiospores and Hyphae emerging between Epidermal Cells — Pycnidiospores fuse with Flexuous Hyphae only — Pycnidiospores incapable of Independent Germination — Pycnidiospores, Oidia, and Pollen Grains — Passage of a Nucleus of a Pycnidiospore down a Flexuous Hypha to a Proto-aecidium — Multiple Fusions and Hybridisation. 2. *Puccinia helianthi.* 3. *P. coronata avenae.* 4. *Gymnosporangium clavipes.*

1. Puccinia graminis. Introduction.—In 1933, Craigie[1] reported fusions between the pycnidiospores and flexuous hyphae of *Puccinia helianthi*, and a little later in the same year Pierson[2] reported similar fusions in *Cronartium ribicola*.

In *Puccinia graminis*, as a result of mixing the nectar of (+) and (−) pycnidia, diploid aecidia are formed on the under surface of the leaves.[3] Moreover, in this species, as a result of mixing pycnidial nectar, it has been found possible to cross races of *P. graminis tritici* differing in pathogenicity and spore colour[4] and also to obtain hybrids from the following varietal crosses: *P. graminis tritici* × *P. graminis*

[1] J. H. Craigie, "Union of Pycniospores and Haploid Hyphae in *Puccinia helianthi* Schw.," *Nature*, Vol. CXXXI, 1933, p. 25.

[2] R. K. Pierson, "Fusion of Pycniospores with Filamentous Hyphae in the Pycnium of the White Pine Blister Rust," *Nature*, Vol. CXXXI, 1933, pp. 728-729.

[3] J. H. Craigie, "Discovery of the Function of the Pycnia of the Rust Fungi," *Nature*, Vol. CXX, 1927, pp. 765-767.

[4] M. Newton, T. Johnson, and A. M. Brown: "Hybridization of Physiologic Forms of *Puccinia graminis tritici*," *Phytopathology*, Vol. XX, 1930, pp. 112-113; and "A Preliminary Study of the Hybridization of Physiologic Forms of *Puccinia graminis tritici*," *Scientific Agriculture*, Vol. X, 1930, pp. 721-731.

agrostidis; *P. graminis tritici* × *P. graminis secalis*;[1] and *P. graminis tritici* × *P. graminis avenae*.[2] The results of these experiments clearly indicate that the haploid mycelium in the pycnidial pustules of *P. graminis* can be diploidised by transferring (+) pycnidiospores to (−) pycnidia or (−) pycnidiospores to (+) pycnidia and that, just as in *P. helianthi*, the pycnidiospores and flexuous hyphae of opposite sex must fuse together, thus providing the means for (+) and (−) nuclei to become associated with one another in conjugate pairs. An account of an investigation which revealed *actual* fusions between the pycnidiospores and flexuous hyphae of *P. graminis* will now be given.

Methods.—Small Barberry bushes with young shoots were inoculated[3] with basidiospores of *Puccinia graminis* and, within a week, pycnidial pustules could be seen upon the leaves. From about ten days to two weeks after inoculation the drops of nectar were sufficiently large for mixing.

The mixing of the nectar was carried out (1) upon leaves attached to the Barberry bush or (2) upon leaves which had first been cut away and floated for a few hours on water in a closed Petri dish. After the nectar on a detached leaf had been mixed, the leaf was put back again into the Petri dish. The Petri-dish technique proved to be very satisfactory. It was found easier to mix the nectar on detached leaves than on leaves left on the bushes. Moreover, the moist air of the Petri dish caused the drops of nectar to increase in volume and, after the nectar had been mixed, prevented the leaves from withering.

In some of the later experiments with detached leaves the technique was improved in that the nectar was mixed before the leaves were floated on water in the Petri dish. Thus time was saved. Sometimes, when the pustules available were few and the nectar small in quantity, the nectar was diluted with about 50 per cent. tap-water.

The operation of mixing the nectar of (+) and (−) pustules was effected by means of a glass tube with a capillary extension at one end. The nectar from about ten pustules was drawn up into the tube, blown out on to a leaf, sucked up again into the tube, and then re-

[1]E. C. Stakman, M. N. Levine, and R. W. Cotter: "Hybridization and Mutation in *Puccinia graminis*," *Phytopathology*, Vol. XX, 1930, p. 113; and "Origin of Physiologic Forms of *Puccinia graminis* through Hybridization and Mutation," *Scientific Agriculture*, Vol. X, 1930, pp. 707-720. Also T. Johnson, M. Newton, and A. M. Brown, "Hybridization of *Puccinia graminis tritici* with *Puccinia graminis secalis* and *Puccinia graminis agrostidis*," *Scientific Agriculture*, Vol. XIII, 1932, pp. 141-153.

[2]T. Johnson and M. Newton, "Hybridization between *Puccinia graminis tritici* and *Puccinia graminis avenae*," *Proceedings of the World's Grain Exhibition and Conference*, Canada, 1933, Vol. II, pp. 219-223.

[3]For the technique of inoculation *vide supra*, p. 97.

deposited on the pustules. Some hours later, usually next day, pieces of leaves bearing pustules were cut away with a pair of scissors; and, immediately thereafter, cross-sections of the pustules were cut with a hand-razor. The sections were mounted in water, covered with a cover-glass, and examined with the microscope. The still-living flexuous hyphae, which could be seen protruding beyond the periphyses of each pycnidium, were then carefully examined to find out whether or not fusions had been formed between them and any of the pycnidio-spores.

Fusions between Pycnidiospores and Flexuous Hyphae described and discussed.—A considerable number of fusions between pycnidio-spores and flexuous hyphae were observed in *Puccinia graminis tritici* and also in *P. graminis avenae*. In general, the fusions resembled those found by Craigie in *P. helianthi*; but, as in *P. graminis* the pycnidiospores are much smaller and the flexuous hyphae less in diameter than in *P. helianthi*, they were more difficult to find and to see clearly.[1]

The fusions observed in the preparations were found: (1) *at the end of a flexuous hypha* which in the part near to and connected with a pycnidiospore was usually bent through a right angle, curved like a crozier, or twisted so as to resemble a goose-neck (Fig. 75); or (2) *at the end of a short lateral branch* of a flexuous hypha, with the branch either very short and peg-like, or short and straight, or longer and bent backwards toward the pycnidiospore through a considerable angle, an angle which sometimes exceeded 180° (Fig. 76). Each pycnidiospore taking part in a fusion had become united with a hypha by one end or near one end and I did not observe any fusion in which the pycnidiospore was united with a hypha exactly between its ends.

The pycnidiospores of *Puccinia graminis* are only 1.3-1.5µ in diameter and the width of the junction between a pycnidiospore and a flexuous hypha is not more than about 1µ. On account of these small dimensions it is only in the most favourable preparations and with the help of an oil-immersion lens that, on examining a fusion, one can feel reasonably certain that an open channel passing from the flexuous hypha into the pycnidiospore has been perceived.

Isolated unfused pycnidiospores in the preparations were seen to have highly refractive yellowish contents, including one or two small

---

[1]The pycnidiospores of *Puccinia helianthi* measure 6-8 × 2.5-3.3µ, and those of *P. graminis* 2-4 × 1.3-1.5µ. In preparations containing a mixture of the two kinds of pycnidiospores, one can see that the length of a pycnidiospore of *P. graminis* is only about equal to the width of a pycnidiospore of *P. helianthi*. *Cf.* A and B in Fig. 44 (p. 145) and A and D in Fig. 48 (p. 150).

drops of oil containing a carotinoid pigment (Fig. 75, M, and Fig. 76, G, *e*). On the other hand, a pycnidiospore which had fused with a flexuous hypha in the course of 24 hours from the time of the mixing of the nectar showed no such contents and stood out in contrast with the unfused pycnidiospores as being colourless, watery-looking, and very transparent.   As a rule, in a fusion examined several hours after

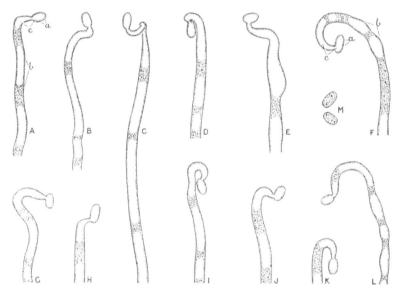

FIG. 75.—*Puccinia graminis*.   Union of the end of a flexuous hypha with a pycnidiospore of opposite sex. The nectar of (+) and of (−) pycnidial pustules was
mixed by hand; and, several hours later, the unions shown in A-L were observed
in living hand-sections: *a*, a pycnidiospore which has retained its original shape
and size; *b*, a flexuous hypha which has united with it; *c*, a vacuole which was
formed in the spore and in the terminal part of the flexuous hypha after the
union was effected.   M, two pycnidiospores which have not fused with a flexuous
hypha. Magnification, 1,000.

the mixing of the nectar, all the protoplasm, with the exception of a very thin layer lining the cell-wall, and the oil-drops had moved out not only from the pycnidiospore but also from the end of the hypha in union with the pycnidiospore, so that both these structures appeared to be empty (Figs. 75 and 76).

A study of the appearance of the fusions suggests that the flexuous hypha plays the major part in the growth process which brings the hypha and the pycnidiospore together.   In the crozier unions (Fig. 75, D, I) it would seem as though a spore had become attached by its gelatinous wall at or very near the end of a hypha and that the hypha had reacted toward the stimulus which the spore had provided by growing at its apex and curving backwards through an angle of about 180° until its tip had come into contact with the spore.   It would also seem that, when a pycnidiospore is situated near to, or has become attached to, the side of the middle part of a flexuous hypha (Fig. 76),

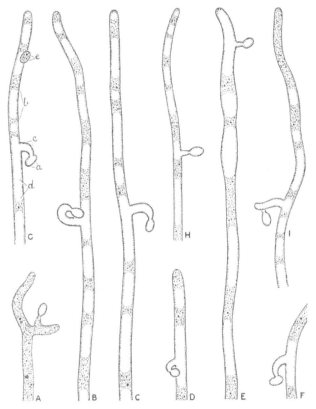

FIG. 76.—*Puccinia graminis*.   Union of a special fusion-branch or *peg* sent out from the side of a flexuous hyphae with a pycnidiospore of opposite sex.   The nectar of (+) and of (−) pycnidial pustules was mixed by hand and, several hours later, the unions shown in A-F were observed in living hand-sections: *a*, a pycnidiospore which has retained its original shape and size; *b*, a flexuous hypha; *c*, a short lateral fusion-branch or *peg* which has grown out from the flexuous hypha toward, and has fused with, the pycnidiospore; *d*, a vacuole which, after union was effected, developed in the spore, the peg, and the shaft of the flexuous hypha; *e*, a pycnidiospore which is attached to the flexuous hypha but has not united with it.   Magnification, 1,000.

the hypha reacts to the stimulus emitted by the spore by sending out a special short lateral branch or *peg* which grows straight forward or curves backward through a considerable angle until it, too, comes into contact with the spore.   The part played by the pycnidiospore in any fusion is probably limited to (1) sending to the hypha some signal of its presence, (2) receiving some signal from the hypha indicating the direction of the hypha's approach, and (3) pushing out a blunt papilla or *peg* to meet and fuse with the end of the approaching hypha.[1] Where the end of a flexuous hypha or the end of one of its ordinary long branches fuses with a pycnidiospore we have a *hypha-to-peg* fusion; and where the side of a flexuous hypha fuses with a pycnidiospore by means of a short special hypha which is developed in response to the presence of the pycnidiospore we have a *peg-to-peg* fusion.[2]

So far, owing to technical difficulties, I have not succeeded in watching a living pycnidiospore and a living flexuous hypha uniting with one another; but there is every reason to suppose that the various stages in the process resemble those already described for numerous hyphal fusions in the living mycelia of Pyrenomycetes, Discomycetes, Hymenomycetes, Gasteromycetes, and Fungi Imperfecti.[3] On the assumption that this is true, the successive steps in the growth processes which result in the union of a peg sent out from the middle of a flexuous hypha with a pycnidiospore of opposite sex have been represented in Fig. 77. At A, after the nectar of a (+) and a (−) pustule have been mixed, a pycnidiospore is adhering by its gelatinous wall to the middle part of a flexuous hypha of opposite sex. At B, in response to a morphogenic stimulus sent out by the pycnidiospore, the flexuous hypha is developing a fusion hypha or *peg*. At C, in response to a chemotropic or other stimulus sent out by the pycnidiospore, the peg is making a growth curvature toward the spore. At D this curvature has become more pronounced, and now the end of the curved peg is stimulating the pycnidiospore so that this is emitting a little papilla from that end of itself which is opposite to the oncoming end of the peg. At E the end of the curved peg and the tiny papilla have come into contact and are about to flatten out against one another. At F fusion has taken place: the double wall between the fusing elements has been dissolved, the outer walls of the curved peg and of the pycnidiospore have been welded together, and protoplasmic continuity has been established. As the tiny papilla of the pycnidiospore is in reality a very short fusion hypha or peg, the fusion itself is of the *peg-to-peg* type. Also the fusion resembles all other fusions in being *end-to-end*:[4] the end of one peg (the curved one) has fused with

[1]*Cf.* observations on stages in the hyphal-fusion process in other fungi, as given in these *Researches*, Vol. V, 1933, pp. 26-40.

[2]For definitions of the terms *hypha-to-peg* and *peg-to-peg* fusions, *vide* these *Researches*, Vol. V, pp. 26, 28-33.

[3]These *Researches*, Vol. V, 1933, pp. 26-40.   [4]*Ibid.*, p. 26.

the end of the other peg (the papilla).   Thus the flexuous hypha and the pycnidiospore have fused together only after mutual stimulation and as a result of physical and chemical co-operation.   Shortly after the fusion was accomplished, a vacuole formed in the pycnidiospore and, by enlargement, soon extended into the curved peg and the adjacent part of the original flexuous hypha.   In this way, as indicated at G, a considerable amount of massive cytoplasm, and doubtless with it the nucleus of the pycnidiospore, has been displaced and has been forced to move down the flexuous hypha in the direction of the ostiole

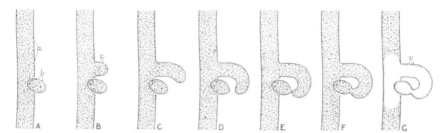

FIG. 77.—*Puccinia graminis*.   Diagram suggesting the successive steps in the growth processes which result in the union of a fusion-branch or *peg*, sent out from the middle part of a flexuous hypha, with a pycnidiospore of opposite sex.   A: *a*, the middle part of a flexuous hypha; *b*, a pycnidiospore of opposite sex which has become attached to the hypha.   B, the pycnidiospore has stimulated the hypha and, in consequence, the hypha is emitting a special fusion branch or *peg*, *c*.   C, the end of the peg is making a growth curvature toward the spore.   D, the curvature has become more marked; and the end of the peg has stimulated the spore, in consequence of which, the spore is now forming a small papilla opposite to the oncoming end of the peg.   E, the end of the peg and the end of the papilla have met and are about to flatten against one another.   F, the peg and the spore have fused together, so that their protoplasm is now in continuity.   G, a large and common vacuole, *v*, has now formed in the spore, the peg, and part of the main shaft of the flexuous hypha.   Magnification, about 2,250.

of the pycnidium.   How the nucleus of the pycnidiospore subsequently makes its way through the cytoplasm to the base of the flexuous hypha and through the intercellular mycelium in the leaf and onwards to the basal cells of a proto-aecidium is at present an unsolved problem.[1]

Particularly good material of *Puccinia graminis avenae* was provided by Barberry leaves which had been inoculated with basidiospores thirteen days previously and whose pycnidial pustules had borne nectar for about six days.   The nectar was mixed on a leaf which had been removed from a bush and had been floated for about two hours

[1]In *Coprinus lagopus*, during the diploidisation process, nuclei of one sex move with surprising speed long distances through a haploid mycelium of opposite sex (These *Researches*, Vol. IV, 1931, pp. 213-229).   Here, too, we are presented with the problem of the means whereby nuclei make their way through cell-contents rapidly from one point to another.

on water in a closed Petri dish.   About 21 hours later the pustules of
this leaf were sectioned with a hand-razor and then many fusions
between the flexuous hyphae and the pycnidiospores were observed.
In the sections of one pustule, which included numerous pycnidia,
about twenty fusions were seen and of this number three were in one
pycnidium and two in another.   It thus became clear that, after the
nectar of diverse pustules has been well mixed, *in a single pustule*
that is young and in good condition there may be formed *several
fusions in each single pycnidium* and *many fusions in all the pycnidia
considered collectively*.

The conclusion just stated was confirmed by further observations
made on *Puccinia graminis tritici*.   A Barberry leaf had been inoculated with basidiospores of this variety fifteen days previously and the
pycnidial pustules had borne nectar for about eight days.   Again the
Petri-dish technique was employed.   About 21-23 hours after the
nectar had been mixed, two pustules were cut into sections and the
sections revealed twenty-two fusions.   In one pycnidium five fusions
were seen and in two other pycnidia three fusions each.

In no instance was a flexuous hypha seen which had fused with two
pycnidiospores, and it may well be that a flexuous hypha which has
fused with one pycnidiospore is incapable of fusing with a second
pycnidiospore.

The total number of fusions between the pycnidiospores and
flexuous hyphae of *Puccinia graminis* observed by me in all the hand-
sections examined was between eighty and one hundred.

From an examination of the appearance of a large number of
fusions, it appears that, in *Puccinia graminis*, the distance grown by
the end of a flexuous hypha or special lateral fusion-branch (peg)
before it meets with and fuses with a pycnidiospore is not more than
10μ and is usually less.   We may conclude that, after nectar has been
mixed, the flexuous hyphae—so far as fusion phenomena are concerned—react only to those pycnidiospores of opposite sex which are
very close to them.

From observations on the progress of the fusion process in other
fungi,[1] it may be supposed that, in *Puccinia graminis*, the time required
to complete a fusion between a flexuous hypha and a pycnidiospore
after the two elements have begun to react toward one another would
not be more than 20-40 minutes.

**Time elapsing between Mixing the Nectar and the Appearance of
Fusions.**—It was found that, in *Puccinia graminis tritici*, after the
nectar had been mixed, fusions between flexuous hyphae and pycnidiospores could be definitely observed at the end of 5.0, 4.5, 3.5, 3.0 hours

[1]These *Researches*, Vol. V, 1933, pp. 52-64.

and even 2 hours and 10 minutes.   It is therefore evident that, after the nectar has been mixed, the flexuous hyphae and pycnidiospores of opposite sex quickly re-adjust themselves to the new conditions and soon begin to react toward one another.

Holton[1] mixed monosporidial lines of opposite sex of *Ustilago avenae* or *U. levis* and spread them out thinly over poured plates of 2 per cent. plain agar.   In one instance he observed that the sporidia had begun to fuse 35 minutes after they had been mixed; but he found that, usually, one or two hours were required before fusions could be observed and, occasionally, three or four hours.   It thus appears that, in *Puccinia graminis* and in *Ustilago avenae* or *U. levis*, after the appropriate cells have been brought together, the time required for fusions to take place is about the same.

Pycnidiospores and Pointed Periphyses.—In *Puccinia graminis*, a union between a pycnidiospore and one of the stiff, tapering, pointed periphyses (ostiolar trichomes) was never observed, and there is no reason to suppose that unions of this kind take place.   However, as has already been recorded,[2] mature periphyses in older pustules sometimes become converted into flexuous hyphae by growing forward at their pointed ends.   In at least one instance I have seen a periphysis which had been converted into a flexuous hypha united with a pycnidiospore by the tip of its flexuous part.   This tip was bent through a right angle and was united with one end of the pycnidiospore in the usual way (Fig. 63, J, p. 185).

In 1933, Miss Allen[3] suggested that, in *Puccinia graminis*, one of the ways of initiating the diploidisation process is by the union of pycnidiospores with pointed periphyses (her paraphyses).   She said: "In active spermogonia new paraphyses continue to form; these fresh young paraphyses may serve as receptive hyphae, the spermatial nuclei entering them and migrating down through them to the cells of the spermogonial wall at their base."   However, Miss Allen did not support her supposition with any direct evidence that the pycnidiospores of *P. graminis* do actually fuse with periphyses, and the flexuous hyphae, described by Craigie in 1933, escaped her notice.

In 1935, Lamb[4] stated that the pycnidia of *Puccinia phragmitis* are "provided with a brush of periphyses round the ostiole," but he

---

[1]C. S. Holton, "Studies in the Genetics and the Cytology of *Ustilago avenae* and *Ustilago levis*," *University of Minnesota Agricultural Experiment Station*, Bull. No. LXXXVII, 1932, p. 12.   [2]*Vide supra*, p. 185.

[3]Ruth F. Allen, "Further Cytological Studies of Heterothallism in *Puccinia graminis*," *Journ. Agric. Research*, Vol. XLVII, 1933, pp. 1-16.

[4]I. M. Lamb, "The Initiation of the Dikaryophase in *Puccinia phragmitis* (Schum.) Körn," *Annals of Botany*, Vol. XLIX, 1935, p. 418.

made no mention of the periphyses being accompanied by flexuous hyphae. However, as I have since observed flexuous hyphae in the pycnidia of eighteen other Pucciniae, it seems very probable that these structures are also present in the pycnidia of *P. phragmitis*.

Lamb[1] mixed the nectar of haploid pustules of *Puccinia phragmitis*, fixed his material 48 hours later, and then sought for fusions between pycnidiospores and periphyses. After examining much material cytologically, he succeeded in finding two cases, and two cases only, in which he thought he saw the fusions sought for. In each case, in his drawings, he represents the pycnidiospore as having put out a germtube[2] and, in one case, he represents the pycnidiospore as having fused with *two* periphyses (with the tip of one and with the side of another). Although the material was fixed 48 hours after the nectar was mixed, Lamb represents his pycnidiospores and adjacent parts of the periphyses with which the spores are supposed to have fused as still containing much protoplasm. In absence of high vacuolation Lamb's fusions differ markedly from the fusions between pycnidiospores and flexuous hyphae observed by myself in *P. graminis* and *P. helianthi* (cf. Figs. 75 and 76 with Lamb's Plate X, Figs. 16 and 17). In one of Lamb's cases (his Fig. 16), the pycnidiospore may have been merely lying in contact with a periphyses and, in the other case (his Fig. 17), one of the two supposed fusions, the one at the tip of a periphysis, does not look like a real fusion but like two bodies in simple contact.[3] In the light of my own extended experience with fusions between the pycnidiospores and flexuous hyphae of *P. graminis* and *P. helianthi*, it seems to me that Lamb has misinterpreted what he saw and that the evidence which he has brought forward cannot be accepted as proving that, in *P. phragmitis*, the pycnidiospores fuse with pointed periphyses. It is to be expected that a further investigation of *P. phragmitis*, made with hand-sections of living material, will reveal not only that this species, like other Pucciniae, possesses flexuous

[1]*Ibid.*, p. 420. In *Puccinia graminis*, after the diploidisation process has been initiated, the flexuous hyphae break down and disappear. Perhaps in Lamb's material, owing to late fixation, the flexuous hyphae had suffered a similar fate.

[2]The germ-tubes, as shown by Lamb, are different from the bridges joining the pycnidiospores with the periphyses. In *P. graminis*, the pycnidiospores which fuse with flexuous hyphae do not put out germ-tubes like those represented by Lamb. For a criticism of the supposition that the pycnidiospores of the Rust Fungi germinate *vide infra*, pp. 229-233.

[3]H. C. I. Gwynne-Vaughan and B. Barnes in the second edition of *The Structure and Development of the Fungi* (Cambridge, 1937, p. 313) have incorrectly reproduced Lamb's Fig. 17 showing a pycnidiospore fused with *two* periphyses, for they have eliminated from it the periphysis with the supposed tip fusion.

hyphae, but that the pycnidiospores fuse with these hyphae in the usual manner.

Savile,[1] like Lamb, has represented in his illustrations what he believed to be cases of fusions between pycnidiospores and "ostiolar filaments" some of which look like periphyses: two cases for *Puccinia sorghi* (his Figs. 68 and 72) and one for *Uromyces fabae* (his Fig. 74). In each of these three supposed fusions the pycnidiospore is very close to the ostiolar filament and the bridge between the spore and the filament is excessively thin and short. The appearance of these fusions, as represented by Savile, is quite unlike those represented by myself for *Puccinia graminis*, *P. helianthi*, *P. coronata avenae*, and *Gymnosporangium clavipes* in this Chapter. He did not see any fusions between pycnidiospores and the bent ends of flexuous hyphae such as are represented in Figs. 75 (p. 218), 78 (p. 236), and 80 (p. 239); nor did he see any fusions between pycnidiospores and what are evidently special fusion branches or pegs sent out from the side of a flexuous hypha as shown in Fig. 76 (p. 219); and his fusions look like those of Lamb which have already been criticised. Savile, as we have seen,[2] in his microtome sections was unable to distinguish clearly between pointed periphyses and longer, cylindrical, bluntly-ending flexuous hyphae and believed that in the species studied by him these two kinds of elements form an intergrading series of "ostiolar filaments." From the conical form of the small pieces of filaments shown in his Figs. 68 and 74 we may suppose that he thought he saw fusions between pycnidiospores and pointed periphyses. My experience with living unpickled material of *Puccinia graminis*, *P. helianthi*, *P. coronata avenae*, and *Gymnosporangium clavipes* has taught me that pycnidiospores fuse not with pointed periphyses but only with flexuous hyphae. On the basis of this discussion I am unable to accept Savile's evidence that, in the species studied by him, he really saw fusions between pycnidiospores and flexuous hyphae or between pycnidiospores and pointed periphyses. *Puccinia sorghi* and *Uromyces fabae* have been proved by experiment to be heterothallic.[3] It is therefore probable that in these species, if the nectar of (+) and (−) pustules were to be mixed and the pycnidia thereafter studied with the help of living hand-sections, real fusions between pycnidiospores and flexuous hyphae would be observed.

[1]D. B. O. Savile, "Nuclear Structure and Behavior in Species of the Uredinales," *American Journal of Botany*, Vol. XXVI, 1939, p. 599.

[2]*Vide* Chapter III, p. 212.

[3]*Vide supra*, Chapter I, p. 85.

Miss Fort,[1] in 1940, reported the results of her study of fixed and stained preparations of *Uromyces scirpi* on *Oenanthe crocata*. In the pycnidia she observed numerous pointed periphyses, but she makes no mention of flexuous hyphae and evidently failed to find any trace of these organs. Her Fig. 9 shows a pointed periphysis with two pycnidiospores sticking to its side, and she remarks that Allen suggested (1933) that periphyses are the means of entrance of spermatial nuclei. It is to be supposed: that living pycnidia of *Uromyces scirpi* resemble the living pycnidia of *Puccinia graminis*, *Uromyces poae*, and other Pucciniaceae in being provided with flexuous hyphae; and that these hyphae, as in *Puccinia graminis*, *P. helianthi*, etc., unite with pycnidiospores of opposite sex.

In concluding this Section it may be stated that, up to the present, there is no satisfactory evidence to support the idea that pycnidiospores of one sex fuse with pointed periphyses of opposite sex.

Pycnidiospores and Hyphae emerging between the Epidermal Cells.—De Bary,[2] in 1884, remarked that, when aecidia are developing in Rust Fungi, some of the mycelial hyphae press to the surface of the host-epidermis at the stomata; but he deprecated the idea that these hyphae have anything to do with a sexual process. He found that the emerging hyphae are not connected with an archicarp and do not seem to have any special relations with the "spermatia"; and he wisely remarked that branches of a mycelium in a leaf "may as well grow outwards through a stoma as inwards into an intercellular space."

In 1896 Richards[3] recorded that, in connexion with the development of the aecidia of *Uromyces caladii*, he had observed occasional hyphae protruding from a stoma, but remarked that these hyphae show no evidence of specialisation and cannot be regarded as trichogynes.

In 1904, Klebahn[4] stated: that, encouraged by his teacher Stahl, he had sought to find out how the "spermatia" of the Rust Fungi might play a part in the development of the aecidia; that he had observed short bluntly-ending hyphae projecting at the mouths of stomata and sometimes in contact with spermatia; that he had not been able to trace any connexion between the protruding hyphae and organs corresponding to ascogonia as Stahl had been able to do be-

[1]Margaret Fort. "A Study of *Uromyces scirpi* Burr.," *Trans. Brit. Myc. Soc.*, Vol. XXIV, 1940, pp. 98-108.

[2]A. de Bary, *Vergleichende Morphologie und Biologie der Pilze, Mycetozoen und Bacterien*, Leipzig, 1884, p. 300.

[3]H. M. Richards, "On some Points in the Development of Aecidia," *Proc. American Acad. of Arts and Sci.*, Vol. XXXI, 1896, p. 257.

[4]H. Klebahn, *Die wirtswechselnden Rostpilze*, Berlin, 1904, pp. 196-197.

tween the trichogynes and ascogonia of collemaceous Lichens; and, finally, that he regarded the contact of the spermatia with the hyphae as accidental.

In 1931, Andrus[1] observed that, in haploid pustules of *Uromyces phaseoli typica* (*U. appendiculatus*) and *U. phaseoli vignae* (*U. vignae*),[2] hyphae come to the surface of the host-leaf through stomata and between epidermal cells. He regarded these hyphae as *trichogynes*, and he made the bold assumption that direct fusions take place between these supposed trichogynes and the pycnidiospores. In 1932, Miss Allen published the results of her studies on *Puccinia triticina*[3] and *P. coronata*.[4] She found in haploid pustules of these Rusts emergent hyphae similar to those described by Andrus and, assuming that the pycnidiospores fuse with them, she called them "receptive hyphae." In 1933, she described receptive hyphae as being present in the haploid pustules of *P. graminis*.[5] However, neither Andrus nor Miss Allen brought forward any evidence that could be taken as proof that the supposed fusions ever take place. In 1933, Miss Rice[6] observed in pustules of *P. sorghi* and *P. violae* finger-like hyphae which arose from hyphal runners under the epidermis and pushed out through stomata. She did not find any fusions between the pycnidiospores and the finger-like hyphae and deprecated calling such hyphae trichogynes. In *Aecidium punctatum* she found finger-like hyphae just below or within the stomata, but she did not observe any which projected through them.

In January, 1933, Craigie[7] announced his discovery of flexuous hyphae in *Puccinia helianthi* and *P. graminis* and brought forward convincing evidence that, in *P. helianthi*, the pycnidiospores and the flexuous hyphae of opposite sex fuse with one another. His discovery of unions between flexuous hyphae and pycnidiospores in *P. helianthi*

[1]C. F. Andrus, "The Mechanism of Sex in *Uromyces appendiculatus* and *U. vignae*," *Journ. Agric. Research*, Vol. XLII, 1931, pp. 559-587.

[2]In employing the names *Uromyces phaseoli typica* and *U. phaseoli vignae* I have followed J. C. Arthur (*Manual of the Rusts in United States and Canada*, Lafayette, U.S.A., 1934, pp. 296-297).

[3]Ruth F. Allen, "A Cytological Study of Heterothallism in *Puccinia triticina*," *Journ. Agric. Research*, Vol. XLIV, 1932, pp. 733-754.

[4]Ruth F. Allen, "A Cytological Study of Heterothallism in *Puccinia coronata*," *Journ. Agric. Research*, Vol. XLV, 1932, pp. 513-541.

[5]Ruth F. Allen, "Further Cytological Studies of Heterothallism in *Puccinia graminis*," *Journ. Agric. Research*, Vol. XLVII, 1933, pp. 1-16.

[6]M. A. Rice, "Reproduction in the Rusts," *Bull. Torrey Bot. Club*, Vol. LX, 1933, pp. 23-54.

[7]J. H. Craigie, "Union of Pycniospores and Haploid Hyphae in *Puccinia helianthi* Schw.," *Nature*, Vol. CXXXI, 1933, p. 25.

greatly weakened the position of those who had imagined that, in other Rust species, the pycnidiospores fuse with interpycnidial surface hyphae.[1]

The hyphae of *Puccinia graminis* which come to the surface of a Barberry leaf between and around the pycnidia and proto-aecidia are, as shown in Miss Allen's illustrations,[2] very short structures which terminate in a slight knob-like swelling at or just beyond the level of the cuticle. In microtome sections (cut 6-12μ thick, stained with safranin and counterstained with fast green), in the lower epidermis I have been able to find hyphae with bluntly-swollen tips pushed out between the guard-cells of certain stomata just like those described by Miss Allen; but, in the upper epidermis which is devoid of stomata, I have so far failed to observe any hyphae which had definitely pene-

---

[1]In December, 1933, Andrus published a second paper on *Uromyces appendiculatus* and *U. vignae* ("Sex and Accessory Cell Fusions in the Uredineae," *Journ. Wash. Acad. Sciences*, Vol. XXIII, pp. 544-557) in which he again described fusions between pycnidiospores and gametophytic hyphae emerging at the surface of the host-leaf at stomata or between epidermal cells. However, the two illustrations which he gives in support of his contention that such fusions take place are very unsatisfactory. In his Fig. 1, A, the pycnidiospore is represented as lying in an intercellular space *below* a stoma. One may ask: how did a *pycnidiospore* get into such a position? And the supposed fusion represented in his Fig. 1, B, looks like an artifact. In all probability flexuous hyphae are present in the nectar which is excreted by the pycnidia of *U. appendiculatus* and *U. vignae*, but Andrus failed to mention them. It may well be that these two Rust species resemble *P. helianthi* and *P. graminis* in that, after the nectar has been mixed, fusions take place between the pycnidiospores and the flexuous hyphae. To determine whether or not this supposition is well based it will be necessary to discard the microtome and to apply the living-material method introduced by Craigie.

Miss Allen in a paper on *Puccinia sorghi* ("A Cytological Study of Hetero-thallism in *Puccinia sorghi*," *Journ. Agric. Research*, Vol. XLIX, pp. 1047-1068), published in December, 1934, made the following assertions. "The entrance of spermatial nuclei into haploid hyphae takes place in several ways. (1) A spermatium placed on a spermogonium becomes attached to a paraphysis and its nucleus passes over into the paraphysis and moves through it to the mycelium within the leaf. (2) A spermatium placed in contact with a stomatal hypha contributes its nucleus to this hypha. (3) Under favourable conditions a spermatium not in contact with any surface hypha can germinate and grow into a slender hypha which enters the nearest stoma and so becomes effective." Miss Allen may well have seen fusions between flexuous hyphae and pycnidiospores; but she admits that she did not actually observe a union between a stomatal hypha and a pycnidiospore, and the somewhat elongated pycnidiospores which she calls "germinating spermatia" (Pl. V, B) may have been elongated when they were first formed and before their extrusion from the pycnidia. Her evidence that pycnidiospores produce slender germ-tubes which enter the host-leaf through stomata seems unconvincing.

[2]Ruth F. Allen, "Further Cytological Studies of Heterothallism in *Puccinia graminis*," *loc. cit.*, Plates I - VI.

trated through the cuticle and had emerged into the outer air. Miss Allen states that in *P. graminis* "as in the other Rusts studied, hyphae reaching the surface are short-lived." When examining hand-sections of living and mature haploid pustules of *P. graminis* I was never able to find any hyphae projecting from the epidermis between the pycnidia, and Dr. Craigie has informed me that his experience with the pustules of both *P. graminis* and *P. helianthi* has been like mine. While the projecting hyphae depicted by Miss Allen have an exceedingly small amount of surface exposed above the epidermis and are short-lived, the flexuous hyphae projecting beyond the periphyses in the drop of nectar have relatively a large amount of exposed free surface and are long-lived. Moreover, while a great many fusions have now been observed between the flexuous hyphae and the pycnidiospores, not a single fusion has been observed between one of Miss Allen's surface hyphae and a pycnidiospore. For all these reasons we may conclude that, even if the tips of the so-called "receptive hyphae" of *P. graminis* are actually exposed at the surface of the Barberry leaf, these hyphae do not fuse with pycnidiospores of opposite sex and are of no interest from the sexual point of view.

Pycnidiospores fuse with Flexuous Hyphae only.—In *Puccinia graminis*, since numerous fusions between pycnidiospores and flexuous hyphae have been observed and none between pycnidiospores and pointed periphyses or between pycnidiospores and hyphae which have emerged between epidermal cells at the surface of the host-leaf, and since surface hyphae have a very restricted area of exposed surface and, after coming into existence, are short-lived, it is justifiable to conclude that the pycnidiospores fuse with the flexuous hyphae and with these alone.

Pycnidiospores incapable of Independent Germination.—The pycnidiospores of *Puccinia graminis* do not germinate in the nectar into which they have been extruded, nor do they germinate in mixed nectar. After nectar has been mixed, the only sign of growth displayed by the pycnidiospores is exhibited by those very few pycnidiospores which actually fuse with flexuous hyphae; and all that a pycnidiospore which is about to fuse with a hypha seems to do is to send out, in response to a stimulus received from the hypha, a very tiny blunt papilla which meets with and fuses with the on-coming hyphal tip. It thus appears that, in mixed nectar, not only isolated pycnidiospores but also those pycnidiospores which unite with flexuous hyphae are incapable of independent germination.

No one, as yet, has produced any conclusive evidence that the pycnidiospores of any of the Rust Fungi germinate independently even to a limited degree, *i.e.* no one has sown pycnidiospores in water, nectar, or any culture medium, has watched the pycnidiospores grow, and has observed the germ-tubes of particular pycnidiospores in various stages of elongation.[1] As noted in Chapter I, the older workers, including Cornu,[2] Plowright,[3] Sappin-Trouffy,[4] and Carleton,[5] who thought they saw pycnidiospores budding, may well have had their culture media (dilute honey, etc.) contaminated with foreign organisms, such as yeasts. The evidence presented by more recent workers including Miss Allen[6,7] and Lamb[8] is just as unconvincing as the older evidence of germination in culture media. A source of error in

[1]In 1934, Miss Allen, when investigating *Melampsora lini* ("A Cytological Study of Heterothallism in Flax Rust," *Journ. Agric. Research*, Vol. XLIX, pp. 765-789), observed branched paraphyses (= flexuous hyphae) protruding from the pycnidia but failed to observe any fusions between them and the pycnidiospores. She concluded that "spermatia placed upon the surface of an infection enter, probably growing in through spermogonia, perhaps also entering through epidermal cells, and then grow into intercellular mycelium the hyphae of which are at first very fine, but later of ordinary appearance." The evidence presented by Miss Allen in favour of the view that the pycnidiospores of *M. lini* actually germinate and that the germ-tubes enter the host-leaf seems unsatisfactory. It may well be that an investigation of *M. lini* by means of hand-cut sections of living material will yet reveal fusions between the pycnidiospores and flexuous hyphae.

[2]M. Cornu and M. E. Roze (Rapport de M. Brogniart), *Comp. Rend. Acad. Sci. Paris*, T. LXXX, 1875, pp. 1464-1468; also M. Cornu in *Bull. Soc. Bot. France*, T. XXIII, 1876, pp. 120-121.

[3]C. B. Plowright, *A Monograph of the British Uredineae and Ustilagineae*, London, 1889, pp. 14-16.

[4]P. Sappin-Trouffy, "Recherches histologiques sur la famille des Urédinées," *Le Botaniste*, Sér. 5, 1896, pp. 106-108.

[5]M. A. Carleton, "Studies in the Biology of the Uredineae. I. Notes on Germination," *The Botanical Gazette*, Vol. XVIII, 1893, p. 451; also "Culture Methods with Uredineae," *Journ. Appl. Micros. and Lab. Methods*, Vol. VI, 1903, pp. 2109-2114.

[6]Ruth F. Allen, "A Cytological Study of Heterothallism in *Puccinia graminis*," *Journ. Agric. Research*, Vol. XL, 1930, p. 600 and Plate XI, A and B. Some of the supposed germinated spermatia look like parts of periphyses and others may have been bits of mycelium of some foreign fungus which had grown over the epidermis of the Barberry leaf.

[7]Ruth F. Allen: "A Cytological Study of Heterothallism in *Puccinia sorghi*," *Journ. Agric. Research*, Vol. XLIX, 1934, pp. 1047-1068, Plate V, B; "A Cytological Study of Heterothallism in Flax Rust," *ibid.*, pp. 765-791, Plate VI, C. In some cytological preparations of *Melampsora lini* (Plates VII and VIII) Miss Allen thought that she had found cases of pycnidiospores germinating on a host-leaf and sending germ-tubes through the outer wall of the epidermal cells. Possibly she had under observation the conidia of some other fungus, which had become mixed with the pycnidiospores.

[8]I. M. Lamb, "The Initiation of the Dikaryophase in *Puccinia phragmitis* (Schum.) Körn," *Annals of Botany*, Vol. XLIX, 1935, p. 415. Lamb has informed me that the percentage of pycnidiospores which seemed to have germinated was in reality very small, and that he did not see the supposed germ-tubes elongating.

searching for germinated pycnidiospores lies in the fact that pycnidiospores are decidedly variable in length. Some pycnidiospores are two or three times as long as others of the same species (Fig. 61, K and L, p. 181). I have occasionally seen in pycnidial nectar of *Puccinia graminis* here and there a large elongated cell, swollen in one part, but full of protoplasm containing carotinoid pigment (Fig. 61, L). It did not look to me as if the cell was germinating, nor was it seen to elongate after it had been first observed. Pycnidiospores which have been supposed to have produced a short germ-tube may well have been pycnidiospores of exceptional size and shape which had not changed in form since they were constricted off their pycnidiosporophores and were extruded from the pycnidial cavities in which they came into existence; and, there is also the possibility that, in some instances, the structures may have been extruded buffer cells or pycnidiosporophores.

Dr. Craigie,[1] in unpublished work carried out in 1930, made a long series of experiments to find out whether or not the pycnidiospores of *Puccinia graminis* and of *P. helianthi* germinate in mixed nectar, but with entirely negative results. He mixed the nectar of (+) and (−) pustules. Then with a pipette, at various intervals of time up to 63 hours afterwards, he withdrew the nectar, placed it on a slide, stained it with safranin or cotton-blue, and examined it carefully with the microscope. He observed that the pycnidiospores varied in length up to two or three times the average length, but in none of his preparations could he find any pycnidiospores which appeared to have germinated.

I myself, on numerous occasions, have mixed the nectar of the pustules of *Puccinia graminis* and, 6-48 hours afterwards, have cut sections through the pustules with a view to finding fusions between pycnidiospores of one sex and flexuous hyphae of the opposite sex. In none of my preparations, which were mounted in water, did I ever see any pycnidiospores, other than those which had fused with flexuous hyphae, show any signs of germination. Thus my own observations on the germination of pycnidiospores in mixed nectar support those of Craigie.

Plowright,[2] Klebahn,[3] Clinton and McCormick,[4] and more recently Lamb[5] all spread pycnidiospores of Rust Fungi over the epidermis of

[1]Personal communication.

[2]C. B. Plowright, *A Monograph of the British Uredineae and Ustilagineae*, London, 1889, p. 20.

[3]H. Klebahn, *Die wirtswechselnden Rostpilze*, Berlin, 1904, p. 387.

[4]G. P. Clinton and Florence A. McCormick, "Rust Infection of Leaves in Petri Dishes," *Connecticut Agric. Exp. Station*, Bull. 260, 1924, p. 491.

[5]I. M. Lamb, *loc. cit.*, pp. 415-416.

the leaves or stems of host-plants, but with negative results, for no infection of the hosts took place. Lamb took nectar of mixed origin from pustules of *Puccinia phragmitis* and applied it to certain marked places on the upper surfaces of a healthy *Rumex crispus* plant which was kept over water under a bell-jar for several days afterwards; but at no time did any trace of infection show itself. It thus appears that Rust pycnidiospores do not germinate on healthy host leaf-surfaces.

Hanna,[1] in 1929, made an attempt to germinate the pycnidiospores of *Puccinia graminis* in a number of liquid and solid media, but without success. The details[2] of Hanna's experiments with culture media, which hitherto have not been published, are as follows:

Pycnidiospores of *Puccinia graminis* were placed in hanging drops of the following culture media: (1) potato-dextrose agar, (2) prune agar, (3) corn-meal agar, (4) malt-extract agar, (5) 1 per cent. sucrose (cane-sugar) solution, (6) 1 per cent. dextrose solution, (7) 1 per cent. malt-extract solution, and (8) 1 per cent. peptone solution. After nine days the pycnidiospores showed no signs of germination.

With a view to providing a larger air supply, Hanna then placed pycnidiospores of *Puccinia graminis* in the following media held in Syracuse dishes: (1) 1 per cent. sucrose solution, (2) 1 per cent. dextrose solution, and (3) malt-extract solution. After two days: in (1) and (2) the pycnidiospores had not germinated; and in (3) no pycnidiospores could be seen, but in the culture there was an abundance of small yeast cells which were budding rapidly.

Hanna also sought to germinate the pycnidiospores of *Gymnosporangium juniperi-virginianae* in hanging drops of: (1) 1 per cent. sucrose solution, (2) 1 per cent. dextrose solution, (3) potato-dextrose agar, and (4) malt-extract agar. After nine days, as with *Puccinia graminis*, the pycnidiospores showed no signs of germination. The same negative result was obtained when the pycnidiospores of the Gymnosporangium were sown in 1 per cent. sucrose solution contained in a Syracuse dish. Hanna's findings with *Gymnosporangium-juniperi-virginianae* confirm those of Reed and Crabill[3] who, without giving details of their experiments, reported that repeated attempts to germinate the pycnidiospores of *G. juniperi-virginianae* had failed.

We may conclude from the evidence at our disposal that the

---

[1]W. F. Hanna, "Nuclear Association in the Aecium of *Puccinia graminis* Pers.," in *Report of the Dominion Botanist for 1929*, Dept. of Agric., Ottawa, 1931, pp. 50-52.

[2]Personal communication to the author.

[3]H. S. Reed and C. H. Crabill, "The Cedar Rust Disease of Apples caused by *Gymnosporangium juniperi-virginianae* Schw.," *Virginia Agric. Experiment Station*, Bull. IX, 1915, p. 48.

pycnidiospores of the Rust Fungi do not germinate: (1) in their own nectar; (2) in mixed nectar, except when very close to a flexuous hypha of opposite sex and then by putting out nothing more than a tiny fusion-papilla; (3) on the surface of host-leaves; and (4) in any culture medium so far tried in a critical manner. There is the possibility that some day, in some as yet untried artificial culture medium, they will be got to put out germ-tubes; but, in any case, it appears fairly certain that, under natural conditions in the open, and in many artificial culture media, the pycnidiospores of the Rust Fungi are incapable of independent germination.

Pycnidiospores, Oidia, and Pollen Grains.—The pycnidiospores of the Rust Fungi, in an earlier geological age, may have germinated just as the oidia of the Coprini and other Hymenomycetes do now; but, if so, they have lost that power. The fact that, at the present day, Rust pycnidiospores are incapable of germinating independently in pycnidial nectar is distinctly advantageous for their function as diploidising agents.

In the open, the pollen grains of Angiosperms, by one means or another, are prevented from germinating until they have been transported to a stigma where their further development may be advantageously undertaken. In the same way, out in the open, the pycnidiospores of the Rust Fungi are prevented from developing any further until they have been transported very close to, or actually in contact with, a flexuous hypha of opposite sex. In a Flowering Plant, the ovule is passive and the pollen grain develops a long pollen-tube which grows down to the ovule and enters it *via* the micropyle; whereas, in a Rust Fungus, a flexuous hypha co-operates with a pycnidiospore by growing toward it, so that all the pycnidiospore has to do to assist the formation of a fusion is to put out a tiny fusion papilla.

Passage of a Nucleus of a Pycnidiospore down a Flexuous Hypha to a Proto-aecidium.—After a flexuous hypha and a pycnidiospore have fused together, the protoplasmic contents of the pycnidiospore, including the nucleus, pass into the flexuous hypha. So much is known; but what happens to the nucleus of a pycnidiospore after its entry into a flexuous hypha and until dicaryotic cells are produced in the proto-aecidia is a matter for conjecture. We may suppose: (1) that the nucleus of the pycnidiospore travels down the flexuous hypha and through the wall of the pycnidium into the mycelium occupying the intercellular spaces of the leaf; (2) that the nucleus there undergoes repeated nuclear division; (3) that the nuclei so produced travel along the mycelial hyphae to the basal cells of all the neighbouring proto-aecidia; and (4) that these nuclei, having reached their final destination, rapidly become associated with nuclei of opposite sex already present in the basal cells with the result that the aecidiosporophores, which are formed in the spore-bed of each young aecidium, all come to contain a pair of conjugate nuclei.

In *Puccinia graminis*, the flexuous hyphae are non-septate and, as Wahrlich[1] observed as long ago as 1893, the mycelium is septate and the septa are each provided with a *small central open pore* through which a thread of protoplasm passes from cell to cell. Thus there is a continuity of protoplasm extending from the end of each flexuous hypha to the hyphal wefts of all the proto-aecidia. A nucleus of a pycnidiospore, after entering a flexuous hypha, can pass freely to the hypha's base; and when, in travelling through the mycelial hyphae on its way toward a proto-aecidium, it (or one of its descendents) encounters a septum, it probably takes the line of least resistance and squeezes through the septum's central pore.

The distance between the end of a flexuous hypha and the basal cells in the mycelial wefts of the nearest proto-aecidia inside the Barberry leaf is somewhat less than 0.5 mm. (*cf*. Fig. 40, p. 130). How long it takes for a pycnidiospore nucleus and its progeny produced by nuclear division to travel this distance is unknown; but, as bearing upon this question, Hanna's work may be cited. Hanna[2] observed that, in *Puccinia graminis tritici*, about 48 hours after the nectar of (+) and (−) pycnidia has been mixed, the nuclei at the base of the hyphal weft of each proto-aecidium become enlarged and stain more deeply and, shortly afterwards, here and there throughout the weft, binucleate cells make their appearance. He also observed that, in sections made through pustules fixed 65 hours after the mixing of the nectar, there can be seen young aecidia with as many as four aecidiospores in some of the spore-chains (*cf*. Figs. 27 and 29, pp. 104 and 108). These data permit us to conclude that the nuclear migration from a flexuous hypha to one or more proto-aecidia takes less than about 48 hours. Possibly the actual time taken in not more than about 24 hours.

Multiple Fusions and Hybridisation.—In many individual pustules multiple fusions between flexuous hyphae and pycnidiospores were observed, and some numerical details concerning these fusions have already been given. The occurrence of multiple fusions readily ex-

[1]W. Wahrlich, "Zur Anatomie der Zelle bei Pilzen und Fadenalgen," *Scripta Botanica Horti Universitatis Imperialis Petropolitanae*, T. IV, 1893, pp. 101-155, Tab. III, Fig. 26. For an account of Wahrlich's work and of the perforate septum of Ascomycetes, Basidiomycetes, and Fungi Imperfecti, *vide* these *Researches*, Vol. V, 1933, pp. 89-97.

[2]W. F. Hanna, "Nuclear Association in the Aecium of *Puccinia graminis*," *Nature*, 1929, p. 267.

plains the fact observed by Newton, Johnson, and Brown[1] that, when the nectar of various races of *Puccinia graminis tritici* is mixed and then placed on a single pycnidial pustule of another race, the aecidia which are formed on the under surface of the pustule often differ from one another in their genetic composition in that some of them give rise to one hybrid race and others to other hybrid races. Newton, Johnson, and Brown also observed that, when the nectar had been mixed in the manner just described, while most of the aecidia resulting from the crossing yielded but one physiologic race each, yet 5 per cent. of the aecidia yielded two or more races each.   When a single aecidium has produced two kinds of hybrid aecidiospores, we must suppose that originally two different nuclei derived from two different pycnidio-spores belonging to two different races entered two different flexuous hyphae of the pustule and that they (or their descendents) both made their way to the spore-bed of the aecidium in the middle layer of the leaf and both succeeded in diploidising haploid cells there, so that in the end the chains of aecidiospores in the mature aecidium came to be of two different kinds.

2. Puccinia helianthi.—My own observations[2] on fusions in *Puccinia helianthi* serve to confirm and extend those made by Craigie in 1933.

Some young foliage leaves of seedling plants of *Helianthus annuus* were inoculated with basidiospores in the Rust Laboratory on May 14, 1938.   Sixteen days later I cut off half a leaf bearing a number of simple rust pustules, mixed the nectar, floated the half-leaf on water in a Petri dish, and left the dish near a window on a laboratory table. About twenty hours later, I cut hand-sections of some of the pustules and mounted them in water.   On examining these sections with the microscope, I had no difficulty in observing a number of fusions between the pycnidiospores and the flexuous hyphae.   The pycnidio-spores of *Puccinia helianthi* are much larger than those of *P. graminis* (*cf.* A and D in Fig. 48, p. 150) and, therefore, fusions are more easily observed in *P. helianthi* than they are in *P. graminis*.   The flexuous hyphae in the particular pustules examined were mostly less than twice the length of the periphyses (ostiolar trichomes), and all the fusions observed, except one which resembled the one shown at H in Fig. 76 (p. 219), had been formed at the ends of the hyphae (*cf.* Fig. 75, p. 218).

---

[1]M. Newton, T. Johnson, and A. M. Brown, "A Preliminary Study on the Hybridization of Physiologic Forms of *Puccinia graminis tritici*," *Scientific Agriculture*, Vol. X, 1930, pp. 728-731.

[2]A. H. R. Buller, "The Flexuous Hyphae of *Puccinia graminis* and Other Uredinales," *Phytopathology*, Vol. XXXI, 1941, p. 4.   An abstract.

In single pycnidia several flexuous hyphae were observed to have each fused with a single pycnidiospore. In one pycnidium six of these fusions were seen (Fig. 78) and in another seven. Thus the observations on multiple fusions made upon *Puccinia graminis* and already recorded have been confirmed by further observations made on *P. helianthi.* It seems likely that, in single pycnidia of the Rust Fungi in general, under favourable conditions, multiple fusions often take place.

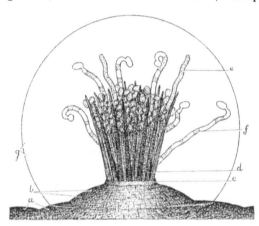

FIG. 78.—*Puccinia helianthi.* Multiple fusions between flexuous hyphae and pycnidiospores in a single pycnidium. The nectar of (+) and (−) pustules was mixed with the help of a pipette. About 20 hours later, the pycnidium here shown was found to have seven fusions (of which six are here represented) between its flexuous hyphae and pycnidiospores presumably of opposite sex: *a,* epidermis of Sunflower leaf; *b,* epipycnidial papilla; *c,* cuticular collar around ostiole of pycnidium; *d,* a periphysis; *e,* an unfused flexuous hypha; *f,* one of the flexuous hyphae which has fused terminally with a pycnidiospore. Magnification, about 600.

Brown[1] studied the interfertility of four strains of *Puccinia helianthi* (1) by employing the nectar-mixing technique and (2) by diploidising haploid pustules with diploid mycelium derived from uredospores; and he found that the strains fall into two groups: A, the strains on *Helianthus annuus* and *H. petiolaris,* and B, the strains on *H. tuberosus* and *H. subtuberosus.* The experiments revealed that the two strains in each group are highly interfertile, and that the two strains of one group are highly intersterile with the two strains of the other group. Brown suggested that each of the two groups A and B represents a variety of *Puccinia helianthi* comparable to the varieties of *P. graminis.*

In Brown's experiments with the nectar-mixing technique, which involved several hundred pustules of the four strains of *Puccinia helianthi,* we may suppose, in the light of my own observations recorded above, that each successful cross involved the fusion of one or more pycnidiospores of one strain with one or more flexuous hyphae of opposite sex of the other strain (*cf.* Fig. 78).

[1]A. M. Brown, "Studies on the Interfertility of Four Strains of *Puccinia helianthi* Schw.," *Canadian Journal of Research,* C, Vol. XIV, 1936, pp. 361-367, and Plate I.

3. Puccinia coronata avenae.—In May, 1938, some small pot plants of *Rhamnus cathartica* were inoculated in the greenhouse with basidiospores of the Oat Rust, *Puccinia coronata avenae*; and, after pustules had been formed, the infected host-plants were exposed so that flies had access to them.

Hand-sections of some of the pustules were cut and mounted in water. In a pycnidium in one of the sections I clearly saw a fusion between the end of a flexuous hypha and a pycnidiospore (Fig. 79, B).

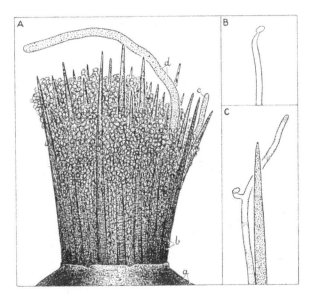

FIG. 79.—*Puccinia coronata avenae.* A, the exposed part of a pycnidium: *a*, epi-pycnidial papilla; *b*, periphyses and pycnidiospores; *c*, a short flexuous hypha of normal thickness; *d*, a long flexuous hypha of about twice the normal thickness. B and C: fusions between flexuous hyphae and pycnidiospores. In B, the hypha and spore, when found, had lost their protoplasmic contents and the wall of the hypha was becoming gelatinous. In C, the pycnidiospore has fused with a special lateral branch or fusion peg sent out by the hypha. A periphysis drawn at the same time as the hypha is also shown. Magnification, 700.

The end of the flexuous hypha was bent round toward its junction with the pycnidiospore, and the spore and adjacent part of the flexuous hyphae had already lost their massive contents and become highly vacuolated.

On June 16, 1939, a few isolated pustules of *Puccinia coronata avenae* were found on shoots of *Rhamnus cathartica* just outside the Rust Laboratory. On that day, I gathered six of the shoots, each bearing one or two leaf-pustules, and set them in water under a bell-jar. On June 17, in the evening, I mixed the nectar of six pustules and, on June 18, I cut hand-sections of the pustules, mounted them in water, and examined them under the microscope. In most of the pustules the flexuous hyphae had already become highly gelatinised and were disappearing. However, in one pustule, which was in better condition, I found a flexuous hypha and a pycnidiospore fused together (Fig. 79, C). The flexuous hypha had put out a special short lateral fusion-branch similar to the fusion-branches already described for *Puccinia graminis* (*cf*. Fig. 76, I). The pycnidiospore and adjacent part of the flexuous hypha showed the usual vacuolisation.

It thus appears that, in *Puccinia coronata avenae*, fusions take place between the flexuous hyphae and the pycnidiospores in the same way as they do in *P. graminis*, *i.e.* at the end of a hypha or laterally by means of a special fusion-branch.

4. Gymnosporangium clavipes.—At Minaki, in western Ontario, on May 20, 1939, I gathered twigs of *Juniperus communis* bearing teleutospore sori of *Gymnosporangium clavipes* and allowed them to dry. On May 27 I soaked some of the twigs in water for about half an hour so that the gelatinous teliospore sori became much swollen, and then I suspended the twigs over some small bushes of *Amelanchier alnifolia* which were growing in a wood on the bank of the Red River, close by the Rust Laboratory. The host-plants were sprayed with water and then covered with a metal tub having a glass top. Three days later the tub was removed.

On June 10, pycnidial pustules had already been formed on the Amelanchier leaves, and an examination of some of them revealed that the pycnidia contained not only pointed periphyses but also typical flexuous hyphae. A twig with leaves bearing simple pustules of the Gymnosporangium was taken to the laboratory and placed in water in a covered glass vessel set in a well-lighted position near a window.

On June 12, the pycnidial pustules on the leaves of the infected twig were covered with large drops of nectar. At 5 p.m. on that day, the nectar and spores of about twelve pustules situated on a single leaf were mixed, then re-deposited on the pustules.

About 3 p.m. on June 13 (about twenty hours after mixing the nectar), hand-sections were made of some of the pustules and these sections were then mounted in water and examined while still living. A number of fusions between the pycnidiospores and the flexuous hyphae were soon found, and it was observed that these fusions resembled those which had been seen in *Puccinia graminis* and *P. heli-*

anthi.   Some of the fusions had been formed at the ends of the flexuous hyphae and others on short lateral branches (Fig. 80).   Just as in *P. graminis*, it appeared that the flexuous hyphae had made growth-curvatures toward the pycnidiospores with which they were about to fuse and it was observed that the pycnidiospore and adjacent part of the fused flexuous hypha had become highly vacuolated.

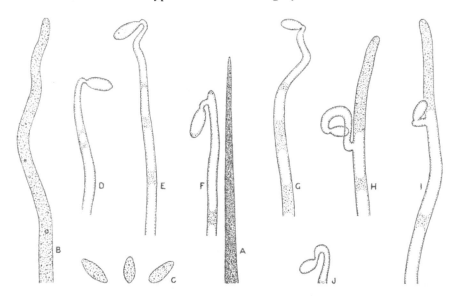

FIG. 80.—*Gymnosporangium clavipes* on *Amelanchier alnifolia*.   The nectar of (+) and (−) pustules on leaves was mixed with the help of a pipette.   Appearance of some of the elements of the pycnidia 20 hours later.   A, an orange periphysis which happened to be close to F, as shown.   B, a young flexuous hypha full of protoplasm.   C, three yellowish pycnidiospores, full of protoplasm.   D-J, seven flexuous hyphae each of which has fused with a pycnidiospore presumably of opposite sex.   In D, E, G, and J, the apical end of the flexuous hypha has fused with a pycnidiospore.   In F, H, and I, a lateral branch or fusion peg has fused with a pycnidiospore.   Each fused pycnidiospore and the adjacent part of the flexuous hypha is vacuolate.   Magnification, 1,050.

The pycnidiospores of *Gymnosporangium clavipes* are more or less spindle-shaped, 6-8μ in length, and about 3μ in width.   On account of their large size, it is easier to perceive a fusion between one of them and a flexuous hypha than it is in *Puccinia graminis*; and, for the same reason, one can readily convince oneself that in *Gymnosporangium clavipes*, when a union has taken place, there is a distinct and continuous channel leading from the pycnidiospore to the flexuous hypha. The channel shown in Fig. 80, F, was used as a demonstration object for fellow workers.

The pycnidiospores in the mixed nectar showed no signs of independent germination. Those which had become fused with a flexuous hypha were attached to the hypha by one of their ends. From the appearance of the fused elements it became clear that, just as in *Puccinia graminis*, the flexuous hypha is attracted towards and grows toward the pycnidiospore with which it is destined to fuse and that the pycnidiospore co-operates in establishing the fusion when the flexuous hyphae is about to touch, or has already touched, one of its ends. Fusions between flexuous hyphae and pycnidiospores in Gymnosporangium differ in no essential particular from those in Puccinia.

In *Gymnosporangium clavipes*, since fusions were not seen in isolated pycnidial pustules in which nectar had not been mixed but were found in abundance in similar pustules within 24 hours after the nectar had been mixed, one may safely infer that *G. clavipes* is heterothallic.

Five of the simple Gymnosporangium pustules whose nectar had been mixed were left intact on an Amelanchier leaf. Within about 48 hours the nectar on these pustules dried up and, during the next two days, aecidia made their appearance. On the other hand, five simple pustules on another leaf of the twig, whose nectar had not been mixed, retained their nectar and did not produce aecidia. These observations, although few in number, support the inference made from the observations on fusions and provide additional evidence justifying the conclusion that *Gymnosporangium clavipes* is heterothallic.

Since (1) *Gymnosporangium globosum*, *G. haraeanum*, and *G. juniperi-virginianae*, by means of the nectar-mixing technique, have all been shown to be heterothallic,[1] since (2) the evidence from fusions, as just set forth, indicates that *G. clavipes* is heterothallic, and since (3) *G. clavariiforme* and *G. juvenescens* have been found to have flexuous hyphae,[2] we may conclude that the genus Gymnosporangium is heterothallic.

[1] *Vide supra*, Chapters I and III.
[2] *Vide supra*, Chapter III.

CHAPTER V

THE PRESENCE OR ABSENCE OF PYCNIDIA AND ASSOCIATED SPORE-FORMS IN CERTAIN UREDINALES

Microcyclic Rust Fungi without Pycnidia — A Microcyclic Rust with Pycnidia only on Certain Hosts — Desirability of Experiments on Microcyclic Rusts provided with Pycnidia — Vestigial Pycnidia in *Coleosporium pinicola* and *Calyptospora goeppertiana* — A Variety of *Puccinia coronata* with Vestigial Pycnidia — Correlation of Facts concerning the Presence or Absence of Pycnidia and Sexuality — Pycnidia and the Production of Uredospores and Teleutospores of *Puccinia graminis* on Barberry Leaves — Short-cycling in *Uromyces fabae* — Supposed Occasional Association of Pycnidia and Uredospore Sori in *Puccinia helianthi* — *Uromyces hobsoni* and its Pycnidia.

Microcyclic Rust Fungi without Pycnidia.—There are a number of microcyclic (short-cycled) Rusts which never produce pycnidia, and as examples of them one may mention *Puccinia malvacearum* (the Hollyhock Rust) and *P. xanthii* (on *Xanthium commune*, etc., in North America). In both these species a basidiospore germinates on a host-leaf and the germ-tube, after boring through the cuticle, develops into a mycelium which gives rise not to pycnidia and protoaecidia, but to a sorus of teleutospores.

In Fig. 81 is shown a semi-diagrammatic drawing of a transverse section of a teleutospore pustule of *Puccinia malvacearum*. We may suppose that the mycelium within the pustule was derived from a single basidiospore which fell upon, and germinated on, the upper surface of the host-leaf. It will be noted that the mycelium has failed to produce any pycnidia and bears teleutospores only. The teleutospores are shown germinating *in situ* and giving rise to basidia and to basidiospores which are being violently discharged.

One may suppose, with Jackson,[1] that such a fungus as *Puccinia malvacearum* was once macrocyclic (long-cycled) like *P. graminis* and

[1]H. S. Jackson, "Present Evolutionary Tendencies and the Origin of Life Cycles in the Uredinales," *Memoirs Torrey Bot. Club*, Vol. XVIII, 1931, pp. 1-108. Jackson holds that microcyclic forms arise in most cases from the haploid generation of heteroecious eu-forms through the replacement of the aecidia by teleutospore sori.

241

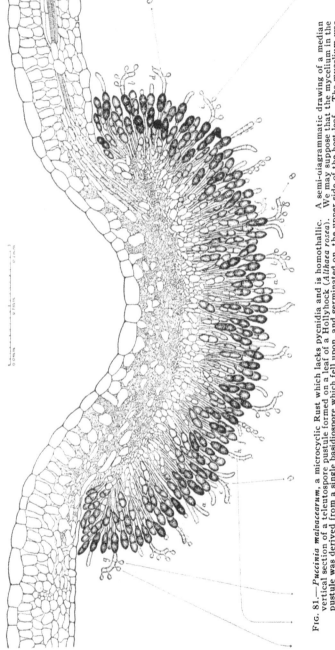

Fig. 81.—*Puccinia malvacearum*, a microcyclic Rust which lacks pycnidia and is homothallic.   A semi-diagrammatic drawing of a median vertical section of a teleutospore pustule formed on a leaf of a Hollyhock (*Althaea rosea*).   We may suppose that the mycelium in the pustule was derived from a single basidiospore which fell upon, and germinated on, the upper side of the host-leaf.   The mycelium was at first uninucleate; but, in the spore-bed of th teleutospore sorus, it spontaneously became binucleate.   It failed to produce pycnidiospores, aecidiospores, and uredospores and it developed teleutospores only, as shown.   Two nuclei fused together in each of the two cells of each teleutospore.   Many of these cells are forming, or already have formed, basidia.   In each basidium the fusion nucleus divided twice, so as to provide the four haploid nuclei destined to pass through four sterigma into as many basidiospores.   For a further description of the pustule and, more particularly, of the basidia and the discharge of the basidiospores *vide* Vol. III, Fig. 216, p. 537.   Magnification, 163.

that, in the course of evolution, its pycnidial, aecidial, and uredinial stages were eliminated with a consequent short-circuiting of the life-history.

In the course of my studies of the structure and sexual processes of the Hymenomycetes I had noticed that, in the genus Coprinus, the heterothallic species, *e.g. C. lagopus, C. macrorhizus,* and *C. niveus,* have haploid oidia on their haploid mycelia (derived from basidio-spores), whereas the homothallic species, *e.g. C. sterquilinus, C. sterco-rarius,* and *C. narcoticus,* produce no oidia whatever. In the belief that the Uredinales are related to the Hymenomycetes and may have been derived from one of their "lower" or ancestral forms, and on the supposition that the pycnidiospores of the Uredinales are equivalent to the oidia of the Hymenomycetes, I came to the conclusion in 1931 that Rusts which are devoid of pycnidia are probably homothallic.[1] The idea that *Puccinia malvacearum* is homothallic was supported by my finding on one of a large number of otherwise entirely rust-free Hollyhocks (*Althaea rosea*) in a garden at Kew a solitary *P. malvacearum* pustule which appeared quite simple (as if derived from a single basidiospore), but which, nevertheless, bore teleutospores.[2]

In 1931, Jackson,[3] working independently, had also come to the conclusion that *Puccinia malvacearum* is homothallic. In his masterly paper on the evolutionary tendencies in the Uredinales he said: "If the pycniospores exercise the function indicated by Craigie's work, then it would appear quite probable that they would be functionless in homothallic species and might soon be dropped from the life cycle. It also seems probable that when they are absent the species is homo-thallic. Preliminary culture experiments conducted by the writer with *Puccinia malvacearum,* a micro-form which does not develop pycnia, strongly indicate that in this species every infection, mono-sporidial or otherwise, develops one or more sori of teliospores. The indications are that this species is homothallic."

The supposition that *Puccinia malvacearum* is homothallic was confirmed by Miss Ashworth[4] who studied the fungus in monosporidial cultures. With the help of a micromanipulator she removed basidio-spores from a spore-deposit and sowed each one of them separately in a drop of water on a host-leaf. In 1000 single-spore inoculations she obtained rust pustules from 18 only; but, in all the 18 pustules, teleutospores developed on the under surface. Moreover, she observed

[1]These *Researches*, Vol. IV, 1931, p. 286.   [2]*Ibid.*
[3]H. S. Jackson, *loc. cit.,* p. 99.
[4]Dorothy Ashworth, "*Puccinia malvacearum* in Monosporidial Culture," *Trans. Brit. Myc. Soc.,* Vol. XVI, 1931, pp. 177-202.

that, after inoculation, teleutospores were formed just as soon in the simple pustules as they had been in compound pustules produced by the sowing of basidiospores *en masse*.   The low percentage of single-spore infections, 1.8, was doubtless due to the technique employed in handling and sowing the basidiospores.[1]

Miss Allen,[2] on the basis of a cytological study of *Puccinia malvacearum*, said: Miss Ashworth "assumed that infection did not take place in the other 982 inoculations.   So far as stated, however, these were not sectioned, and it is possible that many of the inoculations resulted in minute infections that died without becoming large enough to be visible macroscopically.   It is also possible that, while the 18 inoculations resulting in infections with sori were still small, insects transferred conidia or sporidia to them.   On this basis *P. malvacearum* would be heterothallic."   This criticism of Miss Ashworth's conclusion that *P. malvacearum* is homothallic is not well based; for (1) there is no evidence of any value that *P. malvacearum* produces conidia,[3] and (2) there is no evidence whatsoever that insects assist the wind in the dispersal of Rust basidiospores.   It is most unlikely that the 18 pustules bearing teleutospores which Miss Ashworth obtained from single-spore inoculations should have been diploidised in the manner suggested by Miss Allen.

Fig. 82.—*Puccinia malvacearum*, without pycnidia and homothallic, on *Malva rotundifolia*; 12 days after inoculation with single basidiospores. The 10 simple and 2 compound pustules have all produced teleutospores and simultaneously.   Photographed by A. M. Brown.   Natural size.

[1]J. H. Craigie ("An Experimental Investigation of Sex in the Rust Fungi," *Phytopathology*, Vol. XXI, 1931, pp. 1010-1013) tried a micromanipulator technique with *Puccinia graminis*, but the number of infections obtained was so small that he discarded it in favour of allowing the basidiospores to fall on to the Barberry leaves from germinating teleutospores attached to straws suspended above the bushes.

[2]Ruth Allen, "A Cytological Study of *Puccinia malvacearum* from the Sporidium to the Teliospore," *Journ. Agric. Research*, Vol. LI, 1935, p. 813.

[3]Miss Allen alone (*ibid.*, p. 808) has suggested that *Puccinia malvacearum* may produce conidia, but her supposition, put forward doubtfully even by herself, is based on a study of microtome sections only.   Her supposed conidia appear to be nothing more than the ends of hyphae which, after having grown outwards to the surface of a Mallow leaf *via* a stomatic cleft, have afterwards been pinched subterminally by the closing of the guard-cells.

Brown,[1] by permitting the basidiospores of *Puccinia malvacearum* to fall as they do in nature on to leaves of *Althaea rosea* and *Malva rotundifolia*, obtained over 1,000 single-spore infections resulting in pustules. None of these pustules was sterile, but all bore an abun-

Fig. 83.—*Puccinia xanthii*, without pycnidia and homothallic, on *Xanthium commune*. About two weeks after inoculation with single basidiospores. A, the *upper* side of a leaf with three simple pustules without any pycnidia. B, the *under* side of a leaf with two simple and one compound pustule, all of which have produced teleutospores and produced them simultaneously. Photographed by A. M. Brown. Natural size.

dance of teleutospores on their under surface (Fig. 82). It was also observed that, after inoculation and on the same leaf, a simple pustule produces teleutospores at the same time as a compound pustule derived from the fusion of two simple pustules.

Thus Brown's observations confirm those of Miss Ashworth and show conclusively that *Puccinia malvacearum* is homothallic.

[1]A. M. Brown, "The Sexual Behaviour of Several Plant Rusts," *Canadian Journal of Research*, Vol. XVIII, Section C, 1940, pp. 23-24. Also *vide supra*, Chapter I, p. 86.

Brown[1] sowed basidiospores of *Puccinia xanthii* singly on leaves of *Xanthium commune* and obtained 350 rust infections. When from ten to twelve days old, all these infections produced teleutospore sori (Figs. 83 and 84). From the results of his experiments Brown rightly concluded that *Puccinia xanthii*, like *P. malvacearum*, is homothallic.

If a short-cycled Rust such as *Puccinia malvacearum* or *P. xanthii* were heterothallic instead of homothallic, its chances of survival would be very small; for, in the absence of pycnidiospores, flies would not assist these fungi in the initiation of their diploidisation process, and every isolated haploid pustule would perforce remain sterile. From this consideration alone we may conclude that short-cycled Rusts that have lost their pycnidia but have retained their heterothallism probably do not exist. The lack of pycnidia and the consequent impossibility of employing the pycnidial mode of initiating the diploidisation process finds its compensation in homothallism.

A Microcyclic Rust with Pycnidia only on Certain Hosts.—In the Dominion Rust Research Laboratory at Winnipeg, Brown[2] sowed basidiospores of *Puccinia grindeliae* on leaves of *Grindelia squarrosa* and obtained 150 well-separated infections. All these infections, when from ten to twelve days old, produced tele-

Fig. 84.—*Puccinia xanthii* on *Xanthium commune*. Under side of leaf, 10 days after inoculation: the simple pustules have begun to form teleutospores as soon as the compound pustule. Photograph by A. M. Brown. Natural size.

utospore sori. It was thus proved that *Puccinia grindeliae* is homothallic.

Although *Puccinia grindeliae* resembles *P. malvacearum* and *P. xanthii* in being microcyclic and homothallic, it differs from those species

[1] *Ibid.*

[2] *Ibid.* Also *vide supra*, Chapter I, p. 86.

in that it sometimes produces pycnidia although only "on certain hosts, particularly on *Xylorrhiza*."[1] At Winnipeg, on *Grindelia squarrosa*, it produces no pycnidia whatever.

Puccinia grindeliae, as Jackson[2] has pointed out, is a microcyclic Rust which can be correlated with, and has probably evolved from, the heteroecious long-cycled Rust, *P. stipae*, which has aecidia on *Grindelia squarrosa*, etc., and uredospore and teleutospore sori on species of *Stipa* and related grasses. Occasionally in *Puccinia grindeliae*, as Jackson[3] and Brown[4] have observed, aecidiospores and peridial cells occur in the teleutospore sori. This fact indicates very clearly that the fungus once produced aecidia regularly and that these organs have since been suppressed. As pycnidia are produced by *P. grindeliae* only on certain hosts, it would appear that they, too, are being suppressed.

The suppression of the pycnidia in the variety of *Puccinia grindeliae* which occurs at Winnipeg is doubtless correlated with the fact that the variety in question is homothallic. It may be that this homothallic variety was produced by a mutation from the ancestral heterothallic species and that it became homothallic and ceased to produce pycnidia at one and the same time.

The fact that there are varieties of *Puccinia grindeliae* which still produce pycnidia suggests that these varieties are heterothallic. As will be shown shortly, in *P. coronata*, the variety *avenae* with well-developed pycnidia is heterothallic, whereas the variety *elaeagni* with vestigial pycnidia is homothallic. It is therefore possible that the species *P. grindeliae* may be composed of one or more varieties that lack pycnidia and are homothallic and of one or more varieties that have well-developed pycnidia and are heterothallic. Whether or not this suggestion is well based can be decided only by further experiment. Another possibility that cannot be excluded, but which seems less likely to be true, is that the varieties of *P. grindeliae* with pycnidia resemble the variety without pycnidia in being homothallic. However, it has yet to be shown that in any species of Rust in which the pycnidia are well-developed, *i.e.* in which the pycnidia excrete nectar, discharge pycnidiospores freely, and possess flexuous hyphae, the pycnidia are completely functionless. It is known from experiment that the pyc-

[1]J. C. Arthur, *Manual of the Rusts in United States and Canada*, Lafayette, U.S.A., 1934, p. 141.

[2]H. S. Jackson, "Present Evolutionary Tendencies and Origin of Life Cycles in the Uredinales," *loc. cit.*, pp. 33-34.

[3]*Ibid.*, p. 36.

[4]A. M. Brown, personal communication, 1939.

nidia of twenty-two species[1] of Rust Fungi are functional, and we are therefore justified in assuming as true, or at least as a working hypothesis, that *in the Rust Fungi in general, including the microcyclic forms, well-developed pycnidia are functional.*

Desirability of Experiments on Microcyclic Rusts provided with Pycnidia.—It is desirable that the presumption that microcyclic Rusts provided with well-developed pycnidia are heterothallic should be put to the test of nectar-mixing experiment. However, up to the present, this has not been attempted.

Lamb,[2] in 1934, gave an account of his cytological investigation of *Puccinia prostii*, a micro-form with pycnidia, that in Italy and southern France parasitises *Tulipa sylvestris* and *T. celsiana*. Lamb found the fungus attacking *T. sylvestris* in the Royal Botanic Garden at Edinburgh and, apparently, perennating in the bulb by means of a systemic mycelium. It was observed that the dikaryophase is initiated at the bases of the fundaments of the teleutospore sori by cell-fusions and nuclear migrations. Nevertheless, Lamb admitted that *Puccinia prostii* and similar microcylic Rusts may be heterothallic. Unfortunately, Lamb was unable to follow up his cytological findings with nectar-mixing experiments on account of the fact that the teleutospores could not be induced to germinate.

Vestigial Pycnidia in Coleosporium pinicola and Calyptospora goeppertiana.—When, in the course of evolution, a Rust passes from the heterothallic condition to the homothallic, there is the possibility that its pycnidia, instead of becoming entirely suppressed as they have been in *Puccinia malvacearum* or entirely suppressed only on some of the hosts as in *P. grindeliae*, may become degenerate and vestigial, so that traces of them can still be observed. Such vestigial pycnidia have been found in (1) *Coleosporium pinicola* and (2) *Calyptospora geoppertiana*.

(1) The short-cycled Pine Rust, *Coleosporium pinicola* (*Gallowaya pinicola*), is parasitic on *Pinus virginiana* in the eastern part of the United States of America and has been reported on *Pinus cembra* in western Siberia.[3] In this rust, according to Dodge,[4] "vestigial spermogonia are to be found occasionally, but the host tissues above them are not fully ruptured and spermatia are seldom formed."

---

[1] *Vide* the list of heterothallic Rust species given on p. 85.

[2] I. M. Lamb, "On the Morphology and Cytology of *Puccinia prostii* Moug., a Micro-form with Pycnidia," *Trans. Roy. Soc. Edinburgh*, Vol. LVIII, 1933-1934, pp. 143-162. Author's abstract in *Biological Abstracts*, Vol. X, November, 1936, no. 21975.   [3] J. C. Arthur, *loc. cit.*, p. 47.

[4] B. O. Dodge, "Organisation of the Telial Sorus in the Pine Rust, *Gallowaya pinicola* Arth.," *Journ. Agric. Research*, Vol. XXXI, 1925, p. 641 and Plate I, Fig. 1.

Dodge's finding has been confirmed by Miss Hunter.[1] She observed in her microtome section that the pycnidia of *Coleosporium pinicola* are: of the applanate type, lenticular in vertical section, subepidermal, and amphigenous, nearly always centrally located beneath the guard cells in stomatal cavities, and spread out between the mesophyll and the epidermis; and she further observed that, in a single pycnidium, a basal stroma, made up of interwoven hyphae gives rise to attenuate pycnidiosporophores (spermatiophores) that do not produce any pycnidiospores (spermatia). In her own words "Spermatia do not form."

(2) *Calyptospora goeppertiana* (*Pucciniastrum goeppertianum*) is heteroecious. It produces aecidia and inconspicuous pycnidia on the leaves of Abies and teleutospores in the epidermal cells of the stems of Vaccinium. On Vaccinium the mycelium causes the formation of witches brooms in which the upright stems are seen as smooth, swollen, bright-brown, cylindrical structures. I found these brooms in profusion in Blueberry patches (*Vaccinium canadense* and *V. pennsylvanicum*) at Minaki in the extreme western part of Ontario in August, 1943. Similar brooms have also been found at Victoria Beach on Lake Winnipeg in Manitoba and at Lake Waskesiu in Saskatchewan.[2]

Our knowledge of *Calyptospora goeppertiana* was reviewed by Faull[3] in 1939, and from that review the following observations have been taken. Robert Hartig, 1880-1881, proved the connexion of the two phases of the life-history by means of culture experiments: he passed the fungus from *Vaccinium vitis-idaea* to *Abies pectinata* and then in the reverse direction. Cultures on Abies species were also made by Kühn and Bubák in Europe and by Fraser in Canada, but none of these mycologists observed any pycnidia. Arthur made similar cultures and remarked "pycnia rarely if ever found." Weir cultured a rust, which he took to be *Calyptospora goeppertiana*, from *Abies lasiocarpa* to *Vaccinium membranaceum*, and *vice versa*. Faull examined this material, found some pycnidia on the Abies, but expressed some doubt as to the identification of the rust species. Finally, in 1939, Faull[4] reported that he had made many cultures of *Calyptospora goeppertiana* on *Abies balsamea* with inoculum obtained from *Vaccinium*

[1]Lillian M. Hunter, "Morphology of the Spermogonia of the Melampsoraceae," *Journal of the Arnold Arboretum*, Vol. XVII, 1936, p. 139 and Plate CLXXXVIII, Fig. 41.

[2]G. R. Bisby *et al.*, *The Fungi of Manitoba and Saskatchewan*, Ottawa, 1938, p. 63.

[3]J. H. Faull, "A Review and Extension of our Knowledge of *Calyptospora goeppertiana* Kuehn," *Journal of the Arnold Arboretum*, Vol. XX, 1939, pp. 104-113. To this paper the reader is referred for literature citations for the work of Hartig, Kühn, Bubák, Fraser, and Weir.   [4]*Ibid.*

canadense and *V. pennsylvanicum*, with the result that, on the leaves of the Balsam Fir, pycnidia appeared from 13 to 19 days after inoculation. Thus the fact that *Calyptospora goeppertianum* does give rise to pycnidia has been placed beyond doubt.

Miss Hunter,[1] working in Faull's laboratory, with the help of stained microtome sections, made a thorough study of *Calyptospora goeppertiana* on *Abies balsamea* and found that the pycnidia do not form pycnidiospores and do not rupture the cuticle lying above them. Subsequently she[2] examined more than 500 pycnidia in cultured material and found, as before, that they had not formed pycnidiospores and had not ruptured the cuticle.

Since the pycnidia of *Coleosporium pinicola* and *Calyptospora goeppertiana* do not produce pycnidiospores or rupture the overlying epidermis or cuticle of their hosts, it is clear, as Miss Hunter remarks, that they cannot function in the manner described for Puccinia species by Craigie and others. Indeed, it is obvious that they are just as functionless as the vestigial back pair of stamens in a Salvia flower or as the eye-less eye-stalks of crustacea that live in the dark in the waters of certain underground caverns.

Since *Coleosporium pinicola* and *Calyptospora goeppertiana* lack functional pycnidia, it seems highly probable that they are homothallic.

A Variety of Puccinia coronata with Vestigial Pycnidia.—Brown and Craigie[3] have studied *Puccinia coronata elaeagni* Fraser and Ledingham. Fifty-one isolated monosporidial pustules were obtained on leaves of *Elaeagnus commutata*. In the infections no pycnidia or pycnidial nectar could be detected with the aid of a strong hand-lens and, in consequence, the transfer of nectar from one pustule to another was precluded. Within eighteen days aecidia formed spontaneously in forty-nine of the infections, and this clearly indicated that the fungus is homothallic. In three infections studied cytologically, the mycelium was uninucleate and the aecidiospores binucleate, but it was not determined how the binucleate condition arose. Further investigations with the help of microtome sections revealed that, in some of the pustules, there were under the upper epidermis a few minute

---

[1]Lillian M. Hunter, "Comparative Study of Spermogonia of Rusts of Abies," *Botanical Gazette*, Vol. LXXXIII, 1927, pp. 1-23.

[2]Lillian M. Hunter, "Morphology and Ontogeny of the Spermogonia of the Melampsoraceae," *Journal of the Arnold Arboretum*, Vol. XVII, 1936, p. 123.

[3]A. M. Brown and J. H. Craigie, "Studies on the Sexual Behaviour of Plant Rusts," presented to the Canadian Phytopathological Society in June, 1937, *Report by the Dominion Botanist for the years 1935 to 1937 inclusive*, Ottawa, 1938. Also A. M. Brown, "The Sexual Behaviour of Several Plant Rusts," *Canadian Journal of Research*, Vol. XVIII, 1940, pp. 21-24.

pycnidia which had become arrested in their development (Fig. 85). These structures had not opened at their apex, had not developed any periphyses or flexuous hyphae, and had not produced any pycnidiospores. Thus, in *Puccinia coronata elaeagni*, such pycnidia as are formed are undoubtedly nothing but vestigial organs which in the ancestors of this Rust form were once active, but which have now become degenerate and functionless.

Fraser and Ledingham,[1] by culture methods and a study of the characters of the aecidia, found that in *Puccinia coronata*, in the prairie provinces of Canada, there are four varieties:

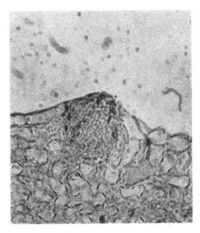

FIG. 85.—A vestigial pycnidium of *Puccinia coronata elaeagni*. Photographed by A. M. Brown and J. H. Craigie. Magnification, about 900.

(1) *Puccinia coronata avenae*, the Crown Rust of oats. Pycnidia large, numerous, with an average diameter of 124 μ (Fig. 86, A).

(2) *Puccinia coronata calamagrostidis*. Pycnidia small, average diameter 103 μ (Fig. 86, B).

(3) *Puccinia coronata bromi*. Pycnidia few, small, average diameter 109 μ (Fig. 86, C).

(4) *Puccinia coronata elaeagni*. Pycnidia rare or wanting (Fig. 85).

Puccinia coronata avenae, as Miss Allen[2] has shown, is heterothallic. *P. c. calamagrostidis*, as I have observed in greenhouse cultures, has pycnidia that develop periphyses, flexuous hyphae, and pycnidiospores and excrete nectar. This indicates that *P. c. calamagrostidis* is heterothallic. It may well be (*cf.* Fig. 86, B and C) that the pycnidia of *P. c. bromi* are also functional. On the other hand, as Brown and Craigie have taught us, the pycnidia of *P. c. elaeagni* are vestigial (Fig. 85) and functionless and *P. c. elaeagni* is homothallic. Thus the four varieties of *P. coronata* enumerated above appear to form a series which passes from the heterothallic to the homothallic condition in association with a diminution in size of, and a final degeneracy and almost total disappearance of, the pycnidia.

[1]W. P. Fraser and G. A. Ledingham, "Studies of the Crown Rust, *Puccinia coronata* Corda," *Scientific Agriculture* (Ottawa), Vol. XIII, 1933, pp. 313-323.

[2]Ruth F. Allen, "A Cytological Study of Heterothallism in *Puccinia coronata*," *Journ. Agric. Research*, Vol. XLV, 1932, pp. 521-523.

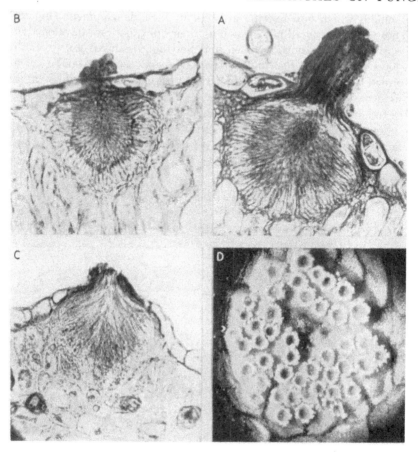

FIG. 86.—Progressive degeneracy of pycnidia in three varieties of *Puccinia coronata*. Sections of pycnidia: A, *P. coronata avenae* on *Rhamnus cathartica*; B, *P. coronata calamogrostis* on *Rhamnus alnifolia*; C, *P. coronata bromi* on *Lepargyraea canadensis*; cf. Fig. 85, *P. coronata elaeagni* on *Elaeagnus commutata*. D, open aecidia on the under side of a pustule on a leaf of *Rhamnus cathartica*. Photographs by W. P. Fraser and G. A. Ledingham (1933). Magnification: A, B, and C, 508; D, about 15.

Puccinia coronata elaeagni produces aecidia and vestigial pycnidia on Elaeagnus and uredospores and teleutospores on a variety of *Calamagrostis elongata*.[1]   It is therefore long-cycled.   As a long-cycled Rust, it is remarkable in being homothallic.   No other such long-cycled Rust is at present known.

Correlation of Facts concerning the Presence or Absence of Pycnidia and Sexuality.—On the basis of our knowledge (1) that *Puccinia malvacearum, P. xanthii, P. grindeliae* (form at Winnipeg),

[1] W. P. Fraser and G. A. Ledingham, *loc. cit.*, p. 322.

and *P. coronata elaeagni* are devoid of functional pycnidia and are homothallic, and (2) that twenty-two other Rust species,[1] *e.g. P. graminis, P. helianthi, P. coronata avenae, Phragmidium speciosum,* and *Gymnosporangium juniperi-virginianae,* have well-developed pycnidia and are heterothallic, we are justified in coming to the conclusion that, throughout the Uredinales, the *presence of well-developed pycnidia is correlated with heterothallism and the absence or imperfect development of pycnidia is correlated with homothallism.* If there are any exceptions to this rule, they have yet to be discovered.

Pycnidia and the Production of Uredospores and Teleutospores of Puccinia graminis on Barberry Leaves.—Newton and Johnson,[2] employing the nectar-mixing technique, selfed certain selected strains of *Puccinia graminis tritici* for several successive generations and thereby obtained rust strains with various abnormal characteristics manifested in all the stages of the life-history: pycnidial, aecidial, uredinial, and telial.

In certain of their inbred strains, in the F_3 generation, the pycnidial mechanism had become weakened, for the pycnidia themselves had become fewer, the nectar scant, and intermixing of the nectar of the pustules (or the transfer to these pustules of nectar from other strains) ineffective so far as the production of aecidia is concerned. The mycelium in the pustules appeared to be vigorous and abundant. Forty-four days after one of the cultures had been made by inoculating the leaves of *Berberis vulgaris* with basidiospores, small uredospore sori[3] were observed on the upper surface of several of the pustules (*cf.* Fig. 87) and "further examination revealed that, in a total of 129 pustules, uredia were present in 50, and that 21 of these contained telia." The authors added: "The uredia and telia may occur on the upper or lower surface of a pustule, but more frequently on the upper surface."[4]

Newton and Johnson had previously observed that, in the F_2, F_3, and F_4 progeny of certain crosses, after the nectar had been intermixed, the pustules had failed to produce aecidia; but never before had they found failure to produce aecidia associated with the production of uredospores and teleutospores.

[1] *Vide supra,* Chapter I, p. 85.

[2] Margaret Newton and T. Johnson, "Production of Uredia and Telia of *Puccinia graminis* on *Berberis vulgaris,*" *Nature,* Vol. CXXXIX, 1937, p. 800. Also T. Johnson and Margaret Newton, "The Origin of Abnormal Rust Characteristics through the Inbreeding of Physiologic Races of *Puccinia graminis tritici,*" *Canadian Journal of Research,* Section C, Vol. XVI, 1938, pp. 38-52. [And see *Mycologia,* XXXIX, 1947, pp. 145-151.]

[3] I had the pleasure of seeing some of these sori on the day that they were discovered.   [4] M. Newton and T. Johnson, 1937, *loc. cit.*

What happened in Newton and Johnson's abnormal F_3 cultures may have been as follows. After the nectar had been mixed, some of the flexuous hyphae fused with pycnidiospores of opposite sex and the pycnidial nuclei made their way down the flexuous hyphae into the leaf-mycelium in the usual way, so that in this mycelium there were both (+) and (−) nuclei. For some reason or other the proto-aecidia

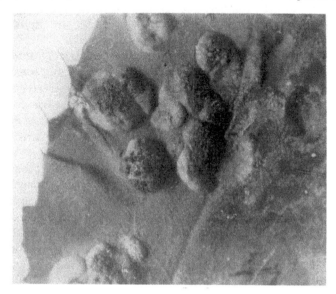

Fig. 87.—Abnormal inbred race of *Puccinia graminis* on *Berberis vulgaris*. Upper surface of part of a leaf showing upwardly swollen pustules (derived from basidiospores) which, after mixing of the nectar, have produced uredospores and then teleutospores. These two kinds of spores are here present together in the central parts of the pustules. On the under side of the leaf there are no aecidia. Photographed by T. Johnson and M. Newton. Several times the natural size.

failed to attract to themselves, or failed to receive, the (+) or (−) nuclei required to effect their diploidisation and so remained sterile. The mycelium in a pustule, having grown well and attained a certain mass, was ready for reproduction and, not having been able to relieve itself by producing aecidiospores, at length found an unusual outlet by producing uredospores and teleutospores. The mycelium in each abnormal pustule had grown old and it produced uredospores on a Barberry leaf at about the same time as, in the normal course of the life-history if aecidiospores had been formed, uredospores would have developed on a wheat-plant or other alternate host.

Johnson and Newton[1] point out that the development of abnormal strains of *Puccinia graminis* is not an inevitable consequence of in-

[1]T. Johnson and M. Newton, 1938, *loc. cit.*, p. 38.

breeding, as many inbred strains show no abnormal characteristics. Inbreeding, therefore, does not always lead to the weakening of the pycnidial mechanism and failure to produce aecidia. Johnson and Newton finally suggest that the abnormal characteristics which they observed in certain of their inbred strains "are, in most cases, the result of recessive mutations that have taken place in the past history of the rust, the part played by the selfing being that of segregating and recombining the mutant factors in a homozygous state under which their effects are manifested in various types of abnormalities."

Johnson and Newton[1] also obtained uredospores and teleutospores on a Barberry leaf by the culture of Race 21 of *Puccinia graminis* collected in the field at Indian Head, Saskatchewan, in 1934. Teleutospores of this culture were produced in the greenhouse in the spring of 1936 and Barberry bushes were infected by these in January, 1937. When pustules developed on the Barberry, it was observed that only about one-half of them produced pycnidial nectar in a normal manner. The remaining pustules were almost white in colour and many of them produced no pycnidia or only rudimentary ones. Some, however, contained scattered pycnidia which eventually produced small quantities of pycnidiospore-bearing nectar. Many of the white pustules began to produce uredospores and teleutospores about six weeks after inoculation. Some of the compound pustules produced aecidia in one component and uredospores and teleutospores in the other component and such a compound pustule is shown in Fig. 88. Here we have pycnidia, aecidia, uredospore sori and teleutospore sori all present together on the under side of a single leaf!

The culture of Race 21, in which the production of aecidia was only partly suppressed, stands midway between a normal culture and the inbred cultures already described in which the production of aecidia was entirely suppressed.

Johnson and Newton found that the uredospores produced in their abnormal cultures on Barberry leaves are incapable of infecting Barberry leaves, but readily infect Wheat seedlings. To that extent the new *Puccinia graminis* strains are still heteroecious. However, we may presume that the teleutospores produced on a Barberry leaf, if properly matured, would germinate and give rise to basidiospores that would again infect a Barberry leaf. Such an infection would demonstrate that the new *P. graminis* strains have become *autoecious* in the sense that they are able to carry through on a single host a short-cycled life-history corresponding precisely to that of a microcyclic Puccinia provided with pycnidia.

[1] *Ibid.*, pp. 47-48.

FIG. 88.—Abnormal culture of Race 21 of *Puccinia graminis* transferred to *Berberis vulgaris*. Under side of a Barberry leaf showing two compound pustules (derived from basidiospores). The pustule with a rounded margin near the apex of the leaf bears pycnidia and it gave rise to a group of elongated aecidia, but these were removed for culture purposes and their position is now seen as a dark central cavity. The central compound pustule shows: (1) *pycnidia, a*, with small dried-up masses of nectar; (2) in one component of the pustule, a few elongated finger-like *aecidia*, b, which, owing to the dryness of the greenhouse atmosphere, have not yet opened; (3) *uredospore sori, c*, still partly covered by epidermis; and (4) sori, *d*, in which *teleutospores* have come up beneath uredospores, so that both uredospores and teleutospores are associated with one another. Photographed by T. Johnson and M. Newton through an appropriate Wratten light-filter. About five times the natural size.

The development in Puccinia graminis of strains that give rise on Barberry leaves to pycnidia, aecidia, uredospore sori, and teleutospore sori is, from the evolutionary point of view, of special interest; for it supports the contentions of those who, like Jackson,[1] hold that in the course of the more recent evolution of the Uredinales there has been a derivation of autoecious species from heteroecious species and of microcyclic species from macrocyclic species.

Short-cycling in Uromyces fabae.—*Uromyces fabae* is a long-cycled heterothallic Rust, and it produces its pycnidia, aecidia, uredospore sori, and teleutospore sori on one and the same host-plant. It attacks various species of Lathyrus, Pisum, and Vicia, and it is well known as a parasite on the Garden Pea, *Pisum sativum*, and on the Broad Bean, *Vicia faba*.

Uromyces fabae tends to short-cycle its life-history, and this was first noticed by de Bary[2] who, in 1863, remarked that in some instances the same mycelium that produces aecidia gives rise to uredospore sori.

A. M. Brown,[3] in the course of his study of *Uromyces fabae* in the Dominion Rust Research Laboratory, observed that while, normally, the mycelium derived from basidiospores gives rise to pycnidia and aecidia only, not infrequently the pycnidia are associated: either (1) with aecidia and uredospore sori, as de Bary had observed, or (2) with uredospore sori only. Some particulars of Brown's observations will now be given.

(1) Basidiospores of the strain of *Uromyces fabae* inhabiting *Lathyrus venosus* were sown on seedling hosts of *L. ochroleucus* and *L. venosus*, and sixty well-isolated simple pustules (each containing a haploid mycelium derived from a single basidiospore) were secured. To fifty of these pustules, when sixteen days old, mixed nectar taken from a number of other haploid pustules was applied. Seven days later, forty-nine of the fifty pustules produced aecidia. Within three days after the appearance of the aecidia seventeen of the pustules produced uredospore sori (Fig. 89), so that in these seventeen pustules *aecidia and uredospore sori became associated.*

In another experiment uredospores were sown near the periphery of each of ten simple pustules. About ten days later, after the diploid mycelia derived from the uredospores had fused with, and had di-

[1]H. S. Jackson, "Present Evolutionary Tendencies and the Origin of Life Cycles in the Uredinales," *Memoirs Torrey Bot. Club*, Vol. XVIII, 1931, pp. 1-108.

[2]A. de Bary, "Recherches sur le développement de quelques champignons parasites," *Ann. Sci. Nat.*, Bot., Sér. 4, T. XX, 1863, pp. 76 and 79.

[3]A. M. Brown, "The Sexual Behaviour of Several Plant Rusts," *Canadian Journal of Research*, Vol. XVIII, Section C, 1940, pp. 19-20.

ploidised, the haploid mycelia derived from basidiospores,[1] all of the ten pustules *produced aecidia and uredospore sori in association and at one and the same time.* It may be added that two haploid pustules to which uredospores had not been applied remained unchanged.

(2) As a result of sowing basidiospores on host-leaves thirty compound pustules, each formed by the coalescence of two simple pustules, came under observation. Some days after coalescence had taken place, seventeen of the thirty compound pustules *produced uredospore sori without any accompaniment of aecidia,* while the other thirteen compound pustules (presumably owing their origin to the fusion of two (+) pustules or two (−) pustules) remained in the haploid condition and produced neither aecidia nor uredospore sori. It may be added that here, as in the production of aecidia in compound pustules of *Puccinia helianthi*, the uredospore sori usually appeared in one component of a compound pustule several days in advance of their appearance in the other component of the pustule.

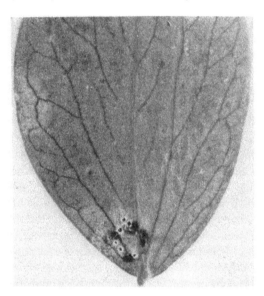

Fig. 89.—*Uromyces fabae.* Aecidia and uredospore sori on the same mycelium. To a haploid pustule derived from the sowing of a single basidiospore mixed nectar was applied. Thereafter the pustule developed six aecidia and three days later four uredospore sori. Photographed by A. M. Brown. Twice the natural size.

It thus appears that Brown has obtained convincing experimental evidence that, in *Uromyces fabae*, short-cycling of the life-history sometimes takes place in such a way as to cause the complete omission of the aecidial stage.

Brown's observations on *Uromyces fabae* give further support to Jackson's[2] contention that, in the course of evolution, the short-cycled Rusts have been derived from the long-cycled Rusts.

[1]For a full discussion of this mode of initiating the sexual process in long-cycled autoecious heterothallic Rusts *vide* Chapter VI under *Mode III.*

[2]H. S. Jackson, *loc. cit.*

Supposed Occasional Association of Pycnidia with Uredospore Sori in Puccinia helianthi.—*Helianthus annuus*, the Sunflower, was introduced into Europe from America. It arrived in Spain toward the end of the sixteenth century and in Russia about 1850. In southern Russia it became important as a source of oil. In 1867, Russian farmers complained that their Sunflower plants were being killed by a mysterious disease supposed to be due to evil effects of the sun, the soil, or the wind. This disease was therefore investigated by Woronin[1] who found that it was due to a Rust Fungus, *Puccinia helianthi*.

Puccinia helianthi is a macrocyclic, autoecious, heterothallic Rust. That it is macrocyclic and autoecious, producing all its spore-forms on the Sunflower plant, was discovered by Woronin, and that it is heterothallic was demonstrated by Craigie.

During three years, 1869-1871, Woronin investigated *Puccinia helianthi* both in the field and in the laboratory. Employing cultural methods he found that the fungus produces pycnidia, aecidia, uredospore sori, and teleutospore sori in succession on one and the same host, *Helianthus annuus*; and he embodied the results of his work in a paper in which, among other things, he described in detail and illustrated in coloured figures: the germination of the teleutospores; the production of basidia and basidiospores; the infection of Sunflower leaves by germ-tubes of basidiospores, aecidiospores, and uredospores; the structure of pycnidia, aecidia, uredospore sori, and teleutospore sori; the haustoria of the mycelial hyphae; and the two pores in the wall of each uredospore. Woronin's paper may be considered as one of the classics of Phytopathology; but, since its contents were unknown to most later workers, it attracted but little attention and was not properly appreciated.[2]

[1] M. Woronin, *Investigations on the Development of the Rust Fungus, Puccinia helianthi, attacking the Sunflower*, in Russian, published by W. Damakowa at St. Petersburg, 1871, pp. 1-35, Plates I and II.

[2] Thus J. C. Arthur (*Manual of the Rusts in United States and Canada*, U.S.A., 1934, p. 268) says that, in *Puccinia helianthi*: "Occurrence of the aecia had not been definitely established up to 1902, when the first part of Sydow's *Monographia Uredinarum* was published, which led the authors of that work to think that the species was a Hemi-puccinia, and the reported production of aecia to be an error." As a matter of fact Woronin, by cultural methods and field observations definitely established the occurrence of aecidia in the life-history in his paper of 1871. P. and H. Sydow (*Monographia Uredinarum*, Vol. I, 1904, pp. 92-93) and Arthur do not cite Woronin's original paper but only Woronin's German abstract of it entitled "Untersuchungen über die Entwicklung des Rostpilzes (*Puccinia helianthi*), welche die Krankheit der Sonnenblume verursacht," published in the *Bot. Zeit.* (Jahrg. XXX, 1872, pp. 677-684, 693-697). In this abstract there are no illustrations and Woronin says very little about the aecidia of *P. helianthi* beyond the fact that, along with spermogonia, they arise as a result of the sowing of basidiospores.

For what follows it is important to note that Woronin, although such a keen observer, in his investigations upon the life-history of *Puccinia helianthi* did not report having seen any anomalies in the succession of the spore-forms.

Carleton and Bailey concluded from their observations that the life-history of *P. helianthi* is sometimes short-cycled so that pycnidia are occasionally associated with uredospore sori instead of aecidia. According to Carleton:[1]

The aecidium occurs rarely in comparison with the occurrence of other stages, but is to be found on a number of hosts and occasionally in considerable abundance. This rarity of its occurrence, together with the occurrence so often with the uredo, may be accounted for by the fact that the uredo is often produced by teleutosporic infection.

And Bailey,[2] following Carleton, in 1923, said:

The normal development of pycnia and aecia does not always follow teliosporic infection. There is a distinct tendency for the rust to omit the aecia and develop uredinia after the production of pycnia. The uredinia produced by short-cycling subsequent to pycnial formation usually had a very distinct appearance. They developed below pycnidia and were confined to definite, light colored, slightly hypertrophied spots. The length of the incubation period also served to distinguish them from ordinary uredinia which usually developed within from five to seven days after inoculation. Short-cycled uredinia, on the other hand, do not develop until two or three days after the pycnia, and then are limited to the region directly beneath them. . . . In our experience, although short-cycling was not frequent, it occurred often enough and was so clear that no doubt remains that it actually does take place.

Jackson[3] in his study of the present evolutionary tendencies in the Uredinales has cited the passages just quoted as evidence that *Puccinia helianthi* has an unstable life-history. An examination of this evidence therefore seems to be desirable.

So far as Carleton's statement is concerned it may be pointed out that, as shown by Brown at Winnipeg, in the course of work on the Sunflower Rust extending over eight years, all the forms of *Puccinia helianthi* studied, after the sowing of basidiospores, always produced both pycnidia and aecidia and, furthermore, since *P. helianthi* is autoecious, it is quite possible for belated basidiospores and aecidiospores or uredospores to be sown naturally together in close succession on one and the same leaf. I also have examined many hundreds of infections and never once have I seen a pustule with pycnidia above and uredospores instead of aecidia below.

[1]M. A. Carleton, "Investigations of Rusts," *U.S. Dept. Agric. Bureau of Plant Industry*, Bull. LXIII, 1904, pp. 1-29.

[2]D. L. Bailey, "Sunflower Rust," *Minnesota Agric. Exp. Station.*, Tech. Bull. XVI, 1923, pp. 6-7.

[3]H. S. Jackson, "Present Evolutionary Tendencies and the Origin of Life Cycles in the Uredinales," *Memoirs Torrey Bot. Club*, Vol. XVIII, 1931, pp. 7-8.

Bailey[1] found that the association of uredospore sori with pycnidia followed no rule, for he said:

Short-cycling seems to occur rather infrequently and the conditions governing it are not at all understood. It was observed only on rather heavily infected leaves which suggested that a nutrient relation might be involved. Attempts to induce short-cycling by adverse host conditions did not yield conclusive results. The phenomenon occurred so erratically that it was difficult to ascertain definitely the cause. It occurred on plants incubated at approximately 60° F. as well as on plants kept dry at 75° F. but it did not do so consistently. Moreover, sometimes there was fully as much short-cycling under conditions which were apparently entirely favorable to the host.

Bailey's usual method of inoculating his Sunflower plants was to smear the teleutospores on moistened leaves and then to incubate for 48 hours in a damp-chamber. This method involves the risk of uredospores being sometimes deposited on the leaves along with the teleutospores. The method used by Craigie, Brown, and myself involves no such risk. From an old well-soaked leaf bearing teleutospores and perhaps some uredospores no spore can fall away, and the basidiospores produced escape from the leaf only because they are shot violently away from their sterigmata. The erratic occurrence of the supposed short-cycling in Bailey's experiments also suggests that his results may have been due to his mode of inoculation. Furthermore, since he found that the uredospores may retain their vitality when stored at 8° - 23° C. and at 20 - 40 relative humidity for at least six months, the risk of some of these spores germinating if any of them were mixed with the teleutospores that he smeared on his leaves was not negligible.

It would be interesting deliberately to sow basidiospores and uredospores of *Puccinia helianthi* on one and the same spot on a Sunflower leaf, and then to observe whether or not, after a young haploid mycelium and a young diploid mycelium had met and fused together, uredospore sori would take the place of aecidia. If so, we should have a case of apparent short-cycling, but not a real one.

Uromyces hobsoni and its Pycnidia.—In 1876, Vize[2] described an Indian Rust growing on *Jasminum grandiflorum* and called it *Uromyces hobsoni*. In 1891 Barclay[3] re-described this species and, believing it to be new, named it *U. cunninghamianus*.

Uromyces hobsoni is autoecious and gives rise to pycnidia, aecidia, and teleutospore sori. Uredospores are absent from the life cycle. Barclay observed that the fungus has three peculiarities: (1) the

---

[1] D. L. Bailey, *loc. cit.*, p. 7.

[2] J. E. Vize, in M. C. Cooke, "Some Indian Fungi," *Grevillea*, Vol. IV, 1876, p. 115.

[3] A. Barclay, "On the Life-History of a Remarkable Uredine on *Jasminum grandiflorum* L. (*Uromyces Cunninghamianus* nov. sp.)," *Trans. Linn. Soc. London*, Second series, Botany, Vol. III, 1891, pp. 141-151, Plates XLIX and L.

aecidiospores germinate on the host that produces them and the mycelia so produced give rise to new aecidia, the aecidiospores thus taking the place of uredospores in disseminating the species; (2) the germ-tubes produced by the aecidiospores in moist chambers become bicellular and then each cell sends out a slender sterile branch; and (3) the teleutospores arise in the exhausted aecidia and the teleutospore sori so formed push to the side the last-formed aecidiospores.

Barclay[1] germinated teleutospores of *Uromyces hobsoni* and observed that the basidia become divided by septa into three or four cells and that each cell may produce a basidiospore; and he sowed basidiospores on host-leaves and found that the young mycelia give rise to pycnidia "of the usual form and structure with a tuft of paraphyses protruding through the mouths to the extent of about 44 μ." The presence of normal pycnidia produced on mycelia derived from basidiospores suggests that the fungus is heterothallic.

In 1931, Ajreker and Parandekar[2] confirmed many of Barclay's observations and, in addition, showed that the aecidiospores and the teleutospores which are present in one and the same mature pustule, as also the mycelium which produces the teleutospores and the hyphae from which the aecidiospores and the intercalary cells are abstricted, are all binucleate. No pycnidia were seen, in all probability because the material was too old and these organs had disintegrated.

In 1939, Thirumalachar[3] gave a further account of *Uromyces hobsoni*. He observed that: (1) the fungus causes hypertrophy of stems, leaves, and flowers; (2) pycnidia, aecidia and teleutospore sori occur side by side on the hypertrophied portions of the host; (3) the production of aecidia from aecidiospores proceeds almost throughout the season; (4) in many cases aecidia and teleutospores were seen to develop within old pycnidia; and (5) the teleutospores germinate without rest and form 3 - 4 binucleate basidiospores.

With respect to the pycnidia of *Uromyces hobsoni* Thirumalachar[4] remarks: "The pycnial primordia arise just below the epidermal layer by the massing of hyphae. Generally in rusts, the pycnia are borne on the gametophytic uninucleate mycelium resulting from germinating sporidia and subsequent host infection. In the case of *Uromyces hobsoni*, however, the pycnia-bearing mycelium is always binucleate

---

[1]*Ibid.*, pp. 146-147.

[2]S. L. Ajrekar and S. A. Parandekar, "Observations on the Life History of the Rust Fungus *Uromyces* species on *Jasminum malabaricum* and its Relation to *Uromyces hobsoni* Vize (*U. cunninghamianus* Barc.) on *Jasminum grandiflorum*," *Journ. Indian Bot. Soc.*, Vol. X, 1931, pp. 195-204, Plates I and II.

[3]M. J. Thirumalachar, "Rust on *Jasminum grandiflorum*," *Phytopathology*, Vol. XXIX, 1939, pp. 783-792.   [4]*Ibid.*, p. 786.

(Fig. 1, C). This condition is unique and, in the knowledge of the writer, is the first case recorded among the rusts."

Thirumalachar further remarks in respect to the pycnidiospores of *Uromyces hobsoni*: "many of the spores that were shed near the ostioles were binucleate. . . . The presence of binucleate mycelium and basal cells in the pycnia is an interesting feature which may be correlated with the binucleate condition of the pycnidiospores."

This remarkable fungus needs some further study. Basidiospores should be sown on leaves of *Jasminum grandiflorum*, and the young mycelia and young pycnidia studied cytologically.

Thirumalachar observed, as we have seen, that "in many cases aecia and telia were seen to develop within the pycnial cup," but it remains to be determined whether the binucleate mycelium from which these aecidia and teleutospore sori developed was present under the pycnidia when the pycnidia were being formed or whether it made its way there subsequently. We need illustrations showing stages in the production of the pycnidiospores from the pycnidiosporophores, such as those given by Colley for *Cronartium ribicola* (*vide* Fig. 101, p. 330). It still seems possible that, after all, the pycnidia of *Uromyces hobsoni* have the usual haploid characteristics with single nuclei in its pycnidiosporophores and pycnidiospores and that, at its first origin, every pycnidium arises on a haploid (uninucleate) mycelium. Possibly the fungus may be heterothallic.

In treating of the basidiospores of *Uromyces hobsoni* Thirumalachar[1] says: "Basidiospores are oval, flattened on one side, and binucleate. The nucleus in the promycelium migrates into the sterigma and becomes diplodized before entering the spore." This last statement, as it stands, is incomprehensible; but, if it is intended to mean that somehow or other each spore comes to contain a ($+$) nucleus and a ($-$) nucleus, it can be said that no evidence is offered by Thirumalachar for any such supposition. Thirumalachar shows an abnormal basidium with one basidiospore including two black structures presumed to be nuclei; but, since normally the basidia are 3-4-celled and bear as many basidiospores, it seems probable that each normal basidiospore receives a single nucleus from the basidium and thus comes to be, as usual, a haploid structure. In some Rusts, as Miss Stevens[2] has shown for *Gymnosporangium juniperi-virginianae*, it often happens that the nucleus of a basidiospore, after entering the spore by passing through the sterigma soon divides, with the result that the mature basidiospore is binucleate. It is possible that this may happen in *Uromyces hobsoni*.

[1]M. J. Thirumalachar, *loc. cit.*, p. 790.

[2]Edith Stevens, "Cytological Features of the Life History of *Gymnosporangium juniperi-virginianae*," *Botanical Gazette*, Vol. LXXXIX, 1930, pp. 398, 400.

CHAPTER VI

MODES OF INITIATING THE SEXUAL PROCESS IN THE RUST FUNGI

The Sexual Process — Modes of Initiating the Sexual Process — Modes I and II, the Normal Modes — Mode III: Fusion of a Diploid Mycelium with a Haploid Mycelium — Critical Remarks concerning Mode I — Mode IV, in Homothallic Rusts — Mode V: Annual Self-diploidisation in Certain Rusts having a Systemic Mycelium — De-diploidisation and Self-diploidisation in *Puccinia minussensis* — Defect of Homothallism — Rust Fungi in which the Sexual Process has become Imperfect or Inoperative — Uninucleate Rusts considered as Self-propagating Haploid Strains derived from Heteroecious Species.

The Sexual Process.—The *sexual process* of a typical heterothallic Rust, *e.g. Puccinia graminis*, resembles the sexual process in a typical heterothallic Agaric, *e.g. Coprinus lagopus*, in that it has five stages: (1) *cell-fusions* between haploid cells of opposite sex; (2) *nuclear migration* accompanied by *nuclear division, i.e.* the passage of one or more nuclei of one sex through a mycelium of opposite sex and an increase in the numbers of the migrating nuclei brought about by karyokinesis; (3) *nuclear association, i.e.* the coming together of a (+) nucleus and a (−) nucleus in intimate association in the cytoplasm of many single cells; (4) the multiplication of the first-formed pairs of associated nuclei by *conjugate nuclear division*; and (5) in each cell of a teleutospore, the *fusion* of a (+) nucleus with a (−) nucleus.

The first three stages of the sexual process in the Uredinales may be included under the term *diploidisation*. When this is done, the sexual process in such a Rust as *Puccinia graminis* may be described as: (1) beginning with diploidisation; (2) continuing with conjugate nuclear division; and (3) terminating with nuclear fusion. The *diploidisation process* will be more fully discussed in subsequent pages. Here it suffices to remark that the diploidisation process is a part, and only a part, of the sexual process.

The sexual process in a heterothallic Rust, as contrasted with the sexual process in Higher Plants and Higher Animals, occupies much time, is associated with long-continued cell-growth, and is consummated not by a single fusion of two nuclei of opposite sex, but by hundreds, thousands, or even millions of such fusions.

In such a Rust as *Puccinia graminis*, after the basal cells of a proto-aecidium have been diploidised, conjugate nuclear division is initiated. The process of multiplying the first-formed pairs of conjugate nuclei begins with the formation of the aecidiospores and is continued: in the mycelia derived from the aecidiospores; during the formation of the first-generation uredospores; in the mycelia derived from the first-generation uredospores; during the formation of the second-generation uredospores; in the mycelia derived from the second-generation uredospores; and so forth for any other generation of uredospores; and, finally, during the formation of the teleutospores. Thus the increase in the numbers of pairs of conjugate nuclei derived from the pairs of conjugate nuclei first formed in a single proto-aecidium may be enormous.

For a long-cycled Rust like *Puccinia graminis* it is not difficult to imagine the sexual potentialities of a single successful (+) or (−) nucleus contained within a pycnidiospore. Such a nucleus, after the pycnidiospore has fused with a flexuous hypha of opposite sex, migrates down the flexuous hypha and through the associated mycelium, multiplying by nuclear division as it moves. Its progeny of nuclei pass to the basal cells of many different proto-aecidia and there effect diploidisation. The conjugate pairs of nuclei in the basal cells of the proto-aecidia undergo conjugate division and so provide a pair of conjugate nuclei for each aecidiospore. The aecidiospores are shot out from the aecidia, are dispersed by the wind and, if fortunate, many of them settle on suitable host-plants, germinate there, and cause infection. As a result, many scores or hundreds of uredospore pustules may be formed, and the uredospores, when set free from their pedicels may be blown by the wind to new hosts, germinate there, and give rise to a second generation of uredospores. A third and a fourth, and even fifth and sixth, generations of uredospores are possible. Finally, on numerous host-plants in hundreds of thousands of pustules there are formed many millions of teleutospores, and in each cell of each teleutospore two nuclei of opposite sex unite. Thus a single (+) or (−) nucleus originally present in a pycnidiospore of a Rust like *P. graminis* may succeed, by multiplication, diploidisation, and conjugate nuclear division, in being represented by its progeny in the consummation of the sexual process in millions of cells, cells scattered far and wide on numerous host-plants but all derived in the first place from a haploid mycelium of opposite sex.

It may well be that the sexual process initiated by the nucleus of a single pycnidiospore in the proto-aecidia of a rust pustule of *Puccinia graminis* on a Barberry bush in southern Minnesota in the United States has often been consummated in the teleutospores on the leaves and straw of wheat-plants and other graminaceous hosts scattered through the States of Minnesota, South Dakota, and North Dakota and through the Canadian Provinces of Manitoba, Saskatchewan, and Alberta.

Modes of Initiating the Sexual Process.—The ways or *modes* of initiating the sexual process in the Uredinales, as here recognised, are five in number, and they will be designated Modes I, II, III, IV, and V.

Modes I and II may be regarded as the original ancestral modes which were employed by those primitive, long-cycled, heterothallic, heteroecious Rusts from which it is believed our present-day Uredinales were evolved.   Modes III, IV, and V appear to have been derived from Mode I.

Modes I and II, the Normal Modes.—In ordinary, heterothallic, long-cycled Rusts, *e.g. Puccinia graminis* which has alternate hosts (Barberry and certain Gramineae) and *P. helianthi* which has one host (Sunflower), there are two normal modes of initiating the sexual process:

Mode I.   By the fusion of a (+) mycelium with a (−) mycelium in a host-leaf, when such mycelia (each derived from a basidiospore) happen to meet and form a compound rust pustule, and

Mode II.   By the fusion of a (+) pycnidiospore with a (−) flexuous hypha, or the fusion of a (−) pycnidiospore with a (+) flexuous hypha, in the nectar of a pycnidium after the nectar of (+) and (−) pustules has been mixed through the agency of insects.

As already set forth in Chapter I, both of these modes were discovered by Craigie,[1] Mode I in 1927, and Mode II in part in 1927 and in part in 1933.

To the illustrations of Mode I already given (Fig. 10, p. 61, and Fig. 11, p. 64) there is here added Fig. 90; and to the illustrations of Mode II already given (Fig. 14, p. 69, and Fig. 15, p. 70) there is here added Fig. 91.

Mode I is relatively simple, but it becomes operative only when two mycelia of opposite sex happen to come into contact with one another.   Mode II leads to the diploidisation of pustules which are isolated from one another and, for its initiation, the visits of insects are generally required.

[1] *Vide supra*, pp. 65-73.

As, under natural conditions in the open, the pustules derived from (+) or (−) basidiospores are far more often isolated from one another than united or, in other words, simple rust pustules are far more numerous than compound, Mode II of initiating the sexual process in heterothallic long-cycled Rusts is of much greater importance than Mode I. No wonder, therefore, that the pycnidia of these Rusts are so beautifully adapted for making use of insects and, in each generation, so soon come into good working condition.

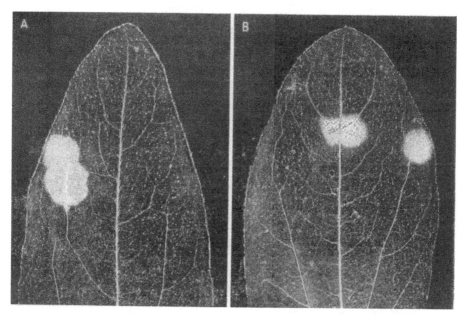

FIG. 90.—Mode I of initiating the sexual process in *Puccinia helianthi*. Under side of two Sunflower leaves upon which basidiospores were sown and which have developed rust pustules. A: a leaf (inoculated Sept. 10, photographed Oct. 11) in which two simple pustules have met and fused but without the development of any aecidia. The pustules were both (+) or both (−). B: a leaf (24 days after inoculation) with a compound pustule astride the mid-rib with aecidia, and a simple pustule without aecidia. Of the components of the compound pustule one was (+) and the other (−). Photographs by J. H. Craigie: A, hitherto unpublished; B, from *Phytopathology*, 1932. Magnification, 1.5.

Mode III: Fusion of a Diploid Mycelium with a Haploid Mycelium.

—In an *autoecious* heterothallic long-cycled Rust, *e.g. Puccinia helianthi*, but *not* in a heteroecious heterothallic long-cycled Rust, *e.g. P. graminis*, in addition to the two normal modes of initiating the sexual process, there is a third possible mode:

Mode III. By the fusion of a diploid mycelium, derived from an aecidiospore or a uredospore, with a haploid mycelium derived from a

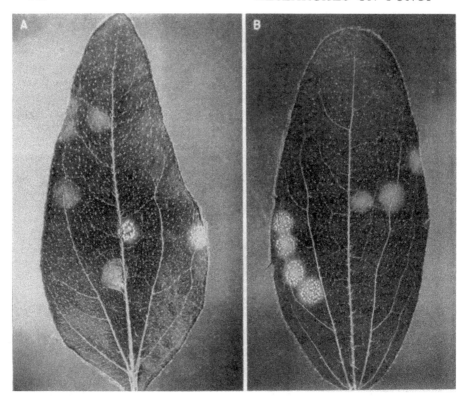

Fig. 91.—Effect of mixing nectar of haploid pustules of *Puccinia helianthi*. Under side of two Sunflower leaves upon which basidiospores were sown and which developed haploid rust pustules bearing pycnidia but no aecidia. A: seven monosporidial pustules developed; on July 29, the nectar of the two pustules on the right of the mid-rib was mixed with the help of a pipette, while that of the five pustules on the left of the mid-rib was stirred separately but not mixed. On Aug. 3 aecidia appeared on the two experimental pustules, but not on the five control pustules. The photograph shows the final condition. B: seven mono-sporidial pustules developed; the nectar of the four pustules on the left of the mid-rib was mixed, while that of the three pustules on the right of the mid-rib was stirred separately but not mixed. The result, as here shown, was that the pustules with the mixed nectar developed aecidia, while those with unmixed nectar did not. Photographs by J. H. Craigie, hitherto unpublished. Magnification, 1.75.

basidiospore. This Mode was discovered by A. M. Brown[1] in the Dominion Rust Research Laboratory, in 1932, as a result of experiment.

Brown was preparing for an M.Sc. examination and, when reading Volume IV of these *Researches*, he came upon the account of experi-

[1]A. M. Brown, "Diploidisation of Haploid by Diploid Mycelium of *Puccinia helianthi* Schw.," *Nature*, Vol. CXXX, 1932, p. 777.

ments which prove that, in *Coprinus lagopus*, a diploid mycelium is able to diploidise haploid mycelia of the same genetic constitution as those two haploid mycelia from which it was compounded.[1] It then occurred to him that experiments similar to those which I had made with *C. lagopus* might be made with *Puccinia helianthi*, an autoecious Rust which produces pycnidiospores, aecidiospores, uredospores, and teleutospores on a single host-plant, namely, the Sunflower. Brown soon proved in the laboratory that his idea had been well conceived.

Brown, after sowing basidiospores of *Puccinia helianthi* on *Helianthus annuus*, obtained: (1) Sunflower plants bearing (+) or (−) haploid pustules; and (2) other Sunflower plants in which the life-history of the fungus was more advanced, so that the leaves bore uredospores. The uredospores are diploid, and each contains a (+) and a (−) nucleus. Brown used these uredospores as diploidising agents in experiments which he[2] described as follows:

Sporidia of *Puccinia helianthi* were sown sparsely on the upper surface of the first two foliage leaves of Sunflower seedlings (*Helianthus annuus* L.). From these inoculations there arose forty-nine haploid pustules.

When three weeks old, none of these pustules bore aecia. Twelve of them were then marked, to serve as controls; and beside sixteen others, at a point just beyond the periphery of each, urediniospores were sown. A week later urediniospores were sown similarly beside twelve other pustules; and a week later still, a sowing of urediniospores was made beside the remaining pustules.

As a result of these inoculations, uredinia (diploid pustules) arose at, or very near to, the margin of each of the thirty-seven haploid pustules. Thus diploid mycelium grew in juxtaposition to haploid mycelium.

From eight to twelve days after the inoculations with urediniospores were made, aecia began to appear on the under side of all the thirty-seven pustules. Usually the aecia appeared in that part of a pustule lying nearest to the uredinial pustules, and later in the parts more remote (Fig. 1) [here reproduced as Fig. 92]. In a few pustules, the first aecia to appear arose rather irregularly spaced over the whole under-surface of each. No aecia appeared in any of the control pustules.

A cytological examination of the pustules which produced aecia has not been made, but it is assumed on the basis of these experiments that, when contact between a haploid and a diploid mycelium of *Puccinia helianthi* is established, the diploidisation of the haploid mycelium is effected by successive nuclear divisions and migrations, as has been described by Buller[3] for *Coprinus lagopus*.

Aecidiospores resemble uredospores in that they each contain a pair of conjugate nuclei. In a further series of experiments Brown[4] found that the aecidiospores of *Puccinia helianthi*, just like the uredo-

[1]A. H. R. Buller, these *Researches*, Vol. IV, 1931, pp. 204-229.

[2]A. M. Brown, *loc. cit.*

[3]Buller, A. H. R., *Nature*, **126**, 686, Nov. 1, 1930; "Researches on Fungi," vol. 4, p. 187; 1931.

[4]A. M. Brown, personal communication of the results of experiments hitherto unpublished.

spores, can be used as diploidising agents.   He sowed aecidiospores on Sunflower leaves close to haploid pustules derived from basidiospores, the diploid mycelium derived from them came into contact with the haploid mycelia in the pustules, and the diploid and haploid mycelia fused together, with the result that the haploid proto-aecidia developed into diploid aecidia.

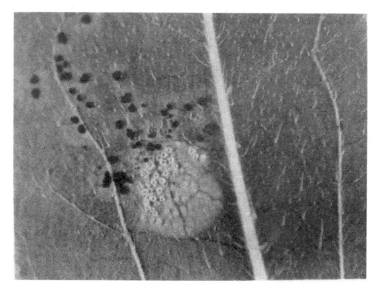

FIG. 92.—Diploidisation of a haploid mycelium by a diploid mycelium in a macro-cylic autoecious Rust.   Part of the under side of a *Helianthus annuus* leaf showing (left of mid-rib) a pustule derived from a single basidiospore of *Puccinia helianthi*.   This pustule, 16 days after inoculation, was still haploid.   Some uredospores were then sown near to it on the under side of the leaf.   They germinated and produced in the leaf a diploid mycelium which, on coming into contact with the haploid mycelium in the haploid pustule, diploidised it, with the result that the proto-aecidia developed into aecidia.   At the same time, the diploid mycelium gave rise to new uredospore sori of which about thirty can be seen in the photograph.   The aecidia have appeared on the side of the haploid pustule which is nearest to the uredospore sori.   Photographed by A. M. Brown 14 days after sowing the uredospores.   Magnification, 4.5.

Brown[1] has also succeeded in diploidising haploid pustules, derived from basidiospores, by means of diploid mycelia, derived from uredo-spores, in two other species, namely, *Uromyces trifolii hybridi*, which he grew on *Trifolium hybridum* and *Uromyces fabae* which he grew on *Pisum sativum*.   In both of these species—which, like *Puccinia heli-*

[1]A. M. Brown: personal communication, 1939; also "The Sexual Behaviour of Several Plant Rusts," *Canadian Journal of Research*, Vol. XVIII, Section C, 1940, pp. 19-20.   Experiments made at the Dominion Rust Research Laboratory.   For details of the experiments on *Uromyces trifolii hybridi vide supra*, Chapter I, p. 88.

anthi, are macrocylic, autoecious, and heterothallic—the uredospores, after being sown near a haploid pustule, germinated and gave rise to a diploid mycelium that produced a fresh crop of uredospore pustules and, at the same time, diploidised the haploid pustule with the result

FIG. 93.—Diploidisation of a haploid mycelium by a diploid mycelium in a macro-cylic autoecious Rust. Part of the under side of a leaf of *Trifolium hybridum*, showing on the mid-rib of the central leaflet a pustule derived from a single basidiospore of *Uromyces trifolii hybridi*. This pustule, 21 days after inoculation, was still haploid. Some uredospores were then sown near to it on the under side of the leaflet. They germinated and produced a diploid mycelium which, on coming into contact with the haploid mycelium in the haploid pustule, di-ploidised it, with the result that the proto-aecidia developed into aecidia except in the central older part of the pustule which was dying. At the same time, the diploid mycelium gave rise to new uredospore sori which can be readily recog-nized in the photograph. Photographed by A. M. Brown about 10 days after sowing the uredospores. Magnification, 2.

that the proto-aecidia developed into aecidia. The result of one of Brown's experiments with *Uromyces trifolii hybridi* is illustrated in Fig. 93.

Working with *Uromyces fabae*, Oliveira[1] has obtained results similar to those of Brown. He too has observed that, in this species, a diploid

[1] B. d'Oliveira, "Aspectos actuais do problema das Ferrugens," Separata de Palestras Agronómicas, Vol. II, Part II, 1939, Lisboa, 1940, pp. 1-77. On p. 37, the result of experiments with *Uromyces fabae* are announced and a paper with details promised.

mycelium derived from a uredospore is able to diploidise a haploid mycelium derived from a basidiospore.

Phragmidium speciosum grows on species of *Rosa* on which it produces pycnidiospores, aecidiospores, and teleutospores, but no uredospores. Brown[1] attempted to diploidise its haploid pustules with diploid mycelia derived from aecidiospores; but he failed because he found that, although the aecidiospores germinated readily in water, the germ-tubes did not cause any infection of the Rose leaves.[2]

Mode III of initiating the sexual process is of considerable theoretical interest, for it proves beyond doubt that, in the Rust Fungi, under conditions that exclude the employment of pycnidiospores, *one mycelium is able to diploidise another mycelium.*

Under natural conditions, in such a fungus as *Puccinia helianthi*, it may be but rarely that an isolated (+) or (−) mycelium, which has escaped diploidisation by means of pycnidiospores transferred by insects, subsequently becomes diploidised through the agency of wind-blown aecidiospores or uredospores. Under natural conditions, therefore, Mode III of initiating the sexual process in autoecious long-cycled Rusts may well be of but very little importance.

Critical Remarks concerning Mode I.—Some critical remarks concerning the reality of Mode I for initiating the sexual process in such Rusts as *Puccinia graminis* and *P. helianthi* will now be made. After Craigie had discovered that the intermixing of the nectar of (+) and (−) pustules leads to the production of aecidia (Mode II), some doubt arose as to Mode I. In a compound pustule in which a (+) mycelium has fused with a (−) mycelium and aecidia have been produced, has the sexual process been initiated: (1) by the fusion of the two mycelia, or (2) by the spontaneous flowing together and comingling of the drops of nectar above the (+) and (−) mycelia?

In favour of initiation by the fusion of the mycelia, it may be said that, in compound pustules, the two mycelia meet in the leaf tissues long before the pycnidial drops of the (+) and (−) pycnidia have developed sufficiently to fuse with one another. Fig. 25 (p. 102) shows

---

[1]A. M. Brown, "The Sexual Behaviour of Several Plant Rusts," *loc. cit.*

[2]Brown observed that *Phragmidium speciosum* is autoecious and that its teleuto-spore sori arise in the same infections that produce the aecidia and not as a result of aecidiospore infection. Whether or not the aecidiospores play any part in the life-history of the fungus remains to be determined. Brown has suggested that, although the aecidiospores do not infect *Rosa blanda* with which he worked, they may be able to infect other Rosa species or possibly a species of another genus. However, there is also the possibility that, when *P. speciosum* shortened its cycle and became autoecious, its aecidiospores survived as merely vestigial and functionless structures. So far as is known at present, *P. speciosum* is dispersed under natural conditions solely by its basidiospores.

the upper side of some compound pustules of *Puccinia helianthi* and an inspection of it permits one to observe that, whereas the mycelia of the two or more components of these pustules are in contact, the adjacent groups of pycnidia are relatively remote from one another.

Also in favour of initiation by the fusion of the mycelia is the fact that, in *Puccinia helianthi*, *Uromyces trifolii hybridi*, and *U. fabae*,

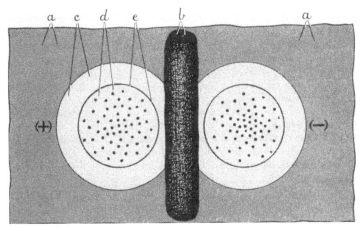

FIG. 94.—Diagram illustrating Brown's barrier experiments: *a a*, a Sunflower leaf bearing two haploid pustules of *Puccinia helianthi*, one (+) and the other (−), which when young were separated above by a barrier of Opaque, *b*; *c*, outer part of pustule; *d*, inner part with pycnidia; *e*, present outer boundary of drop of nectar. As the two pustules grow toward one another in the leaf, they meet and fuse, but their nectar cannot intermingle. The fact that, under these conditions, aecidia are produced by both pustules on the under side of the leaf proves that the (+) mycelium and the (−) mycelium in the pustules are able to diploidise each other. Drawn by A. M. Brown and A. H. R. Buller. Pustules and barrier several times the natural size.

under conditions that preclude the intermixing of (+) and (−) nectar, a diploid mycelium is able to diploidise a haploid mycelium.

Finally, all doubt about the reality of Mode I has been removed by a series of *ad hoc* experiments which Brown[1] carried out in the Dominion Rust Research Laboratory and which are here illustrated in the diagram reproduced in Fig. 94. Working with *Puccinia helianthi* he obtained haploid pustules on seedling plants of *Helianthus annuus*. He then concentrated his attention on pairs of pustules in which the two components were at first separated from one another by a distance of 2-4 mm. Brown's further account of his work may be given in his own words:

[1]A. M. Brown, "A Study of Coalescing Haploid Pustules in *Puccinia helianthi*," *Phytopathology*, Vol. XXV, 1935, pp. 1085-1090.

Care was taken to protect the plants bearing the pustules from any agency that might accidently mix the pycnial nectar.

As a further precaution to ensure that all the pustules used in the experiment were haploid, the pustules were allowed to develop until they were 2 weeks or more old. If any individual pustule was of bisporidial origin and consisted of mycelia of opposite sex, it would most probably have produced aecia by the time it had reached this age. Only those pairs in which the two individual pustules were at this time separated by a narrow strip of green tissue, or slightly chlorotic tissue in which no pycnia had yet developed, were included in the experiment proper.

In order to prevent the spontaneous intermixing of the nectar of each pair of pustules when later the two components would coalesce to form a compound pustule, a barrier was placed on the upper leaf-surface between the two pustules of each pair. For this purpose, Eastman's Opaque No. 1 preparation was found quite satisfactory. It adhered readily to the leaf, dried quickly, and caused no perceptible injury. The barrier, or wall, was placed at right angles to a line passing through the centres of the two pustules comprising each pair and extended sufficiently far on either side of this line to assure its always exceeding any diameter that the pustules might later attain. The experiment was carried out during the winter of 1933-34. At this season of the year in Winnipeg the days are short and the light intensity is rather low, conditions under which the production of nectar is scanty and confined mostly to the central part of the pustule. The plants at this time of the year are almost free from insects. Hence the possibility of any intermixing of pycnial nectar was practically eliminated.

A total of 288 pairs of pustules were included in the experiment. The 2 pustules of each pair were separated by a barrier of Opaque, as described above. Finally they coalesced. The barrier appeared to have no adverse effect on the growth of the pustules: the radial expansion of the pustules, as observed on the under side of the leaves was quite uniform. Of the 288 compound pustules that arose through the coalescence of the individual pairs of simple pustules, 110 produced aecia.

Thus Brown by experiment conclusively proved that, in *Puccinia helianthi*, "when two pustules of opposite sex coalesce to form one compound pustule, diploidisation of the mycelia of the components of the compound pustule can be effected without the intervention of pycniospores."[1] Mode I, therefore, is a true mode of initiating the sexual process in heterothallic Rust Fungi.[2]

In the course of his studies of compound pustules of *Puccinia helianthi* in which the (+) component and the (−) component had

[1]*Ibid.*, p. 1090.

[2]Gilbert M. Smith, in Volume I of his *Cryptogamic Botany* (New York and London, 1938, pp. 497-498), no doubt inadvertently, has given a very erroneous account of the manner in which the sexual process is initiated in *Puccinia graminis*. Among other things he says: "Unlike most other heterothallic Basidiomycetes, *P. graminis* has no conjugation when two haploid mycelia of opposite sex come in contact with each other. Instead, the diplophase is initiated by a transfer of spermatia from a spermogonium of one sex to another of the opposite sex (Craigie, 1927, 1927 A) or by a spermatium coming in contact with a vegetative hypha of the opposite sex (Allen, 1933)." After Brown's critical investigation of coalescing pustules of *P. helianthi*, published in 1935, the last possible doubt as to the sexual

been separated above by a barrier of Opaque, Brown[1] observed that, after the mycelia of the two components had come into contact and presumably had fused with one another, it almost invariably happened that aecidia appeared in one component first and afterwards in the other component and that aecidial production was completed in the

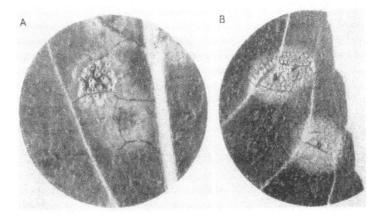

FIG. 95.—*Puccinia helianthi*. Mutual diploidisation in compound pustules with one component producing aecidia earlier than the other. A: a compound pustule in which aecidia have been formed in one component but not yet in the other. B: a simple pustule without aecidia and a compound pustule with aecidia. In one component of the latter the aecidia are well developed, whereas in the other component the aecidia are only beginning to appear. Photographed by A. M. Brown (*Phytopathology*, 1935). Magnification, 3.

one component before it became evident in the other (Fig. 95, A). Brown also observed that in both components aecidia developed first, although at different times, in the areas adjacent to the line of fusion and latest in the areas most remote from that line. Fig. 95, B, shows a simple (+) or (−) haploid pustule without aecidia and a compound pustule consisting of a (+) component and a (−) component. In one of these components the development of aecidia is about complete, whilst in the other component the development of aecidia has progressed most near the line of fusion. Thus Brown's experiments upon *P. helianthi* indicate that, in heterothallic Rust Fungi, in a compound

interaction of the (+) and (−) components in a compound pustule of a heterothallic Rust, such as *P. helianthi* and *P. graminis*, was removed. Moreover, Miss Allen only supposed, but did not prove, that in *P. graminis* the pycnidiospores fuse with hyphae emerging through the epidermis; and, in the same year as her publication, 1933, Craigie showed that in *P. graminis* and *P. helianthi* the pycnidia are provided with flexuous hyphae and that in *P. helianthi* the pycnidiospores fuse with these hyphae. Of the existence of flexuous hyphae in *P. graminis* Smith failed to make any mention.

[1] *Ibid.*, p. 1087.

pustule made up of a (+) component and a (−) component: (1) the diploidisation process is initiated in the areas where the hyphae of the two mycelia first intermingle; and (2), as a rule but not always, the process runs its course in one component before it makes any noticeable progress in the other component.

The progressive diploidisation of one component of a compound pustule consisting of a (+) component and a (−) component was illustrated by Brown[1] in a photomicrograph of a microtome section here reproduced as Fig. 96, A. The section was cut in a direction passing

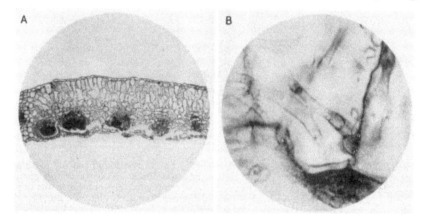

Fig. 96.—*Puccinia helianthi.* A, a cross-section through the right-hand component of a compound pustule on a Sunflower leaf. The five proto-aecidia are becoming progressively diploidised by nuclei from the left-hand component which are travelling from left to right. Already the two left-most proto-aecidia have been converted into aecidia (aecidiospores deeply stained). The central proto-aecidium has been diploidised on its left side but not on its right. The two other proto-aecidia are still undiploidised. B, a hyphal fusion found in the region of coalescence of a compound pustule. Photographed by A. M. Brown (*Phytopathology*, 1935). Magnification: A, 25; B, about 460.

through the centres of the two component pustules which had coalesced, and Fig. 96, A, shows what we may consider to be the right-hand component only. The region of coalescence is just to the left of the part shown in the photograph and diploidisation of the proto-aecidia is progressing from left to right. If we number the proto-aecidia from left to right, Nos. 1, 2, 3, 4, and 5, it is apparent that: (1) the proto-aecidia Nos. 1 and 2 have both been diploidised and are now aecidia, for both are producing chains of aecidiospores (represented by the deeply-stained areas); (2) the left-hand portion of proto-aecidium No. 3 has been diploidised, for this portion is proceeding with the formation

[1]*Ibid.*, p. 1088, Figs. 2, A.

of aecidiospores although progress in this operation is not so advanced as in Nos. 1 and 2; (3) the right-hand portion of proto-aecidium No. 3 is still haploid; and (4) proto-aecidia Nos. 4 and 5 are still in the haploid condition, for as yet they show no signs of aecidiospore production.

That the (+) component and the (−) component of a compound pustule of a Rust frequently, but not always, diploidise one another at unequal rates is, in view of the complexity of the diploidisation process, not surprising, and it finds a parallel in the unequal rates of diploidisation which I have often observed in two mycelia of opposite sex of the Hymenomycete, *Coprinus lagopus*, after the mycelia have been paired on a plate of dung-agar. Even in a single mycelium of *C. lagopus* the rate of diploidisation in all directions is not uniform. Thus, as may be seen from an examination of Fig. 128 (p. 222) and Fig. 129 (p. 225) in Volume IV, the diploidisation of a large haploid mycelium that is being diploidised by a very small one may proceed more rapidly on one side than on the other.

An attempt will now be made to visualise what happens to the nuclei when the sexual process is being initiated by Mode I, *i.e.* when a (+) mycelium and a (−) mycelium of such a fungus as *Puccinia helianthi* or *P. graminis* come into contact in the intercellular spaces of a host-leaf; and, in making this attempt, it will be borne in mind that, as a result of experiments and simple deductions, it has been discovered that in certain Hymenomycetes, *e.g. Coprinus lagopus*, when two mycelia of opposite sex have come into contact and have formed hyphal fusions with one another, the nuclei of one sex, on entering the mycelium of the opposite sex, are able to travel long distances through the hyphae[1] and may undergo a great many nuclear divisions *en route* to their final destinations.[2]

As soon as a (+) mycelium and a (−) mycelium of a heterothallic Rust, such as *Puccinia helianthi* or *P. graminis*, come into contact, they form hyphal fusions (*cf.* Fig. 96, B). Through these fusions an exchange of nuclei takes place: a few (+) nuclei pass into the (−) mycelium and a few (−) nuclei pass into the (+) mycelium. The (+) nuclei which have entered the (−) mycelium travel along in the cytoplasm of the (−) hyphae, undergo repeated divisions, and make their way to all the (−) proto-aecidia where they become paired and conjugately associated with the (−) nuclei present in what is to become the aecidiospore-bed. Similarly, the (−) nuclei which have entered the (+) mycelium travel along in the cytoplasm of the (+) hyphae, undergo repeated divisions, and make their way to all the (+) proto-

[1]These *Researches*, Vol. IV, 1931, pp. 213-241, Figs. 124-132.
[2]*Ibid.*, pp. 241-243.

aecidia where they become paired and conjugately associated with the (+) nuclei present in what is to become the aecidiospore-bed.

Finally, it may be asked: in *Puccinia graminis* and other heteroecious Rust Fungi, in effecting by Mode I the diploidisation of, say, a (−) proto-aecidium, does one (+) nucleus or do a number of (+) nuclei pass through the (−) intercellular mycelium to the proto-aecidium? In attempting to answer this question, we shall consider: (1) the structure of a proto-aecidium when becoming converted into an aecidium, and (2) the implication of the results of certain hybridisation experiments made with the help of Mode II.

(1) An examination of the structure of a proto-aecidium of *Puccinia graminis* when the proto-aecidium is becoming diploidised and is beginning to produce aecidiospores (Figs. 26, 27, and 29; pp. 103, 104, and 108) enables one to perceive that the aecidiosporophores are developed not as branches from a single diploidised basal cell, but from many diploidised basal cells. It is not difficult, therefore, to imagine that a number of (+) nuclei may pass through the (−) mycelium to a (−) proto-aecidium to initiate the diploidisation of the many haploid (−) cells which are destined to develop into aecidiosporophores.

(2) Newton, Johnson, and Brown[1] applied mixed pycnidiospores of eight physiological races of *Puccinia graminis* to the pycnidia of a single race and then genetically analysed 276 of the resultant aecidia by transferring the aecidiospores to differential-host wheat seedlings, and they found that, while 262 of the 276 aecidia yielded but one hybrid race each, yet 13 aecidia yielded two races each, and one aecidium three races. It is obvious: (1) that the 13 aecidia which yielded two races each at their origin must each have received nuclei from at least two genotypically different pycnidiospores; (2) that the aecidium which yielded three races at its origin must have received nuclei from at least three genotypically different pycnidiospores; and (3) that the nuclei received must have travelled (perhaps dividing on the way) down flexuous hyphae and through the intercellular mycelium to the basal cells of the proto-aecidia destined to be diploidised and to develop into aecidia. Since, when the sexual process is being initiated by Mode II, two or three or more nuclei may pass through the intercellular mycelium to a single proto-aecidium, we may suppose that a similar migration of nuclei may also take place when the sexual process is initiated by Mode I.

[1]M. Newton, T. Johnson, and A. M. Brown, "A Preliminary Study on the Hybridization of Physiologic Forms of *Puccinia graminis tritici*," *Scientific Agriculture*, Vol. X, 1930, pp. 728-729.

After one, two, three, or more of the invading nuclei have made their way to the basal cells of a proto-aecidium, it seems probable that they pass along the hyphae through the central pores of septa and through hyphal fusions, dividing as they go, and so, in the end, provide a sufficient number of nuclei to diploidise all the basal cells which subsequently elongate and give rise to chains of aecidiospores.

It was formerly thought by Blackman, Christman, Colley, and other cytologists that the nuclei which, by migration or cell-fusion, come together in conjugate pairs at the base of a proto-aecidium are sister nuclei or nearly related nuclei that arose locally by the division of a common ancestral nucleus; but, in the light of Craigie's experimental work on *Puccinia helianthi* and *P. graminis* and of Newton and Johnson's work on the production of hybrids in *P. graminis*, we can now discard that idea, in so far as it applies to heterothallic Rust species.

A modern interpretation of Colley's cytological observations on cell-fusions at the base of the young aecidium of *Cronartium ribicola* will be given in Chapter IX.

Mode IV, in Homothallic Rusts.—In homothallic short-cycled Rusts, *e.g. Puccinia malvacearum*, *P. xanthii*, and *P. grindeliae* (form at Winnipeg[1]) which produce no pycnidia and in which each mycelium derived from a basidiospore gives rise to teleutospores only, diploidisation is effected not by Modes I, II or III, but by another mode:

Mode IV. By an internal change in each mycelium derived from a basidiospore, as a result of which every mycelium passes spontaneously from the haploid condition in which the nuclei are separate into the diploid condition in which the nuclei are in conjugate pairs.

Why the nuclei of a mycelium of *Puccinia malvacearum* and of other homothallic Rusts should at first be separate from one another and subsequently come to be arranged in conjugate pairs is a problem which has not as yet been solved. A similar unsolved problem is presented by homothallic Hymenomycetes, *e.g. Coprinus sterquilinus* and *C. stercorarius*.

Since it has been proved by experiment that the *micro*-forms *Puccinia malvacearum* and *P. xanthii* which have no pycnidia are homothallic, we may assume that all other similar *micro*-forms are also homothallic. Therefore, in what follows it has been assumed that the following species are homothallic: *Puccinia adoxae*, *P. buxi*, *P. transformans*, *Uromyces scillarum*, and *U. ficariae*.

Cytological investigations have revealed that in the *micro*-forms of

[1]For the evidence that these three species are indeed homothallic, *vide supra*, Chapters I and V.

the Rust Fungi there are two distinct types.   In Type I, represented
by *Puccinia adoxae*[1] and *Uromyces scillarum*,[2] the germ-tubes of the
basidiospores give rise to a diploid or binucleate mycelium which
spreads between the host-cells and produces binucleate teleutospore
sori and binucleate teleutospores directly without the formation of any
special cell fusions.   In Type II, represented by *Puccinia transformans*,[3]
P. malvacearum[4] and *P. buxi*,[5] the germ-tubes of the basidiospores give
rise to a haploid or uninucleate mycelium.   This spreads between the
host-cells and produces uninucleate rudiments of teleutospore sori.
Then, under the epidermis, uninucleate cells fuse together in pairs and
thus binucleate cells are formed.   These grow in length and so produce
short chains of binucleate cells and, finally, from these chains of cells
the binucleate teleutospores are formed.   According to Mme Moreau,
in *Uromyces ficariae*,[6] which otherwise resembles the Rusts of Type II,
the cell-fusions take place at the base of the uninucleate primordium
of the teleutospore sorus and the number of binucleate cells produced
before teleutospore formation is greater than in *Puccinia malvacearum*,
etc.   Thus she regards *Uromyces ficariae* as being intermediate between Type I and Type II.   Kursanov[7] made a further examination
of *U. ficariae* and in 1914 reported that in his material the binucleate
phase appeared earlier in the mycelium than had been observed by

[1]V. H. Blackman and Helen C. I. Fraser, "Further Studies on the Sexuality of
the Uredineae," *Annals of Botany*, Vol. XX, 1906, p. 43, Plate IV, Figs. 24 and 25.

[2]*Ibid.*; also Mme F. Moreau, "Les phénomènes de la sexualité chez les Urédinées,"
Le Botaniste, T. XIII, 1914, pp. 194-196.   Mme Moreau confirmed the observations
of Blackman and Fraser on *Puccinia scillarum* and clearly distinguished between the
two types of *micro*-forms (p. 195).

[3]E. W. Olive, "Sexual Cell Fusions and Vegetative Nuclear Divisions in the
Rusts," *Annals of Botany*, Vol. XXII, 1912, pp. 340-341, Plate XXII, Figs. 39 and 40.

[4]E. Werth and K. Ludwigs, "Zur Sporenbildung bei Rost- und Brandpilzen,"
Ber. d. D. Bot. Gesellschaft, Bd. XXX, 1912, pp. 525-527, Taf. XV, Figs. 14-39;
Mme F. Moreau, *loc. cit.*, pp. 190-191, Plate XXII, Fig. 1-9; T. Lindfors, "Studien
über den Entwicklungsverlauf bei einigen Rostpilzen aus zytologischen und ana-
tomischen Gesichtspunkten," *Svensk Bot. Tidskrift*, Bd. XVIII, 1924, pp. 34-37,
Fig. 13; Ruth F. Allen, "A Cytological Study of *Puccinia malvacearum* from the
Sporidium to the Teliospore," *Journ. of Agric. Research*, Vol. LI, 1935, pp. 810-812,
Plate VIII; Blackman and Fraser (*loc. cit.*, p. 42, Plate IV, Figs. 19-23) found that
the teleutospore sori arise on a uninucleate mycelium, but left the question of the
origin of the binucleate condition unsolved.   Dorothy Ashworth ("*Puccinia malvacearum* in Monosporidial Culture," *Trans. Brit. Myc. Soc.*, Vol. XVI, 1931, pp. 191-
192, 200, Plate IX, Figs. 3-7) did not observe any cell fusions and brought forward
evidence that the diplophase is initiated "both by nuclear migration and by nuclear
division unaccompanied by wall formation."

[5]Mme F. Moreau, *loc. cit.*, pp. 191-192, Plate XXII, Figs. 10-17.

[6]*Ibid.*, pp. 193-194, 195.

[7]L. Kursanov, "Recherches morphologiques et cytologiques sur les Urédinées,"
Bull. Soc. Nat. Moscou, T. XXXI, 1917, pp. 53-54.

Mme Moreau; and he remarked that where the initiation of the binucleated stage is not definitely associated with specialised cells, it is of little importance whether it takes place in the mycelium earlier or later.

From the cytological observations that have just been recorded it appears that, in *micro*-forms of Rusts, the association of nuclei in conjugate pairs takes place in some species at a very early stage in the development of the haploid mycelium and in other species at a much later stage and shortly before the formation of the teleutospores. In those species, *e.g. Puccinia malvacearum*, in which the association takes place shortly before the formation of the teleutospores and with an accompaniment of cell fusions, the fusion of uninucleate cells in the rudiment of each teleutospore sorus, just as in heterothallic Rusts, permits of the (+) and (−) nuclei finding their conjugate mates.

In 1924, Lindfors[1] described the nuclear phenomena connected with the life-history of *Puccinia arenariae*, a *micro*-form devoid of pycnidia. He found that the life-history takes the following course. Two nuclei are present in each of the two cells of the young teleutospore and fusion nuclei are formed in the usual way. In a basidium the fusion nucleus divides and a septum is formed, so that the basidium becomes divided into two uninucleate cells. The two nuclei then divide so that each of the two cells becomes binucleate; but *no further septa are formed*. Two basidiospores are formed, one by each cell, and two nuclei pass up into each basidiospore. The binucleate basidiospores produce binucleate mycelia and these give rise to the binucleate teleutospore sori. Thus, in *P. arenariae*, the association of nuclei in pairs takes place in the two basidiospores and not as a result of cell fusions in the mycelium.

Lindfors[2] observed that, in addition to *Puccinia arenariae*, the following *micro*-forms or *lepto*-forms have a binucleate mycelium only: *Puccinia albulensis, P. epilobii, P. gigantea, P. holboelii, P. saxifragae,* and *Uromyces solidaginis*, but in these species whether the association of (+) and (−) nuclei takes place in the basidiospores or in the germ-tubes of the basidiospores was not investigated. In any case it is evident that in these and other similar species there is no necessity for any cell fusions to be formed at the base of the teleutospore sorus.

In 1926, Dodge[3] investigated the short-cycled Pine Rust *Coleo-*

---

[1] T. Lindfors, "Studien über den Entwicklungsverlauf bei einigen Rostpilzen, etc." *loc. cit.*, pp. 43-46, Text-fig. 16 and Taf. IV, Figs. 37-49.

[2] *Ibid.*, pp. 46-48.

[3] B. O. Dodge, "Organization of the Telial Sorus in the Pine Rust, *Gallowaya pinicola* Arth.," *Journ. Agric. Research*, Vol. XXXI, 1925, pp. 641-651, Text-fig. 1 and Plates I and II.

sporium pinicola (*Gallowaya pinicola*). This form gives rise to teleuto-spore sori and to vestigial pycnidia. Since the pycnidia do not fully rupture the host tissues above them and seldom form any pycnidio-spores, it is probable that *C. pinicola* is homothallic. Dodge found that its mycelium derived from one or more basidiospores is uninucleate and that the binucleate condition is initiated by lateral cell-fusions between cells in chains of cells composing the primordium of the teleutospore sorus. These fusions point to a sorting-out process in which (+) nuclei and (−) nuclei seek one another just as they must do in a heterothallic Rust. It is desirable that it should be determined experimentally whether *C. pinicola* is homothallic or heterothallic, as such a determination would be of assistance in interpreting the cell fusions which Dodge observed.

In 1927, Ruth Walker[1] investigated four *micro*-forms: without pycnidia, *Puccinia xanthii* and *P. asteris*; with pycnidia, *P. fusca* and *P. cryptotaeniae*; and she found that the binucleate condition preceding the production of teleutospores originated in various ways. Her results were as follows.

In *Puccinia xanthii*, the uninucleate condition passes into the binucleate by cell fusions in the rudiment of the teleutospore sorus. Thus *P. xanthii* resembles *P. malvacearum* (as described by Werth and Ludwig, Mme Moreau and Miss Allen) and other *micro*-forms of Type II.

In *Puccinia asteris*, the mycelium consists of both uninucleate and binucleate cells, and cells with 3, 4, or 5 nuclei are not infrequent. No cell fusions were observed and Miss Walker suggested that "the binucleate condition is brought about in a vegetative manner, *i.e.* by simple mitotic division followed by cell division." If this is really what happens in *P. asteris* during the establishment of conjugate pairs of nuclei, then in this species (+) nuclei and (−) nuclei do not have to seek one another, and nuclear migrations and cell-fusions have been rendered unnecessary.

In *Puccinia fusca*, the mycelium is uninucleate and it becomes binucleate at the base of the teleutospore sorus by cell fusions. Since *P. fusca* has pycnidia, it may well be heterothallic. If monobasidio-sporic pustules were found to remain sterile unless their nectar had been mixed, we should be obliged to conclude that, in each of the first pairs of conjugate nuclei formed in the sorus rudiment, one nucleus must be of local origin and the other must have come from without.

[1]Ruth I. Walker, "Cytological Studies of Some of the Short-cycled Rusts," *Trans. Wisconsin Acad. of Sci., Arts, and Letters*, Vol. XXIII, 1927, pp. 567-582, Plates VIII-X.

In *Puccinia cryptotaeniae*, the binucleate condition "arises at the base of the sorus and in the vegetative mycelium by nuclear division without a subsequent cell division and rarely by nuclear migration." Miss Walker says that pycnidia "are rarely present," but Arthur[1] remarks that they are "amphigenous, numerous, preceding or among the telia." The presence of pycnidia in *P. cryptotaeniae* suggests that the fungus is heterothallic. If *P. cryptotaeniae* should be found to be heterothallic, Miss Walker's solution of the problem of the origin of its binucleate condition could no longer be accepted.

Before any *micro*-form of Rust bearing pycnidia is studied cytologically to discover the mode of origin of its binucleate condition it should be examined experimentally to determine whether it is heterothallic or homothallic; for, if it is heterothallic, the cytological preparations should yield evidence of the migration of nuclei from without as these seek their conjugate mates in the rudiments of the teleutospore sori.

Mode V: Annual Self-diploidisation in Certain Rusts having a Systemic Mycelium.—There are a number of Rusts whose mycelium is *systemic, i.e.* is non-localised and grows widely through the host tissues. Systemic mycelia often persist in the host tissues for years. There are various types of systemic mycelia.

In *Cronartium ribicola* the systemic mycelium in a canker on a White Pine arises from a basidiospore and is haploid (uninucleate). It grows for years, remains uninucleate, and annually produces a new crop of pycnidia and proto-aecidia.

In *Gymnosporangium clavariiforme* the systemic mycelium in a branch of *Juniperus communis* arises from an aecidiospore and is diploid (binucleate). It grows for years, remains diploid, and annually produces a crop of teleutospores.

In *Puccinia minussensis* the systemic mycelium in *Lactuca pulchella* arises from an aecidiospore or a uredospore and is at first diploid (binucleate).[2] It perennates in the diploid condition in the rhizomes of its host; but, in the new shoots, it produces haploid (uninucleate) branches which bear pycnidia and proto-aecidia. There are other Rusts whose systemic mycelium resembles that of *Puccinia minussensis* and one of them is *P. suaveolens*. It is in systemic mycelia of the *P. minussensis* type that the phenomenon of annual self-diploidisation is possible.

[1]J. C. Arthur, *Manual of the Rusts in United States and Canada*, Lafayette, U.S.A., 1934, p. 321.

[2]A. M. Brown: personal communication of unpublished data, 1940; also "Studies on the Perennial Rust *Puccinia minussensis*," *Canadian Journal of Research*, Vol. XIX, Section C, 1941, pp. 75-79.

By the term *annual self-diploidisation* is meant the phenomenon in which a diploid systemic mycelium annually self-diploidises its own haploid branches, thus enabling these branches to develop aecidia or uredospore sori.

De-diploidisation and Self-diploidisation in Puccinia minussensis.— *Puccinia minussensis* is a macrocylic autoecious Rust with a systemic mycelium that gives rise to pycnidia, aecidia, uredospore sori, and teleutospore sori on *Lactuca pulchella* and *L. canadensis*. On *L. pulchella* (the Blue Lettuce) it is not uncommon in the neighbourhood of

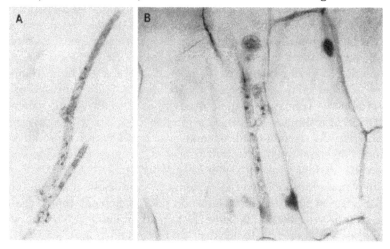

FIG. 97.—Conjugate nuclei in diploidised mycelia of two Basidiomycetes. A, *Clitocybe illudens* (Agaricaceae), a branched hypha with three pairs of conjugate nuclei one of which, in a developing clamp-connexion, is undergoing division. B, *Puccinia minussensis* (Uredineae), a branched hypha in a resting over-wintering rhizome of *Lactuca pulchella*, with three pairs of conjugate nuclei. A, photographed by W. F. Hanna, magnification 750. B, photographed by A. M. Brown, magnification about 1,400.

the Dominion Rust Research Laboratory. Its life-history has been studied in this Laboratory by A. M. Brown.

Lactuca pulchella is a herbaceous perennial. It persists through the winter in the form of subterranean rhizomes. In the spring these rhizomes send up aerial shoots that bear leaves, flowers, and seeds and then die down.

Brown[1] found that, in the winter, the mycelium of *Puccinia minussensis*, living within the rhizomes and up to the top of the stumps of the old died-down aerial shoots, contains conjugate nuclei (Fig. 97, B) and is therefore *diploid*.

[1] A. M. Brown, *loc. cit.*

He has also observed that, in the spring, the mycelium grows out from the rhizomes into every new aerial shoot and, upon the leaves and sometimes on the stems, always first produces pycnidia which on the under side of each leaf are scattered over the general surface including the mid-rib (Fig. 98, A, *cf. C*). The mycelium newly grown into the young leaves is *haploid*, as is shown by its producing pycnidia and by having its nuclei isolated from one another and not in conjugate pairs.

In many of the leaves, shortly after the production of pycnidia, proto-aecidia are formed in and about the mid-rib and sometimes between the mid-rib and the leaf-margin. The mycelium making up these proto-aecidia, as Brown observed, is haploid, without any trace of conjugate nuclei.

It appears, therefore, that, in *Puccinia minussensis*, in the spring, the diploid mycelium that has lived in the rhizomes through the winter becomes *de-diploidised* in that it sends up into the new stems and leaves haploid branches. It seems reasonable to suppose that these haploid branches are of two kinds, one (+) and one (−); but how these branches are distributed in the new shoots is at present unknown. It may be that both (+) and (−) branches grow into one and the same shoot or leaf or that a shoot or leaf contains either a (+) branch or a (−) branch.[1]

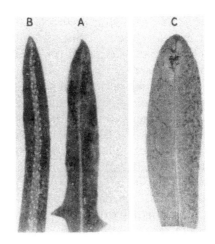

FIG. 98.—*Puccinia minussensis* on *Lactuca pulchella*. Under side of leaves, all bearing numerous scattered pycnidia invisible in A and B, appearing as minute black dots in C. A, as yet, with pycnidia only. B, with aecidia on and about the mid-rib. C, with teleutospore sori near the tip. Photographed by A. M. Brown. A and B, slightly reduced; C, about natural size.

The pycnidia produced on a Blue Lettuce leaf come into existence in a series from below upwards on the leaf. Later, on and about the mid-rib and sometimes between the mid-rib and the leaf-margin, the proto-aecidia are converted into aecidia (Fig. 98, B). Some leaves

[1]For a general discussion of the phenomenon of *de-diploidisation* in Hymeno-mycetes, Uredinales, etc., *vide* A. H. R. Buller, "The Diploid Cell and Diploidisation in Plants and Animals with Special Reference to the Higher Fungi," *The Botanical Review*, Vol. VII, 1941, pp. 411-414.

that produce aecidia produce nothing more, while other leaves, after producing aecidia, produce uredospore sori and then teleutospore sori, or teleutospore sori alone. It also happens that a leaf which has produced pycnidia thereafter produces nothing but teleutospore sori (Fig. 98, C).

The observations just recorded indicate that, in *Puccinia minussensis*, a haploid mycelium that has produced pycnidia and proto-aecidia, afterwards, by some means or other, becomes diploidised, at least at all those points where aecidia, uredospore sori, or teleutospore sori are developed. Thus *de-diploidisation* of the original diploid mycelium that persisted in the rhizome is followed by *re-diploidisation* of at least parts of that mycelium's haploid branches.

Brown observed that this re-diploidisation of the haploid branches of an old diploid mycelium takes place in a greenhouse under *conditions that exclude the visits of flies to the pycnidia*. Therefore, if we consider the mycelium in a Blue Lettuce plant as a whole (with a diploid part in the rhizome and haploid parts in the new stems and leaves), it seems certain that the diploidisation of the haploid branches can be effected by a process of *self-diploidisation*, *i.e.* the diploidisation of the haploid branches can be effected by nuclei which migrate either from the older diploid part of the mycelium or from haploid branches of opposite sex, ($+$) branches receiving ($-$) nuclei and ($-$) branches ($+$) nuclei.

Brown sowed some aecidiospores of *Puccinia minussensis* on a bud at the end of a healthy rhizome of *Lactuca pulchella* and then planted the rhizome about one inch below the surface of the soil. The bud grew into an aerial shoot and, thereafter, it was observed that, on the leaves, pycnidia first appeared, then aecidia, and finally teleutospore sori. This experiment was repeated again and again during three years and always with the same result.

Brown's experiment was carried out in the winter in the absence of insects and without the nectar being mixed by hand; and it can be interpreted as follows. A diploid mycelium, derived from aecidiospores, underwent local de-diploidisation and then re-diploidised its haploid branches by the process of self-diploidisation.

In respect to such a fungus as *Puccinia minussensis* it may be asked: when once a diploid mycelium has been established in a host-plant, why should such a diploid mycelium develop haploid branches and spend so much material and energy in producing pycnidiospores? These pycnidiospores, as we have seen, are not required for the diploidisation of the haploid branches. How, therefore, do they function? Are they or are they not of some benefit to the species which produces them? An answer to these questions will now be attempted.

The production of pycnidia annually on a systemic mycelium: (1) increases the chances that a unisexual haploid mycelium produced from a basidiospore on a new host-plant will have its proto-aecidia diploidised through the agency of insects; and (2) increases the chances that aecidiospores may be produced as a result of crossing between different strains.

Although every year in *Puccinia minussensis* self-diploidisation of the haploid mycelia in the new shoots takes place, it is possible that the diploidisation of these mycelia here and there on the leaves, may be often effected by the intermixing of nectar through insect agency, *i.e.* by *cross-diploidisation*.[1] Self-diploidisation, which takes place automatically after a certain time, does not necessarily prevent prior cross-diploidisation by means of transferred pycnidiospores.

Crossing between different strains in a single Rust species may be assumed to be favourable to the production of new variations and therefore favourable to evolutionary survival. If the perennating mycelium of such a fungus as *Puccinia minussensis* did not produce pycnidia annually, the chances that aecidiospores of diverse genetic composition would be formed would be greatly diminished.

We may therefore conclude that, in *Puccinia minussensis* and in similar Rust species, the de-diploidisation process that takes place in each old mycelium annually and the consequent production of an annual crop of pycnidiospores may be looked upon as of distinct benefit to the species as a whole.

That, in *Endophyllum sempervivi*, the pycnidia are not required for the annual production of aecidia on the leaves was proved by Miss Ashworth[2] who killed all the young pycnidia which appeared by stabbing them with a hot needle. *E. sempervivi*, therefore, like *Puccinia minussensis* is able to self-diploidise its proto-aecidia.

Cirsium arvense, like *Lactuca pulchella*, is a herbaceous perennial. The mycelium of its parasite, *Puccinia suaveolens*, persists in the roots during the winter and, in the spring, enters all the buds which develop on these roots. The buds develop into aerial shoots and, as a rule, every leaf on a shoot first bears pycnidia and afterwards uredospore sori and teleutospores. As will be shown in Chapter X the processes of de-diploidisation and re-diploidisation which we have recognised as taking place annually in the systemic mycelium of *P. minussensis* also take place annually in the systemic mycelium of *P. suaveolens*.

---

[1] For the further employment of the terms *de-diploidisation, re-diploidisation, self-diploidisation,* and *cross-diploidisation vide* the discussion of the sexual process of *Puccinia suaveolens* in Chapter X.

[2] Dorothy Ashworth, "*Puccinia malvacearum* in Monosporidial Culture," *Trans. Brit. Myc. Soc.*, Vol. XVI, 1931, p. 253.

In Chapter IX it will be suggested that annual self-diploidisation may take place in cankers of *Cronartium ribicola* that have already borne aecidia.

In systemic Rusts like *Puccinia minussensis*, the diploidisation of the haploid branches of a mycelium by nuclei which migrate from the older diploid part of the mycelium finds a close parallel in Brown's discovery that, in *P. helianthi*, a diploid mycelium derived from an aecidiospore or a uredospore can be used experimentally to diploidise a haploid mycelium derived from a basidiospore. In other words, Mode V of initiating the sexual process in the old mycelia of certain systemic Rusts is very similar to Mode III observed by Brown in a non-systemic Rust.

Defect of Homothallism.—In the Uredinales and the Hymenomycetes, the homothallic species are constantly self-diploidising themselves and thus becoming more and more homozygous, whereas the heterothallic species, owing to cross-diploidisation, remain highly heterozygous and variable.

Just as in the Plant Kingdom and the Animal Kingdom in general cross-fertilisation between the individuals of a species is the rule and self-fertilisation the exception, so in the Uredinales and the Hymenomycetes in general cross-diploidisation is the rule and self-diploidisation the exception. From this we may deduce the fact that, in the Uredinales and the Hymenomycetes, in the course of evolution heterothallism has proved superior to homothallism.

There is every reason to believe that, in the Uredinales and the Hymenomycetes, the original primitive sexual condition was one of heterothallism and that the homothallic species have all been derived from heterothallic species that are now extinct or are still living. Thus, for example: (1) in the genus Puccinia it seems clear from the arguments adduced by Jackson[1] that the short-cycled homothallic species have been derived from the long-cycled heterothallic species; and (2) in the genus Coprinus in which a few species, *e.g. C. narcoticus, C. stercorarius*, and *C. sterquilinus*, are homothallic, whereas far more species, *e.g. C. comatus, C. curtus, C. lagopus, C. niveus, C. macrorhizus, C. radians*, and *C. rostrupianus*, are heterothallic, one seems compelled to admit that the homothallic species have been derived from heterothallic species and that the genus at its first origin was heterothallic.

For the successful operation of the evolutionary process variability is a *sine qua non* and, since homothallism involves much less variability

[1]H. S. Jackson, "Present Evolutionary Tendencies and the Origin of Life Cycles in the Uredinales," *Memoirs Torrey Bot. Club*, Vol. XVIII, 1931, pp. 1-108.

than heterothallism, from the evolutionary point of view we may regard homothallism as *a defective sexual process*. When in the course of evolution a species of the Uredinales or Hymenomycetes changes from heterothallism to homothallism, it takes a short cut that leads, very truly, to the certain production of spores by the individual and the avoidance of the risk of non-diploidisation and sterility but ends, unfortunately, in non-adaptability to new conditions and in evolutionary stagnation.

Rust Fungi in which the Sexual Process has become Imperfect or Inoperative.—In the Uredinales there are a number of aberrant types so far as the behaviour of the nuclei in the aecidiospores, teleutospores, and basidia are concerned and, in some of them, the sexual process has become either imperfect or entirely inoperative. A few examples of these aberrant types will now be considered.

Sappin-Trouffy[1] in 1896 showed that, in *Endophyllum euphorbiae-sylvaticae*, the mature teleutospore contains two nuclei which, as the teleutospore germinates, migrate—one following the other—into the germ-tube. Here they divide and so form four nuclei. The germ-tube then becomes a basidium. Three septa are formed cutting off one nucleus in each cell. Each of the four cells then produces a sterigma and a basidiospore. Sappin-Trouffy, on the basis of these observations, came to the conclusion that in *E. euphorbiae-sylvaticae* there are *no nuclear fusions whatever*, a conclusion that has been confirmed by the more recent work of the Moreaus.[2]

In a normal long-cycled Rust, *e.g. Puccinia graminis* or *P. helianthi*, the sexual process has three stages: (1) nuclear association in the basal cells of the proto-aecidium; (2) a series of conjugate nuclear divisions; and (3) nuclear fusion followed by nuclear reduction. In a well-established perennial mycelium of *Endophyllum euphorbiae-sylvaticae* the sexual process presumably passes through the first two stages but not the third and is therefore imperfect.

Kunkelia nitens is a short-cycled Orange Rust which attacks Rubus species, and it appears to have been derived from a long-cycled Orange Rust, *Gymnoconia peckiana* (*G. interstitialis*). As in Endophyllum, its mycelium is perennial in its host-plant and its aecidiospore-like

---

[1]P. Sappin-Trouffy, "Recherches histologiques sur la famille des Urédinées," *Le Botaniste*, T. V, 1896, pp. 184-187.

[2]F. Moreau et Mme F. Moreau, "L'écidiospore de l'*Endophyllum Euphorbiae-sylvaticae* (DC) Winter est-elle le siège d'une Karyogamie?," *Bull. Soc. Myc. France*, T. XXXIII, 1918, pp. 97-99; and "Les Urédinées du groupe Endophyllum," *Bull. Soc. Bot. France*, T. LXVI, 1919, pp. 27-33.

teleutospores develop directly into basidia.[1]   Dodge and Gaiser[2] have found a form of *Kunkelia nitens* which, in its nuclear behaviour, resembles *Endophyllum euphorbiae-sylvaticae*.   This form of *Kunkelia nitens* has pycnidia, cell-fusions in the aecidial (telial) spore-bed, and 4-celled 4-spored basidia; but there are *no nuclear fusions in the life-history*.   The two nuclei in the aecidiospore migrate into the germ-tube and divide.   Then a septum is formed between the two pairs of nuclei and then two more septa, so that the basidium is divided into four cells. The nuclei in two basidiospores formed at one end of the basidium must therefore be "the direct descendents by vegetative division of one of the two nuclei entering the fusion cell at the origin of the spore chain." Here again, the sexual process is not carried through to the normal climax of nuclear fusion and nuclear reduction and is therefore imperfect.

[1] *Gymnoconia peckiana* is an aut-opsis form, and its pycnidia and caeomoid aecidia, borne on a systemic and perennial mycelium, and its teleutospore sori are found on various Brambles (Rubus species) in Europe, Asia, Japan, and North America.   In Saskatchewan it has been collected on *Rubus acaulis* and *R. arcticus* and in Manitoba on *R. triflorus* (G. R. Bisby *et al.*, *The Fungi of Manitoba and Saskatchewan*, Ottawa, 1938, p. 64), and in England, on mycological excursions, I have often observed it on Blackberry bushes.   The caeomoid aecidia "are chiefly hypophyllous on dwarfed leaves and shoots and they are crowded over all the under surface of the leaflets, often confluent and tortuous, somewhat pulverulent, and bright golden-orange when fresh" (J. C. Arthur, *Manual of the Rusts of United States and Canada*, Lafayette, 1934, p. 96).   Owing to their extent and brilliant colour, the caeomoid aecidia attract the eye, with the result that the fungus has been called the Orange Rust.

Kunkelia nitens, a microcyclic form of *Gymnoconia peckiana*, was discovered by L. O. Kunkel in 1913 ("The Production of a Promycelium by the Aecidiospores of *Caeoma nitens* Burrill," *Bull. Torrey Bot. Club*, Vol. XL, 1913, pp. 361-366; also "The Nuclear Behaviour of the Promycelia of *Caeoma nitens* Burrill and *Puccinia Peckiana* Howe," *Am. Journ. Bot.*, Vol. I, 1914, pp. 37-47).   He made his demonstration with aecidiospores gathered in the parks of New York City.   The fungus was referred to by Kunkel, Dodge, and others as *Caeoma nitens*.   But *Caeoma* is merely a form-genus. Therefore, in 1917, Arthur placed Kunkel's fungus in a new genus, *Kunkelia*, and named it *Kunkelia nitens* (Schw.) Arth. (J. C. Arthur, *vide* his *Manual, loc. cit.*, p. 97). H. S. Jackson ("Evolutionary Tendencies, etc.," *loc. cit.*, p. 55) remarks that B. O. Dodge clearly recognised that the forms of *Kunkelia nitens* have been derived from the haploid generation of *Gymnoconia peckiana* and that Kunkelia is properly interpreted as an *endo*-form with the aecidia having the gross morphology of a caeoma.

Of the genus Gymnoconia from which the genus Kunkelia has doubtless been derived Jackson (*loc. cit.*, pp. 96-97) says: "It is entirely possible that this is really an *-opsis* Phragmidium which in the process of simplification has reverted to the original two-celled teliospore type as did Teloconia."

[2] B. O. Dodge and L. O. Gaiser, "The Question of Nuclear Fusions in the Blackberry Rust, *Caeoma nitens*," *Journ. Agric. Research*, Vo.. XXXII, 1926, pp. 1008-1012, 1022-1023.

In another form of *Kunkelia nitens*, Dodge[1] found that: there are no pycnidia; no cell-fusions are formed in the aecidial (telial) sporebed; the cells are all uninucleate and devoid of conjugate nuclei; and the basidia are 2-celled and 2-spored. In a well-established perennial mycelium of this form of *K. nitens*, therefore, the sexual process passes through none of its three stages and is entirely inoperative.

In *Endophyllum euphorbiae-sylvaticae* var. *uninucleatum*, as discovered by Mme Moreau,[2] and in a form of *E. centranthi-rubri* according to Poirault,[3] the "aecidiospores" arise without cell-fusion and from the first have one nucleus only. When a spore germinates, a single nucleus enters the germ-tube, and in *E. centranthi-rubri* a two-celled basidium is formed. Here again, the sexual process is no longer operative.[4]

The problem of sexuality in such a Rust as *Endophyllum euphorbiae-sylvaticae* has so far been attacked by the cytologist only, and the cytologist has investigated well-established mycelia exclusively. The work of the cytologist ought now to be supplemented by that of the experimenter; for only by experiment can we learn whether or not *E. euphorbiae-sylvaticae* is heterothallic. If the fungus should be found to be heterothallic, then it would be necessary to find out whether or not the two nuclei in each of the aecidiospores formed by the interaction of a (+) mycelium and a (−) mycelium fuse together. On the assumption that *E. euphorbiae-sylvaticae* is heterothallic, the sexual problem for the experimenter involves the following questions.

In *Endophyllum euphorbiae-sylvaticae*, if two basidiospores, one (+)

[1]B. O. Dodge, "Uninucleated Aecidiospores in *Caeoma nitens* and Associated Phenomena," *Journ. Agric. Research*, Vol. XXVIII, 1924, pp. 1045-1058; also Dodge and Gaiser, *loc. cit.*

[2]Mme F. Moreau: "Sur l'existence d'une forme écidienne uninucléée," *Bull. Soc. Myc. France*, T. XXVII, 1911, pp. 489-493; "Les phénomènes de la sexualité chez les Urédinées," *Le Botaniste*, T. XIII, 1914, pp. 177-188; and "Note sur la variété uninucléée de l'*Endophyllum Euphorbiae* (DC) Winter," *Bull. Soc. Myc. France*, T. XXXI, 1915, pp. 68-70.

[3]G. Poirault: "Sur quelques Urédinées nouvelles," *Bull. Mens. des Naturalistes de Nice et Alpes-Maritimes*, T. 1, 1913, pp. 105-108; and "Sur quelques champignons parasites rares ou nouveaux observés dans les Alpes-Maritimes," *Riviera Scientifique: Bull. Assoc. des Naturalistes de Nice et Alpes-Maritimes*. T. II, 1915, pp. 7-19. Cited from H. S. Jackson's "Present Evolutionary Tendencies, etc.," *loc. cit.*

[4]Unfortunately, Mme Moreau had great difficulty in getting the aecidiospores of her Endophyllum to germinate. The germ-tubes shown in her illustrations (*Le Botaniste, loc. cit.*, Pl. XXI, Figs. 10-15) look very abnormal. Some of them are divided by two or three septa and some of the cells have a blunt process projecting from them. The processes failed to develop any basidiospores. Presumably the germ-tubes were developing into basidia; but whether or not, if they had developed normally, they would have had four cells or two it is impossible to decide.

and the other (−), germinated close to one another on a leaf of *Euphorbia sylvatica* and gave rise to mycelia which met in a compound pustule, would the sexual process become operative in the same way as it does in *Puccinia graminis*?   In other words, would the (+) mycelium and the (−) mycelium fuse together, exchange nuclei, and diploidise one another's proto-aecidia?   Would every aecidiospore come to contain a (+) nucleus and a (−) nucleus?   And would or would not these two nuclei fuse together before or after entering the basidium?

Further, in *Endophyllum euphorbiae-sylvaticae*, let us suppose that some basidiospores have germinated at a little distance from one another on some leaves of a *Euphorbia sylvatica* plant and that in consequence there have been formed several isolated pycnidia-bearing pustules, some (+) and others (−).   In the *absence* of insects, would these isolated pustules remain haploid so that the proto-aecidia would not develop into aecidia?   And, in the *presence* of insects which would mix the nectar of (+) and (−) pustules, would (+) pycnidiospores unite with the flexuous hyphae of (−) pycnidia and (−) pycnidiospores unite with the flexuous hyphae of (+) pycnidia and so initiate the sexual process in the same way as in *Puccinia graminis*?

The pycnidia of *Endophyllum euphorbiae-sylvaticae* are produced in great numbers every spring on the perennial mycelium of an infected host-plant.   It seems not unlikely that, under the conditions set forth in the last question, they would function in a normal manner.   Are the pycnidia of *E. euphorbiae-sylvaticae* vestigial organs or are they functional?   There, in brief, is a problem which the experimenter alone can solve.

The problem of sex in the Orange Rusts of Rubus species should also be investigated experimentally.   In the form of *Kunkelia nitens* with pycnidia and resembling *Endophyllum euphorbiae-sylvaticae*, described by Dodge and Gaiser, it might be found that (+) and (−) mycelia, newly developed from basidiospores, would, on meeting in a compound pustule or after exchanging pycnidiospores, interact in a normal manner, *i.e.* would diploidise each other's proto-aecidia.   It might then be found that in the first crop or generation of aecidiospores thus produced the two nuclei in each aecidiospore would fuse together instead of remaining separate as they do in aecidiospores formed on mycelia several years old.   If, on the other hand, it were to be found that the nuclei in the first-generation aecidiospores do not fuse, then we might conclude that the sexual process in Dodge and Gaiser's form of *K. nitens* is always imperfect and never carried to the normal climax.

Haploid aecidia with uninucleate aecidiospores have been observed not only by Mme Moreau and B. O. Dodge, but also by Kursanov.

Kursanov,[1] in 1917, found in three different places near Moscow a form of *Aecidium punctatum* (a stage of *Tranzschelia pruni-spinosae*) on *Anemone ranunculoides* which differs from the normal form on the same host in that its aecidia are haploid: in every aecidium the uninucleate condition is displayed by the aecidiosporophores, the aecidiospores, the cells of the peridium, and the intercalary cells of the spore-chains and of the peridium.[2] Kursanov[3] also observed a form of *Aecidium leucospermum* (a stage of *Ochrospora ariae*) on the same host, *Anemone ranunculoides*, in which the aecidia contained chains of uninucleate aecidiospores mixed with a few chains (up to 5 per cent.) of binucleate aecidiospores. In these two forms, just as in *Endophyllum euphorbiae-sylvaticae* var. *uninucleatum* and the form of *Kunkelia nitens* with haploid aecidia, the nuclear condition of the aecidia indicates an imperfection of the sexual process in respect to nuclear association.

The development of uninucleate aecidiospores in *Aecidium punctatum*, *Aecidium leucospermum*, etc., also teaches us that the development of aecidiospores is not always dependent on their nuclei being associated in conjugate pairs.

There are known not only uninucleate aecidiospores but also uninucleate uredospores and teleutospores.

Recently, Newton and Johnson,[4] working in the Dominion Rust Research Laboratory, have observed in certain abnormal selfed races of *Puccinia graminis*, propagated on wheat-plants, uredospore sori in some of which the uredospores are all uninucleate while in others some of the uredospores are uninucleate and others binucleate. I have had the privilege of seeing Newton and Johnson's preparations and thus observing the uninucleate uredospores there present.

In 1911 Olive[5] described a variation of *Uromyces rudbeckiae*, a *micro*-form without pycnidia, aecidia, or uredospore sori, in which the binucleate stage is entirely wanting and the young teleutospores are uninucleate.

[1]L. Kursanov, "Recherches morphologiques et cytologiques sur les Urédinées," *Bull. Soc. Nat. Moscou*, n.s. (1917) Vol. XXXI, 1922, pp. 15-16.

[2]L. Kursanov, "Über die Peridienentwicklung in Aecidium," *Ber. d. D. Bot. Gesellschaft*, Bd. XXXII, 1914, p. 320.

[3]L. Kursanov, "Recherches morphologiques, etc.," *loc. cit.*, pp. 17-18. Kursanov's observations on *Aecidium leucospermum* have been confirmed by E. O. Callen ("Examination of *Aecidium leucospermum* D.C. from Scotland," *Trans. Brit. Myc. Soc.*, Vol. XXIV, 1940, pp. 109-111, Fig. 1).

[4]Margaret Newton and T. Johnson, personal communication, 1939.

[5]E. W. Olive, "The Nuclear Conditions in Certain Short-cycled Rusts," *Science*, Series II, Vol. XXXIII, 1911, p. 194.

In 1927 Jackson[1] confirmed Olive's observations on *Uromyces rudbeckiae* and extended them by germinating the teleutospores. He found that each teleutospore, uninucleate from its first origin, gives rise to a two-celled basidium: the nucleus of the teleutospore passes into the basidium and divides once; a septum is laid down; from each of the two uninucleate cells so formed a sterigma develops; and, finally, the contents of each cell pass through the sterigma into the developing basidiospore which thus becomes uninucleate. Said Jackson: "There can be little question that this species is uninucleate throughout." It is obvious that, in the form of *U. rudbeckiae* investigated by Olive and Jackson, the sexual process is entirely inoperative.

From the observations on *Puccinia graminis* and *Uromyces rudbeckiae* just recorded it appears that the development of uredospores and teleutospores, like the development of aecidiospores, is not dependent, or at least not always dependent, on the nuclei being associated in conjugate pairs.

Uninucleate Rusts considered as Self-propagating Haploid Strains derived from Heteroecious Species.—Such Rusts as (1) *Endophyllum euphorbiae-sylvaticae* var. *uninucleatum* observed by Mme Moreau, (2) the uninucleate *Endophyllum centranthi-rubri* observed by Poirault, (3) the uninucleate form of *Kunkelia nitens* observed by Dodge, (4) the uninucleate form of *Aecidium punctatum* observed by Kursanov, and (5) the uninucleate *Uromyces rudbeckiae* observed by Olive and Jackson all appear to be spore-producing and self-propagating haploid strains of what originally were, or still are, heterothallic species. The uninucleate Rusts, on this view, exhibit the phenomenon that I[2] have called *haploidy*[3] and Kniep[4] *somatogenous apomixis*. This phenomenon is exhibited by many Hymenomycetes and, owing to the similarity of the sexual process in the Hymenomycetes and the Uredinales, *a priori* one might expect it to occur in some of the Rust Fungi.

The phenomenon of haploidy in the Higher Fungi was reviewed by the writer[5] in 1941 and, here, for the sake of comparison with the

[1]H. S. Jackson, "Present Evolutionary Tendencies and the Origin of Life Cycles in the Uredinales," *Memoirs Torrey Botanical Club*, Vol. XVIII, pp. 23-24.

[2]A. H. R. Buller, "The Diploid Cell and the Diploidisation Process in Plants and Animals, with Special Reference to the Higher Fungi," *Botanical Review*, Vol. VII, 1941, p. 366.

[3]The word *haploidy* means the condition of being haploid. Haploidy is a mutant condition of the sporophytes of Higher Plants and is known in a number of species, *e.g. Datura stramonium* and *Nicotiana tabacum*.

[4]H. Kniep, *Die Sexualität der niederen Pflanzen*, Jena, 1928.

[5]A. H. R. Buller, *loc. cit.*, pp. 366-371.

uninucleate Rusts, some of the facts respecting haploidy in the Hymenomycetes given in the review will be brought forward again.

In the Hymenomycetes, a number of heterothallic species are known in which: (1) dikaryotic diploid fruit-bodies are formed on dikaryotic diploid mycelia, and (2) haploid (monocaryotic) fruit-bodies are formed on haploid mycelia. *Coprinus lagopus*, under natural conditions on horse dung, is typically heterothallic and produces dikaryotic diploid fruit-bodies. Yet in 1938 Hanna[1] succeeded in cultivating it from single haploid basidiospores for ten successive generations. The mycelia, sporocarps, and basidiospores were all haploid, and the basidiospores proved to be of the same genetic constitution as that of the original spore from which the first haploid mycelium originated.

Among other Hymenomycetes which, under experimental conditions, have been shown by various workers[2] to be able to develop haploid as well as diploid fruit-bodies are: *Armillaria mellea, Collybia tuberosa, Collybia velutipes, Coprinus ephemerus, Fomes pinicola, Panaeolus campanulatus, Peniophora ludoviciana, Schizophyllum commune,* and *Typhula erythropus.*

Haploid fruit-bodies of certain heterothallic Hymenomycetes, as well as dikaryotic diploid fruit-bodies, are known to occur under natural conditions in the open. Thus Bauch[3] found four-spored and two-spored forms of *Camarophyllus (Hygrophorus) virgineus* growing in fairy rings near to one another. Each fairy ring was composed entirely of four-spored fruit-bodies or entirely of two-spored fruit-bodies. In the four-spored form the cells of the fruit-body and the mycelium attached to its base bore clamp-connexions and contained conjugate pairs of nuclei; whereas, in the two-spored form, the cells of the fruit-body lacked clamp-connexions and were all uninucleate. In a basidium of the four-spored form, two haploid nuclei fused together, and this was followed by meiosis and the passage of the four resulting haploid nuclei up the four sterigmata into the four basidiospores; whereas, in a basidium of the two-spored form, there was at first only one haploid nucleus present so that no fusion was possible, the single

[1]W. F. Hanna, "Sexual Stability in Monosporous Mycelia of *Coprinus lagopus*,' *Annals of Botany,* Vol. XLII, 1928, pp. 378-379.

[2]For citations the reader is referred to the author's review of "The Diploid Cell, etc.," *loc. cit.*

[3]R. Bauch, "Untersuchungen über zweisporige Hymenomyzeten. Haploid Parthenogenesis bei *Camarophyllus virgineus*," *Zeitschrift für Botanik,* Bd. XVIII, 1926, pp. 337-387. The passage here cited is taken from the author's review of "The Diploid Cell, etc.," *loc. cit.,* pp. 368-369.

haploid nucleus divided once and, finally, the two haploid daughter-nuclei passed up the sterigmata into the two basidiospores." Also in the genus Mycena, *e.g.* in *Mycena galericulata, M. alkalina*, and *M. polygramma*, it has been shown by Kühner[1] and Smith[2] that, under natural conditions, haploid as well as diploid fruit-bodies are developed.

In the light of the facts concerning the Hymenomycetes just presented we have strong support for the view, brought forward at the beginning of this Section, that the uninucleate Rusts are spore-producing and self-propagating haploid strains of what originally were, or still are, heterothallic species. Thus *Endophyllum euphorbiae-sylvaticae* var. *uninucleatum* may be regarded as a haploid strain of the normal *E. euphorbiae-sylvaticae*, and *Aecidium punctatum* with haploid aecidia may be regarded as a haploid strain of *Tranzschelia pruni-spinosae*.

Uromyces rudbeckiae, as observed by Olive and Jackson, appears to be a species that now consists of haploid generations comparable with Hanna's ten successive haploid generations of *Coprinus lagopus* and having a two-spored basidium comparable with the two-spored basidium of the haploid forms of *Camarophyllus (Hygrophorus) virgineus*. Possibly *Uromyces rudbeckiae*, as it exists now, was originally derived from a single (+) or a single (−) strain. If there are both (+) and (−) self-propagating strains of *U. rudbeckiae* in existence today (comparable in distribution with (+) and (−) *Mucor mucedo* strains as found by Blakeslee), perhaps, if they were to come together under natural conditions or were brought together experimentally, they would fuse and give rise to the dikaryotic diploid phase of the life-history.

A possible explanation of the cytological condition of the uninucleate Rusts, alternative to the one already advanced, is that the nuclei of the uninucleate Rusts are all diploid, like the fusion nuclei in the teleutospores of Rusts with a normal life-history. In support of such an alternative explanation, so far as the writer knows, there is no evidence of any value. We may therefore accept the view that the nuclei of the uninucleate Rusts are haploid and not diploid and that the uninucleate Rusts are spore-producing and self-propagating strains of what originally were, or still are, heterothallic species.

[1] R. Kühner, "Étude cytologique de l'hymenium de *Mycena galericulata* Scop.," *Le Botaniste*, T. XVIII, 1927, pp. 168-176.

[2] A. H. Smith, "Investigations of Two-spored Forms in the genus Mycena," *Mycologia*, Vol. XXVI, 1934, pp. 305-331.

CHAPTER VII

COMPARISON OF THE SEXUAL PROCESSES IN THE UREDINALES WITH THOSE IN THE HYMENOMYCETES

Dikaryotic and Synkaryotic Diploid Cells in the Hymenomycetes and Uredinales — Uredinales and Hymenomycetes as Closely Related Groups — *Puccinia graminis* and *Coprinus lagopus* as Types for Comparison — Modes of Initiating the Sexual Process — Mode I — Mode II — Mode III — Mode IV — Mode V — Appendix.

Dikaryotic and Synkaryotic Diploid Cells in the Hymenomycetes and Uredinales.—For a number of years prior to 1941, the writer and other geneticists working with the Hymenomycetes and the Uredinales had referred to mycelia and cells containing conjugate nuclei as being diploid.

In 1941, in an extended review concerned with the nuclear condition of the cells of plants and animals but with special reference to the Higher Fungi, the writer[1] maintained that a cell containing conjugate nuclei is a diploid cell and he defined a *diploid cell* as a *cell that contains a complete set of pairs of homologous chromosomes.* Two kinds of diploid cells were recognised: (1) the *dikaryotic diploid cell* which contains two haploid nuclei, (n) + (n), one derived from one parent and the other from the other parent; and (2) the *synkaryotic diploid cell* which contains a single diploid nucleus, (2n), formed by, or derived from a nucleus resulting from, the fusion of two parental nuclei.

In support of the view that the dikaryotic cells of Hymenomycetes, Rust Fungi, etc., are to be regarded as diploid cells, it was pointed out[2] that: (1) a pair of conjugate nuclei provides the cell in which it lies with a complete set of homologous chromosomes; (2) the chromosomes of both nuclei affect the cell protoplasm and influence mycelial development; and (3) certain hybrid mycelia and fructifications containing conjugate nuclei exhibit the phenomenon of Mendelian dominance and recessiveness. Eight cases of such Mendelian dominance were

---

[1] A. H. R. Buller, "The Diploid Cell and the Diploidisation Process in Plants and Animals, with Special Reference to the Higher Fungi," *The Botanical Review*, Vol. VII, 1941, pp. 335-431. [2] *Ibid.*

described. The hybrids of the Higher Plants and of animals were called *synkaryotic hybrids* and those of the Hymenomycetes and Rust Fungi, etc., *dikaryotic hybrids*.

Both kinds of diploid cells are to be found in the Hymenomycetes and Uredinales.

In heterothallic Hymenomycetes, *e.g. Coprinus lagopus*, after two haploid mycelia of opposite sex have been mated, say (*Ab*) and (*aB*), the cells of the mycelium so formed contain pairs of conjugate nuclei, (*Ab*) + (*aB*). These cells are dikaryotic diploid cells. The young basidia contain a pair of conjugate nuclei, (*Ab*) + (*aB*), and are therefore dikaryotic diploid cells; but, as soon as these two nuclei fuse to form a diploid nucleus, (*AaBb*), the basidia become synkaryotic diploid cells.

In heterothallic Uredinales, *e.g. Puccinia graminis*, after two haploid mycelia have co-operated sexually, the basal cells of the proto-aecidium form aecidiosporophores, each of which contains a pair of conjugate nuclei. These aecidiosporophores and the aecidiospores to which they give rise are dikaryotic diploid cells. The two cells of a young teleutospore each contain a pair of conjugate nuclei, say (*A*) and (*a*), and are therefore dikaryotic diploid cells; but, as soon as the two nuclei of each pair fuse to form a diploid nucleus, (*Aa*), the cells of the teleutospore become synkaryotic diploid cells.

Uredinales and Hymenomycetes as Closely Related Groups.— Both the Uredinales and the Hymenomycetes produce basidia, and the basidiospores of species of both these groups are developed asymmetrically on the ends of conical sterigmata and are violently discharged by the drop-excretion mechanism.[1] Moreover, in both groups: (1) there is an absence of any organs that can be regarded as male or female; and (2) nuclear association is followed by a long series of divisions of conjugate nuclei. There can be no doubt that the Uredinales and the Hymenomycetes are related to one another, and it may well be that the Uredinales—a relatively small group of highly specialised parasites—were evolved from one of the lower forms of the Hymenomycetes—a much larger group of more generalised saprophytes. The view that the Uredinales and the Hymenomycetes are related to one another is supported by the fact that the modes of initiating the sexual process in the Uredinales are essentially similar to those in the Hymenomycetes.

Puccinia graminis and Coprinus lagopus as Types for Comparison.—For a comparison of the modes of initiating the sexual process in heterothallic Rust Fungi and heterothallic Hymenomycetes,

[1]But see the paper by A. E. Prince in *Farlowia*, Vol. I, 1943, pp. 79-93.

Puccinia graminis and *Coprinus lagopus* may be considered as typical representatives of their respective groups.

Coprinus lagopus is a saprophyte that grows on horse dung. Its haploid mycelia, derived from basidiospores, produce aerial oidiophores which bear oidia embedded in mucilaginous drops (Volume IV, Figs. 114 and 115, p. 198). These drops are eagerly sought by flies.[1] The oidiophores and oidia of *C. lagopus* correspond in their haploid nature and in their functions with the pycnidiosporophores and the pycnidiospores in the pycnidia of *Puccinia graminis*. In *Coprinus lagopus*, the oidiophores are produced in large numbers all over the surface of the mycelium on the outside of horse-dung balls; whereas, in *Puccinia graminis*, the pycnidiosporophores are collected together in special receptacles known as pycnidia.

A haploid mycelium of *Coprinus lagopus* differs from a haploid mycelium of *Puccinia graminis*, not only in not having its haploid spores (oidia) assembled in pycnidia, but also in not forming any organs corresponding to proto-aecidia. By forming proto-aecidia at an early stage in its development a haploid mycelium of *P. graminis* fixes within the host-leaf the exact positions in which, if the diploidisation process should be initiated, the aecidia will be subsequently developed. The dorsi-ventral structure and the thickness of the host-leaf are fairly constant, and this permits of the parasite forming proto-aecidia at even distances apart in the middle and lower layers of the mesophyll. The formation of proto-aecidia, therefore, appears to be an adaptation to a parasitic mode of life. On the other hand, the substratum of *Coprinus lagopus*—horse-dung masses—is very variable both in form and size and, as diploidisation of the haploid mycelia usually takes place inside the dung and at an early stage in development, there would be no advantage in the haploid mycelia producing any organs in the dung corresponding to proto-aecidia. What actually happens is that: the haploid mycelia of *C. lagopus* in each dung-ball soon become diploidised, the diploid mycelia grow to the surface of the dung and there produce fruit-body rudiments, and then one or a very few of the rudiments—those which happen to have been formed at the most favourable points for development—grow up into mature spore-producing fruit-bodies, while all the other rudimentary fruit-bodies stop growth and undergo abortion. By forming proto-aecidia, such a Rust as *Puccinia graminis* saves time in much the same way as does a Phanerogam by forming ovules. Everything is prepared for the coming of nuclei of opposite sex, so that, as soon as nuclear associa-

[1] H. J. Brodie, "The Oidia of *Coprinus lagopus* and their Relation with Insects," *Annals of Botany*, Vol. XLV, 1931, pp. 315-344.

tion or nuclear fusion has taken place, aecidiospores or seeds may begin their development at the earliest possible moment.

Modes of Initiating the Sexual Process.—In Chapter VI five modes of initiating the sexual process in such a heterothallic Rust as *Puccinia graminis* were described. These modes will now be used as a basis for comparing the sexual processes in the Uredinales with those in the Hymenomycetes.

Mode I.—In both *Puccinia graminis* and *Coprinus lagopus*, when two mycelia of opposite sex meet (in a leaf or in a dung-ball) they form hyphal fusions at their points of contact, exchange nuclei, and thus diploidise one another.

While Mode I in essentials is the same in *Puccinia graminis* and in *Coprinus lagopus*, it differs in these two fungi in ultimate details. In *Puccinia graminis*, the only parts of the mycelium that are truly diploidised are the basal cells of the proto-aecidia, for here alone are conjugate pairs of (+) and (−) nuclei formed. It is true that the branched intercellular (+) or (−) parts of the compound mycelium receive nuclei of opposite sex; but, as shown by the cytological researches of Allen, Hanna, Savile, and others, there is no evidence that in these parts of the mycelium definite conjugate pairs of nuclei are formed. It may well be that the few (+) nuclei which have entered a (−) mycelium and the few (−) nuclei which have entered a (+) mycelium use these mycelia merely as paths leading toward the proto-aecidia and as places in which they may multiply *en route* to the groups of basal cells where their presence is required. In all probability the following structures never receive any nuclei of opposite sex: (1) the peripheral hyphae of the (+) and (−) mycelia; (2) the wall-cells, ostiolar hairs (periphyses), and pycnidiosporophores of the pycnidia; and (3) the pseudoparenchymatous parts of the proto-aecidia which lie below the basal cells and are destroyed during the growth of the young aecidium; for why should hyphae or hyphal cells of one sex receive nuclei of another sex when they are destined neither directly nor indirectly to produce aecidiospores? Is it not likely that the law of least action will be followed and that no more nuclear divisions and nuclear migrations will take place than are necessary for the diploidisation of the basal cells of the proto-aecidia?

In 1930, I introduced the term *diploidisation*[1] to designate the process by which a haploid cell is converted into a diploid cell, or a

---

[1]A. H. R. Buller: "The Biological Significance of Conjugate Nuclei in *Coprinus lagopus* and other Hymenomycetes," *Nature*, Vol. CXXVI, 1930, pp. 686-689; these *Researches*, Vol. IV, 1931, pp. 190 and 302; and "The Diploid Cell, etc.," 1941, loc. cit., p. 398.

haploid mycelium into a diploid mycelium, by the formation of conjugate nuclei within the cell or within the mycelium.   In such Hymenomycetes as *Coprinus lagopus* the diploidisation process is completed when a haploid mycelium has been converted into a diploid mycelium. Each cell previously contained an (n) nucleus; but, at the end of the process, it contains a pair of conjugate nuclei, (n) + (n), and enclosed by the membranes of these two nuclei is the full complement of chromosomes required for the formation of the (2n) fusion nucleus in each basidium.   I also introduced the verb to diploidise.   Thus a haploid or diploid cell is said to *diploidise* a haploid cell when, through its agency, the latter becomes converted into a diploid cell containing conjugate nuclei (n) + (n); and a haploid or diploid mycelium is said to *diploidise* a haploid mycelium when, through its agency, the latter becomes converted into a diploid mycelium containing conjugate pairs of nuclei, (n) + (n).   Such terms as *fertilisation* and *conjugation* which have been used to describe certain well-known sexual processes in other plants and in animals cannot be successfully employed in the quite different sexual process which occurs in the Hymenomycetes.[1]

Now when a (+) mycelium and a (−) mycelium of *Puccinia graminis* or any similar Rust Fungus fuse and exchange nuclei, the intercellular parts of these mycelia come to have lying within their cytoplasm a mixture of (+) and (−) nuclei; but, apparently, they do not become diploidised in the sense that pairs of conjugate nuclei come to be everywhere established within them.   We may therefore say that the intercellular parts of the two mycelia which have fused have been *heterokaryotised* but not truly diploidised.   The term "hetero-

---

[1] The useful term *conjugate nuclei* was introduced by Poirault and Raciborski in 1895 (*vide* Chapter I).   In 1902 Maire referred to a pair of nuclei in a young aecidium as a *synkaryon* (*vide* Chapter I), but in 1912 (*Myc. Centralblatt*, T. I, p. 214) he proposed that such a pair of nuclei should be called a *dikaryon* and that the term synkaryon should be restricted to the fusion nucleus in the teleutospore.

Instead of the terms *conjugate nuclei, diploid cell, diploid mycelium, diploidise,* and *diploidisation* as used by myself, one may use the terms *dikaryon, dikaryotic cell, dikaryotic mycelium, dikaryotise,* and *dikaryotisation.*   I prefer the first set of terms because they imply not merely that two nuclei are involved but that the two nuclei are paired (are not alike but are of opposite sex or pole) and that, when present in a cell, they provide the cell with the complete set of chromosomes, (n) + (n), required for the formation of the (2n) fusion nucleus in a basidium.   The terms *diploidise* and *diploidisation* have been adopted: for the Hymenomycetes by A. Quintanilha, H. J. Brodie, and others; and for the Uredinales by I. M. Lamb and A. M. Brown.

For a fuller discussion of terminology connected with the sexual process in Hymenomycetes, Uredinales, etc., *vide* A. H. R. Buller, "The Diploid Cell, etc.," 1941, *loc. cit.*, pp. 398-404.

karyotised," introduced here for the first time,[1] may be explained more
fully as follows. A mycelium becomes heterokaryotised when it receives nuclei of opposite sex or pole, without this leading to the formation within it of definite pairs of conjugate nuclei. A (+) or a (−)
mycelium may be said to *heterokaryotise* another mycelium of opposite
sex or pole when it sends into the latter some of its own nuclei, without
these nuclei forming conjugate pairs with the nuclei already present in
the invaded hyphae. And, just as we can speak of the diploidisation
of a haploid mycelium of *Coprinus lagopus* or of the basal cells of the
proto-aecidia of *Puccinia graminis* so, too, we can speak of the *heterokaryotisation* of the intercellular parts of the (+) mycelium and the (−)
mycelium which have fused together in a compound pustule of *P. graminis* formed on a Barberry leaf. Heterokaryotisation apparently
without strict diploidisation also takes place when the (+) and (−)
strains of certain Pyrenomycetes are mated. This has been proved
by Dodge[2] for *Neurospora tetrasperma, N. sitophila, Gelasinospora
tetrasperma,* and *Pleurage anserina.*

To sum up this discussion, what apparently happens in such a
fungus as *Puccinia graminis,* after a (+) mycelium and a (−) mycelium
have met, fused, and exchanged nuclei, is as follows: (1) the *basal cells
of all the proto-aecidia* alone are truly *diploidised,* in that conjugate
nuclei come to be established within them; (2) the general *intercellular
mycelium,* at least temporarily, is *heterokaryotised* in that it comes to
contain both (+) and (−) nuclei, although not in equal numbers, not
everywhere in all its hyphae, and not anywhere, perhaps, in conjugate
pairs; and (3) the *wall-cells of the pycnidia,* the *pycnidiosporophores,*
and the *ostiolar hairs (periphyses),* as well as the *pseudoparenchymatous*

[1]For employment by others and by myself of the terms *heterokaryosis* (hetero-
caryosis) and *heterokaryotic* (heterocaryotic) in connexion with the supposed mixing
of nuclei of different strains in *Botrytis cinerea, Hypochnus solani,* and species of
Fusarium and Helminthosporium as a result of vegetative fusions between mycelia of
different strains *vide* these *Researches,* Vol. V, 1933, pp. 73-74. The terms *heterokaryotise* and *heterokaryotisation* here introduced into the text by myself are merely
adaptations of the terminology of heterokaryosis to fit the special case of a mixture
of nuclei of two different sexes or poles at the beginning of the sexual process in the
Uredinales.

The existence of sexually *heterokaryotic* mycelia in certain Pyrenomycetes,
e.g. Neurospora tetrasperma, Pleurage anserina, and *Gelasinospora tetrasperma* and in
certain Uredinales, *e.g. Puccinia graminis,* was recognised and discussed in my review,
"The Diploid Cell, etc.," 1941, *loc. cit.,* pp. 395-397.

[2]*Cf.* B. O. Dodge: "The Mechanics of Sexual Reproduction in Neurospora,"
Mycologia, Vol. XXVII, 1935, pp. 422-430; and "Spermatia and Nuclear Migrations
in *Pleurage anserina,*" *Mycologia,* Vol. XXVIII, 1936, pp. 288-291.

cells of the proto-aecidia, are presumably *neither diploidised nor hetero-karyotised.*

In *Coprinus lagopus*, on the other hand, as my own experiments have shown, when two mycelia of opposite sex meet, fuse together, and exchange nuclei, the diploidisation process (indicated by the formation of clamp-connexions) is not localised as it is in *Puccinia graminis*, but extends to all the leading hyphae growing radiately outwards away from the centre of the mycelium.[1] Experiment shows that, in *Coprinus lagopus*, when a large mycelium has been diploidised by a small one of opposite sex, while all the peripheral hyphae of the large mycelium become diploidised, the older, more central, non-growing part of the large mycelium remains in the haploid condition;[2] but it is certain that in *C. lagopus*, in mating experiments made under favourable conditions, all those hyphae which have the potentiality of forming a fruit-body or fruit-bodies at a later stage of development of the mycelium as a whole, come to contain definite pairs of conjugate nuclei.

Diploidisation by Mode I secures that all those parts of the two interacting mycelia which sooner or later may produce spores shall be diploidised. This end is accomplished in *Puccinia graminis* by the diploidisation process being confined to the basal cells of the proto-aecidia and in *Coprinus lagopus* by the process being extended to every leading, elongating, branching, peripheral hypha.

Mode II.—In both *Puccinia graminis* and *Coprinus lagopus*, Mode II of initiating the sexual process provides for those haploid mycelia which are physically separated from one another and in which, therefore, mutual diploidisation by Mode I cannot be effected.

In *Puccinia graminis*, as Craigie[3] proved, insects intermix the nectar of (+) and (−) pustules and so initiate the sexual process. The pycnidiospores do not germinate like ordinary conidia, but are able to fuse with flexuous hyphae of opposite sex.

In *Coprinus lagopus*, as Brodie[4] proved, insects can carry oidia of a haploid mycelium derived from a basidiospore to a haploid mycelium of opposite sex derived from another basidiospore, with the result that the mycelium which receives the nuclei of the oidia becomes diploidised. But the oidia, unlike the pycnidiospores of *Puccinia graminis*, can and do germinate and they diploidise the haploid mycelia derived from

---

[1] *Vide* these *Researches*, Vol. IV, 1931, pp. 204-234.   [2] *Ibid.*, pp. 229-231.

[3] J. H. Craigie, "An Experimental Investigation of Sex in the Rust Fungi," *Phytopathology*, Vol. XXI, 1931, p. 1023.

[4] H. J. Brodie, "The Oidia of *Coprinus lagopus* and their Relation with Insects," *Annals of Botany*, Vol. XLV, 1931, pp. 315-344.

basidiospores *by first producing a germ-tube or a mycelium* and then, by means of these structures, forming hyphal fusions with the haploid mycelia. In reality, one haploid mycelium derived from an oidium fuses with another haploid mycelium of opposite sex derived from a basidiospore.

The adaptations for making use of insects for transportation of haploid spores are much more highly evolved in the Uredinales than in the Hymenomycetes. In both *Coprinus lagopus* and *Puccinia graminis*, the haploid spores are contained in mucilaginous drops which are eagerly sucked up by flies and which prevent the spores from being dispersed by the wind. The bright colours and the attractive aroma associated with the pycnidiospores of the Uredinales are not present in connexion with the oidia of the Hymenomycetes; and whether or not sugar, which is contained in the nectar of the Rust Fungi, is also present in the drops which crown the oidiophores of the Coprini remains to be determined by chemical analysis.

The oidia of *Coprinus lagopus* which float in the mucilaginous drop in which they are set free after abscission from the oidiophore never germinate *in situ* and, in this respect, they resemble the pycnidiospores of the Rust Fungi. Germination *in situ* in the drops would result in the formation of a tangle of mycelium, and this would be fatal to the employment of the oidia or the pycnidiospores as diploidising agents. But, after transference by insects, the oidia will germinate on horse-dung balls where mycelia of opposite sex may be growing and be ready for diploidisation, and the pycnidiospores will unite with flexuous hyphae of opposite sex. It is not difficult to imagine that ages ago the pycnidiospores of the Rust Fungi germinated like the oidia of the Hymenomycetes do now and that, later on in their evolution, they became so specialised for their diploidising function that they ceased to show any signs of further development except when extremely close to flexuous hyphae of opposite sex and then only by putting forth a minute fusion papilla.

Mode III.—In 1931, in Volume IV of these *Researches*, I showed that, in *Coprinus lagopus*, where there are four groups of basidiospores, (AB), (ab), (Ab), and (aB), there are two kinds of diploid mycelium, $(AB) + (ab)$ and $(Ab) + (aB)$, and I further showed that a diploid mycelium $(AB) + (ab)$ will diploidise the haploid mycelia (AB) and (ab), and that a diploid mycelium $(Ab) + (aB)$ will diploidise the haploid mycelia (Ab) and (aB).[1] In 1937, Quintanilha[2] called the

[1]These *Researches*, Vol. IV, 1931, pp. 209-212.

[2]A. Quintanilha, "Contribution à l'étude génétique du phénomène de Buller," *Compt. Rend. Acad. Sci. Paris*, T. CCV, 1937, p. 745.

phenomenon of the diploidisation of a haploid mycelium of a Hymeno-mycete by a diploid mycelium the *Phénomène de Buller*.

The Buller phenomenon has been investigated experimentally by: Quintanilha (1933-39), in *Coprinus lagopus* (his *C. fimetarius*); Chow (1934), a pupil of Dangeard, in *C. lagopus*; Dickson (1934-1936), in *C. sphaerosporus* and *C. macrorhizus*; Noble (1937), in *Typhula trifolii*; and Oikawa (1939), in *Stropharia semiglobata* and *Galera tenera*.[1] We now know that the Buller phenomenon is not confined to the Hymeno-mycetes but may also occur in heterothallic, *autoecious*, long-cycled Uredinales, for Brown[2] has demonstrated it experimentally in *Puccinia helianthi* (1932), *Uromyces trifolii hybridi* and *U. fabae* (1940).

The Buller phenomenon cannot take place in the heterothallic, *heteroecious* long-cycled Rust, *Puccinia graminis* because the aecidio-spores and the uredospores do not germinate on the leaves of Barberry bushes and cause infections there, in other words because it is not possible for a haploid mycelium and a diploid mycelium to come into contact with one another in the tissues of one and the same host-leaf.

In one of the three Rust species with which Brown experimented, namely, *Puccinia helianthi*, as was indicated by Craigie's investigations, there are only two sexual groups of basidiospores. In this respect *P. helianthi* resembles certain Hymenomycetes, *e.g. Coprinus radians* and *C. rostrupianus*. We may here refer to the two groups of basidio-spores in *Puccinia helianthi* not as $(+)$ and $(-)$, but as (A), and (a). As already recorded in detail in Chapter VI, Brown has shown that, in *P. helianthi*, a diploid mycelium, $(A) + (a)$, derived either from an aecidiospore or from a uredospore, if brought into contact in a Sun-flower leaf with a haploid mycelium, either (A) or (a), is able to diploidise the basal cells of the haploid mycelium's proto-aecidia and so cause the development of diploid aecidiosporophores and the pro-duction of chains of diploid aecidiospores, each of which contains a pair of conjugate nuclei, $(A) + (a)$.

It is obvious that Mode III of initiating the sexual process is the same in both *Coprinus lagopus* and *Puccinia helianthi*.

There can be but little doubt that Mode III is very frequently employed by *Coprinus lagopus* under natural conditions in the open. Let us imagine that a freshly extruded horse-dung ball contains 500 basidiospores of *C. lagopus*, that all the spores germinate, and that the mycelia so produced fuse together as they come into contact with one

---

[1]For a citation of the papers of A. Quintanilha, C. H. Chow, H. Dickson, Mary Noble, and K. Oikawa *vide* A. H. R. Buller, "The Diploid Cell, etc.," *loc. cit.*, 1941, pp. 414-429.

[2]A. M. Brown, for details of experiments *vide* Chapter VI.

another. Then, any diploid mycelia which may be formed will diploidise surrounding haploid mycelia of suitable genetic constitution with which they happen to come into contact; and thus the whole mass of mycelium in the dung-ball will become diploidised and eventually give rise to diploid fruit-bodies.

On the other hand, it seems unlikely that Mode III is much employed by any Rust fungus. The diploidisation of the haploid pustules of *Puccinia helianthi*, doubtless, is usually effected by Modes I and II; and, so far, diploidisation by Mode III has been observed only under experimental conditions in the laboratory.

Mode IV.—*Coprinus sterquilinus, C. stercorarius*, and *C. narcoticus*, like *Puccinia malvacearum, P. xanthii*, and *P. grindeliae* (form on *Grindelia squarrosa* at Winnipeg), are homothallic; but how the change in a mycelium derived from a single basidiospore takes place so that this mycelium passes from the haplophase to the diplophase automatically has not as yet been satisfactorily explained.[1]

Mode V.—In 1941, I introduced the term *de-diploidisation*[2] and defined it as follows: *in Basidiomycetes and Ascomycetes, the production of haploid cells or hyphae by a dikaryotic diploid mycelium or by a dikaryotic diploid cell.*

The de-diploidisation of the systemic perennial mycelium of *Puccinia minussensis* and *P. suaveolens* on entering the leaves of new shoots in the spring, treated of in Chapter VI, finds its parallel in the de-diploidisation of the diploid mycelium of certain Hymenomycetes.

Thus, in the agaric *Pholiota aurivella*, it was observed by Vandendries and Martens[3] (1932) that: a dikaryotic mycelium produces dikaryotic oidiophores and that these give rise either to dikaryotic diploid oidia or to monokaryotic haploid oidia. In the latter case: a pair of conjugate nuclei (a dikaryon) enters a fusiforme oidium: a cross-wall (septum) is then formed in the oidium, dividing it into two uninucleate cells; and, finally, the uninucleate cells break apart and become haploid oidia. The oidia on germination give rise to haploid mycelia.

Again, in another agaric, *Collybia velutipes*, it was observed by Brodie[4] (1934) that: a dikaryotic mycelium gives rise not only to di-

[1]For a discussion of homothallism in the Higher Fungi *vide* A. H. R. Buller, "The Diploid Cell, etc.," 1941, *loc. cit.*, pp. 348-364.

[2]A. H. R. Buller, "The Diploid Cell, etc.," 1941, pp. 411-414.

[3]R. Vandendries and P. Martens, "Oidies haploïdes et diploïdes sur mycelium diploïde chez *Pholiota aurivella* Batsch," *Bull. Acad. Roy. Belg.*, T. XVIII, 1932, pp. 468-472.

[4]H. J. Brodie, "The Occurrence and Function of the Oidia in the Hymenomycetes," *American Journ. of Bot.*, Vol. XXIII, 1934, pp. 321-322.

karyotic oidiophores and oidia, but also to uninucleate oidiophores and oidia; the uninucleate oidiophores and oidia are borne on uninucleate branches of the dikaryotic mycelium; the origin of the uninucleate branches is due to the separation of the two nuclei of a conjugate pair; and half of the oidia (those borne on certain haploid branches) are of one sex and the other half (borne on other branches) of opposite sex. The oidia, on germination, give rise to haploid mycelia that can be used to diploidise haploid mycelia of opposite sex.

While cases of de-diploidisation of the mycelium are known in both the Hymenomycetes and the Uredinales, in none of the Hymenomycetes do the annual changes take place that have been observed in *Puccinia minussensis* and *P. suaveolens, i.e.* no diploid mycelium of a Hymenomycete produces annually haploid branches which, later on, are re-diploidised by self-diploidisation. However, the uninucleate oidia produced on haploid branches of the diploid mycelium of such an agaric as *Collybia velutipes* germinate and produce (+) and (−) haploid mycelia which are able to diploidise one another, so that here we have a re-diploidisation not of the haploid branches of the original diploid mycelium but of the products of those branches, *i.e.* a re-diploidisation which is delayed and finally carried out in a second generation of mycelia.

An instance of de-diploidisation resulting in the formation of haploid basidiospores in the Uredinales and comparable with de-diploidisation resulting in the formation of haploid oidia in such Hymenomycetes as *Pholiota aurivella* and *Collybia velutipes* is to be found in such Uredinales as *Endophyllum euphorbiae-sylvaticae* and *Kunkelia nitens*. The cytological facts were brought to light: for the Endophyllum, by Sappin-Trouffy,[1] in 1896; and for the Kunkelia, by B. O. Dodge and Miss Gaiser,[2] in 1926. As set forth in Chapter VI, in both these species: the basidium develops directly from an aecidiospore; the two nuclei of the conjugate pair in the aecidiospore migrate into the young basidium; there they both divide once, thus giving rise to four haploid nuclei; and then the four nuclei pass through the four sterigmata into the four basidiospores. Thus the nuclei for the basidiospores are provided not by nuclear fusion followed by miosis, but by the de-diploidisation of an unfused pair of conjugate nuclei.[3]

[1] P. Sappin-Trouffy, "Recherches histologiques sur la famille des Urédinées," Le Botaniste, T. V, 1896, pp. 184-187.

[2] B. O. Dodge and L. O. Gaiser, "The Question of Nuclear Fusions in the Blackberry Rust, *Caeoma nitens*," *Journ. Agric. Research*, Vol. XXXII, 1926, pp. 1008-1012, 1022-1023.

[3] It is perhaps not without interest to note here that de-diploidisation of a mycelium followed by the production of spores is known not only in certain Hymeno-

APPENDIX: *Origin and Meaning of the Terms Haploid
and Diploid**

In a review concerned with the diploid cell, particularly in respect
to the dikaryotic condition of cells in the Higher Fungi, the writer (1),
in 1941, recorded briefly the history of the origin of the terms haploid
and diploid.   On account of the great interest attaching to these terms,
what was there said will here be expanded.

In 1884, Nägeli (6), when advancing his mechanistic theory of
inheritance, invented the word *idioplasm* as a term for the special
organic structure-determining part of the protoplasm, supposed to
consist of a network of fine fibrillae traversing the substance of the cell,
passing from cell to cell, and pervading the whole body.   Then, in
1891, Weismann (10, 11), in developing his germ-plasm theory of
inheritance, invented: (1) the word *id* as a term for an hereditary unit,
supposed to be made up of "determinants" and to be present in granular form in chromosomes; and (2) the word *idant* as a term for a serial
complex of ids, such as might make up, or be present in, a chromosome.

The *idio-* in Nägeli's term *idioplasm* is derived from *idios* (ἴδιος),
a Greek word meaning "peculiar."   *Id* is a noun of Greek origin invented by Weismann as a short alternative term for a unit of what he
had previously designated as "ancestral germ-plasm."   In introducing
the word "ids," he remarked that the term "recalls the 'idioplasm' of
Nägeli" (10, 11).   It is therefore clear that the word *id* was derived
from *idios*.

The derivation of *id* from *idios* is given by the Hendersons (3) in

mycetes and Uredinales, but also in the Ustilaginales. W. F. Hanna ("Studies in
the Physiology of *Ustilago zeae* and *Sorosporium reilianum*," *Phytopathology*,
Vol. XIX, 1929, p. 433) floated portions of Corn leaves heavily infected with the
diploid mycelium of *U. zeae* on distilled water in a Petri dish and found that branches
of the mycelium grew upwards through the epidermal cells and gave rise to uninucleate
haploid sporidia. Monosporidial cultures were made from these oidia; but, when
injected separately into Corn plants, they failed to produce galls. In the Corn Smut,
therefore, the diploid mycelium may sometimes become de-diploidised. R. Bauch
("Über *Ustilago longissima* und ihre Varietät *macrospora*," *Zeitschr. f. Bot.*, Bd. XV,
1923, pp. 241-279) states that the diploid or fusion sporidia of *U. longissima* which
are formed by the conjugation of two haploid sporidia of opposite sex may, in pure
culture, produce once more haploid sporidia of the two kinds; and S. Dickinson
("Experiments on the Physiology and Genetics of the Smut Fungi—Hyphal fusion,"
Proc. Roy. Soc. London, B, Vol. CI, 1927, pp. 126-136) found that the same process
occurs in cultures of *U. hordei* and *U. levis*. Here, again, in the Ustilaginales, we
have evidence of de-diploidisation.

*This paper, found amongst Dr. Buller's manuscripts, was evidently intended for
independent publication in some Journal. It seems appropriate to include it here
as an appendix, exactly as it was in his revised copy.

their Dictionary of Scientific Terms; but Jackson (4), in his Glossary of Botanic Terms, refers *id* to the suffix *-ides*. This suffix implies paternity and was commonly used in Greek literature in words such as *Pelopides* (son of Pelops) and *Atreides* (son of Atreus). Since *id* has a connotation of ancestry and descent, it is possible that Weismann, when inventing it, was thinking not only of the *idios* of Nägeli's *idioplasm* but also of the Greek suffix *-ides*.

The term *diploid* has been invented twice, once in Crystallography and again in Biology, and in two different ways.

Diploid in Crystallography, as stated in 1897 in Volume III of Murray's New English Dictionary published at Oxford (5), is derived from the Greek words *diploos* and *eidos* (διπλό-ος double + εἶδος form), means "a solid belonging to the isometric system, contained within twenty-four trapezoidal planes," and is equivalent to *diplohedron*. It may be added that a *diploid* or *diplohedron* is a hemihedral form of the hexoctahedron and that, as shown in an illustration in Webster's Dictionary (9), its twenty-four trapezoidal faces are similar and arranged in pairs. Doubtless it was this pair-wise arrangement of the faces that suggested the terms *diplohedron* and *diploid* for the solid body or crystal which they designate.

In Volume VII of the Oxford Dictionary (5), published in 1909, the suffix *-oid* in English is given as derived from modern Latin *-oīdēs* and Greek *-οειδής*, *i.e.* -o- of the preceding element or connective + ειδής "having the form or likeness of," "like," from εἶδος form, and it is pointed out that *-oid* is extensively used in scientific terms. Of such terms the following may be cited as examples: *globoid, trapezoid, sigmoid, scorpioid, fungoid, amoeboid,* and *alkaloid*.

As will now be shown, the *-oid* of the biological terms *haploid* and *diploid* had an origin different from that of the *-oid* in the crystallo-graphical term *diploid* and in such terms as *globoid* and *trapezoid*.

The words *haploid* and *diploid*, as used in Biology, were invented by Strasburger (7), in 1905, at the end of a discussion of typical and allotypical nuclear division. As a contribution toward the construction of a terminology that could be used for nuclear phenomena throughout the realm of living organisms, he said: "Finally, it would perhaps be desirable if alongside the terms gametophyte and sporophyte, which can be used only for plants with the single and double chromosome number, there were set such other terms as would be suitable also for the animal kingdom. For this purpose I therefore permit myself to propose the words haploid and diploid, respectively haploidic and diploidic generation (bezw. haploidische und diploidische Generation)."

In 1907, in his review of the ontogeny of the cell, Strasburger (8) explained how he came to construct the words proposed in 1905. He said that the *id* in *haploid* and *diploid* served to connect the new terms with the *idioplasm* of Nägeli and the *id* and *idant* of Weismann. This historical fact is not recorded in text-books on Genetics and does not seem to be generally known.

Haploos (ἁπλόος) in Greek is equivalent to the Latin *simplex* and means "simple" or "single" ("one-fold"), whilst *diploos* (διπλόος) in Greek is equivalent to the Latin *duplex* and means "double" or "two-fold."

In origin, therefore, Strasburger's haploid (from *haploos* and *id*) and diploid (from *diploos* and *id*) are three-syllable words: ha-plo-id and di-plo-id; and respectively they mean, literally, "simple-number-of-ids" or "single-number-of-ids" and "double-number-of-ids" and, as defined by Strasburger, "simple-number-of-chromosomes" or "single-number-of-chromosomes" and "double-number-of-chromosomes."

In French, the words haploid and diploid are spelled *haploïde* and *diploïde*, with a diaeresis over the letter i, and they are therefore written and pronounced as three-syllable words.

In German, the words haploid and diploid are spelled without any diaeresis over the letter i, but they are pronounced as three-syllable words.

In English, the last two syllables of haploid and diploid, when spoken, are usually run together, and thus the words are pronounced with two syllables instead of three. On this account, the etymological origin of the terms haploid and diploid has become obscured.

Some errors concerning the derivation of the biological terms haploid and diploid, as found in standard works of reference, may now be noted.

Jackson (4), in 1928 gives *haploid* as being derived from ἁπλόος, meaning single, and εἶδος, meaning resemblance, and *diploid* as being derived from διπλόος, meaning two-fold. The Hendersons (3), in 1929, give *haploid* as derived from the Greek *haploos*, simple, and *eidos*, form, and *diploid* as being derived from the Greek *diploos* meaning double. These derivations should be corrected in future editions of the works in which they occur, as the *id* in the words *haploid* and *diploid* is the *id* invented by Weismann directly or indirectly from *idios* meaning peculiar, and not from *eidos* meaning form.

In the Supplement to Murray's Oxford Dictionary (5) edited by W. A. Craigie and C. T. Onions and published in 1933, the biological terms *haploid* and *diploid* are introduced for the first time. No derivation is given for the term *diploid* and the reader is referred to the crystallographic word *diploid* in an earlier volume published in 1897. The derivation for this term, as we have seen, is there given as from *diploos* and *eidos*. For the entirely new word *haploid* the derivation is given as from the Greek ἁπλόος single + εἶδος form. Here, again, for the derivation of the biological terms haploid and diploid, *eidos* should be removed and replaced by *id* as invented by Weismann directly or indirectly from the Greek word *idios*.

Very naturally, with the advance in our knowledge of cell-structure and inheritance, to the terms invented by Strasburger there have been added others, such as *tetraploid* and *octaploid*. The whole series of terms can be used adjectivally in the description of generations, tissues, cells, and nuclei. Finally, with the urge toward descriptive brevity, the terms have come to be employed as nouns; so that, as occasion demands, we now speak of certain plants and animals as *haploids*, *diploids*, *tetraploids*, *octaploids*, *polyploids*, and so forth. Thus the lead in terminology given to cytologists and geneticists by Strasburger nearly forty years ago has been logically and successfully followed.

In accordance with Strasburger's definition of the terms haploid and diploid, a *diploid cell* may be defined as *a cell containing the double number of chromosomes*. If this definition be accepted, then we can describe as diploid not only the cells of the sporophytes of Mosses, Ferns, and Flowering Plants and the body cells of Higher Animals, all of which contain a diploid nucleus, but also those cells of the Higher Fungi (Basidiomycetes and Ascomycetes) that contain a pair of conjugate haploid nuclei. To distinguish between cells that contain a single diploid nucleus, (n + n), and cells that contain two conjugate haploid nuclei, (n) + (n), the writer (1) has designated the former as *synkaryotic diploid cells* and the latter as *dikaryotic diploid cells*.

Some biologists, having but "small Latin and less Greek," have supposed that, while *diploid* means "*double*-the-number-of-chromosomes," *haploid* means "*half*-the-number-of-chromosomes"; and this entirely erroneous interpretation of the word haploid has led them to regard Strasburger's terms as ill-chosen and confusing! The actual fact, of course, is that, as we have seen, *haploid* means "*simple*-number-of-chromosomes" or "*single*-number-of-chromosomes," and not "*half*-the-number-of-chromosomes." The Greek word *haploos*, it is true, resembles the English word *half* in that it begins with the same two letters *ha*, but there the similarity ends. The Greek prefix for *half*, as used in composition, is *hemi-* (ἡμι-), as in hemisphere, and not *haplo-*.

It is important that writers of text-books, when treating of Cytology, should avoid language which might suggest to students the erroneous interpretation of the word haploid. In one text-book (2) the author, in his introduction to the process of sexual reproduction, says that the product of nuclear fusion has "a double or *diploid* number" of chromosomes and that, as a result of meiosis, there is distributed to each of the four nuclei "a halved or *haploid* number" of chromosomes. Since *haploid* does not mean "half," the statement concerned with the word haploid should be emended: "halved" should be struck out and replaced by "simple"; or "a halved or *haploid* number" should be made to read "one-half of the diploid number and, therefore, the simple or *haploid* number."

Finally, it may be remarked that the recently introduced word *monoploid*, as an alternative for *haploid*, is of faulty construction. *Mono-*, truly, means "single" (although not "simple"), but there is no Greek word *monoploos* corresponding to *haploos* and *diploos*. The word *haploid* is an excellent one, as a study of its origin shows; and, therefore, the substitute for it, *monoploid*, should be rejected as a gratuitous and superfluous coinage.

BIBLIOGRAPHY OF APPENDIX

1. Buller, A. H. R. "The Diploid Cell and the Diploidisation Process with special reference to the Higher Fungi." Bot. Review **7**: 335-431 (1941).
2. Darlington, C. D. *Recent Advances in Cytology*, ed. 4. Philadelphia (1937).
3. Henderson, I. F. and Henderson, M. A. *A Dictionary of Scientific Terms*, ed. 2. London (1929).
4. Jackson, B. D. *A Glossary of Botanic Terms with their Derivation and Accent*, ed. 4. London (1928).
5. Murray, J. A. H. *A New English Dictionary on Historical Principles*. Oxford (1888-1933).
6. Nägeli, C. V. *Mechanisch-physiologische Theorie der Abstammungslehre*. München und Leipzig (1884).
7. Strasburger, E. "Typische und allotypische Kernteilung. Ergebnisse und Eröterungen." Jahrb. f. wiss. Bot. **42**: 1-71 (1905).
8. ――― "Die Ontogenie der Zelle seit 1875. Progressus Rei Botanici," **1**: 1-138 (1907).
9. Webster. *New International Dictionary*, ed. 2. Springfield, Mass. (1940).
10. Weismann, A. *Amphimixis, oder die Vermischung der Individuen*. Jena (1891).
11. ――― *The Germ-Plasm. A Theory of Heredity*. Eng. trans. by W. N. Parker and H. Rönnfeldt. London (1893).

CHAPTER VIII

A REVIEW OF CYTOLOGICAL WORK ON THE SEXUAL PROCESS IN HETEROTHALLIC UREDINALES

Introduction — Period I — Period II — Period III — Living Fungi and Dead Preparations — Conclusion.

Introduction.—The great twentieth-century advance in our understanding of the genetics of animals and plants in general has been largely due to the fact that experiment and cytological observation have gone hand in hand and have given one another so much mutual assistance.

In the study of the long-cycled heteroecious Uredinales experiments, made by de Bary and others, demonstrated the succession of the various spore-forms. Then cytological observations made by Dangeard, Sappin-Trouffy, and others yielded the fact that, in Rusts of the *Puccinia graminis* type, from the nuclear point of view there are two alternating phases: (1) the *diplophase*, with conjugate nuclei, beginning in the young aecidia and ending in the teleutospores where the two nuclei of each pair fuse together; and (2) the *monokaryophase*, with isolated nuclei, beginning with the basidiospores and the pycnidiospores and ending in the young aecidia.[1]

In the long interval between the publications of Sappin-Trouffy (1896) and Craigie (1927-1933), rust cytologists endeavoured to elucidate the manner in which the diplophase is initiated at the base of the young aecidium and, lacking the sure guide of experimental fact, interpretations of their preparations were influenced—as was natural—by pre-conceived theories.

In reviewing briefly the history of cytological investigations on the initiation of the sexual process in long-cycled Rusts such as *Puccinia graminis*, we may recognise three periods: *Period I*, from Blackman's investigation made in 1904 up to 1927 when Craigie published his first

[1]For an historical account of this early work and citation of particular papers *vide* Chapter I.

account of heterothallism in Rust Fungi; *Period II*, from 1927 to 1933 in which year Craigie showed that pycnidiospores fuse with flexuous hyphae; and *Period III*, from 1933 to the present day.

Period I.—Blackman and Christman and their followers[1] for some twenty years, at a time when it was thought that the sexual process must be initiated by the co-operation of nuclei of two adjacent cells, found nothing but adjacent cells co-operating with one another either by cell-fusion or nuclear migration, and none of the preparations ever suggested to the observers that nuclei of outside origin were invading the "fertile cells" of the rudiments of the aecidia.

Period II.—But as soon as Craigie[2] in 1927 had shown that long-cycled Uredinales are heterothallic and that mixing the nectar of (+) and (−) pustules initiates the sexual process in a way not fully understood, cytological investigation took a new turn:

(1) Andrus[3] supposed that the pycnidiospores fuse with "trichogynes" that come to the surface of a leaf *via* stomata, and he had no difficulty in finding "male" nuclei that had migrated into the leaf mycelium by the trichogyne route and were on the way to "oogonia" situated at the base of each rudimentary aecidium.

(2) Following the lead of Andrus, Miss Allen paid particular attention to "receptive hyphae" which had pushed their way into stomatic clefts and outwards to the surface of the host-leaves, and she came to the conclusion that "the initiation of the sporophyte by the entrance of spermatial nuclei into hyphae at the surface of the host tissue" takes place not only in *Uromyces appendiculatus* and *U. vignae* as found by Andrus, but also in *Puccinia triticina*, *P. coronata*, and *P. graminis.*[4]

In 1933, Miss Allen[5] suggested that in *Puccinia graminis* there

[1]For the citation of papers by Blackman, Christman, and their followers *vide* Chapter I, pp. 48-52.

[2]J. H. Craigie, "Discovery of the Function of the Pycnidia of the Rust Fungi," *Nature*, Vol. CXX, 1927, pp. 765-767.

[3]C. F. Andrus: "The Mechanism of Sex in *Uromyces appendiculatus* and *U. vignae*," *Journ. Agric. Research*, Vol. XLII, 1931, pp. 559-587; and "Sex and Accessory Cell Fusions in the Uredineae," *Journ. Washington Acad. Sciences*, Vol. XXIII, 1933, pp. 544-557. For a further critical account of the findings of Andrus, *vide supra*, Chapter IV.

[4]Ruth F. Allen: "Heterothallism in *Puccinia triticina*," *Science* (n.s.), Vol. LXXIV, 1931, pp. 462-463; "A Cytological Study of Heterothallism in *Puccinia triticina*," *Journ. Agric. Research*, Vol. XLIV, 1932, pp. 733-754; "A Cytological Study of Heterothallism in *Puccinia coronata*," *Journ. Agric. Research*, Vol. XLV, 1932, pp. 513-541; and "Further Studies of Heterothallism in *Puccinia graminis*," *Journ. Agric. Research*, Vol. XLVII, 1933, pp. 1-16, especially pp. 10-11.

[5]Ruth F. Allen, "Further Studies, etc.," 1933, *loc. cit.*, p. 8.

might be an additional mode of initiating the sexual process; for she said "paraphyses may serve as receptive hyphae, the spermatial nuclei entering them and migrating down through them to the spermagonial wall at their base." By paraphyses Miss Allen meant the same cellular organs that in this work are called periphyses or ostiolar trichomes.

(3) On the basis of having found in mixed nectar of *Puccinia graminis* a few pycnidiospores which looked as though they might have put out short germ-tubes, Hanna[1] suggested that, after nectar has been mixed, the pycnidiospores germinate and the germ-tubes grow into the host-leaf and there produce mycelia which become active in initiating the diploidisation process; but, in his cytological preparations, he did not recognise any hyphae that might be presumed to have been derived from pycnidiospores.

(4) Miss Allen,[2] following the new lead given by Hanna and proceeding on the supposition that pycnidiospores do really germinate, found cytological evidence of such germination in *Melampsora lini* and *Puccinia sorghi* and even thought that, in *M. lini*, within two hours after nectar has been mixed, a pycnidiospore had bored its way through the cuticle and outer epidermal wall. Moreover, she believed, supporting her belief with a series of illustrations,[3] that such a pycnidiospore, intact and surrounded by its wall-membrane, can pass into the lumen of an epidermal cell, be carried in a living streaming protoplasmic bridle to the inner epidermal wall and, finally, can put out a germ-tube which pierces this inner wall and then develops into a mycelium. On the basis of what she saw in her preparations she supposed that the hyphae of the mycelium produced from the germ-tube of a pycnidiospore come to interdigitate with the hyphae of the aecidial rudiment already present in the leaf and so prepare the way for a conjugation of (+) and (−) hyphae.

Period III.—As soon as Craigie,[4] in 1933, had made his third and last important discovery, namely, that pycnidiospores of one sex fuse

---

[1]W. F. Hanna, "Nuclear Association in the Aecium of *Puccinia graminis*," *Nature*, Vol. CXXIV, 1929, p. 267; also *Report of the Dominion Botanist for 1929*, Dept. of Agric., Ottawa, 1931. For further remarks on these papers, *vide infra*, Chapter III.

[2]Ruth F. Allen: "A Cytological Study of Heterothallism in Flax Rust," *Journ. Agric. Research*, Vol. XIX, 1934, pp. 765-789; and "A Cytological Study of Heterothallism in *Puccinia sorghi*," *ibid.*, pp. 1047-1068. For further critical remarks on these papers, *vide* Chapter IV.

[3]Ruth F. Allen, *loc. cit.* (Flax Rust), Plates VII and VIII, pp. 775-780.

[4]J. H. Craigie, "Discovery of the Function of the Pycnia of the Rust Fungi," *Nature*, Vol. CXX, 1927, pp. 765-767.

with flexuous hyphae of opposite sex, it became evident that the cyto-logical interpretations of Blackman, Christman, Andrus, Hanna, and Miss Allen as to the manner in which the sexual process in the hetero-thallic Rusts is initiated were all erroneous, and the course of cyto-logical investigation in the Rusts took still another turn:

(1) Lamb[1] investigated *Puccinia phragmitis* and, having failed to find any flexuous hyphae, imagined that in this species the pycnidio-spores must fuse with the pointed periphyses, and he duly illustrated supposed fusions of this kind. However, for reasons set forth else-where,[2] I am of the opinion: that Lamb failed to find flexuous hyphae because he employed an unsuitable technique; that his fusions were no true fusions at all; and that pycnidiospores in the Rusts in general never fuse with pointed periphyses but exclusively with flexuous hyphae.

(2) After I[3] had supported Craigie's finding by publishing illustra-tions of many fusions between pycnidiospores and flexuous hyphae seen in living material of *Puccinia graminis*, Savile[4] reported the results of his investigations on *P. sorghi*, *Uromyces fabae*, etc. Savile admitted that his pycnidia produced "ostiolar filaments" but thought that the pointed periphyses and the flexuous hyphae form an intergrading series, and he gave illustrations of a few supposed fusions some of which appear to be fusions between pycnidiospores and pointed peri-physes. I am strongly inclined to believe: (*a*) that Savile's failure to distinguish flexuous hyphae from pointed periphyses was due to the fact that he did not take the trouble to study pycnidia in the living condition and through all their stages of development; and (*b*) that the three fusions which he illustrated were not true fusions at all but simply appearances which he misinterpreted.[5]

Savile believed that he was able to distinguish by their appearance nuclei which are travelling down an ostiolar filament or are travelling in hyphae at the base of a proto-aecidium from nuclei of local origin: the travelling nuclei are much contracted and the stationary nuclei are much expanded and provided with an "ectosphere." If, in due course, this distinction between proto-aecidial and pycnidiosporal

[1] I. M. Lamb, "The initiation of the Dikaryophase in *Puccinia phragmitis* (Schum.) Körn," *Annals of Botany*, Vol. XLIX, 1934, pp. 1047-1068.

[2] *Vide supra*, Chapters III and IV.

[3] A. H. R. Buller, "Fusions between Flexuous Hyphae and Pycnidiospores in *Puccinia graminis*," *Nature*, Vol. CXLI, 1938, p. 33.

[4] D. B. O. Savile, "Nuclear Structure and Behavior in Species of the Uredinales," *American Journal of Botany*, Vol. XXVI, 1939, pp. 585-609.

[5] For an expansion of these critical remarks *vide supra*, Chapter III, p. 212, and Chapter IV, p. 225.

nuclei should be confirmed by other workers, it will doubtless be of much help in future cytological work on the initiation of the sexual process in the heterothallic Rust Fungi.

Living Fungi and Dead Preparations.—Before a cytologist begins to pickle material of a Puccinia or Uromyces with a view to embedding it in wax and slicing it up with a microtome he would do well to study both his Rust Fungus and the invaded host-tissues in the living condition. He should first make himself acquainted with the difference between pointed periphyses and flexuous hyphae in sufficiently old but still undiploidised haploid pustules. Then he should mix the nectar of haploid pustules and, thereafter, by means of hand-cut sections, find true fusions between pycnidiospores and flexuous hyphae whilst these elements are still alive and fully turgid. A knowledge of the appearance of such fusions thus gained would stand him in good stead subsequently when seeking for fusions in dead, non-turgid, and more or less distorted material seen in sections cut with the microtome.

In such Hymenomycetes as *Coprinus lagopus*, when one haploid mycelium is diploidising another of opposite sex, the invading nuclei travel through a network of hyphae formed by radial hyphae already interconnected by means of peg-to-peg, hypha-to-peg, and hypha-to-hypha fusions, and no *new* fusions are formed to assist the invading nuclei in taking short cuts to their destinations. It is desirable that the rust cytologist should study the extent to which hyphal fusions in heterothallic Rusts: (1) are normally present between basal cells (fertile cells) in *fully mature haploid proto-aecidia* obtained by sowing individual basidiospores; and (2) come into existence in haploid proto-aecidia some 24 or more hours after the nectar of the pustules containing the proto-aecidia has been mixed. An investigation of this kind would teach us whether the nuclei invading a proto-aecidium of opposite sex make use of *hyphal fusions* between adjacent rows of fertile cells *already in existence* or whether they stimulate the formation of new fusions.

Cytologists, in studying the initiation of the diplophase of the life-history at the base of a young aecidium seem to have contented themselves with vertical sections only. A study of the basal cells (fertile cells) of a young aecidium in *transverse sections* as well as in vertical sections might enable the investigator to construct a three-dimensional illustration of the group of basal cells as a whole and thus to indicate the extent to which any one basal cell becomes fused laterally with the several basal cells by which it is surrounded. It is possible that three-dimensional studies of young aecidia would reveal far more multiple fusions than have so far been admitted.

In 1933, in Volume V of this work, I[1] called attention to Wahrlich's discovery, made in 1893, that the septa of the mycelium of the Higher Fungi, including the Uredinales, are each provided with a small central open pore; and I described and illustrated the perforated septum in species of Pyrenomycetes, Discomycetes, Hymenomycetes, and Fungi Imperfecti. Since then, Savile[2] alone has referred to the pores in the septa of the Rust Fungi; but, unfortunately, having studied them only in dead hyphae, he has failed to understand them. He says: "In preparations steeped first in tannic acid and then in ferric chloride to give intensively stained walls, the septal pores in the older, vacuolate hyphae are seen to be filled by small plugs of undetermined composition. These are presumably destroyed or displaced by the migrating nuclei." Savile does not cite the work of Wahrlich or of myself which shows that in the Higher Fungi the septal pores are open and not closed by plugs. I myself and others have observed, in living hyphae of Pyrenomycetes, Discomycetes, Hymenomycetes, and Fungi Imperfecti, the cytoplasm flowing freely from cell to cell; and, in *Pyronema confluens*, on one occasion, I observed a stream of cytoplasm flowing through the pores of the septa of a series of 161 cells that stretched for a distance of about 1.6 cm.[3] The transportation of cytoplasm from the mycelium into fruit-bodies of the Higher Fungi depends on vacuolar pressure compressing the cytoplasm and forcing it from cell to cell through the perforations of the septa.[4] In living hyphae of the Higher Fungi in general, the pores are open and not plugged. As I have shown for *P. confluens*, etc., plugs are formed when the cells are suddenly killed,[5] and I have no doubt, therefore, that the plugs which Savile saw in his preparations were artifacts and not structures present when his hyphae were alive and functioning. There is, therefore, no need to presume that septal plugs are "destroyed or displaced by the migrating nuclei." It seems very likely, as Savile has supposed, that, during the diploidisation process, the nuclei of one sex invading the mycelium of another sex do pass through the pores of the septa, thus moving in the line of least resistance; but rust cytologists have yet to prove definitely that this is so. Savile's Figure 90 in which a nucleus is supposed to be migrating from one cell to another is, in my judgment, far from convincing.[6]

[1]These *Researches*, Vol. V, 1933, pp. 89-96.   [2]D. B. O. Savile, *loc. cit.*, p. 602.
[3]These *Researches*, Vol. V, 1933, pp. 119-121.
[4]*Ibid.*, pp. 122-125.   [5]*Ibid.*, pp. 130-134.

[6]For experimental and histological evidence that in *Gelasinospora tetrasperma*, after two mycelia of opposite sex have been mated, nuclei of one sex migrate from cell to cell in the mycelium of opposite sex *via* the septal pores *vide* Eleanor S. Dowding and A. H. R. Buller, "Nuclear Migration in Gelasinospora," *Mycologia*, Vol. XXXII, 1940, pp. 483-484.

As a preliminary to a study of the pores in the septa of the mycelium of a Rust, a rust cytologist would do well to study the pores in the septa of the living mycelium of such fungi as *Pyronema confluens*, *Pleurage anserina*, etc., and to witness the transportation of cytoplasm through them and from cell to cell. The technique for doing this is described in Volume V of these *Researches*.

The phenomenon of the flow of cytoplasm from cell to cell through septal pores in certain Pyrenomycetes having wide hyphae and granular cytoplasm is so clear that it can be demonstrated by photography. Thus, in 1939, at New York, my former pupil, Dr. E. S. Keeping,[1] exhibited moving pictures of growing hyphae of *Gelasinospora tetrasperma* in which cytoplasm could be seen "streaming toward the tips of the hyphae and passing through perforations in the transverse septa."

It may here be remarked that, in the Hymenomycetes, it is not to be supposed that nuclei of one sex migrating through a mycelium of opposite sex are propelled on their courses by vacuolar pressure. The movement of these migrating nuclei seems to be quite independent of the mass flow of protoplasm. How a nucleus or nuclei can travel through a mycelium of *Coprinus lagopus* for a distance of 6 cm. at a rate of more than 1.5 mm. per hour, and probably at about 3.0 mm. per hour, as was calculated in a particular instance,[2] is a profound mystery. The means by which, in a heterothallic Rust Fungus, a pycnidiosporal nucleus, after entering the tip of a flexuous hypha, travels down the flexuous hypha, through the cells in the wall of the pycnidium, and through vegetative hyphae situated between the leaf-cells to the lower basal cells in a proto-aecidium of such a Rust as *Puccinia graminis* is also a profound mystery. Whether or not cytologists will be able to throw any light on the means whereby migrating nuclei travel rapidly and in specific directions through the hyphae of a mycelium of opposite sex remains to be seen.

Conclusion.—In Period I, when it was believed that the sexual process in Rusts of the *Puccinia graminis* type was initiated by the co-operation of pairs of cells at the base of each young aecidium, one of the difficulties in admitting this solution of the problem was to imagine how, in a large aecidial rudiment like that of a Cronartium, there should come into existence in the fertile layer approximately equal numbers of gametic or pseudogametic cells of opposite sex or mating power arranged in such a way that each (+) cell was next to a

[1]Eleanor Silver Keeping, "Nuclear Migration in Gelasinospora," Third International Congress for Microbiology, New York, 1939, *Report of Proceedings*, 1940, p. 541.

[2]These *Researches*, Vol. IV, 1931, pp. 213-229.

(−) cell.   But, as soon as Craigie initiated Period III by announcing his discovery that pcynidiospores fuse with flexuous hyphae of opposite sex, it became evident: (1) that the supposed chess-board-like arrangement of (+) and (−) fertile cells in a proto-aecidium had been a completely erroneous conception; (2) that the fertile cells in any particular proto-aecidium are either all (+) or all (−); and (3) that the diploidisation of a proto-aecidium is effected by nuclei of opposite sex that invade the proto-aecidium from without.   To verify these conclusions by means of cellular and nuclear studies is the task with which the rust cytologist interested in the sexual process is now confronted.

Now that it has been established by observation and genetic experiment that the sexual process can be initiated: in *heteroecious* heterothallic Rusts, *e.g. Puccinia graminis,* in two different ways (1) by the exchange of nuclei between (+) and (−) mycelia which have formed a compound pustule, and (2) by the fusion of a pycnidiospore of one sex with a flexuous hypha of opposite sex; and in *autoecious* heterothallic Rusts, *e.g. P. helianthi,* in both those ways and in yet a third way, (3) the diploidisation of a haploid mycelium (derived from a basidiospore) by a diploid mycelium (derived from an aecidiospore or a uredospore), a firm basis has been secured for many new cytological investigations on nuclei travelling to proto-aecidia and there associating themselves with conjugate mates.   In seeking to elucidate the sexual process in heterothallic Rust Fungi the experimenter now needs the assistance of the cytologist.

CHAPTER IX

CRONARTIUM RIBICOLA AND ITS SEXUAL PROCESS

The White Pine Blister Rust Disease — Mode of Infection — Incubation Period for Canker Formation — Successive Formation of Pycnidia and Aecidia — Morphology and Cytology of an Old-established Canker — Comparison of Systemic Mycelia — Simple and Compound Cankers — Heterothallism of *Cronartium ribicola* — Initiation of Diploidisation in a Simple Canker — Initiation of Diploidisation in a Compound Canker — Cell-fusions in the Basal Cells of the Proto-aecidia — Conclusion.

The White Pine Blister Rust Disease.—*Cronartium ribicola*, the White Pine Blister Rust, economically is one of the most important of the Uredinales. It ravages the White Pine (*Pinus strobus*) and other five-needle Pines in eastern and western Canada; and on this account, in 1928, in my Presidential Address to the Royal Society of Canada,[1] I treated of the White Pine Blister Rust Disease as an example of a disease of forest trees due to fungi. The description of the disease given in the Address was compiled from information contained in many technical papers and expressed in simple language; and it is here condensed as a general introduction to what follows:

The White Pine is the tallest and most stately conifer in eastern Canada and the northeastern part of the United States and, where it occurs, it is the most valuable tree of the forest. It grows rapidly, has excellent wood, its yield of wood per acre is high, and it is very adaptable to forest management. In the United States, White Pine is grown in rotations of 40 - 60 years and, in many sections, it yields a profit of $5 to $10 per acre per year. Not only are White Pines wonderful producers of merchantable timber, but they are a delight to the eye and, owing to their high ornamental value, are often planted as lawn trees in parks and private grounds.

The White Pine blister rust is possibly of Asiatic origin. It was first noticed in Europe on White Pines introduced from the United

---

[1]A. H. R. Buller, "The Plants of Canada Past and Present," a Presidential Address, *Trans. Roy. Soc. Canada*, Vol. XXII, 1928, Appendix A, pp. 34-58.

States. Owing to the outbreak of the disease in Europe, the growing of White Pines in that continent for forestry purposes has largely been given up.

The blister rust disease was accidentally introduced into the United States and Canada from Germany on diseased White Pine stock, and it has now become established in both the eastern and western regions.

The blister rust disease kills the bark of White Pines and thus causes their death. The fungus is peculiar in its demands on its host-plant, for it limits its attack to five-needle pines, such as *Pinus strobus* (White Pine) and *P. monticola* (the Western White Pine) and leaves untouched all two-needle pines.

The blister rust fungus infects the bark of White Pines through the needles. Late in the summer or in the autumn, a basidiospore settles on a needle and germinates there. The germ-tube makes its way through the epidermis into the interior of the leaf and, as a result, a yellow leaf-spot is developed. Once inside the leaf, the parasite grows down the needle into the stem. It then vegetates in the bark, spreading out from the point of attack and forming a canker. The canker usually completes its ripening in the third year after infection has taken place and then, from April to June, orange-yellow blisters about the size of a small bean burst through the diseased bark (Fig. 99). The blisters (aecidia) contain millions of aecidiospores, which are carried off by the wind and scattered far and wide.

The bark ruptured by the fungus dies. The mycelium continues to grow in the healthy bark bordering the canker and, each successive spring, produces new blisters or spore-sacs until the limb or trunk on which it is situated has been killed. Young trees may be girdled and killed in a few years. Older trees may survive for a longer time. On older trees the first signs of damage appear as scattered dead and dying branches sometimes called "flags." From these diseased branches the fungus grows back into the trunk. After the tree has been girdled, the portion of the tree above the canker dies and in course of time may break off. The fungus continues to grow down the tree through the healthy bark with the inevitable result that, in the end, the whole tree is killed. Where a canker occurs on a trunk, the growth of annual rings of wood is greatly retarded. White Pines weakened by blister rust become attacked by other fungi and by insects, and the dead trees are soon destroyed by these agencies.

In the spring, the aecidiospores are blown by the wind, but they can infect only the leaves of Ribes. The mycelium vegetates in the internal tissues of the leaf for some three weeks. At the end of this

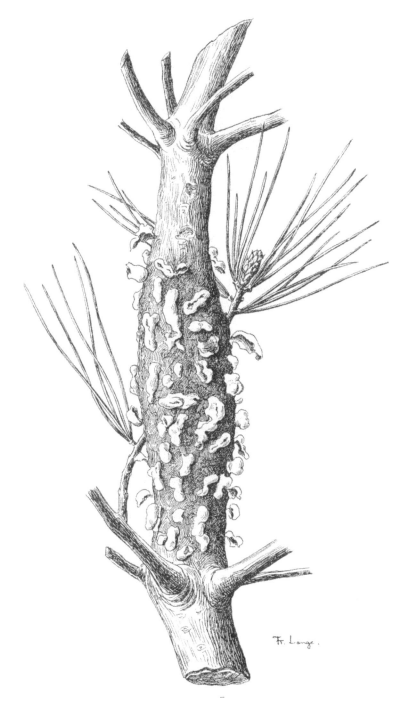

FIG. 99.—*Cronartium ribicola* on *Pinus strobus.* Erumpent from the bark are numerous bladdery aecidia, each with an inflated peridium. The surrounding zone of pycnidia is not indicated. Natural size. From E. Rostrup's *Plantepatologi* (1902).

time, small orange-yellow uredospore sori break open and liberate a great number of uredospores which spread the disease on currants and gooseberries from leaf to leaf and from bush to bush, rapidly producing new infections. This is repeated several times during the growing season.

From late June until the foliage dies, brownish hairlike outgrowths of the fungus, the teleutospore sori appear on the under side of the diseased currant and gooseberry leaves. The teleutospores liberate delicate basidiospores which cannot re-infect currant and gooseberry leaves but which can infect the leaves of the White Pine and other five-needled pines.

The aecidiospores produced on the White Pines are long-lived and spread the rust for miles from infected pines to healthy currant and gooseberry bushes. On the other hand, the basidiospores are short-lived and spread the rust from currants and gooseberry bushes to White Pines for a few hundred feet only.

The short infecting range of the basidiospores makes it possible to control the rust locally by the eradication of all the Ribes growing within 900 feet of the Pine trees. The cost of the currant and goose-berry eradication work on 3,450,000 areas in the United States has averaged 25 cents an acre, and these control measures have been found to be reasonable and profitable. In Canada and the United States it is impossible to eliminate the blister rust disease entirely; but, through our knowledge of the life-history of the causal fungus, it is now possible to protect any stand of White Pine.

In the West, the blister rust fungus has passed from Canada to the United States and is attacking all the five-needled pines which it meets in its progress, *e.g., Pinus monticola* (the Western White Pine) and *P. lambertiana* (the Sugar Pine). Already it has become a serious disease in a very valuable and extensive western forest area.

Mode of Infection.—Klebahn,[1] in 1905, and Tubeuf,[2] in 1914, inoculated the needles of *Pinus strobus* with basidiospores of *Cronartium ribicola* and obtained typical infection spots in the needles but left undecided how these spots were connected with infection in the bark. There was the possibility that entrance to the bark was by direct infection of the young stems. McCubbin,[3] in 1917, from field

[1]H. Klebahn, "Kulturversuche mit Rostpilzen. XII. Bericht (1903 and 1904)," *Zeitschrift für Pflanzenkrankheiten*, Bd. XV, 1905, pp. 86-92.

[2]C. von Tubeuf, "Über das Verhältnis der Kiefern-Peridermien zu Cronartium. II. Studien über die Infection der Weymouthkiefer," *Naturw. Zeitschr. Forst- u. Landw.*, Bd. XV, 1917, pp. 274-307.

[3]W. A. McCubbin, "Contributions to our Knowledge of the White Pine Blister Rust," *Phytopathology*, Vol. VII, p. 96.

observations, concluded that most infections occur in the bases of the dwarf shoots (leaf fascicles). In 1919, Clinton and McCormick[1] demonstrated that the germ-tubes of the basidiospores enter through the stomata of the needles and that the typical mode of infection of the bark is accomplished by the mycelium growing from an infection spot on a needle down the needle and short stem of the dwarf shoot and thus into the bark of the shoot of unlimited growth.

Clinton and McCormick[2] were unsuccessful in their attempts to infect the bark of pines directly, but held that direct infection may occur to a slight extent in bark that is young enough to have stomata through which the germ-tubes of the basidiospores may enter. Spaulding,[3] in 1922, also thought that young bark might occasionally be infected directly. However, the long-continued field observations of Lachmund,[4] published in 1933, indicate that in *Pinus monticola*, as in other White Pine species, the mycelium derived from a basidiospore normally gains entrance to the bark *via* a needle. Said he[5]: "The first characteristic yellow to orange discoloration or swelling in the bark, which marks the appearance of an incipient canker, almost invariably develops symmetrically around a single needle bundle. Aside from its central location in the discolored area this bundle is generally distinguished by a marked swelling at its base. An examination of the needles of this bundle will usually disclose at least one definite needle infection spot. Other needle bundles may occur on the discolored area; but, except where there is evidence that the rust has entered from more than one needle bundle, their bases are generally not swollen. Frequently the central bundle, through which infection entered, persists considerably longer than the rest. Where all needles have fallen the enlarged scar of this bundle is generally conspicuous, or at least is recognizable, in the center of the discoloration." Lachmund[6] also observed that incipient cankers are formed only on branches that still bear needles and that, in long branches, except for the terminal series of nodes and internodes of the new shoots developed during the

[1]G. P. Clinton and F. A. McCormick, "Infection Experiments of *Pinus strobus* with *Cronartium ribicola*," *Conn. Agric. Experiment Station Bull. 214*, 1919, pp. 428-459.

[2]*Ibid.*, p. 441.

[3]P. Spaulding, "Investigations of the White-Pine Blister Rust," *U.S. Dept. of Agric. Bull. 957*, 1922, pp. 26-27.

[4]H. G. Lachmund, "Mode of Entrance and Periods in the Life Cycle of *Cronartium ribicola* on *Pinus monticola*," *Journal of Agric. Research*, Vol. XLVII, 1933, pp. 791-805.

[5]*Ibid.*, p. 792.

[6]*Ibid.*, pp. 792-793.

infection season, upon which the needles appear to be resistant,[1] most of the cankers form upon younger yearly series of nodes and internodes where the needles are most numerous and are retained longest.

Incubation Period for Canker Formation.—In *Puccinia graminis*, after a basidiospore has been sown on a Barberry leaf, there can be detected on the leaf: (1) a *discoloration spot* on the fifth or sixth day; (2) *functional pycnidia* on about the eighth day; and (3), in compound pustules where the mycelial components are of opposite sex or in simple pustules to which mixed nectar has been promptly applied, *aecidia* in from two to three weeks. In *Cronartium ribicola* the corresponding development requires from many months to several years. However, whereas the pustules of *P. graminis* on a Barberry leaf are localised and temporary, those of *Cronartium ribicola* on a pine branch, when once established, are systemic and may persist for many years.

Lachmund,[2] working in British Columbia, observed that for *Cronartium ribicola* on *Pinus monicola* (the Western White Pine) the *incubation period*, *i.e.* the period from infection to the appearance of discoloration or swelling in the bark sufficient for an accurate macroscopic diagnosis, varies according to the time of infection and seasonal conditions and is longer at higher elevations, where the growing season is shorter, than at lower elevations, where growth activity is greater and extends over a longer period. In pine-trees more than 3 feet high and 8 years old, the usual incubation period was found to vary from about 20 months to 26 months. After infection on such trees the minimum period required for cankers to form is generally not less than 16 months. An extreme minimum period of 9.5 months and a maximum period of not less than 35 months or possibly 41 months were also observed. It is not probable that the incubation period ever extends beyond 4 years. On the youngest trees, up to 4 or 5 years of age, the incubation period is about a year shorter than on older trees. Incipient cankers have been observed to form on such young trees after a possible minimum of 6 months.

Successive Formation of Pycnidia and Aecidia.—Lachmund[3] observed in British Columbia that incipient cankers on *Pinus monticola*, produced in any one year, if formed sufficiently early in the year produce pycnidia the same year, but if formed later in the year do not produce pycnidia until the next summer. The period from leaf-

[1]H. G. Lachmund, "Resistance of the Current Season's Shoots of *Pinus monticola* to Infection by *Cronartium ribicola,*" *Phytopathology*, Vol. XXIII, 1933, pp. 917-922.

[2]H. G. Lachmund, "Mode of Infection and Periods in the Life Cycle of *Cronartium ribicola* on *Pinus monicola,*" *Journal of Agric. Research*, Vol. XLVII, 1933, pp. 793-798. [3]*Ibid.*, p. 800.

infection to pycnidial production was found to be 1-10 months longer than the incubation period.

In British Columbia it was also observed by Lachmund: that, in warmer localities near the sea level, the formation of pycnidia usually begins in June, is heaviest in July and August, and subsides when the weather turns cooler usually in September; but that periods of warm or hot weather may induce a considerable resumption of activity in pycnidial production in October or even November. At higher levels the season of pycnidial production is much shortened. It thus appears that, in a single summer season, the nectar of the pycnidia of *Cronartium ribicola* is exposed to the visits of insects and possible mixing during a period of many weeks.

It is important to note that in a mycelium of a canker formed by *Cronartium ribicola*, just as in the mycelium of a pustule of *Puccinia graminis* on a Barberry leaf, the production of pycnidia always precedes the production of aecidia.

Lachmund[1] observed that in the majority of cankers aecidia are usually produced in the spring of the year following that in which the cankers first bore pycnidia, but that frequently a large proportion and sometimes a majority of the cankers do not produce aecidia until the second year after that in which they produced pycnidia.

Lachmund also observed: that young cankers which fail to produce aecidia in the year following that of pycnidia production usually produce pycnidia again the next summer and aecidia the next spring; and that, sometimes some of the cankers do not produce aecidia until the third or even the fourth year after that in which they first produced pycnidia.

It may be presumed that, typically, the mycelium in each canker has been derived from a single basidiospore and is haploid. We may therefore suppose that, in well-developed cankers, after the formation of the first pycnidia, the delay of more than one year in the formation of aecidia is due to lack of diploidisation of the proto-aecidia, and this we may associate with a failure of insects to carry to the pycnidia pycnidiospores of opposite sex.

Tuberculina maxima is a common parasite of *Cronartium ribicola* on *Pinus strobus* in Europe,[2] and it was also observed by Lachmund

[1] *Ibid.*, pp. 800-803.

[2] E. Lechmere, "*Tuberculina maxima* Rost., ein Parasit auf dem Blasenrost der Weymouthskiefer," *Naturw. Zeitschr. Forst- u. Landw.*, Bd. XII, 1914, pp. 491-498. Lechmere observed that the mycelium of *T. maxima* does not invade the tissues of *Pinus strobus* or kill the mycelium of *Cronartium ribicola*, but confines its attacks solely to the aecidia and pycnidia. Hence *T. maxima* does not prevent the Blister Rust extending its cankers year by year.

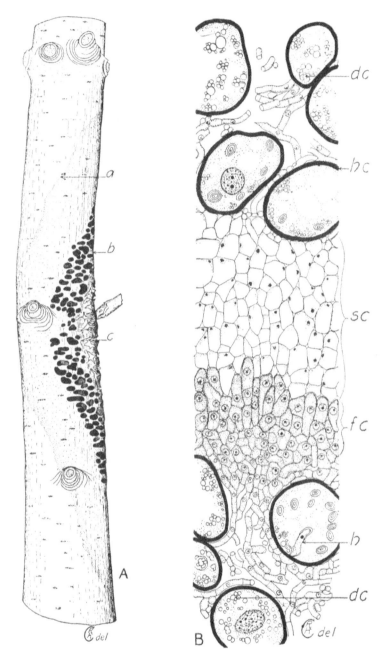

Fig. 100.—*Cronartium ribicola* on *Pinus strobus*. "A, drawing of an infected 12-year-old main stem. The infection entered the main stem along the small branch, the stub of which is shown at the right of the Figure: *a* the advancing edge of the infection; *b*, the pycnial area. The black dots are the pycnial spots; *c*, the aecial area on which the bark is cracked and broken. In another season the aecial area would spread over the pycnial area (*b*), and the pycnial area would be advanced as far as the boundary (*a*) under normal conditions. The boundary (*a*) would be proportionately advanced also. The specimen from which the drawing was made was collected in August, 1917. X ½.

"B, drawing of a section through part of a young stem showing the relation of the fertile cells with their denser protoplasmic contents to the overlying sterile cells, in which the cytoplasm and nuclei have begun to go to pieces. The manner in which the adjacent host-cells are forced apart by the fungus cells is also shown: *dc*, decomposition products in the host cells; *hc* host cell-wall; *sc*, sterile cells; *fc*, fertile cells; *h*, haustorium. The elliptical bodies in the host-cells represent starch grains. X 400." The fungal tissues *sc* and *fc* are parts of a haploid uninucleate proto-aecidium borne on a uninucleate haploid mycelium. Both A and B drawn by R. H. Colley (1918).

sporadically attacking the pycnidia and aecidia on *Pinus monticola*.
There is therefore the possibility that *Tuberculina maxima* by destroying
pycnidia may often prevent the diploidisation of the proto-aecidia and
thus prevent the development of aecidia; and to the attacks of the
parasite Lachmund[1] attributes the sterility of many particular cankers
that came within his observation.

Morphology and Cytology of an Old-established Canker.—For an
excellent and well-illustrated account of the morphology and cytology
of *Cronartium ribicola* and of the relations of the mycelium with the
host-tissues we are indebted to Colley[2] who published his findings in
1918.

In an old-established infection of *Cronatrium ribicola*, in a normal
canker on the bark of *Pinus strobus*, in the summer, there is always:
(1) an outer sterile *zone of discoloration* bounded by the advancing edge
of the infection (Fig. 100, A, *a*); (2) an intermediate *zone of pycnidia* in
which are numerous, broad, flat, somewhat blistery pycnidia that
excrete nectar laden with pycnidiospores (A, *b*); and an inner *zone of
broken bark and bladdery aecidia* (*A, c*). Each year these three zones
move forward centrifugally, so that a new discoloration zone is developed externally, new pycnidia are developed in the old discoloration
zone, and new aecidia are developed in the old zone of pycnidia.

The pycnidia are subcortical and they arise in considerable numbers,
side by side, below the outermost cell-layers of the bark. As represented by Colley in black in Fig. 100, A, zone *b*, one-half the natural
size, they are flattened structures, rounded or oval in outline and extending individually in the bark over an area that varies from about
one to several square millimetres.

A single pycnidium arises from intercellular hyphae which form a
thin mycelial mat parallel to, and a little way beneath, the outermost
layer of the bark. As the mat develops, the covering layer of bark is
forced outwards and thus slightly raised. At maturity and in vertical
section, a pycnidium (Fig. 101) is seen to consist of: (1) a *basal layer
of hyphae* which appears as a pseudoparenchyma with the upper cells
directed toward the exterior surface of the bark; and (2) the *hymenium*.
The hymenium is composed of a palisade layer of many thousands of
orange, closely appressed, slender, slightly tapering *pycnidiosporophores* and of a number of *flexuous hyphae*, long, slender, parallel-sided,
non-septate organs that arise here and there between the pycnidiosporophores and project far beyond them.

[1]H. G. Lachmund, *loc. cit.*, p. 802.

[2]R. H. Colley, "Parasitism, Morphology and Cytology of *Cronartium ribicola*,"
Journal of Agricultural Research, Vol. XV, 1918, pp. 626-631.

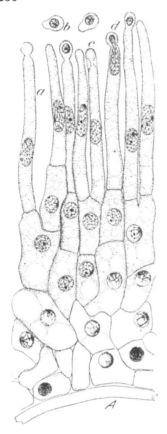

FIG. 101.—*Cronartium ribicola.*
"The elements of the pyc-
nium. The cells at the base
are almost empty. Above
them are the short branching
trunks which bear the sporophores: *a*, a sporophore;
b, pycniospores in sectional
view; *c*, the protoplasm is
constricted just beneath the
spore; *d*, the nucleus is
dividing. X 1,700." Drawn
and described by R. H. Colley
(1918).

There are no structures in the hymenium of a pycnidium of *Cronartium
ribicola* corresponding to the ostiolar
trichomes (periphyses or paraphyses) of
such Rusts as *Puccinia graminis* and
Gymnosporangium juniperi-virginianae.

Each pycnidiosporophore abstricts
from its free end a series of orange,
somewhat pear-shaped, uninucleate
pycnidiospores whose outer walls, as in
other Rusts, swell up and become gelatinous. As more and more pycnidiospores are abstricted from the pycnidiosporophores, they come to form a
dense thick orange layer between the
upper surface of the hymenium and
the covering layer of bark; and, as the
spore-layer increases in thickness, the
covering layer of bark is forced farther
outwards and thus placed under a
considerable mechanical strain.

The flexuous hyphae are not shown
in Colley's illustration of the structure
of a pycnidium (Fig. 101), and they
were first recognised for what they
really are, *i.e.* organs which may unite
with pycnidiospores of opposite sex and
thus take part in the initiation of the
sexual process, by Pierson, in 1933.
Pierson[1] described them shortly after
Craigie had announced his discovery of
the existence and mode of functioning
of the flexuous hyphae of *Puccinia
helianthi.* I have seen flexuous hyphae
similar to those of *Cronartium ribicola*
in hand-cut sections of living pycnidia
of *C. cerebrum* and *C. fusiforme* in
material sent to me from the United States.

Owing to the outward pressure exerted by a pycnidium through its
basal layer, its hymenium, and its thick layer of gelatinous pycnidio-

[1]Royale K. Pierson, "Fusion of Pycniospores with Filamentous Hyphae in the
Pycnium of the White Pine Blister Rust," *Nature*, Vol. CXXXI, 1933, pp. 728-729.

spores, the thin layer of bark covering the pycnidium is bulged outwards and finally split, so that the pycnidium, at least in part, becomes
exposed to the air.   Then through the cracked bark there is excreted
an abundance of nectar and, along with the nectar, vast numbers of
pycnidiospores.   Soon the pycnidiospore-bearing nectar comes to hang
from the swollen bark in drops, as may be seen in Fig. 102.

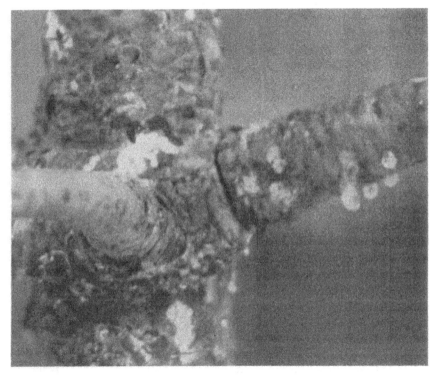

FIG. 102.—Drops of pycnial nectar of *Cronartium ribicola* on *Pinus strobus*, slightly
    enlarged.   Photograph by E. E. Honey on June 27, 1939, in Polk County,
    Wisconsin.

The flexuous hyphae from the first are very much longer than the
pycnidiosporophores and, in a mature pycnidium exposed to the air,
they extend outwards through the thick layer of pycnidiospores into
the drop of nectar, there to await the coming of pycnidiospores of
opposite sex with which they may fuse and co-operate in the initiation
of the sexual process. Thus a drop of nectar excreted from a pycnidium
of *Cronartium ribicola* has associated with it, just as in *Puccinia graminis*, both pycnidiospores and flexuous hyphae.

The drops of nectar hanging from the bark of a canker of *Cro-*

nartium ribicola were first illustrated by Tubeuf.[1] Tubeuf inoculated a young white Pine (*Pinus strobus*) with basidiospores on September 11, 1914, and he photographed the resulting canker on May 31, 1916. The photograph, as reproduced in his communication, shows a pine stem about a quarter of an inch thick with a gall one inch long bearing about a dozen drops of nectar. The somewhat similar photograph, reproduced in Fig. 102, was taken on June 27, 1939, in Wisconsin, by Dr. E. E. Honey.[2]

Klebahn[3] observed that the nectar of *Cronartium ribicola* is sweet to the taste, odoriferous, and attractive to insects; and, concerning the allied species *Cronartium quercuum*, in Japan, Shirai[4] has recorded that the pycnidial nectar in the form of viscid drops are known as *Matsumitsu* (Pine-honey) and are eaten by boys and girls when they happen to find them. There can be no doubt that the drops of pycnidial nectar excreted by *Cronartium ribicola* and other species of Cronartium, like the drops of nectar in flowers, serve as a bait for insects.

Owing to the slow development of *Cronartium ribicola* in its coniferous host, it would require several years work to demonstrate that *C. ribicola* is heterothallic; but that this fungus actually is heterothallic there can be but little doubt. Among the pieces of evidence suggesting heterothallism is the fact that the mycelium in a canker, year after year, produces numerous large pycnidia that excrete drops of nectar that are attractive to insects.

The aecidia of *Cronartium ribicola* are developed from broad flat proto-aecidia (Colley's *young aecia*). The proto-aecidia (Fig. 100, B) are composed of basal cells (Colley's *fertile cells*) and of an upper layer of rather empty cells which Colley calls *sterile cells* and which correspond to the pseudoparenchymatous cells in a proto-aecidium of *Puccinia graminis*. Every year new proto-aecidia are formed beneath the pycnidia in the pycnidial zone; and these proto-aecidia, after being diploidised, are destined to develop and liberate aecidiospores in the year following.

As Colley has shown, in *Cronartium ribicola* the following structures are all composed of uninucleate cells and are therefore haploid: (1) the vegetative mycelium between the cells of the cortex, phloem, medullary

---

[1]C. von Tubeuf, "Über das Verhältnis der Kiefern-Peridermien zu Cronartium," *Naturw. Zeitsch. f. Forst- und Landw.*," Jahrg. XV, 1917, pp. 268-307, Abb. 2.

[2]Dr. Honey was engaged in Blister Rust control work, and he kindly sent me the photograph here reproduced.

[3]H. Klebahn, *Die wirtswechselnden Rostpilze*, Berlin, 1904, p. 387.

[4]M. Shirai, "On the Genetic Connection between *Peridermium giganteum* (Mayr) Tubeuf and *Cronartium quercuum* (Cooke) Miyabe," *Bot. Mag.*, Tokyo, Vol. XIII, 1899, p. 76.

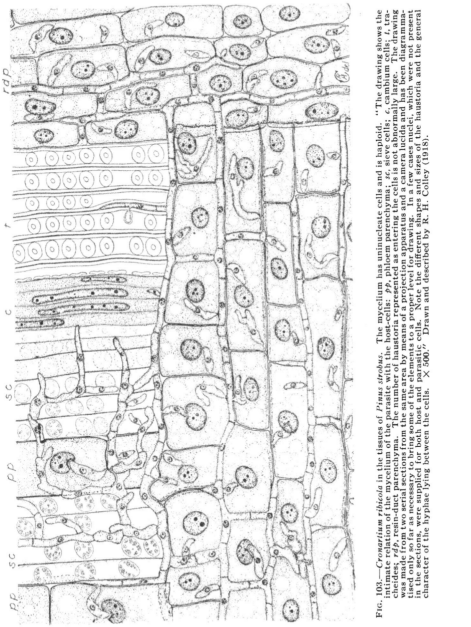

FIG. 103.—*Cronartium ribicola* in the tissues of *Pinus strobus*. The mycelium has uninucleate cells and is haploid. "The drawing shows the intimate relation of the mycelium of the parasite with the host-cells: *pp*, phloem parenchyma; *sc*, sieve cells; *t*, tracheides; *rdp*, resin-duct parenchyma. The number of haustoria represented as entering the cells is not abnormally large. The drawing was made from two serial sections from the same area by means of a projection apparatus and a camera lucida and has been diagrammatised only so far as necessary to bring some of the elements to a proper level for drawing. In a few cases nuclei, which were not present in the sections, were supplied for both host and parasitic cells. Note the different shapes and sizes of the haustoria and the general character of the hyphae lying between the cells. × 500." Drawn and described by R. H. Colley (1918).

rays, and resin-duct parenchyma of the host (Fig. 103); (2) the numerous mycelial haustoria (Fig. 103); (3) the mycelium immediately below the pycnidia and proto-aecidia (Fig. 100, B); (4) the pycnidia (Fig. 101); and (5) the proto-aecidia (Fig. 100, B).

According to Colley, the basal cells of each proto-aecidium (Fig. 100, B, *fc*) fuse in pairs or in greater numbers (Fig. 104) and thus become diploidised, and then the proto-aecidia develop into aecidia with the binucleate aecidiosporophores producing chains of binucleate aecidiospores and intercalary cells (Fig. 105).[1] The intercalary cells are much elongated. Eventually these disjunctor cells break down and then the aecidiospores are left behind in the aecidium in the form of a powder.

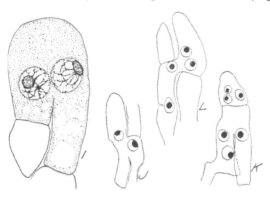

FIG. 104.—*Cronartium ribicola*. "I, an aecial basal cell resulting from the fusion of two adjacent cells of the fertile layer. X 1,700. J, a diagram of a basal cell resulting from the fusion of two cells from different levels. X 850. K, a diagram of a trinucleate irregular basal cell from the tip of which a trinucleate aeciospore initial has been cut off. X 850. L, a diagram of part of an irregular compound fusion cell. X 850." Drawn and described by R. H. Colley (1918).

Comparison of Systemic Mycelia. — In *Cronartium ribicola*, the mycelium in every simple canker on the bark of a pine-tree originates from a *basidiospore* and never from an aecidiospore or a uredospore. Consequently, at its origin, the systemic mycelium of *C. ribicola* is *haploid*. As shown by Colley, this mycelium continues in the haploid (uninucleate) state for years and during the whole period of its growth. Since it never becomes diploid (binucleate) it cannot ever become de-diploidised.

On the other hand, it has been proved experimentally that in *Puccinia minussensis* on *Lactuca pulchella* the systemic mycelium can be initiated by *aecidiospores* or *uredospores*,[2] and that in *Puccinia suaveolens* on *Cirsium arvense* the systemic mycelium can be initiated by *uredospores*.[3] Consequently, in the hosts of these Rusts the

[1] R. H. Colley, *loc. cit.*, pp. 630-631.   [2] A. M. Brown, *vide* Chapter VI.
[3] A. H. R. Buller and A. M. Brown, "Urediospores as the Origin of Systemic Mycelia in *Puccinia suaveolens*," *Phytopathology*, Vol. XXXI, 1941, an abstract, p. 4. A detailed account in this volume, Chapter X.

systemic mycelium that has originated from aecidiospores or uredo-spores is at its origin *diploid* (binucleate). As shown by cytological observations, the systemic mycelia of *Puccinia minussensis* in the rhizomes of the Blue Lettuce and of *P. suaveolens* in the perennial roots of the Creeping Thistle are also diploid. This permits of these mycelia undergoing partial de-diploidisation in the new shoots that come up in the spring and thus producing annual crops of pycnidia and proto-aecidia.

Thus the systemic mycelium of *Cronartium ribicola*, in both its origin and nuclear condition, differs much from the systemic mycelia of *Puccinia minussensis* and *P. suaveolens*.

Cronartium is a genus of the Melampsoraceae and all its species attack Pines. In other members of the Melampsoraceae, *e.g.* Uredinopsis, Milesia, Hyalos-pora, Pucciniastrum, Melampsoridium, and certain species of Melampsora that parasitise Pinaceae, the mycelium derived from a basidiospore remains local-ised in the leaf-needle and is there fertile, producing both pycnidia and aecidia. Cronartium differs from these related Rusts in that the mycelium derived from one of its basidiospores: (1) does not remain localised in the leaf-needle, but grows into the bark where it becomes systemic; and (2), in the leaf-needle, remains completely sterile.

Where, as in Uredinopsis, Milesia, etc., the haploid mycelium is purely foliicolous, the aecidia formed on the coniferous leaves are necessarily small and all produced in a single year; but where, as in Cronar-tium, the haploid mycelium grows from a coniferous leaf into the cortex and cambial region of a branch or tree-trunk before producing pycnidia and proto-aecidia, the aecidia are very large and may be formed annually for many years. In developing a systemic

FIG. 105.—*Cronartium ribicola*. "An aeciospore chain in section view: *a*, the basal cell; *b*, an aeciospore initial; *c*, an intercalary cell; *d*, a young aeciospore. The nuclei in the upper intercalary cells are degenerating. X 850." Drawn by R. H. Colley (1918). The "basal" cell or aecidiosporophore, which has been derived from two fertile cells which have fused, contains two nuclei. Colley supposed that these two nuclei were derived from the two cells which fused; but, if *C. ribicola* is heterothallic, one of the nuclei must be of local origin and the other must have come from without.

mycelium which bears large aecidia perennially, Cronartium exhibits an evolutionary advance on Uredinopsis, Milesia, and other related genera of the Melampsoraceae.

In Uredinopsis, Milesia, etc., since the haploid mycelium is purely foliicolous, the mycelium is quickly and easily established and quickly produces aecidia. This advantage is lost to Cronartium, for, in this genus, the systemic mycelium is established very slowly and with difficulty, and one, two, or more years may pass before it produces any aecidia.

Sterility of the mycelium derived from a basidiospore in an infected host-leaf is known not only in *Cronartium ribicola* but also in *Endophyllum euphorbiae-sylvaticae*, another Rust whose haploid mycelium becomes systemic. Plowright[1] observed: (1) that the basidiospores of this Endophyllum, by means of their germ-tubes, bore through the epidermal cells of a leaf of *Euphorbia sylvatica*, enter the leaf-parenchyma, and there produce a richly branched and widely extending intercellular mycelium; (2) that this mycelium in the leaf remains sterile; and (3) that, if the mycelium by growing down into the rhizomes becomes systemic, it perennates in the rhizomes and produces its first pycnidia and aecidia on the leaves of the next-year shoots.

Thus, in respect to its origin from a basidiospore, its sterility on an infected leaf, and its long-delayed fertility, a systemic mycelium of *Cronartium ribicola* resembles a systemic mycelium of *Endophyllum euphorbiae-sylvaticae*.

There are some Cronartium species which differ from *C. ribicola* in that their Peridermium stage repeats itself on its coniferous hosts by means of aecidiospores. In these species, therefore, the systemic mycelium on a pine may be derived (1) as in *C. ribicola*, from one or more *basidiospores*, and also (2) from one or more *aecidiospores*.

In Europe, *Peridermium pini* (alternate host unknown) causes bark cankers on *Pinus sylvestris*. In 1914, Haack[2] discovered that this rust has repeatng aecidia. He sowed aecidiospores obtained from *Pinus sylvestris* on *Pinus sylvestris* and obtained aecidia; and in 1918 this result was confirmed by Klebahn.[3] Another rust on *Pinus sylvestris*, discovered by York (in 1926)[4] in the State of New York, is now known

[1]C. W. Plowright, *A Monograph of the British Uredineae and Ustilagineae*, London, 1889, p. 227.

[2]G. Haack, "Der Kienzopf (*Peridermium Pini* (Willd.) Kleb.). Seine Uebertragung von Kiefer zu Kiefer ohne Zwischenwirt," *Zeitschr. Forst- und Jagdw.*, Bd. XLVI, 1914, pp. 3-46.

[3]H. Klebahn, "*Peridermium Pini* (Willd.) Kleb. und seine Uebertragung von Kiefer zu Kiefer," *Flora*, Bd. CXI-CXII, 1918, pp. 194-207.

[4]H. H. York, "A *Peridermium* New to the Northeastern United States," *Science* (n.s.), Vol. LXIV, 1926, pp. 500-501.

as the *Woodgate Rust*. It is distinct from *Peridermium pini* and is associated with large woody galls resembling those of *Cronartium cerebrum*. It was found by York[1] and confirmed by True (1938)[2] that the Woodgate Rust is autoecious in that it repeats itself on *Pinus sylvestris* by means of aecidiospores. In 1916 and 1920, in California, on certain American Pine species, Meinecke[3] demonstrated a similar repetition by means of aecidiospores in *Cronartium harknessii* and *Peridermium cerebroides*.[4]

True[5] investigated the Woodgate Peridermium. He sowed aecidiospores obtained from *Pinus sylvestris* on the bark of the current season's shoots of *P. sylvestris* and observed that: the germ-tubes penetrated into the epidermis; the mycelium formed a gall; pycnidia were formed two years after inoculation; and aecidia were formed in ensuing seasons. He also observed that the systemic mycelium in a gall is *uninucleate* and not binucleate. This seems to indicate that the mycelium derived from a binucleate aecidiospore, soon after its entrance into its coniferous host, undergoes de-diploidisation. A similar de-diploidisation process may take place in the mycelium derived from aecidiospores during the formation of cankers in *Cronartium harknessii* and all the other Cronartium species that have repeating aecidia. Further cytological investigations on the mycelium of these species present in galls known to have been formed after the sowing of aecidiospores is desirable.

Simple and Compound Cankers.—*Cronartium ribicola* cankers on the bark of White Pines, like the pustules of *Puccinia graminis* on Barberry leaves, may be *simple* or *compound*. A simple canker is one that contains a single haploid mycelium derived from a single (+) or (−) basidiospore. A compound canker is formed by the fusion of two cankers which have arisen very close to one another. Lachmund,[6] in a passage already cited, incidentally refers to compound pustules that he had observed; but it is evident that such cankers are rare.

[1]H. H. York, "The Woodgate Peridermium," in MS., *vide* True (next citation), p. 42.

[2]R. P. True, "Gall Development on *Pinus sylvestris* attacked by the Woodgate Peridermium, and Morphology of the Parasite," *Phytopathology*, Vol. XXVIII, 1938, pp. 24-46, with 3 Plates.

[3]E. P. Meinecke: "*Peridermium Harknessii* and *Cronartium Quercuum*," *Phytopathology*, Vol. VI, 1916, pp. 225-240; "Facultative Heteroecism in *Peridermium cerebrum* and *Peridermium Harknessii*," *Phytopathology*, Vol. X, 1920, pp. 279-297; and "Experiments with Repeating Rusts," *Phytopathology*, Vol. XIX, 1929, pp. 327-342.

[4]J. C. Arthur in his *Manual of the Rusts in United States and Canada*, includes both of these species in *Cronartium coleosporioides* (1934, p. 29).

[5]R. P. True, *loc. cit.*, p. 41.

[6]H. G. Lachmund, *vide supra*, p. 326.

Heterothallism of Cronartium ribicola.—As noted, it can be safely assumed that *Cronartium ribicola* is heterothallic and thus resembles *Puccinia graminis* and *P. helianthi*. In favour of this assumption are: (1) the fact that the fungus produces large pycnidia which excrete an abundance of nectar that, as Klebahn[1] observed, is sweet to the taste, odoriferous, and attractive for insects; (2) in ordinary simple cankers, the mycelium forms its first pycnidia at least one growing season before it produces its first aecidia; and (3) Pierson's observation that the pycnidia are provided with flexuous hyphae and that, after the nectar of the pycnidia of different pustules has been mixed, just as in *P. graminis* and *P. helianthi*, fusions take place between pycnidio-spores and the flexuous hyphae.

Initiation of Diploidisation in a Simple Canker.—How, in *Cronartium ribicola*, in a simple canker containing a mycelium derived from a single basidiospore, is the diploidisation of the haploid basal cells of the proto-aecidia initiated?

Colley's answer to this question was: by the "fusion of the gamete cells in pairs" or in greater numbers, and the coming together of the nuclei to form a dikaryon or more than one dikaryon in a common mass of cytoplasm.[2]   But, in view of (1) Craigie's discoveries of the Modes I and II by which the sexual process is initiated in *Puccinia helianthi* and *P. graminis* and (2) Pierson's[3] observation that, in *Cronartium ribicola*, 48 hours after the nectar of different pustules had been mixed by hand, eleven cases of the fusion of a pycnidiospore with the end of a filamentous (flexuous) hypha projecting beyond the pycnidiosporophores were observed whereas in control pycnidia in which the nectar had not been mixed no such fusions could be found, Colley's answer to our question must be considered incomplete and unsatisfactory.   Colley[4] actually observed the flexuous hyphae, for he said "occasional long hyphal filaments grow some distance out beyond the tips of the sporophores," but he did not illustrate them (*vide* Fig. 101), and he dismissed the pycnidiospores (Fig. 101) with the remark that "they appear to be completely non-functional."

Granted that *Cronartium ribicola* is heterothallic we can assume that its mature proto-aecidia, like those of *Puccinia helianthi* and *P. graminis*, undergo no important change unless they become di-ploidised by nuclei of opposite sex which have travelled to them through the mycelium.   In other words, we can assume that in a proto-aecidium left to itself the nuclei do not come together in pairs to form dikaryons.

[1]H. Klebahn, *Die wirtswechselnden Rostpilze*, Berlin, 1904, p. 387.
[2]R. H. Colley, *loc. cit.*, pp. 643 and 646.
[3]R. K. Pierson, *loc. cit.*   [4]R. H. Colley, *loc. cit.*, p. 630.

The mycelium in a simple canker has been derived from a single basidiospore and is therefore haploid and unisexual. The mycelium in a young (+) canker is composed solely of (+) cells containing (+) nuclei and bears (+) pycnidia and pycnidiospores and (+) proto-aecidia; and the mycelium in a young (−) canker is composed solely of (−) cells containing (−) nuclei and bears (−) pycnidia and pycnidiospores and (−) proto-aecidia.

For convenience in discussion we may now divide simple cankers into: (1) *virgin cankers* in which the mycelium has never yet been diploidised; and (2) *cankers that have produced aecidia in at least one growing season.*

(1) *Virgin cankers.* In simple isolated virgin cankers there is only one way in which the diploidisation process can be initiated, and that is through the mixing of the pycnidial nectar by insects and the fusion of pycnidiospores with flexuous hyphae of opposite sex. If the virgin canker is (+), then (−) pycnidiospores must be brought from one or more (−) cankers to the (+) flexuous hyphae of the (+) pycnidia; and if the virgin canker is (−), then (+) pycnidiospores must be brought from one or more (+) cankers to the (−) flexuous hyphae of the (−) pycnidia.

If in the first summer a virgin canker produces pycnidia diploidisation of the proto-aecidia is initiated by the flexuous hyphae receiving nuclei of opposite sex, the canker will produce aecidia and aecidiospores in the ensuing spring; but, otherwise, at this time, the canker will still be virgin and unfruitful. As we have seen, Lachmund observed that, while the majority of virgin cankers examined in British Columbia formed aecidia in the spring following the first production of pycnidia in the previous summer, yet in many cankers the formation of aecidia was delayed until the second spring after the production of pycnidia and, in a few cankers, the delay extended to the third or even the fourth spring. These delays may well be attributed to the failure of insects efficiently to mix the pycnidial nectar of the cankers until the year preceding the formation of the aecidia.

(2) *Cankers that have produced aecidia in at least one growing season.* Let us now consider the initiation of the diploidisation process in a simple canker that has already been diploidised once, twice, or more times and in consequence has produced aecidia in one year, two years, or several successive years. In such a canker, as is well shown in Fig. 100, A, in the summer there is a new zone of active pycnidia. Beneath these pycnidia new proto-aecidia (Fig. 100, B) are forming or are shortly to be formed.

The diploidisation of the proto-aecidia of our no-longer-virgin canker may be initiated: (1) as a result of insect visits, by the *mixing of the nectar* and fusions between the flexuous hyphae and pycnidio-spores of opposite sex as already described for virgin cankers; and (2), in the absence of insect visitors and consequent lack of contact between the flexuous hyphae and pycnidiospores of opposite sex, by *self-diploidisation, i.e.* by (+) nuclei in a (−) canker or (−) nuclei in a (+) canker travelling from the older parts of the canker, where aecidia were produced in the previous spring and where both (+) and (−) nuclei must be present, through the mycelium to the basal cells of the new proto-aecidia.

Self-diploidisation, as we have seen, takes place in the systemic mycelium of *Puccinia minussensis*; and it probably takes place also in *P. suaveolens*. It is on this basis that self-diploidisation has been suggested as a means of effecting diploidisation in the systemic myce-lium of old-established cankers of *Cronartium ribicola*. Up to the pre-sent, however, the suggestion has not been put to the test of experiment.

If self-diploidisation can take place in a simple canker that has already borne aecidia, it should take place regularly every year after that in which aecidia were first formed. Thus a simple canker that has once formed aecidia should, without any fresh mixing of the pycnidial nectar, produce aecidia annually.

Initiation of Diploidisation in a Compound Canker.—We may suppose that a young compound canker of *Cronartium ribicola* with two mycelial components of which one is (+) and the other (−) behaves like a corresponding compound pustule of *Puccinia graminis* on a Barberry leaf, *i.e.* the two mycelia meet, fuse, exchange nuclei, and so diploidise one another's proto-aecidia.

In a compound canker of the type just indicated about half of the pycnidia would be produced by the (+) mycelial component and about half by the (−) mycelial component. Insects visiting such a com-pound canker would readily mix the nectar of the (+) and (−) pycnidia and thus could provide for diploidisation of the proto-aecidia.

Thus, in a compound canker with a (+) component and a (−) component that has just produced its first pycnidia and proto-aecidia, there are two possible modes of diploidisation: (1) by the *fusion of the mycelia* and the exchange of nuclei; and (2) by the *receipt of pycnidio-spore nuclei of opposite sex via the flexuous hyphae*.

If a young canker of *Cronartium ribicola* with (+) and (−) com-ponents behaves sexually like a corresponding compound pustule of *Puccinia helianthi*,[1] then diploidisation of the first proto-aecidia is

---

[1] That in a compound pustule of *Puccinia helianthi* with (+) and (−) components diploidisation is initiated by mycelial fusions and not by the mixing of nectar has been proved conclusively by means of barrier experiments by A. M. Brown. *Vide supra*, Chapter VI.

initiated by mycelial fusion and exchange of nuclei and not by the mixing of nectar and fusions of pycnidiospores with flexuous hyphae of opposite sex.

In an older compound canker with (+) and (−) components that has already produced aecidia it is possible that the proto-aecidia of each of the component mycelia are diploidised by *self-diploidisation* in the manner already suggested for a simple canker that has already produced aecidia.

Cell-fusions in the Basal Cells of the Proto-aecidia.—Colley observed that the cells at the base of each proto-aecidium of *Cronartium ribicola* fuse together in pairs and not infrequently in greater numbers (Fig. 104), and he rightly interpreted these fusions as a step toward the formation of the binucleate aecidiosporophores (Fig. 105); but it is not necessary to infer from Colley's observations that the two nuclei which come together in each aecidiosporophore are nuclei which have been derived from the nucleus of a single basidiospore.

To elaborate the statement just made, let us suppose that a particular undiploidised proto-aecidium is a (+) structure, *i.e.* it has been produced on a (+) mycelium derived from a (+) basidiospore. Then all the nuclei in the basal cells of our proto-aecidium, prior to its diploidisation, are (+) nuclei. It can be imagined that the diploidisation of the basal cells of such a (+) proto-aecidium proceeds as follows. One or more (−) nuclei derived from the nucleus of a (−) pycnidiospore which has fused with a (+) flexuous hypha, or one or more (−) nuclei which have been derived from a (−) nucleus in an older part of the mycelium, travel through the (+) haploid mycelium to the base of the (+) proto-aecidium (Fig. 100, B) and enter the basal cells (Fig. 100, B, *fc*) and there seek for (+) mates required for the formation of conjugate pairs of nuclei. The passage of the invading (−) nuclei (which probably divide and so multiply their numbers as they move) from (+) cell to (+) cell is accomplished by these nuclei creeping through the central pores of the septa in the cell-chains and moving through the broad openings between adjacent (+) cells formed by the process of cell-fusion (Fig. 104). It is true that it is the original (+) cells which fuse together; but the process of fusion takes place in the (+) mycelium either (1) in the last stages of development of the mycelium in preparation for diploidisation should this take place at some future time, or (2) as the result of an urgent stimulus given to the cytoplasm of the (+) cells by the invading (−) nuclei after these have entered the mass of (+) basal cells and have begun to find (+) mates. With the microscope it is impossible to tell a (+) nucleus from a (−) nucleus and what in a microtome section of a (+) proto-aecidium a cytologist may suppose to be two (+) nuclei coming together from two fused (+) cells may be in reality, and very probably often are, a local (+) nucleus and an invading (−) nucleus. While in a (+) proto-aecidium becoming diploidised a (−) nucleus is attracted to a (+) nucleus, two unmated (+) nuclei which happen to be together in two (+) cells which have just fused may quickly move apart from one another. Thus a sorting-out process takes place and, in the end, there comes into existence a series of aecidiosporophores each of which contains a pair of conjugate nuclei of which one nucleus is (+) and the other (−).

Fromme,[1] in treating of the origin of aecidia in *Melampsora lini*, called attention to triple-cell fusions formed by the basal cells of the proto-aecidia and remarked that the cells below the "fertile layer" often contribute to the formation of multinucleate fusion cells. In the same species similar lateral fusions of basal cells, found after the nectar of (+) and (−) pustules had been mixed, were observed and illustrated by Ruth Allen.[2] In one of her illustrations, she shows four basal cells containing five nuclei, all fused together in one direction. It seems quite likely that some or all of these four cells were fused with other cells situated at right angles to the plane of section. There is even the possibility that the single layer of basal cells in *M. lini*, by means of lateral fusions, becomes converted into one more or less continuous network of intercommunicating elements. Through such a network invading nuclei of opposite sex, seeking conjugate mates, might freely pass.

The fusions in developing proto-aecidia of *Cronartium ribicola* are very similar to those described by Fromme and Miss Allen in *Melampsora lini*; for Colley[3] has stressed the fact that, in *Cronartium ribicola*,

---

[1]F. D. Fromme, "Sexual Fusions and Spore Development of the Flax Rust," *Bull. Torrey Bot. Club*, Vol. XXXIX, 1912, pp. 113-131.

[2]Ruth F. Allen, "A Cytological Study of Heterothallism in Flax Rust," *Journ. Agric. Research*, Vol. XLIX, 1934, pp. 782-783. Miss Allen, on the basis of cytological evidence, mistakenly supposed that the pycnidiospores of *Melampsora lini* germinate, the germ-tubes pass into the leaf and there give rise to a mycelium, and this mycelium takes part in forming the young aecidia; and she further supposed that "following spermatisation each cell fusion in the aecium is between a hypha of sporidial and a hypha of spermatial origin." However, Craigie's observations have taught us that the pycnidiospores of the Rust Fungi do not germinate but fuse with flexuous hyphae of opposite sex (*vide* this volume, Chapter I). We are now sure that the nuclei of pycnidiospores of *Puccinia helianthi*, *P. graminis*, *Gymnosporangium clavipes*, etc., find their way to the proto-aecidia by migration through flexuous hyphae (*vide* Chapters III and IV), and there is every reason to suppose that a similar migration takes place through the flexuous hyphae of *Melampsora lini*. Miss Allen's supposition that the fusions in *M. lini* take place between (+) and (−) hyphae must therefore be rejected.   [3]R. H. Colley, *loc. cit.*, p. 646.

the basal cells of the proto-aecidia fuse not only in pairs but also in greater numbers. Said Colley: "It is quite certain that the multiple fusions observed are regular occurrences in aecia of all sizes and shapes whether on roots or stems." The formation of multiple fusions between basal cells in a proto-aecidium of a heterothallic Rust Fungus during the maturation of the proto-aecidium or at the time the proto-aecidium is becoming diploidised is to be regarded as a normal physiological process which facilitates the passage of invading nuclei from one basal cell to another and thus in the end facilitates the formation of pairs of conjugate nuclei in the young aecidiosporophores.

Conclusion.—The theoretical solution of the problem of the initiation of the sexual process in a canker of *Cronartium ribicola*, involving as it does a re-interpretation of Colley's observations on cell-fusions at the base of a proto-aecidium undergoing diploidisation, serves to harmonise the work of a cytologist with the experimental work of Craigie on the initiation of the sexual process in Puccinia and also with Lachmund's field observations and Pierson's experiments on *C. ribicola* itself.

CHAPTER X

PUCCINIA SUAVEOLENS AND ITS SEXUAL PROCESS

Introduction — The Name *Puccinia suaveolens* — Host-plant — Can *Puccinia suaveolens* parasitise the Dandelion? — Geographical Distribution of Host and Parasite — Cause of Absence of *Puccinia suaveolens* from Central Canada and its Artificial Introduction — Growth and Reproduction of *Cirsium arvense* — The Effect of *Puccinia suaveolens* on the Structure and Physiology of its Host — Economic Value of *Puccinia suaveolens* — Effect of External Conditions on the Production of Secondary Uredospores and Teleutospores — Source of Fungus Material — Primary and Secondary Uredospores — Localised and Systemic Mycelia — Forms of Localised Mycelia — Forms of Systemic Mycelia — Localised Mycelia derived from Uredospores — Localised Mycelia derived from Basidiospores — Establishment of Systemic Mycelia in *Endophyllum euphorbiae-sylvaticae*, *Aecidium leucospermum*, and *Puccinia minussensis* in Herbaceous Perennials — Establishment of Systemic Mycelia of *Puccinia suaveolens* in *Cirsium arvense* Seedlings — Experiments with Root-buds — Systemic Mycelium in Thistle Roots — Binucleate Condition of the Systemic Mycelium in Stems and Roots — Systemic Mycelia in Leaves and the Production of Pycnidia, Primary Uredospores, and Teleutospores — De-diploidisation, Re-diploidisation, Self-diploidisation, and Cross-diploidisation — Biological Significance of Pycnidia on Systemic Mycelia — Systemic Mycelia bearing Secondary Uredospores and Teleutospores only — Systemic Mycelia bearing Pycnidia only — *Puccinia suaveolens* is Heterothallic — Experiments proving that Systemic Mycelia may arise from Uredospores under Natural Conditions — Uredospores and the Presistence of *Puccinia suaveolens*.

Introduction.—To field mycologists in Europe *Puccinia suaveolens* is one of the best-known Rusts; for it commonly occurs on the Creeping Thistle, *Cirsium arvense*, in pastures and waste places. Its mycelium, when systemic, so alters the appearance of the shoots of the host that infected thistles can be easily distinguished from healthy ones at a distance (Fig. 106). Moreover, the very numerous pycnidia of the fungus give out so strong a scent that a group of infected thistles in a pasture may be detected by their odour before one has observed them with the eye.

The Creeping-Thistle Rust produces pycnidia, uredospore sori, and teleutospore sori, but no aecidia. It is therefore a *brachy*-form.

FIG. 106.—Young shoots of *Cirsium arvense*: healthy on the left, containing systemic mycelium and bearing pycnidia of *Puccinia suaveolens* on the right. Two thirds natural size.

In accordance with Jackson's theory[1] of the origin of *brachy*-forms in general, we may suppose that *Puccinia suaveolens* has been derived from a *eu*-form by a dropping-out of aecidia from the life-history and by the proto-aecidia developing directly into a sorus of uredospores. Thus the proto-aecidia have become what we may now call *proto-uredospore sori* or, following Arthur's terminology, *proto-uredinia*.

[1]H. S. Jackson, "Present Evolutionary Tendencies and the Origin of Life Cycles in the Uredinales," *Memoirs Torrey Bot. Club*, Vol. XVIII, 1931, pp. 74-77.

In North America, according to Jackson,[1] there are about 139 Rust species that are *brachy*-forms, and of these there are included: on Rosales, 11; on Leguminosae, 68; and on Asterales (Compositae), 21. Most of these *brachy*-forms "develop a haplont of limited growth."[2] The others, of which *Puccinia suaveolens* is a typical example, have a systemic mycelium whose haploid branches bear pycnidia and proto-uredospore sori in great numbers on every leaf of an infected host-shoot.

Puccinia suaveolens was investigated cytologically by Olive (1913) and Kursanov (1922); its effect on its host was worked out by Czechinsky (1929?); and its secondary uredospore pustules were obtained by de Bary (1863), Rostrup (1874), and others by sowing primary uredospores on host leaves. In what follows the findings of all these workers will be reviewed.

In 1913, Grove[3] suggested that the systemic infections of new shoots of *Cirsium arvense* seen in the spring are *new* infections initiated by mycelium derived from basidiospores borne on basidia produced by overwintered teleutospores; and, in keeping with this idea, he held that "the hibernation of the mycelium in the rhizome, which is stated by Plowright,[4] has not been proved." Also, as late as 1931, Cunningham[5] said: "Opinions differ as to whether the rust overwinters by means of a systemic mycelium in the rootstocks (Rostrup) or whether it is perpetuated by overwintering teleutospores."

The systemic infections of new thistle shoots do not necessarily all arise as a result of the germination of *basidiospores*, as Grove seems to have supposed; for, as will shortly be shown, systemic infections can be established as a result of the sowing of *uredospores*. Preliminary announcements of this discovery were made by Buller and Brown[6] in 1941 and 1943.

The fact that the mycelium of *Puccinia suaveolens* does actually hibernate in thistle roots was clearly indicated by the field-work of Rostrup (1874) and Kursanov (1922), and by the cytological observa-

[1]*Ibid.*, pp. 86-87.   [2]*Ibid.*, p. 75.

[3]W. B. Grove, *The British Rust Fungi*, Cambridge, 1913, pp. 145-146.

[4]C. B. Plowright, *A Monograph of the British Uredineae and Ustilagineae*, London, 1889, p. 183. Plowright cites Rostrup as his authority for an account of the life-history of *P. suaveolens*.

[5]G. H. Cunningham, *The Rust Fungi of New Zealand*, printed privately by John McIndoe, Dunedin, N.Z., 1931, p. 187.

[6]A. H. R. Buller and A. M. Brown: "Urediospores as the Origin of Systemic Mycelia in *Puccinia suaveolens*," abstract, *Phytopathology*, Vol. XXXI, 1941, p. 4; and "De-diploidisation of the Mycelium of *Puccinia suaveolens*," abstract, *Trans. Roy. Soc. Canada*, Vol. XXXVII, 1943, *Proceedings*, Appendix C, p. 127.

tions of Olive (1913);[1] and it has now been proved conclusively by means of greenhouse experiments and field observations made by A. M. Brown and myself at Winnipeg.[2]

With a view to giving a more complete account of the life-history and sexuality of *Puccinia suaveolens* than has hitherto been possible the writer, in association with A. M. Brown, has undertaken an experimental investigation in the Dominion Rust Research Laboratory. It has been sought: (1) to develop in the greenhouse all the various states of *P. suaveolens* on its host; (2) to solve the problem of the origin of the systemic mycelium; and (3) to determine whether the fungus is homothallic or heterothallic.

The Name Puccinia suaveolens.—The Creeping-Thistle Rust is often referred to as *Puccinia obtegens*; but, as will now be explained, this designation is erroneous.

The first description of this fungus was given by Persoon[3] in 1799. He noted its peculiar scent and named it *Uredo suaveolens*. In 1816, Link[4] re-named the fungus *Uredo obtegens*.[5] He cited Persoon's name "*Ur. suaveolens* Pers." and explained why he had rejected it and replaced it by another name. Said he:[6] "Although during the past year I have seen the fungus in great abundance, yet I detected no odour in it. Hence I have changed a name which might cause doubt."

Link's change of Persoon's specific epithet was not justified, not only as regards priority, but also physiologically; for mycologists who have smelled *Puccinia suaveolens* in the pycnidial stage in the field all agree that the fungus gives out a strong odour and that *suaveolens* is a better descriptive epithet than *obtegens*. Moreover, between 1816 and 1825, Link himself must have detected the odour; for, in 1825, in a new edition of the *Species Plantarum*, he[7] states that the odour of *P. suaveolens* is "potius nauseosum quam suavem." Before and up to the year 1816, he may have smelled the infected host-leaves too late in the year, after the scented pycnidial nectar had dried up, the

---

[1]For a citation of the papers of Rostrup, Kursanov, and Olive, *vide infra*.

[2]*Vide infra* in the Section *Systemic Mycelium in Thistle Roots.*

[3]C. H. Persoon, *Observationes mycologicae*, Pars II, 1799, p. 24. He also used the name *U. suaveolens* in his *Synopsis methodica fungorum*, Gottingae, 1801, p. 221, at the start of the present nomenclature of the Uredinales.

[4]H. F. Link, "Observationes in Ordines plantarum naturales Diss. II," *Magaz. d. Ges. naturf. Freunde zu Berlin*, Vol. VII, 1816, p. 27.

[5]*Suaveolens* means *sweet-smelling* and *obtegens* means *covering* (presumably referring to the fact that the fungus covers the surface of the host-leaves with pycnidia and uredospore sori).

[6]H. F. Link, *loc. cit.*, pp. 27-28.

[7]H. F. Link, *Linné Species Plantarum*, Ed. 4, Berolini, 1825, p. 221.

pycnidia had degenerated, and the odourless uredospore and teleutospore pustules had developed.[1]

Most of the older uredinologists, *e.g.* Rostrup,[2] Jacky, Karsten,
Winter, De Toni, Oudemans, Plowright, Massalongo, and Magnus,
used Persoon's specific epithet in combination with the generic name
Puccinia thus: *Puccinia suaveolens* (Pers.) Rostr. Unfortunately,
however, in 1904, the Sydows,[3] in the first volume of their great
Monographia Uredinearum, cited *Puccinia suaveolens* as *Puccinia
obtegens* (Lk.) Tul. and added as a synonym, "*Caeoma obtegens* Lk.
Obs. II, p. 27 (1791)." They also remarked that "recently this species
very incorrectly has been designated *P. suaveolens* (Pers.) Rostr.
Nevertheless, priority belongs to the name *Puccinia obtegens* (Lk.) Tul."

In 1903, Magnus[4] pointed out that the Sydows had made a serious
error in supposing that Link[5] in his *Observationes mycologicae* of 1791
had published the name *Caeoma obtegens*. At that time Link made
no mention of this name; and, as we have seen, he named the fungus
for the first time in 1816. Thus it became clear that Persoon's specific
epithet *suaveolens* preceded Link's specific epithet *obtegens* by seventeen
years. In 1904, the Sydows[6] in an *Appendix* to their Volume I admitted their error and explained that it had been made when Link's
writings had been inaccessible. Grove,[7] in 1913, in his *British Rust
Fungi* (perhaps having been influenced by the Sydows' original terminology), unfortunately introduced the erroneous name *Puccinia obtegens*
into usage in England.[8]

From the above discussion it is clear that mycologists should give
up the name *Puccinia obtegens* and return once more to the use of the
much pleasanter Persoonian name *Puccinia suaveolens* (Pers.) Rostr.

Host-plant.—*Puccinia suaveolens* as a parasite is a pronounced
specialist; for, apparently, it lives on a single host, *Cirsium arvense
(Carduus arvensis)*, the Creeping Thistle.

[1]Secondary uredospore pustules developed on a local mycelium derived from one
or more uredospores are just as odourless as primary uredospore pustules developed
on a systemic mycelium.

[2]F. G. E. Rostrup in 1874 (*Bot. Zeit.*, XXXII, p. 556) was the first to use the
combination *Puccinia suaveolens* and he said: "Tulasne named *Uredo suaveolens
Puccinia obtegens* Lk. (in Fuckel *Puccinia obtegens* Tul.) which I cannot approve of."

[3]P. and H. Sydow, *Monographia Uredinearum*, Lipsiae, Vol. I, 1904, pp. 53-54.

[4]P. Magnus, "Bemerkungen zur Benennung einiger Uredineen in P. and H.
Sydows Monographia Uredinearum," *Hedwigia*, Bd. XLII, 1903, *Beiblatt*, pp. 305-306,

[5]H. F. Link, in "Botanische Bemerkungen" in his *Annalen der Naturgeschichte*,
one issue only, published at Gottingen, 1791.

[6]P. and H. Sydow, *loc. cit.*, pp. 855-856.

[7]W. B. Grove, *The British Rust Fungi*, Cambridge, 1913, p. 145.

[8]J. C. Arthur uses the name *Puccinia obtegens* in his *Manual of the Rusts in United
States and Canada*, but his reasons for doing so are not clear.

Can Puccinia suaveolens parasitise the Dandelion?—In their works on the Uredinales the Sydows,[1] Grove,[2] and Arthur[3] all cite *Cirsium arvense* as the sole host of *Puccinia suaveolens* and they make no mention of the fact that de Bary believed that he had been able to transfer *P. suaveolens* from the Creeping Thistle to the Dandelion.

In 1863, in his classic paper called *Recherches sur le développement de quelques champignons parasites* in which he laid the foundation for his future discoveries concerning the life-history of *Puccinia graminis*, de Bary[4] recorded the results of his experiments with the uredospores of *P. suaveolens*. He sowed uredospores obtained from *Cirsium arvense* on the leaves of healthy plants of *Cirsium arvense, Taraxacum officinale, Tragopogon pratensis*, and *Tragopogon porrifolius*. The uredospores readily germinated, and he observed that the germ-tubes entered the stomata of the leaves of all the four Compositae. About two weeks later rust pustules appeared on *Cirsium arvense* and *Taraxacum officinale*, but not on the two species of Tragopogon. On both *Cirsium arvense* and *Taraxicum officinale* uredospores were produced first and then teleutospores, but these spore-forms were never accompanied by pycnidia.

From the results of de Bary's experiments it would appear that *Puccinia suaveolens* is able to parasitise not only *Cirsium arvense* but also *Taraxacum officinale*.

At Winnipeg, employing uredospores of an English strain of *Puccinia suaveolens*, Brown and I have endeavoured to repeat de Bary's experiment, but the outcome has been entirely negative. We could readily transfer the fungus from Creeping Thistle to Creeping Thistle but, in spite of three separate trials, never from a Creeping Thistle to a Dandelion.

De Bary, as everyone recognises, was a first-class experimenter and observer; and it is impossible to prove that he made a mistake. Can it be that his Dandelions somehow or other became infected with the uredospores of *Puccinia hieracii* (*P. taraxaci*), a Rust that parasitises *Taraxacum officinale* but not *Cirsium arvense*? Or did de Bary employ a strain of *Puccinia suaveolens* different from that used by Brown and myself, and really transfer the fungus from a Creeping Thistle to a Dandelion? So far as I know, there is no evidence that would indicate that, where Dandelions and Creeping Thistles are

[1]P. and H. Sydow, *loc. cit.*, pp. 53-54.

[2]W. B. Grove, *loc. cit.*, p. 145.

[3]J. C. Arthur, *Manual of the Rusts in United States and Canada*, Lafayette, U.S.A., 1934, p. 347.

[4]A. de Bary, "Recherches sur le développement de quelques champignons parasites," *Ann. Sci. Nat.*, Bot., Sér. 4, T. XX, 1863, pp. 84-85, 89, 92, Pl. XI, Figs. 11 and 12.

growing together in the open, *P. suaveolens* passes freely from one species to the other. This being so and in view of the failure of Brown and myself to repeat de Bary's experiment successfully, it would seem best, for the present at least, to conclude that *P. suaveolens* does not parasitise *Taraxacum officinale* and has but a single host, namely, *Cirsium arvense*.

Geographical Distribution of Host and Parasite.—*Cirsium arvense* is indigenous throughout Europe, Asia, and Northern Africa, and it has spread across Siberia, through China, to Japan. It has become a naturalised plant in South Africa (Natal, Transvaal, eastern Cape Province); and it became naturalised in the Province of Victoria, Australia, about 1887. It was introduced into Canada with impure farm seeds; and, in the State of Vermont, U.S.A., it was officially recognised by law as early as 1795.[1]

Puccinia suaveolens is known to be present not only in Europe, including the British Isles, but also in North America where it must be considered as an introduced species, India, China, Japan, and New Zealand.[2] The tendency has been for *P. suaveolens* to follow its host wherever its host has gone.

The introduction of *Puccinia suaveolens* into North America and New Zealand was probably through human agency: by means of uredospores or teleutospores attached to stems and leaves of *Cirsium arvense* included in packing straw, etc.

In the United States of America and in Canada *Cirsium arvense*, although originally introduced from Europe, is known as the *Canada Thistle*. On this weed, in the U.S.A., according to Arthur,[3] *Puccinia suaveolens* has been found, from east to west, in the following States: Maine, Vermont, Massachusetts, Rhode Island, New Jersey, New York, Pennsylvania, Ohio, Michigan, Wisconsin, South Dakota, Montana, Utah, Washington, and California.

In Canada, according to Arthur,[4] *Puccinia suaveolens* has been recorded in Nova Scotia,[5] Ontario, and Quebec. The parasite also occurs in British Columbia, but, so far, it has never been found in the prairie regions of Manitoba and Saskatchewan.

Freshly gathered specimens of *Puccinia suaveolens* on *Cirsium*

---

[1] Ada Hayden, "Distribution and Reproduction of Canada Thistle in Iowa," *American Journal of Botany*, Vol. XXI, 1934, p. 345.

[2] J. C. Arthur, *Manual of the Rusts in United States and Canada*, Lafayette, U.S.A., 1934, p. 347.

[3] and [4] *Ibid.*

[5] W. P. Fraser ("The Rusts of Nova Scotia," *Trans. Nova Scotian Inst. of Sci.*, Vol. XII, 1913, p. 396) states that in Nova Scotia *Puccinia suaveolens* is common and widely distributed.

arvense have been sent: to A. M. Brown from Victoria in British Columbia (July, 1932); and to me by Ivan H. Crowell from Macdonald College in the Province of Quebec (Sept. 24, 1940). Near London in the Province of Ontario, on June 11, 1943, whilst on a botanical excursion with Dr. John Dearness, I found shoots of *C. arvense* infected with the systemic mycelium of *Puccinia suaveolens* in several places in a low-lying thistley pasture and in two patches of thistles growing on the banks of a small stream.

Cause of Absence of Puccinia suaveolens from Central Canada and its Artificial Introduction.—In view of the fact that *Puccinia suaveolens* has been recorded from such north-central States as Michigan, Wisconsin, South Dakota, and Montana, it is of interest to enquire why this parasite should be absent from the Provinces of Manitoba and Saskatchewan.

It has been suggested by Bisby[1] that the absence of *Puccinia suaveolens* from the prairie region of Canada may be due to climatic factors. In the laboratory it has been found that *P. glumarum*[2] and *Uromyces betae*,[3] both of which are absent from Manitoba and eastern Saskatchewan, cease to develop above certain temperatures that are still favourable to their host-plants; and it may therefore be argued that these parasites are cool-summer fungi and, on this account, are absent from central Canada. Whether or not this deduction is correct, three field experiments, now to be recorded, seem to indicate that the absence of *Puccinia suaveolens* from the prairie region of Canada is not due to climate.

(1) Some leaves on two thistles, *Cirsium arvense*, growing in a patch of thistles along one side of a greenhouse of the Dominion Rust Research Laboratory were inoculated with primary uredospores of *Puccinia suaveolens* in the summer of 1940, with the result that second-

---

[1] G. R. Bisby with the collaboration of A. H. R. Buller, John Dearness, W. P. Fraser and R. C. Russell, *The Fungi of Manitoba and Saskatchewan*, published by the National Research Council of Canada, Ottawa, 1938, p. 18.

[2] Margaret Newton and T. Johnson, "Stripe Rust, *Puccinia glumarum*, in Canada," *Canadian Journal of Research*, Vol. XIV, 1936, pp. 89-108. The authors found that: "*P. glumarum* is extremely sensitive to environmental conditions, particularly temperature. The optimum for uredospore germination is 10° to 12° C., and for rust development 13° to 16° C. Varieties (of wheat) susceptible at from 10° to 16° C. developed resistance at higher temperatures, becoming extremely resistant at 25° C. On account of the sensitiveness of this rust to high temperatures it seems improbable that it will ever become thoroughly established in Manitoba and Saskatchewan, as in these two provinces the summer temperature is probably too high to permit its development."

[3] Margaret Newton and B. Peturson, "*Uromyces betae* in Western Canada," 1943, as yet unpublished.

ary uredospores and teleutospores soon developed on the hosts in the usual manner.   In the spring, in 1941, 1942, and 1943, after the thistles in the patch had sent up their aerial shoots, it was observed that some of the shoots were infected with the systemic mycelium of *P. suaveolens* and had given rise to pycnidia and primary uredospores.   Moreover, it was found that the parasite had spread to new host-plants.   In the spring of 1943, on May 12, out of some 400-500 thistle shoots in the patch 35 were found infected with the systemic mycelium of the parasite and to be producing pycnidia and primary uredospores, and it was observed that the infected shoots were scattered along the side of the greenhouse for a distance of 25 feet.   In the autumn of 1943, on October 2, a further examination of the patch of thistles revealed that all of the shoots that had been free from the fungus in the spring had become infected and now bore pustules of secondary uredospores or of teleutospores, thus indicating that *P. suaveolens* by means of uredo-spores may spread throughout the summer in central Canada just as it does in England and on the continent of Europe.

(2) Similar results were obtained in a second patch of thistles growing in the Laboratory garden: infected plants came up in 1941 and 1942; but, owing to weeding by the gardener, no infected plants appeared in the patch in 1943.

(3) A third, longer established, much larger, and more vigorous patch of thistles was growing among grass by the side of a fence about one hundred yards from the Rust Laboratory.   In the summer of 1941, a single infected thistle-shoot with the leaves bearing an abundance of primary uredospores was removed from the first patch of thistles just outside the greenhouse and was thrown into the patch of thistles by the fence, with the result that infection took place.   In the spring, in 1942 and 1943, in this third patch of host-plants, a number of the new shoots came up characteristically altered in their appearance and mode of growth by the systemic mycelium of *Puccinia suaveolens* and with their leaves bearing in succession pycnidia and primary uredospores.   In the summer of 1943, owing to the construction of a road, the patch of thistles was almost destroyed, so that no further observations could be made upon it.

The results of the three experiments just described clearly indicate that, in Manitoba, when *Puccinia suaveolens* has once been introduced into a patch of thistles, it can spread from the shoots of diseased plants to the shoots of healthy plants and also, by means of its perennial systemic mycelium, can maintain itself on some of its hosts from year to year.   We may therefore conclude that climate is not responsible for the absence of *P. suaveolens* from the prairie region of central Canada.[1]

The Selkirk Settlers did not arrive in what is now Manitoba until 1812, and the development of agriculture on the prairies of western Canada on a large scale, followed by the gradual spread of the Creeping Thistle as a farm weed, is a comparatively recent historical development which took place subsequent to 1886. In that year on Dominion Day, July 1, on the newly constructed transcontinental Canadian Pacific Railway, there passed through Winnipeg the first through train from Montreal to Vancouver.[2] With the opening of the C.P.R. railway, overseas markets became accessible to western farmers and, thereafter, millions of acres of virgin prairie-land were turned with the plow. Thus favourable seed-beds were prepared not only for cereals, but also for farm weeds, among them the Creeping Thistle. This pest had been introduced into Manitoba long before 1886, but subsequently it spread, slowly but surely, far and wide, on to farm-land and railway tracks and along the sides of roads and ditches.

Batho,[3] in 1937, in treating of the history of the Creeping Thistle in Manitoba, said: "The spread of the Canada Thistle in Manitoba has been very steady and gradual. As early as 1884 an official publication of the Manitoba Department of Agriculture, speaking of these thistles, said they 'abound.' Still at that time, and for many years afterward, Canada Thistles could not be said to be distributed in a widespread way throughout Manitoba. The fact is that though they appeared early in the agricultural history of the Province, the seed probably being somehow brought in in the early seed grain and feed, they were not very widespread for quite a long time afterwards." Today, dense patches of thistles can be seen on many Manitoban farms, and photographs of two such patches, one taken in the "park country" about 60 miles north-west of Brandon and the other taken in the prairie country south-east of Brandon, are reproduced by Batho in his publication.

In view of the results of the field experiments already recorded and the history of the development of the prairie region of Canada, it seems

[1][A. M. Brown, in a letter of Aug. 1947, reported that by 1946 the rust had spread some thirty feet from the point of infection in the one remaining clump of thistles of the original three that were inoculated. In the fall of 1946 this clump also had to be destroyed.]

[2]A. H. Reginald Buller, *Essays on Wheat*, New York, 1919, pp. 1-34.

[3]George Batho, *The Canada Thistle*, Circular No. 104, ed. 2, 1937, Manitoba Dept. of Agric. and Immigration.

probable that the absence of *Puccinia suaveolens* from central Canada hitherto has been due, not to climatic factors, but to the parasite not yet having had time to follow its host on its migration to its extreme range in the north-central region of the North American continent.

Growth and Reproduction of Cirsium arvense.—As a preliminary to a study of the life-history of *Puccinia suaveolens* it is necessary to know the chief facts concerning the growth and reproduction of its host. These have been elucidated by Miss Hayden[1] and others.

Cirsium arvense, like *Lactuca pulchella* mentioned previously, is a herbaceous perennial. When winter comes, the aerial shoots die away, leaving behind, not a rhizome as with *L. pulchella*, but a series of horizontal and vertical *roots* which persist through the winter. In the following spring, the horizontal roots give rise to adventitious buds which grow upwards through the soil and form new green shoots. By means of its perennial root system the Creeping Thistle gradually extends its area of occupation. The vertical roots, under favourable conditions, may extend downwards in the soil seven or more feet, and a depth of eighteen feet has been recorded by Malvez[2] (1931) in the black earth of the cotton regions of Russia.

The adventitious buds arise on the horizontal roots at a distance of a few inches to a foot or more beneath the surface of the ground. The subterranean part of the stem of an aerial shoot bears scale leaves, axillary buds, and roots at the nodes, and it may be considered as a *vertical rhizome.* If a green shoot is cut off at or just below the surface of the ground, one or more of the axillary buds on the rhizome soon become active and form new aerial shoots. If an aerial shoot bearing the systemic mycelium of *Puccinia suaveolens* is severed at or just below the surface of the ground, as a rule the new aerial shoot or shoots formed from the axillary buds of the rhizome all bear leaves covered with pycnidia.

Cirsium arvense is dioecious. Some plants bear staminate florets only and other plants pistillate florets only. Staminate and pistillate plants grow to about the same height, but they can be readily distinguished in the field by differences in the appearance of their capitula.

The receptacle of a capitulum bears about 100 tubular reddish-purple (rarely white) florets and numerous soft brownish bristles or hairs; and the florets and the hairs are surrounded by a green ovoid involucre made up of many imbricating bracts. The differences

[1]Ada Hayden, "Distribution and Reproduction of Canada Thistle in Iowa," *American Journal of Botany*, Vol. XXI, 1934, pp. 355-373.

[2]*Ibid.*, p. 370.

between staminate and pistillate capitula, as observed by myself at Winnipeg, may be summed up as follows.

In a capitulum of a *staminate* plant: (1) the involucre is relatively short (12-14 mm. in length); (2) the five corolla-lobes of each floret are relatively long (5 mm.) and broad and, as the floret opens, they become widely divergent; (3) the filaments (1 mm. long) and anthers (4 mm. long) are well-developed, pink, and together are of the same length as the corolla-lobes; (4) the end of the style is cylindrical and not bifid, so that there is no trace of two stigmas; (5) as the style elongates, its brush pushes out from the staminal tube an abundance of pollen which clings to the style until it is removed by insects; (6) the florets do not all open simultaneously but in succession in four age-groups on four successive mornings, the oldest group at the periphery of the capitulum opening on the first morning and the youngest group in the centre of the capitulum opening on the fourth morning; and (7) the pappus of an old floret does not protrude beyond the top of the involucre, and an old dead unopened capitulum is crowned not by a brush of pappus but by spreading pale-brown corolla-lobes and staminal tubes.

On the other hand, in a capitulum of a *pistillate* plant: (1) the involucre is relatively long (15-20 mm. in length); (2) the five corolla-lobes of each floret are relatively short (3 mm.) and narrow and, as the floret opens, they diverge but slightly: (3) the anthers are blackish and abortive, and the filaments and anthers together are only about 1 mm. in length; (4) the end of the style is divided into two stigmatic lobes; (5) as there is no pollen, the style, during its elongation, does not push any pollen out of the staminal tube; (6) the florets of a capitulum all open simultaneously on a single morning; (7) an old capitulum, before opening to allow the plumose fruits to escape, is crowned by a brush of brown pappus that protrudes 5-8 mm. beyond the top of the involucre and conceals the dead corollas and styles of the florets.

As we have seen, both the staminate and the pistillate florets possess stamens and a pistil; but the staminate flowers never develop seeds while the pistillate florets, owing to the abortion of the stamens, never produce any pollen grains. Therefore, staminate plants are strictly male in function and pistillate plants strictly female in function.

Staminate plant colonies, whether mixed with pistillate colonies or not, never develop any seeds. Isolated pistillate colonies also bear no seed; but, when pistillate colonies are near to or are mixed with staminate colonies, the pistillate colonies bear abundant seed. This is due to the fact that, for the production of seed, it is necessary for insects to transfer pollen from the flowers of staminate colonies to those of pistillate colonies.

Miss Hayden,[1] in Iowa, found that about 50 per cent of the flowers in the heads of pistillate plants bore seed, and that the viability of these seeds, when six months old, varied from 10 to 27 per cent. Some freshly gathered seeds collected in August gave 95 per cent. germination.

At Winnipeg, in mixed colonies, the pistillate colonies produce numerous seeds many of which are viable. Seeds collected in October, 1939, produced but few seedlings at the time of collection and a much higher percentage of seedlings five months later (in March). Seeds collected in July, 1940, and at once set on wet blotting paper showed about 66 per cent. germination.

When a Creeping-Thistle seed germinates, the seedling emerges from the earth by elongating its hypocotyl and thrusting its two cotyledons from the soil into the air. The cotyledons are at first appressed and covered with the pericarp and testa; but soon they rid themselves of these protective coverings, separate from one another, and become exposed to the sun. The radicle develops into a slender tap-root which gives off fibrous branches, and on the tap-root adventitious buds arise which develop into aerial shoots. The main vertical root produces secondary roots which grow out horizontally and sometimes arch downward or upward. On these horizontal roots arise adventitious buds. These develop into shoots that push their way up to the surface of the soil and there form rosettes. Thus from a single seed there may develop a seedling and then a colony of staminate or pistillate plants. Should a seedling become infected with the systemic mycelium of *Puccinia suaveolens*, there is the possibility that, *via* the horizontal root-system and the adventitious buds, the fungus may spread to some, or many, or all of the green shoots of the colony derived from the seedling.

The Effect of Puccinia suaveolens on the Structure and Physiology of its Host.—In 1889, Plowright[2] remarked that *Cirsium arvense* plants bearing the systemic mycelium of *Puccinia suaveolens* "appear sooner than the healthy ones, have a sickly pale-green colour, and do not bear flowers."

While it is true that the systemic mycelium of *Puccinia suaveolens* does usually prevent a host-thistle from flowering, there are occasional exceptions to this rule and then the mycelium makes its way into the capitula and florets.

[1]Ada Hayden, *loc. cit.*, p. 365.
[2]C. B. Plowright, *A Monograph of the British Uredineae and Ustilagineae*, London, 1889, p. 183.

About 1929, Czechinsky,[1] one of Kursanov's pupils, gave a detailed account of the effect of the Creeping-Thistle Rust upon its host, and his findings will now be summarised.

At first glance, one can see that a sick thistle as compared with a sound one, shows a general depression in growth and development. A healthy thistle has foliage more tufted, its internodes are shorter, its leaves are larger and characteristically lobed, and sometimes their surface area is triple that of the leaves of a sick plant. The lamina of a leaf of a sick plant is usually more or less entire, rarely with slight teeth.

The change in the form of the leaf, which from compound-lobed becomes entire, and the diminution of the surface of the lamina are so characteristic for plants invaded by the systemic mycelium of *Puccinia suaveolens* that these plants can be distinguished immediately in the midst of a thick growth of healthy plants.

Usually, very few of the plants attacked by the fungus attain to full flowering and seeds are formed but rarely. The seeds actually gathered were all empty and attempts to germinate them failed.

In a rusted thistle bearing flowers, every part of each capitulum, including the florets, are traversed by the systemic mycelium; but on a capitulum the fungus fructifies almost exclusively on the exterior parts. Sometimes tiny teleutospore pustules are formed on the petals of the florets. The hyphae of the parasite have been found in the staminal filaments as well as in the anthers where the pollen has an abnormal appearance and, apparently, is not always completely developed. The presence of the fungus has also been demonstrated in the ovule, in the integument, and sometimes in the immediate neighbourhood of the nucleus of the embryo-sac. Not infrequently ovules that had been already fertilised were strongly infected with the fungus. Such ovules soon cease their development. The fungus always penetrates into the capitula and into the ovaries of the florets from below, so that each floret becomes infected in the earliest stage of its development. However, in a capitulum a few florets may escape infection. The mycelium found in infected capitula was always binucleate.

Czechinsky reported that, in 1926, Kursanov[2] had made experi-

[1] N. I. Czechinsky, "De l'influence du champignon *Puccinia suaveolens* sur la structure anatomique de son hôte," a reprint from an unspecified journal, undated (?1929), pp. 101-138, in French with a Russian summary.

[2] A. L. Kursanov: "De l'influence de l'*Ustilago tritici* sur la respiration du froment," en russe, *Journal Morbi plantarum*, Courrier de la section de Phytopath. du Jardin Bot., 1926 (cited from Czechinsky, *loc. cit.*, p. 106); and "De l'influence de l'*Ustilago tritici* sur les fonctions physiologiques du froment," *Revue Generale de Bot.*, T. IV, 1928, pp. 277-302, 343-371. He records his experiments with *Puccinia suaveolens* on pp. 301-302.

ments on *Cirsium arvense* which showed: that the respiration of sick plants is more energetic than that of healthy plants; and, in particular, that plants with the mycelium in the pycnidial stage respire more actively than plants with the mycelium in the uredospore stage.

Czechinsky found that: the osmotic pressure is weaker in infected plants than in healthy plants; and, in particular, that the osmotic pressure is weaker in plants attacked by *Puccinia suaveolens* in the uredospore stage than in plants attacked in the pycnidial stage.

Sick plants wither more quickly and more easily than do healthy ones.

At first the development of a sick plant, including the maturing of its tissues, is more rapid than that of a healthy plant; but a sick plant ceases to grow sooner than a healthy one.

So far as anatomical developmental differences are concerned, Czechinsky found that in the *stem* of a sick plant, as compared with the stem of a healthy plant:

(1) There is relatively less conducting tissue in all stages of development; (2) the sclerenchymatous tissue exterior and interior to a vascular bundle is less well developed, especially on the interior side of the bundle; but the mechanical tissue around the vascular bundles is formed earlier; (3) the wood vessels and the woody sclerenchymatous fibres are formed earlier; (4) the protoxylem is less well developed; (5) the collenchyma is much less well developed; (6) the central cavity of the stem is formed sooner and develops much more energetically; (7) the cells of the epidermis and of the cortical parenchyma are somewhat enlarged, and the cells of the parenchyma of the pith become very elongated (in the direction of the long axis of the stem); (8) the osmotic pressure in the parenchyma of the cortex and the pith is lower.

Czechinsky also found that in a *leaf* of a sick plant, as compared with the leaf of a healthy plant:

(1) There is an increase in the total number of stomata per unit of leaf area, and this is due to the formation of "supplementary" stomata in the upper epidermis; (2) there is a considerable increase in the dimensions of the guard-cells of the stomata in the upper epidermis; (3) the height of the palisade cells is greater in the upper leaves and much less in the lower leaves by comparison with corresponding leaves of healthy plants; (4) the relative quantity of intercellular spaces in the mesophyll is less, and the spongy parenchyma is more compact in structure; (5) the system of veins is less well developed, and the length of the veins per unit of surface is less; (6) in cross-section, the area of the xylem in the vascular bundles of the median vein of a leaf is considerably increased while the area of the phloem is diminished; (7) the

osmotic pressure of the cell-sap is weaker in all infected tissues (palisade layer, spongy parenchyma, and mid-rib).

Economic Value of Puccinia suaveolens.—*Cirsium arvense* is a troublesome weed; and, when attacked by the systemic mycelium of *Puccinia suaveolens*, as we have seen, it usually fails to produce any flowers. Nevertheless, the fungus is of but very little help to farmers in keeping down the Creeping Thistles in their fields. It is true that, when the rust becomes established in a patch of thistles, it persists for many years and tends to prevent seed formation; but, unfortunately, it merely attacks certain of the thistle plants and does not eradicate the weed colonies. For this reason, as Cunningham[1] in his account of the Rusts of New Zealand has remarked: "its use as a controllant of this thistle cannot be recommended."

Effect of External Conditions on the Production of Secondary Uredospores and Teleutospores.—A systemic mycelium of *Puccinia suaveolens*, as a rule, produces in succession pycnidia, uredospores, and teleutospores, and the uredospores, when sown on a new host, give rise to uredospores and teleutospores. A uredospore pustule may produce uredospores alone, or uredospores followed by teleutospores and, finally, there may be formed teleutospore pustules that are entirely free from uredospores.

Waters[2] in 1928 treated of the conditions which control the production of uredospores and teleutospores in various Rust Fungi, including *Puccinia suaveolens*; and his observations on *P. suaveolens* will now be summarised.

Waters potted wild thistles, *Cirsium arvense*, and then sprayed them with uredospores obtained in the field. The incubation period for the formation of new uredospore pustules varied from six to ten days. The optimum temperature for the germination of the uredospores was found to be 18° - 22° C.

"Cirsium rust, in nature, forms an abundance of teliospores throughout the growing season. Teliospores may be found in badly infected plants at all times, even in the primary systemic infections early in the summer."

"The uredinia appear almost always on the upper surface of the leaf, although the primary uredinia appear usually on the lower."

"Under optimum conditions of temperature and moisture, the rust may not produce teliospores for more than a month after infection.

[1]G. H. Cunningham, *The Rust Fungi of New Zealand*, printed privately by John McIndoe, Dunedin, N.Z., 1931, p. 187.

[2]C. W. Waters, "The Control of Teliospore and Urediniospore Formation by Experimental Methods," *Phytopathology*, Vol. XVIII, 1928, pp. 156-213.

In most cases, however, the uredinia are found to contain teliospores as early as two weeks after infection. The teliospores appear on the lower leaves first and then extend progressively up the plant. The rust will not propagate itself in the greenhouse, but must be re-inoculated about every 30 days in order to keep it in the uredinial stage. Under conditions of high temperature and low humidity, the regions surrounding the uredinia will become dry and the rust will die."

"Telial formation is hastened by subjecting the infected plant to conditions unfavourable for the metabolism of the hosts," namely, *low temperature* and *darkness*.

The number of days after inoculation with uredospores for the formation of teleutospores was: (1) for a check plant kept at 20° C. and fully exposed to daylight, 32 days; (2) for an experimental plant, kept for seven days at 19° C. in the dark, 28 days; and (3) for another experimental plant, kept for seven days at 7° C. in the dark, 24 days.

"In all cases the placing of infected plants in the dark either at low or room temperature was sufficient to inhibit further formation of urediniospores."

"During winter days when the light intensity was low and the hours of daylight few, it was very difficult to maintain a stock supply of urediniospores. When plants were infected, in many cases the first pustules that broke through contained 50 per cent. of teliospores. In no case was it possible to maintain the uredinial stage for more than two weeks after the first infection. The difficulty was partially overcome by installing seven 100-watt electric lamps in the experimental room of the greenhouse. These were suspended about three feet above the pots and were supplied with reflectors to concentrate the rays of light. With the aid of this auxiliary illumination, which was turned on from about four o'clock in the afternoon until eleven at night, it was possible to secure infection and prolong the uredinial stage to three weeks or more."

From the observations of Waters, with which the later observations of Brown and myself are in agreement, it is evident that *Puccinia suaveolens* is a Rust species that produces teleutospores very readily and in great abundance, especially under conditions unfavourable to the metabolism of the host. The ultimate fate of the teleutospores and the part which they may play in the life-history of the parasite will be treated of in a later Section.

Source of Fungus Material.—During several successive summers prior to 1939, I made it my business to attempt to exterminate the Creeping Thistles on a cousin's farm near Banbury, England, by repeated hoeing. The thistles in each field were hoed down three or four times during the growing season and, in the course of three or four years, heavily infested permanent pastures were practically freed from the pest.[1] Whilst engaged in this work I often had occasion to observe thistles infected with the systemic mycelium of *Puccinia suaveolens*. They were particularly noticeable in the early summer. Later in the year, many other thistles on the farm were found that were spotted with pustules bearing secondary uredospores and teleutospores. Also at Kew and in other parts of England infected thistles were examined.

The material used for a first study of the structure of living pycnidia was obtained on waste ground near Birmingham, England.

In the spring of 1939, I procured from England some thistle shoots bearing primary uredospores and then, in the Dominion Rust Research Laboratory at Winnipeg, these uredospores were used to infect (1) mature thistles and (2) seedling thistles. Thereafter, the English strain of *Puccinia suaveolens* was kept constantly available for further work.

In the late summer of 1939, I procured from England thistle shoots bearing teleutospores, but attempts to germinate these spores proved ineffective.

In November, 1941, at Winnipeg, teleutospores that had appeared on inoculated thistles in the open were collected and placed in a refrigerator. In March, 1942, they were removed from the refrigerator, moistened with water, and subjected to room temperatures; whereupon they germinated in the usual manner. The basidia each gave rise to four basidiospores, and the basidiospores were successfully employed as an inoculum for seedling thistles.

Primary and Secondary Uredospores.—In view of what is to follow it should be noted that a uredospore sorus formed on a haploid (uninucleate) mycelium is known as a *primary uredospore sorus*, while a

[1]My experience with the hoeing method of exterminating thistles in permanent cow-pastures at Banburry is summed up in a rhyme.

THE CREEPING THISTLE
IN THE PASTURE

Hoe one and all,
Both great and small;
Hoe them in June,
'Tis not too soon;
And in July
The hoe apply;
In August too,
And then they'll be few.
When three years have been,
The fields will be clean.

uredospore sorus formed on a diploid (binucleate) mycelium is known as a *secondary uredospore sorus*.

Localised and Systemic Mycelia.—The mycelium of *Puccinia suaveolens* can be either *localised* or *systemic*. Localised mycelia are temporary and disappear with the dead leaves of the host. Systemic mycelia enter the perennial root-system of the host, grow up into new green shoots annually, and may persist for years.

Forms of Localised Mycelia.—Localised mycelia arise on the leaves of a host-plant: (1) as a result of the germination of uredospores, and also (2) as a result of the germination of basidiospores.

(1) Primary or secondary uredospores germinate and form local temporary pustules on new host-leaves. These pustules give rise to *secondary uredospores* and *teleutospores* only. This condition of the fungus is common on the Creeping Thistle in the late summer and autumn, and it was obtained experimentally by de Bary[1] in 1863.

(2) It was supposed on theoretical grounds by Brown and myself and then verified experimentally and by field observation that the basidiospores of *Puccinia suaveolens* derived from teleutospores germinate on thistle leaves in the spring and form local temporary pustules comparable with the pustules formed by *P. graminis* on the leaves of *Berberis vulgaris*.

Forms of Systemic Mycelia.—Systemic mycelia which perennate in the root-system of the host and invade new green shoots annually are formed: (1) as the result of uredospores germinating on seedling thistles and on very young root-buds, as found by Brown and myself;[2] and probably also (2) as a result of basidiospores germinating on seedling thistles.

After a systemic mycelium has become established in the perennial root-system of a Creeping Thistle, it may behave in the new aerial shoots in any one of three ways:

(1) Usually, it gives rise to *pycnidia*, then to *primary uredospore sori* (Fig. 107), and finally to *teleutospores*. This behaviour of the systemic mycelium is commonly seen in the field in the spring and summer, and it was in part recognised by Persoon[3] as long ago as 1799.

(2) Less frequently, the systemic mycelium gives rise to *secondary uredospore sori* and *teleutospores only*. This mode of behaviour was

---

[1]A. de Bary, "Recherches sur le développement de quelques champignons parasites," *Ann. Sci. Nat.*, Bot., Sér. 4, T. XX, 1863, pp. 85.

[2]*Vide infra.*

[3]D. C. H. Persoon, *Observationes mycologicae*, Lipsiae et Lucerne, Pars II, 1799, p. 24.

FIG. 107.—*Cirsium arvense* bearing primary uredospore sori of *Puccinia suaveolens.* Natural size.

first described by Olive[1] in 1913, and it has been obtained experimentally by Brown and myself.[2]

(3) Rarely, the systemic mycelium gives rise to *pycnidia only.* In 1939, Mr. R. Eric Taylor of the University of Cambridge, told me that he had observed this mode of behaviour in the field and, in 1940, Brown and I[3] realised it under experimental conditions.

[1]E. W. Olive, "Intermingling of Perennial Sporophytic and Gametophytic Generations in *Puccinia podophylli, P. obtegens,* and *Uromyces glycyrrhizae,*" *Annales Mycologici,* Vol. XI, 1913, p. 304.   [2] and [3] *Vide infra.*

Localised Mycelia derived from Uredospores.—The local patches of temporary mycelium formed on new host-plants, that develop secondary uredospore sori and teleutospores, come into existence somewhat late in the year. Rostrup[1] found them in July, and referred to them as belonging to the *second generation* to distinguish them from the perennial mycelia which he referred to as belonging to the *first generation*. Second-generation mycelia doubtless owe their origin to the germination of primary or secondary uredospores; and, in support of this statement, may be cited the experiments, already described, recorded by de Bary[2] in 1863 and also confirmatory experiments made by Rostrup[3] about 1873, by C. R. Orton[4] in 1912, by A. M. Brown[5] in 1934, and by myself[6] in 1939. Rostrup sowed some primary uredospores on Creeping-Thistle leaves and observed that, from a week to two weeks later, uredospore sori appeared on the infected spots and that, still later, teleutospores became mingled with the uredospores. Orton's experiments were made in the United States of America and Brown's and my own in Canada.

Olive[7] found that the mycelium of the second generation consists solely of *binucleate hyphae*, which is what might be expected in view of the origin of the mycelium from diploid uredospores.

The secondary uredospore sori produced on a second-generation mycelium differ from the primary uredospore sori produced on a perennial mycelium in that they are developed from the first from binucleate hyphae and not from a uninucleate proto-uredospore sorus.

In England in the late autumn, as I have noticed, Creeping-Thistle plants with secondary infections consisting of local scattered confluent uredospore and teleutospore sori are very common and greatly outnumber the plants which bear a perennial mycelium. This indicates

---

[1]F. G. E. Rostrup, "Om et ejendommeligt Generationsforhald hos *Puccinia suaveolens* (Pers.)," *Forhandl. ved de skandinav. Naturforsekeres*, Kjöbenhavn, 1874, pp. 338-350. Cited from an Abstract under the title "Ein eigenthumliches Generations-verhältniss bei *Puccinia suaveolens* (Pers.)" in *Bot. Zeit.*, Jahrg. XXXII, 1874, pp. 556-557.

[2]A. de Bary, *vide supra*, p. 362.

[3]F. G. E. Rostrup, *loc. cit.*, p. 557.

[4]C. R. Orton, in J. C. Arthur's *Manual of the Rusts in the United States and Canada*, Lafayette, U.S.A., 1934, p. 347. Arthur refers to the primary uredospores as *uredinoid aeciospores*.

[5] A. M. Brown, personal communication, 1939. The fungus was collected in British Columbia in July, 1934, and the uredospores were sown on young plants of *Cirsium arvense* in September in the Dominion Rust Research Laboratory at Winnipeg.

[6]At Winnipeg. Inoculation material obtained in England was sown on Canadian Creeping Thistles.   [7]E. W. Olive, *loc. cit.*, p. 303.

that a mycelium that is localised in origin does not, as a rule, succeed in growing down into the roots of the host and thus becoming perennial.

Rostrup[1] remarked that the primary uredospore sori bear relatively few teleutospores but that the secondary uredospore sori bear many. In England the teleutospores in the secondary uredospore sori can be readily found from September to November. These teleutospores doubtless over-winter, produce basidia in the spring, and so give rise to basidiospores which on new green Thistle shoots germinate and cause the development of haploid leaf-pustules; but, up to the present, no British or European field mycologist has recorded having observed such pustules.

The only trace of the sexual process which can be observed in the diploid mycelium of the second generation is the existence of the nuclei in conjugate pairs and the multiplication of these nuclei by conjugate divisions. It is possible that the diploid uredospores might be used to diploidise haploid pustules derived from basidiospores by means of experiments like those successfully carried out on *Puccinia helianthi* by A. M. Brown.[2]

Localised Mycelia derived from Basidiospores.—Localised mycelia derived from the basidiospores of *Puccinia suaveolens* have been observed by Brown and myself for the first time, both (1) experimentally and (2) under natural field conditions.

Our observations were prompted by the reflection that there was every reason to suppose that the teleutospores of *Puccinia suaveolens* resemble the teleutospores of other Rust Fungi in being able to germinate and give rise to basidia and basidiospores. The failure of field-mycologists to observe haploid pustules of *P. suaveolens* derived from basidiospores hitherto may be explained as follows. In the past, in fields in the spring, attention has very naturally been focused on the conspicuous, sickly-looking, pale-green thistles whose systemic mycelium bears thousands and tens of thousands of orange pycnidia, with the result that the observer has not thought of examining the foliage of near-by, otherwise perfectly healthy, green thistles in the hope of finding tiny, inconspicuous, pycnidia-bearing patches of mycelium derived from germinating basidiospores.

(1) *Experiments.* Mr. R. Eric Taylor of the University of Cambridge, England, told me in January, 1939, that he had not been able to get the teleutospores of *Puccinia suaveolens* to germinate. In the autumn of 1939, I procured teleutospores from England; but, during the ensuing winter and spring, Brown and I did not succeed in germi-

[1] F. G. E. Rostrup, *loc. cit.*, pp. 556-557.

[2] *Vide supra*, p. 268.

nating them. Subsequently, however, successful germination was obtained with teleutospores that had developed in the open at Winnipeg.

The teleutospores of *Puccinia suaveolens* are readily detachable and they break off from their pedicels with the greatest ease, thus differing from the teleutospores of *P. graminis*. The pore of the upper cell is terminal and that of the lower cell very near to the pedicel. When germinating on the surface of water, both cells of a teleutospore often send out germ-tubes simultaneously and, on account of the position of the pores, in opposite directions. When germinating on the surface of agar, etc., the germ-tube of each cell becomes curved and develops into a typical four-celled, four-spored basidium. The basidiospores, when ripe, are shot away from their sterigmata in the normal manner; and, after their discharge, if kept moist, they soon begin to germinate.

Teleutospore pustules that had ripened on thistles growing outside the Rust Research Laboratory were collected on November 21, 1941, placed in a refrigerator, and kept there until March, 1942. Then they were taken into the laboratory and used for cultures.

Teleutospores from some of the pustules that had been over-wintered in the refrigerator were scraped off on to glass slides, then floated on water for two days and, finally, put on the very young leaves of four seedling thistles. After 10-12 days it was observed that on each of the four seedlings, many pycnidial pustules had developed and that these pustules resembled pustules of similar origin developed by *Puccinia graminis* on *Berberis vulgaris*.

Also, some teleutospore pustules attached to a thistle shoot that had been in the refrigerator was soaked in water in the usual way and then suspended over a glass slide in a moist Petri dish. The teleutospores germinated and, in the course of 24 hours, the basidia discharged many basidiospores which fell on to the glass slide. A drop of water was placed on the slide and then the basidiospores were smeared on to the first foliage leaves of two seedling thistles. After about ten days, on the inoculated leaves of each of the two host-plants, there had developed several haploid pustules bearing pycnidia.

Unfortunately, owing to war conditions, the experimental seedling thistles used in the experiments just described could no longer be attended to and they died without any further observations being made upon them. Had all gone well, since *Puccinia suaveolens* is heterothallic (*vide infra*), doubtless some of the compound pustules—those formed by the coalescence of (+) and (−) simple pustules—would have given rise to primary uredospore pustules on their under surface.

(2) *Field observations.* On April 20, 1943, in the patch of thistles infected with *Puccinia suaveolens* just outside the Rust Laboratory greenhouse,[1] some two hours were spent in a search for local haploid pustules bearing pycnidia, that might have been derived from naturally produced basidiospores of *P. suaveolens.* One such pustule was definitely found. It was on a small spinescent leaf of an otherwise healthy thistle shoot that had just produced a rosette of three leaves at the surface of the ground, and it was situated on the margin of the leaf about half way up the lamina. It was about 5 mm. in length and less in breadth and in its central part, on both sides of the leaf, it bore orange-coloured pycnidia. It had not produced any primary uredospores and was therefore in the typical haploid condition.

Thus limited but convincing evidence was obtained that for *Puccinia suaveolens*, at Winnipeg, under natural conditions in the open: (1) the teleutospores can survive the winter, germinate in early spring, and produce basidia and basidiospores; (2) the discharged basidiospores can settle on young *Cirsium arvense* leaves and germinate there; and (3) the germ-tubes so produced can initiate the development of haploid pustules bearing pycnidia.

Establishment of Systemic Mycelia of Endophyllum euphorbiae-sylvaticae, Aecidium leucospermum, and Puccinia minussensis in Herbaceous Perennials.—The work accomplished by various mycologists in obtaining a systemic mycelium of certain Rusts in herbaceous-perennial hosts served as a guide to Brown and myself in obtaining systemic mycelia of *Puccinia suaveolens* in the Creeping Thistle. Here, therefore, that work will be reviewed.

Before a Rust can establish itself in the subterranean parts of a herbaceous-perennial host-plant, it is necessary for its mycelium to grow from an aerial leaf or stem, where it originated from one or more spores, downwards through an aerial stem into the rhizome or perennial root-system. The distance from a point of primary infection on a leaf to a rhizome or root is often several inches or more; and the time during which mycelial growth may take place is limited, for in the autumn the aerial shoot dies down and disappears. Thus a problem presents itself: how does a systemic mycelium of a perennial Rust species growing on a herbaceous-perennial Flowering Plant find its way into the subterranean parts of its host?

The establishment of a subterranean perennial mycelium has taken place under experimental conditions with: (1) *Endophyllum euphorbiae-sylvaticae* on *Euphorbia amygdaloides*; (2) *Aecidium leucospermum* on *Anemone nemorosa*; and (3) *Puccinia minussensis* on *Lactuca pulchella*.

[1]For the history of this patch of thistles, *vide supra*, p. 351.

(1) *Endophyllum euphorbiae-sylvaticae* on the Wood Spurge, *Euphorbia amygdaloides*. In the spring a Euphorbia plant infected with the perennial mycelium of its Rust resembles a Creeping-Thistle plant infected with the perennial mycelium of *Puccinia suaveolens*, in that its infected shoots are longer and paler green than is normal and in that its leaves, which are shorter and thicker than normal leaves, bear pycnidia in great abundance. The pycnidia are followed by aecidia; and the germ-tubes of the aecidiospores develop directly into basidia.

Plowright[1] sowed aecidiospores of the Endophyllum on healthy Euphorbia leaves of well-established plants and observed that they germinated and gave rise to basidia and basidiospores[2] and that the basidiospores, by means of their germ-tubes, bored through the epidermal cells and entered the leaf-parenchyma where they soon produced a richly branched and widely extending intercellular mycelium; and he also observed that "if the entrance has been effected in an old leaf, the further development of the parasite ceases when the leaf falls."

"I have always failed," said Plowright, "in permanently infecting old plants of Euphorbia; no matter what the age of the leaves may be, in the ensuing spring the foliage has always been healthy. But if a young seedling be infected shortly after it has come up—that is while not more than a month or two old—the mycelium produced in its leaves readily gains an entrance into the stem. The foliage and shoots sent up by it in the following year are pervaded by the perennial mycelium, and produce aecidia abundantly during the spring; but the late summer and autumn foliage differs little from healthy foliage, excepting that the leaves are somewhat shorter. The next vernal foliage is, however, aecidiiferous. The affected plants seldom flower."

Thus Plowright succeeded in obtaining perennial mycelia in his Euphorbia by inoculating *seedlings* instead of mature plants. In seedlings, as compared with mature plants, the tissues are relatively tender and the distance to be traversed from a primary point of infection to the subterranean organs of the host is relatively short.

The perennial mycelium formed in the rhizome of a young Euphorbia plant under conditions similar to those in Plowright's experiments has been derived from (+) and (−) basidiospores which germinated on the leaves. It would be of interest to find out whether the perennial mycelium has isolated nuclei and is made up of a mixture of (+) and (−) mycelia or whether it has conjugate nuclei and is diploid.

[1] C. B. Plowright, *A Monograph of the British Uredineae and Ustilagineae*, London, 1889, p. 227.

[2] For my own illustrations of the basidia of *Puccinia suaveolens vide* these *Researches*, Vol. III, Fig. 206, p. 511.

In the latter case the perennial mycelium in a Euphorbia rhizome would resemble the perennial mycelium of *Puccinia minussensis* in the rhizome of *Lactuca pulchella* as observed by Brown.

(2) *Aecidium leucospermum* on *Anemone nemorosa*. *Ochropsora ariae* (*O. sorbi*) is a heteroecious Rust with pycnidia and aecidia on *Anemone nemorosa* and uredospores and teleutospores on Sorbus. The stage on the Anemone is known as *Aecidium leucospermum*.

In 1893, Soppit[1] showed that *Aecidium leucospermum* is able to repeat itself on *Anemone nemorosa*. His attempts to infect mature Anemone plants with aecidiospores were unsuccessful;[2] but, on sowing aecidiospores on seedling plants, he obtained aecidia on these plants in the following year.

(3) *Puccinia minussensis* on the Blue Lettuce, *Lactuca pulchella*. A. M. Brown,[3] working in the Dominion Rust Research Laboratory, succeeded in obtaining the perennial mycelium of *Puccinia minussensis* in the rhizomes of its host in two ways: (*a*) by sowing uredospores on the foliage leaves of very young plants derived from seeds; and (*b*) by sowing uredospores in the buds of rhizomes of old established plants at a time when the buds are still buried beneath the ground and shortly before they begin to develop into aerial shoots. Like Plowright with Endophyllum, Brown failed to obtain a perennial mycelium in the rhizome of an old Lactuca plant whenever he sowed uredospores on the leaves of well-developed aerial shoots.

Brown found that the perennial mycelium in old established rhizomes of *Lactuca pulchella* which lie dormant during the winter contains conjugate nuclei and is therefore diploid (Fig. 97, B, p. 284).

Cryptomycina pteridis, one of the Ascomycetes, attacks the Bracken Fern (*Pteridium latiusculum*) and causes a leaf-roll of the young fronds and gives rise to abundant black stromatic areas on the lower surface of the pinnules between the veinlets. *C. pteridis* resembles the three Rust Fungi just treated of in that its host is herbaceous and perennial and in that it produces both localised and systemic mycelia. It further resembles those Rusts in the manner in which its systemic mycelium becomes established; for, in 1940, Sara Bache-Wiig[4] found that: (1) inoculation of young fronds or immature portions of older

[1]H. J. Soppit, *"Aecidium leucospermum DC."* *Journal of Botany*, Vol. XXXI, 1893, p. 273-274.

[2]E. O. Callen (Examination of *Aecidium leucospermum* DC. from Scotland, *Trans. Brit. Myc. Soc.*, Vol. XXIV, 1940, p. 111) also failed to infect mature plants of *Anemone nemorosa*.

[3]A. M. Brown, personal communication, 1939. Also *vide supra*, p. 286.

[4]Sara Bache-Wiig, "Contributions to the Life History of a Systemic Fungus Parasite, *Cryptomycina Pteridis*," *Mycologia*, Vol. XXXII, 1940, pp. 214-250.

fronds with conidia was followed by infections resulting in *localised* lesions only; whereas (2) inoculation of *young bracken sporophytes* with conidia was followed by infections resulting in *systemic infection* and typical leaf-roll symptoms.

Establishment of Systemic Mycelia of Puccinia suaveolens in Cirsium arvense Seedlings.—The aerial shoots of the Creeping Thistle die down in the autumn just like those of *Euphorbia amygdaloides* and *Lactuca pulchella*; but, whereas in the Euphorbia and the Lactuca the perennating organs left beneath the ground are rhizomes, in the Cirsium they are horizontal roots.

The root-system of an old Creeping-Thistle plant established in a pasture is so deep underground that it seems unlikely that uredospores would be washed down to it by rain and thus be placed in a position to infect it directly.

Since, as I myself and others have observed in pastures, in the spring the number of Creeping Thistles infected with a perennial mycelium of *Puccinia suaveolens* is always very small compared with the number which in the previous autumn bore localised secondary infections, it is clear that, as a rule, the mycelium in a secondary infection on old foliage does not spread to the perennial parts of the host-plant. A consideration of this fact indicated that the production of a perennial mycelium on a new host-plant must take place in very young shoots of old-established plants or in the shoots of seedlings.

Since Plowright succeeded in obtaining a perennial mycelium of *Endophyllum euphorbiae-sylvaticae* in Euphorbia seedlings by means of *basidiospores* which germinated on the leaves, and since Brown succeeded in obtaining a perennial mycelium of *Puccinia minussensis* in Lactuca seedlings by means of uredospores which germinated on the leaves, it seemed likely that a perennial mycelium of *P. suaveolens* might be obtained by inoculating seedlings of *Cirsium arvense* either with basidiospores or with uredospores.

And since Brown succeeded in obtaining a perennial mycelium of *Puccinia minussensis* in the rhizomes of old established Lactuca plants by sowing uredospores in the resting buds of the rhizomes buried beneath the ground, it seemed not unlikely that a perennial mycelium of *P. suaveolens* might be obtained in the perennial root-system of Creeping Thistles if uredospores were sown: in (*a*) the adventitious buds which grow out from the roots in the spring, or (*b*) in the axillary buds of vertical rhizomes whose aerial shoots have been cut away.

On the basis of the theoretical considerations just set out, successful attempts to establish systemic mycelia of *Puccinia suaveolens* by sowing uredospores on seedlings of *Cirsium arvense* were made by A. M. Brown and myself in 1939 and 1940.

The first experiments were made on seedling thistles that bore several foliage leaves and with some success, but the later and still more successful experiments were made on seedlings strictly in the cotyledonary stage. Some particulars of three series of experiments will now be given.

(1) *First foliage leaves inoculated. Systemic mycelium obtained in axillary buds.* In May, 1939, some uredospores of *Puccinia suaveolens* were placed on moistened first-foliage leaves of four Creeping-Thistle seedlings growing in pots in a greenhouse. The pots were kept in a damp-chamber for two days and were then transferred to a greenhouse bench. The ensuing weather was very hot and the air of the greenhouse relatively dry. After about a fortnight from the time of inoculation uredospore pustules developed on the leaves, but these pustules were small and localised. On some leaves the pustules developed along the sides of the mid-rib near to the stem. The first rosette of leaves in each seedling plant ceased to grow and died, and new shoots developed from axillary buds. In the winter the infected plants were kept in the greenhouse in a cool place and, during this time, they remained in a quiescent state. In February, 1940, the plants were repotted, whereupon the buds on the old shoots began to develop. About a week later, it was observed that all the leaves produced from a single bud on each of two plants and from two buds on another plant bore numerous orange pycnidia crowned with little drops of nectar. The pycnidia were present on both sides of each leaf, and their nectar emitted a strong perfume. Thus in three of the four inoculated seedling thistles a systemic mycelium of *P. suaveolens* had become established.

(2) *Cotyledons inoculated. Systemic mycelium obtained in axillary buds.* In the spring of 1940, four thistle seeds were sown in as many separate pots; and, as the seedlings came up, their cotyledons were inoculated with uredospores. Infection took place and localised pustules of uredospores were soon formed. From two to three months after the seedlings had been inoculated one or more axillary buds on each plant grew into shoots bearing a systemic mycelium. This mycelium affected the appearance of the shoots and produced on the leaves great numbers of pycnidia.

(3) *Cotyledons inoculated. Systemic mycelium obtained in the terminal bud.* More numerous experiments were then undertaken with a view to obtaining systemic mycelium in terminal buds. The cotyledons of seedlings down to the axils were inoculated with uredospores as soon as they had opened out. Infection took place and, on the upper surfaces of the cotyledons, secondary uredospore pustules were soon formed. In from one to two months after the cotyledons had been

inoculated they had withered away and the seedlings had formed several foliage leaves. In about a dozen plants it was then seen that the upper foliage leaves produced by the terminal bud had become invaded with systemic mycelium, for they had developed, or were developing, pycnidia in great abundance. The particulars concerned with one seedling, which may serve as a type for the other similar seedlings, are as follows. After two months, the plant had developed nine foliage leaves. The first six appeared healthy, but the seventh, eighth, and ninth were obviously invaded with systemic mycelium for they were bearing numerous pycnidia particularly on their under surface (Fig. 108). A little later, the fifth and sixth foliage leaves also began to produce pycnidia. It would appear that the mycelium developed from the uredospores had grown down the cytoledons into the terminal bud of each of the seedlings under discussion and, after entering the growing-plant, had grown out into the rudiments of each of the later-formed foliage leaves.

Other series of experiments made upon seedlings gave results confirmatory of those already obtained. Altogether, some fifty seedling thistles that had been inoculated with uredospores became infected with systemic mycelium either in one or more of their axillary buds or in their terminal bud.

The experiments just described demonstrate that the systemic mycelium of *Puccinia suaveolens* can come into existence as a result of the inoculation of seedling thistles with uredospores; and they also indicate that: the younger a seedling, the more easily does a systemic mycelium become established in it.

As we have seen, the seeds of *Cirsium arvense* are able to germinate as soon as they have been formed. In the open, under suitable weather conditions, seedling Creeping Thistles may spring up from the middle of July until late autumn. Now, during this period, primary and secondary uredospores of *Puccinia suaveolens* are abundantly present on old infected thistles. Under natural conditions, therefore, it would appear that, at any time from July to late autumn: the wind may carry uredospores from old thistles to seedling thistles; the uredospores may germinate on the cotyledons and first foliage leaves; and the mycelium so produced may grow into the seedling stems and thus become systemic.

It seems likely that, if the cotyledons of Creeping-Thistle seedlings were inoculated with the basidiospores of *Puccinia suaveolens*, the mycelium resulting therefrom would become systemic in the seedling stems; but Brown and I have not yet put this suggestion to the test of experiment.

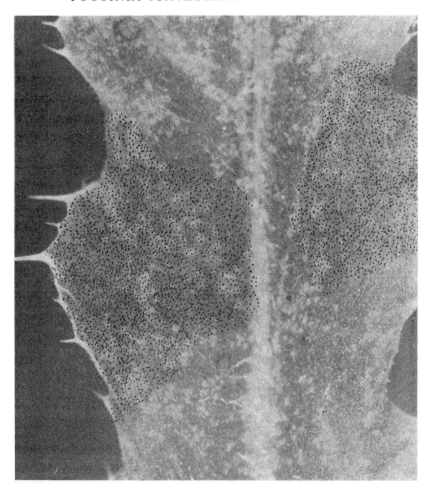

FIG. 108.—Lower surface of a young leaf of *Cirsium arvense* showing pycnidia of *Puccinia suaveolens*. Magnification 6.

Dried seeds of *Cirsium arvense* retain their vitality for six months or longer and, under natural conditions, many Creeping-Thistle seedlings come up in the spring. At this time, too, we may suppose that the over-wintered teleutospores of *Puccinia suaveolens* germinate and give rise to basidiospores. In the spring, therefore, it may be that systemic mycelia are established in seedling thistles as a result of basidiospores germinating on the cotyledons and first foliage leaves.

Experiments with Root-buds.—Some horizontal roots of *Cirsium arvense*, 2-3 mm. thick, bearing tiny adventitious buds, were dug up in a patch of thistles, and pieces of them, 3-4 inches long, were placed in a damp-chamber in the greenhouse. In the course of a day or two, the buds began to swell and the bud-scales developed stomata. Thereupon, some uredospores of *Puccinia suaveolens* were pushed in between the bud-scales. Then the roots with their inoculated buds were set in moist sand in pots, at a depth of about two inches. At the end of ten days, some of the root-buds had developed into aerial shoots. Two of the shoots were found to be infected with perennial mycelium. One bore pycnidia scattered over the under surface of each leaf in the usual manner. The other bore no pycnidia whatsoever, but soon developed uredospore pustules in abundance. In this second shoot, the mycelium, on entering the leaves, had failed to become de-diploidised and thus the shoot came to resemble similar shoots that have occasionally been observed in the field.

The results of the experiment just described suggest that, under natural conditions in the open, a systemic mycelium can become established in the root-shoots of old thistle-plants, providing that uredospores have access to the buds just as they are opening, or just after they have opened, at the surface of the ground. Therefore, in a patch of wild thistles in a pasture, it is easily conceivable that, in the absence of seedlings, some of the shoots may come to be infected with systemic mycelium. It is further conceivable that such infected root-shoots would develop horizontal roots into which the mycelium would penetrate, so that, in the end, the systemic mycelium would become perennial.

In a field, a few days after *Cirsium arvense* shoots have been hoed or otherwise destroyed to the level, or to just below the level, of the ground, some of the axillary buds on the vertical rhizomes still left in the soil begin to develop into aerial shoots. It may well be that these young shoots sometimes become infected with systemic mycelium derived from uredospores and that the mycelium eventually enters the root-system and thus becomes perennial.

Systemic Mycelium in Thistle Roots.—Olive[1] and Kursanov[2] observed that the systemic mycelium of *Puccinia suaveolens* is present in the roots of thistles whose green shoots bear systemic mycelia. Experiments made by Brown and myself confirm this finding.

At Winnipeg in 1939, a seedling thistle that had been inoculated

[1]E. W. Olive, "Intermingling of perennial sporophytic and gametophytic generations in *Puccinia podophylli, P. obtegens*, and *Uromyces glycyrrhizae*," Annales Mycologici, Vol. XI, 1913, pp. 295-311.

[2]L. Kursanov, "Recherches morphologiques et cytologiques sur les Urédinées," *Bull. Soc. Nat. Moscou*, n.s. (1917), Vol. XXXI, 1922, p. 58.

with uredospores of *Puccinia suaveolens* produced from an axillary bud a pale-green shoot invaded by a systemic mycelium and bearing pycnidia. In the spring of 1940, the plant was re-potted in a large pot. It soon sent out lateral roots, and on these roots were formed adventitious buds. Two of these buds, each situated at a distance of about four inches from the older seedling stem, developed into shoots bearing the systemic mycelium. From these observations we may conclude that the mycelium derived from uredospores grew: (1) down the young foliage leaves of the inoculated seedling into the young stem; (2) then from the stem into the primary root; (3) then from the primary root into two horizontal secondary roots; (4) then along these roots into two adventitious buds developed on the roots; and, finally, (5) into the stems and green leaves of the new aerial shoots developed from the adventitious buds.

Some ten or eleven other plants similar to the one just described have been observed in the greenhouse. In all of them shoots derived from root-buds and infected with systemic mycelium came up in the pot about four inches away from the seedling plant originally inoculated with uredospores. In every plant the mycelium derived from uredospores must have made its way from the stem of the seedling shoot into the primary root and then into one or more secondary horizontal roots.

In infected shoots, as Kursanov found, the systemic mycelium is present in the youngest tissues of the growing-points and so is able to enter the rudiment of every new foliage-leaf. Probably, also, in infected roots the systemic mycelium is present in the growing-points, but this has yet to be determined by precise investigation.

Binucleate Condition of the Systemic Mycelium in Stems and Roots.—In 1913 Olive[1] maintained, primarily on cytological evidence, that throughout the tissues in the host-plants parasitised by *Puccinia suaveolens* (his *P. obtegens*), *P. podophylli*, and *Uromyces glycyrrhizae* there is an intermingling of two kinds of mycelium which are independently and simultaneously perennial: (1) *uninucleate* mycelium, belonging to the "gametophytic generation" and bearing pycnidia only; and (2) *binucleate mycelium*, belonging to the "sporophytic generation" and bearing aecidiospores or uredospores and, finally, teleutospores.

Olive's hypothesis, from the first, was regarded by mycologists with suspicion; and more recent and more critical investigations made on *Puccinia podophylli* and *P. suaveolens*, have led to its rejection.

---

[1]E. W. Olive, *loc. cit.*

Puccinia podophylli parasitises the May-apple, *Podophyllum peltatum*, in the United States and eastern Canada, and on its host it produces pycnidia, aecidiospores, and teleutospores. It is therefore an *opsis*-form.

In 1925, Whetzel, Jackson, and Mains,[1] by means of tissue examination and experiment proved conclusively that Olive's conception of the state of the mycelium in the tissues of *Podophyllum peltatum* is erroneous. It was shown that: (1) the mycelium of *Puccinia podophylli* is not perennial and never becomes systemic; (2) the early crop of teleutospores which appear on the bud scales, stems, and sepals of the new aerial organs of the May-apple in the spring (very rarely accompanied by pycnidia or by pycnidia and aecidia) arise directly from the mycelium produced by basidiospores derived from overwintered teleutospores; (3) the pycnidia and aecidia which develop on the peltate laminae of the leaves in the spring also arise from the mycelium produced by basidiospores derived from overwintered teleutospores; and (4) the late or summer crop of teleutospores is produced on mycelium developed from infection by aecidiospores.

In respect to *Puccinia suaveolens*, Olive, on the basis of cytological observations made on *Cirsium arvense* leaves, supposed that not only in the leaves, but also in the stems and roots, uninucleate and binucleate mycelia intermingle systemically. This intermingling of two kinds of mycelium was admitted by Kursanov[2] for the young leaves but not for the stems and the underground organs of thistles; and he pointed out that Olive's illustration of a mycelium in a thistle rootstock[3] shows this mycelium in the binucleate condition.

Kursanov[4] observed that the systemic mycelium of *Puccinia suaveolens* in the stems of *Cirsium arvense*, as a rule, is composed of binucleate cells, *i.e.* that the mycelium in the stems is diploid.

Among the binucleate cells in the stems Kursanov sometimes found trinucleate or uninucleate cells. In sufficiently large longitudinal stem-sections, especially among the large pith-cells, he succeeded in following hyphae composed of a number of cells and in observing deviations from the normal binucleate condition. The number of nuclei in successive cells of some of these exceptional hyphae (found in older parts of host-stems) were:

---

[1] H. H. Whetzel, H. S. Jackson, and E. B. Mains, "The Composite Life History of *Puccinia podophylli* Schw.," *Journal of Agricultural Research*, Vol. XXX, 1925 pp. 65-79.

[2] L. Kursanov, *loc. cit.*, p. 60.

[3] E. W. Olive, *loc. cit.*, Plate XV, Fig. 3.

[4] *Ibid.*, pp. 58-60.

(1) 3 - 1 - 2;
(2) 2 - 2 - 2 - 3 - 2 - 2 - 2 - 1;
(3) 2 - 3 - 3 - 3 - 3 - 2 - 2;
(4) 2 - 2 - 1;
(5) 3 - 2 - 1.

Kursanov explained the deviations as follows. Where trinucleate cells are next to uninucleate cells, they are formed by the unequal distribution of the nuclei during the division of the cells (3 in one cell, 1 in the other). If, however, there is no such association of trinucleate and uninucleate cells, the cause of the formation of trinucleate cells may well be a break in the integrity of the dikaryon and a division of one only of the two nuclei. The probability of the correctness of this last explanation is increased by the observation that the cells of the mycelium concerned were very long and the nuclei were resting almost at the opposite ends of the cells.

Kursanov also remarked: that the number of nuclei in the binucleate mycelium of a thistle stem never exceeds three; that uninucleate cells are much rarer than trinucleate; and that such uninucleate cells, owing to their position in a mycelium otherwise generally binucleate, should be regarded as of secondary origin.

As might be expected, the mycelium of *Puccinia suaveolens* in *Cirsium arvense* roots resembles cytologically the mycelium in the stems, *i.e.* it is binucleate. It appears from Kursanov's nuclear investigations and from the experiments of Brown and myself recorded in the last Section that the binucleate mycelium in infected thistle stems grows into some of the horizontal roots produced by the stems, maintains itself there without change, and so passes into the adventitious root-buds and into the stems of the shoots that develop from these buds in the next spring. Therefore it may be assumed that, as a rule, the systemic mycelium of *Puccinia suaveolens* in the stems and roots of its host is binucleate and diploid.

As we shall see in the next Section, as a binucleate mycelium of *Puccinia suaveolens* passes from a stem into a young leaf, it undergoes a remarkable change.

Systemic Mycelia in Leaves and the Production of Pycnidia, Primary Uredospores, and Teleutospores.—The behaviour of the mycelium which perennates in the roots and subsequently produces pycnidia, primary uredospore sori, and teleutospores on the leaves of new shoots was followed by Kursanov.[1] In the spring the mycelium enters each new shoot and makes its way into the growing-points of

[1] L. Kursanov, *loc. cit.*, pp. 58-66.

the buds. The mycelium in a growing-point, like that in a mature stem, is made up of binucleate cells and is therefore *diploid*. This diploid mycelium grows into the base of each very young leaf; but soon, under the epidermis of the leaf whilst this organ is still enclosed in the bud, hyphae with uninucleate cells come into existence and form pycnidia. The mycelium with uninucleate cells is obviously *haploid*. At the stage of the leaf's growth just described, binucleate and uni-nucleate mycelia are present in the leaf tissues in close proximity. Later on, as the leaf extends, the uninucleate mycelium increases in amount and its hyphae come to exceed those of the binucleate mycelium in number. However, in a longitudinal section through a mid-rib, in connexion with the large parenchymatous cells, one finds binucleate hyphae only. The two kinds of mycelium had been seen by Olive[1] in an older leaf and he had supposed that they accompany one another throughout the host-plant; but Kursanov, having found in the growing-points of stems and in very young leaves a binucleate mycelium only, suggested that the uninucleate hyphae in leaves more advanced in development may be secondary products (produits secondaires). It therefore appears that, in the spring, an overwintered diploid my-celium, after growing out of the roots into the stems and leaf-rudiments of the new shoots of the host-plant, produces haploid branches in the growing leaves, and thus, at least in part, becomes *de-diploidised*.

Kursanov's cytological observations and the experiments of Brown and myself in which we obtained systemic mycelia from uredospores indicate: (1) that the systemic mycelia in thistle roots and stems is binucleate (dikaryotic) only; (2) that, in the leaf rudiments of buds, the binucleate mycelium gives rise to uninucleate hyphal branches; and (3) that, therefore, in an older leaf the pycnidia-bearing uninucleate mycelium and the binucleate mycelium along the mid-rib, etc., are both merely derivatives of the binucleate mycelium that entered the leaf rudiment. It remained for Kursanov to envisage the idea that the uninucleate hyphae in a leaf are but *secondary products* and for me now to express the same idea in other words by saying that the uni-nucleate hyphae owe their origin to *de-diploidisation*.[2]

From Kursanov's further observations and illustrations it appears that the uninucleate mycelium which has come to be present in a young thistle leaf develops two kinds of haploid organs: *pycnidia* and

[1]E. W. Olive, *loc. cit.*, p. 304.

[2]*Cf.* A. H. R. Buller, "The Diploid Cell and Diploidisation in Plants and Animals with Special Reference to the Higher Fungi," *The Botanical Review*, Vol. VII, 1941, pp. 411-414. The term *de-diploidisation* is there defined as: "in Basidiomycetes and Ascomycetes, the production of haploid cells or hyphae by a dikaryotic diploid mycelium or by a dikaryotic diploid cell."

proto-uredospore sori (his very young primary *uredo-sorus*). The proto-uredospore sori (proto-uredinia), when very young, are made up of uninucleate cells which grow up under the epidermis and form a palisadal complex.[1] In a complex the cells are all "fertile" and none "sterile."

According to Kursanov, a proto-uredospore sorus becomes converted into a primary uredospore sorus in one of two different ways. Usually, (1) binucleate hyphae grow up into a proto-uredospore sorus and produce uredospores there after pushing aside and causing the degeneration and death of the uninucleate cells;[2] while, sometimes, in a proto-uredospore sorus (2) the uninucleate cells are "fertilised" by a nucleus moving into them from below through an opening in the cell-wall[3] in the manner described by Blackman[4] for the aecidium of *Phragmidium violaceum*, and then the "fertilised" cells form uredospores.

The primary uredospore sori of *Puccinia suaveolens*, like the aecidia of *P. helianthi*, *P. graminis*, and long-cycled Rusts in general, are always formed a little later than the pycnidiospores. They are "hypophyllous, occupying the whole under surface of the leaf, minute, crowded, often confluent, pulverulent, reddish-brown, then darker."[5] Later still, teleutospores make their appearance among the uredospores. The fact that the primary uredospore sori develop shortly after the pycnidia suggests that the pycnidiospores may have something to do with the origin of the uredospore sori from the proto-uredospore sori.

Sections of Creeping-Thistle leaves which bore young pycnidia crowned with drops of nectar were cut with a hand-razor, mounted in water, and examined under the microscope; and it was observed that the pycnidia are provided with numerous pointed periphyses and also with *flexuous hyphae*. I was able to demonstrate the existence of these hyphae to the late W. B. Grove.

The pycnidia of *Puccinia suaveolens* are not only provided with flexuous hyphae, but are orange-coloured, emit scent, excrete pycnidiospore-bearing nectar, and are visited by insects.[6] A consideration of

---

[1]L. Kursanov, *loc. cit.*, Plate II, Fig. 37.

[2]*Ibid.*, Plate III, Figs. 38 and 39.   [3]*Ibid.*, Plate III, Figs. 41 and 42.

[4]V. H. Blackman, "The Fertilization, Alternation of Generations, and General Cytology of the Uredineae," *Annals of Botany*, Vol. XVIII, pp. 338-341, Plate XXIII, Figs. 66-70.

[5]W. B. Grove, *The British Rust Fungi (Uredinales) their Biology and Classification*, Cambridge, 1913, p. 145.

[6]E. Ráthay ("Untersuchungen über die Spermogonien der Rostpilze," *loc. cit.* in Chapter I) observed eleven species of insects visiting the pycnidia of *Puccinia suaveolens*.

these facts leads to the supposition that *P. suaveolens*, like *P. helianthi* and *P. graminis*, is heterothallic; and, shortly, we shall see that this supposition has been justified by experiment.

We may now ask: how do the proto-uredospore sori produced by a haploid mycelium that is systemic in the leaves of an infected Creeping Thistle become diploidised?

To answer this question we must take into account the following facts: (1) *Puccinia suaveolens* is heterothallic; and (2) Kursanov's cytological observations which led him to conclude that nuclear migration in proto-uredospore sori from one cell to the next cell with resultant diploidisation of uninucleate "fertile" cells sometimes takes place.

We may assume that the diploidisation of the proto-uredospore sori is initiated in either of two ways: (1) by the mixing of the nectar of (+) and (−) pycnidia through insect agency; and (2) by the migration of (+) mycelial nuclei to (−) proto-uredospore sori and of (−) mycelial nuclei to (+) proto-uredospore sori.

(1) Mode I: mixing of nectar. As we shall see, Brown and I have proved by experiment that the mixing of pycnidial nectar of systemic mycelia results in the diploidisation of proto-uredospore sori and the production of primary uredospores. This being so, there is every reason to believe that, under natural conditions, the mixing of the nectar of (+) and (−) pycnidia of systemic mycelia by insects may also lead to the diploidisation of the proto-uredospore sori.

(2) Mode II: migration of mycelial nuclei. Early in 1940, in the greenhouse at Winnipeg, in the absence of flies, Brown and I observed that systemic mycelia of *Puccinia suaveolens* which appeared on shoots of plants inoculated in 1939 formed pycnidia first and then uredospore sori. Under the conditions described, diploidisation may have been effected: by (+) nuclei migrating from (+) haploid branches of the mycelium or from diploid (dikaryotic) branches of the mycelium to (−) proto-uredospore sori; and by (−) nuclei migrating from (−) haploid branches of the mycelium or from diploid (dikaryotic) branches of the mycelium to (+) proto-uredospore sori.

The two modes of diploidisation of haploid proto-uredospore sori produced on the systemic mycelia of *Puccinia suaveolens* as just described correspond to the two modes previously described for the haploid proto-aecidia produced on localised mycelia of *P. helianthi* and *P. graminis*.

Mode I for *Puccinia suaveolens*, as described above, is well-based on both probability and experiment.

Mode II is more theoretical and possibly may not be true. To prove Mode II satisfactorily, it would be necessary to keep insects entirely away from infected thistle plants, so that the nectar of the pycnidia could not be mixed, and then to study cytologically the development of the proto-uredospore sori. If it were found that some or all of these structures developed in the same way as the proto-aecidia in fertile compound pustules of *Puccinia helianthi* or *P. graminis, i.e.* that the proto-uredospore sori became diploidised by migrating nuclei, then it could be said that Mode II had been definitely established.

The alternative possibility so far as Mode II is concerned is that, in the absence of insects and without any mixing of pycnidial nectar, the proto-uredospore sori are invaded by diploid hyphae and suppressed, so that in the end the uredospores are produced not by diploidised proto-uredospore sori, but by the invading diploid hyphae.

That proto-uredospore sori on systemic mycelia of *Puccinia suaveolens* are frequently suppressed by the invasion of diploid uredospore-producing hyphae, as we have seen, was asserted by Kursanov, on the basis of a careful cytological study. Moreover, Kursanov's observations on *P. suaveolens* appeared to confirm similar observations made by Olive[1] on *P. podophylli.* However, the suppression of proto-uredospore sori by diploid hyphae is a surprising idea and, before it is accepted without reservation, it is desirable that it should be supported by further investigations.[2]

Assuming that Kursanov's observations were correctly made, we must admit that the proto-uredospore sori on a systemic mycelium of *Puccinia suaveolens* sometimes function as in other Rusts but that, frequently, they are functionless owing to their displacement by diploid hyphae which have been derived from the perennial mycelium and which take over the task of forming the uredospores.

The uredospores theoretically or actually produced by a haploid proto-uredospore sorus of *Puccinia suaveolens* after diploidisation by nuclei of opposite sex which have migrated to the sorus from another part of the systemic mycelium and the uredospores produced by diploid hyphae of the mycelium which, as described by Kursanov, have invaded and suppressed a haploid proto-uredospore sorus *would be identical genetically*; for the (+) nucleus and the (−) nucleus in the uredospores formed in both ways would all be the descendents by

[1]E. W. Olive, *loc. cit.*, p. 301 and Plate XV, Fig. 1.

[2]Early in 1939, Mr. R. Eric Taylor of Cambridge University told me that he had investigated *Puccinia suaveolens* cytologically but, so far, had been unable to confirm Kursanov's observations on the invasion of proto-uredospore sori by diploid (dikaryotic) mycelia.

division of the (+) nuclei and the (−) nuclei of the original diploid mycelium that became systemic. Thus, from the point of view of genetics, the formation of uredospores on a systemic mycelium by diploid hyphae which displace the palisadal complex of haploid cells of a proto-uredospore sorus would be a matter of interest, but of no real importance. The one mode of uredospore formation would give exactly the same result as the other.

De-diploidisation, Re-diploidisation, Self-diploidisation, and Cross-diploidisation.—The terms de-diploidisation, re-diploidisation, self-diploidisation, and cross-diploidisation, already introduced in a discussion of the behaviour of the systemic mycelium of *Puccinia minussensis* growing on *Lactuca pulchella*,[1] can be conveniently used in connection with the behaviour of the systemic mycelium of *Puccinia suaveolens*.

In *Puccinia suaveolens*, a systemic mycelium derived from uredospores sown on a seedling thistle is at first diploid (dikaryotic); but, as soon as this diploid mycelium grows out of the growing-point of a new thistle shoot into the young leaves, as we have seen, it produces haploid (monokaryotic) branches. These soon give rise to haploid pycnidia and haploid proto-uredospore sori. In producing haploid branches the original diploid mycelium may be said to undergo partial *de-diploidisation*.

If and when, in *Puccinia suaveolens*, in the absence of insects and without any mixing of the nectar of the pycnidia, the haploid proto-uredospore sori on a haploid branch of a diploid mycelium become diploidised by nuclei which have migrated from haploid branches of the mycelium of opposite sex or from diploid branches, we can say that a haploid part of the original diploid mycelium has undergone *re-diploidisation* and that *re-diploidisation* has been effected by *self-diploidisation*.

As we have seen from the discussion in the last Section, in *Puccinia suaveolens* it yet remains to be demonstrated that re-diploidisation by self-diploidisation actually takes place.

In *Puccinia suaveolens*, in Mode I of initiating the diploidisation of the proto-uredospore sori on a systemic mycelium, diploidisation is initiated by means of pycnidiospores of opposite sex transported by insects from other infected plants and deposited on or very near to the flexuous hyphae of pycnidia associated with the proto-uredospore sori. Here, if we wish, we can speak of the diploidisation process as being effected by *cross-diploidisation*.

[1] *Vide supra*, Chapter VI, p. 284.

Thus, just as we can speak of *self-pollination* and *cross-pollination* in the Phanerogamia, so in the Uredinales with systemic mycelia having haploid branches we can speak of *self-diploidisation* and *cross-diploidisation*.

Biological Significance of Pycnidia on Systemic Mycelia.—A diploid systemic mycelium of *Puccinia suaveolens*, established in the root-system of a thistle colony, in its annual production of haploid branches bearing pycnidia on the leaves of new shoots expends much material and energy. Yet, as we have seen, the pycnidiospores and flexuous hyphae of these pycnidia are not required for the formation of the uredospores: in the absence of insects and without any mixing of the nectar, the proto-uredospore sori either become diploidised by self-diploidisation or, as Kursanov holds, frequently become invaded and displaced by binucleate hyphae which form uredospores. Therefore this question arises: of what use to *P. suaveolens* are the pycnidia produced annually in such large numbers on its systemic mycelia?

The answer to our question is in essence the same as that already given to a similar question raised in connexion with the pycnidia on the systemic mycelia of *Puccinia minussensis*.[1] However, a slight modification of the answer is necessary owing to the fact that, whereas *P. minussensis* is a *eu*-form of Rust so that the diploidisation process is effected in its proto-aecidia, *P. suaveolens* is a *brachy*-form so that the diploidisation process is effected in its *proto-uredospore sori*.

In *Puccinia suaveolens*, the production of pycnidia annually on a systemic mycelium: (1) increases the chances that an isolated haploid (+) or (−) pustule derived from a basidiospore and located on a new host-thistle will be diploidised through the agency of insects; and (2) makes it possible for some or many of the primary uredospores to be produced as a result of crossing between different genetic strains. A discussion of the second of the two points here raised will now follow.

In *Puccinia suaveolens*, although every year in the proto-uredospore sori on the haploid branches of a systemic mycelium, in the absence of insects, uredospores can be formed either by self-diploidisation or by invading diploid hyphae, it is possible and very likely that, here and there on the leaves, following insect visits and the mixing of nectar, diploidisation of the proto-uredospore sori is effected as the result of pycnidiospores of one sex fusing with flexuous hyphae of opposite sex. Thus self-diploidisation or its substitute (the formation of uredospores by diploid hyphae) which takes place after a certain time automatically, may be forestalled and replaced by cross-diploidisation effected by means of transferred pycnidiospores.

[1] *Vide supra*, Chapter VI, p. 287.

Crossing between different strains in a single Rust species may be assumed to be favourable to the production of new varieties and, therefore, favourable to evolutionary survival. In *Puccinia suaveolens*, crossing between different strains is favoured by the production of pycnidia annually on systemic mycelia and we may therefore argue that this production of pycnidia is advantageous to *P. suaveolens* as a species.

If a perennating systemic mycelium of *Puccinia suaveolens* did not become partially de-diploidised annually and did not produce pycnidia annually, the uredospores in its uredospore sori would all be genetically identical.[1] It is de-diploidisation and the annual production of pycnidia alone that makes it possible for one and the same systemic mycelium to produce uredospores of diverse genetic constitution.

Systemic Mycelia bearing Secondary Uredospores and Teleutospores only.—It was observed in the field by Olive[2] that systemic mycelia of *Puccinia suaveolens* bearing secondary uredospores and teleutospores are much less common than systemic mycelia bearing pycnidia, primary uredospores, and teleutospores.

A systemic mycelium which on the leaves of infected Creeping Thistles produces nothing but secondary uredospore sori and teleutospores, as Olive[3] observed, is borne on a mycelium which, throughout the whole of the host-plant, consists of binucleate hyphae only. Unlike the type of systemic mycelium described in the previous Section, after entering the young leaves it does not give off haploid branches or, in other words, it does not become partially de-diploidised. In remaining diploid in the leaf-tissues it resembles exactly the mycelia of the localised second-generation infections derived from uredospores and, like these second-generation mycelia, it gives rise to diploid organs only, *i.e.* secondary uredospore sori in which there appear first of all secondary uredospores and subsequently teleutospores.

Brown and I, in 1940, succeeded in obtaining experimentally a systemic mycelium of the type here under discussion in three plants.

(1) A seedling thistle was inoculated with uredospores in the summer of 1939, and the mycelium became systemic in a shoot developed from an axillary bud. This shoot bore pycnidia followed by primary uredospore sori. In the spring of 1940, the plant was placed in a large pot, and its lateral roots soon grew to the edge of the pot. On one of these roots an adventitious bud was formed, and this bud grew up into a green aerial shoot. The shoot was infected with the systemic mycelium. On every leaf the mycelium produced a great abundance of uredospore sori but no pycnidia were present.

[1] *Vide supra*, p. 382.   [2] E. W. Olive, *loc. cit.*, p. 303.   [3] *Ibid.*

What happened in the plant just described seems to have been as follows. The uredospores placed on the first foliage leaves of the seedling germinated and gave rise to a mycelium which grew down the leaves into the stem. From the stem the mycelium grew into an axillary bud; and, as the bud developed into a green shoot, the mycelium in it became partially de-diploidised so that the leaves of the shoot came to bear pycnidia and then primary uredospores. The systemic mycelium in the seedling plant also penetrated into the primary root and thence into a secondary horizontal root. From this root the mycelium grew into an adventitious bud borne on the root and then, as the bud grew into a green aerial shoot, the systemic mycelium on entering the leaf rudiments *failed to become de-diploidised* and, therefore, in all the leaves remained in the diploid (dikaryotic) state. Not having become partly de-diploidised, the mycelium in the leaves had no haploid branches and so did not produce any pycnidia; and owing to its having remained in the diploid condition, it produced diploid secondary uredospore sori. Whether or not teleutospores subsequently developed in these sori was not determined.

(2) In August, 1940, another potted thistle, with an early history similar to that of the thistle already described, sent up from one of its roots an aerial shoot infected with systemic mycelium. This mycelium, without producing any pycnidia, on the backs of all the leaves gave rise to numerous well-developed uredospore sori.

The observations of Brown and myself just recorded teach us that, in *Puccinia suaveolens*, one and the same systemic mycelium derived from uredospores: in one aerial shoot may act as is usual in the field and produce pycnidia, primary uredospores, and teleutospores; and yet in another aerial shoot may produce secondary uredospores followed (presumably) by teleutospores. But why a systemic mycelium in one shoot should become partly de-diploidised and in another shoot of the same plant should not remains an unsolved problem.

(3) As already recorded in an earlier Section,[1] one of two root-buds that had been inoculated with uredospores came to be infected with a systemic mycelium that developed uredospore sori, without previously developing pycnidia. In this case, the dikaryotic diploid mycelium derived from one or more uredospores completely failed to become de-diploidised. In the other root-bud de-diploidisation took place in the usual manner. Here, again, we are presented with an unsolved problem, for the reason why the mycelium in one of the two shoots should have become de-diploidised and in the other should not is not apparent.

[1]Experiments with Root-buds. *Vide supra*, p. 374.

Presumably, de-diploidisation is dependent on a reaction of the dikaryotic diploid mycelium to some stimulus given by the young leaves of the host-plant. We may suppose that the stimulus is a chemical one and that it acts upon the conjugate nuclei in a hypha in such a way as to cause them to separate and become independent of one another; so that, eventually, they and their descendents come to be the nuclei in (+) and (−) haploid hyphae formed by a mycelium that, at first, was dikaryotically diploid. We may further suppose that, if the stimulus postulated were to become very weak or the resistance to it were to become very strong, de-diploidisation would not take place and the dikaryotic diploid mycelium would continue its development unchanged.

Systemic Mycelia Bearing Pycnidia Only.—As already mentioned, Mr. R. E. Taylor of Cambridge University informed me in 1939 that he had observed in the field thistles whose systemic mycelium produced pycnidia only; but exact details of his observations were not communicated to me.

In 1940, at Winnipeg, Brown and I succeeded in obtaining in the greenhouse a systemic mycelium which bore pycnidia only.

In the summer of 1939, a seedling thistle bearing young foliage leaves was inoculated with uredospores and the mycelium became systemic. After being re-potted in February, 1940, the plant sent up a shoot which was infected with a systemic mycelium; but this particular mycelium produced *pycnidia only* and, subsequently, no uredospore sori whatsoever.

The shoot that bore pycnidia only contrasted strikingly with several other shoots on other infected plants that produced pycnidia first and, shortly afterwards, an abundance of uredospore sori.

What happened in the shoot that bore pycnidia only seems to have been as follows. The diploid mycelium derived from the uredospores used for inoculating the seedling plant entered the young stem and then became de-diploidised in such a way that one of its haploid branches, either (+) or (−), entered a bud and grew out into all the leaves. As the systemic mycelium in the shoot was haploid and entirely (+) or entirely (−), it produced haploid pycnidia and haploid proto-uredospore sori and nothing more. Strong evidence that the mycelium was actually entirely (+) or entirely (−) is afforded by the result of diploidisation experiments which will be described in the next Section.

Puccinia suaveolens is Heterothallic.—The fact that the pycnidia on the systemic mycelia of *Puccinia suaveolens* are similar in all essentials to the pycnidia of *P. helianthi* and *P. graminis* is in itself sufficient to suggest that the Creeping-Thistle Rust, like *P. helianthi* and *P. graminis*, is heterothallic.

It may be noted that the pycnidia of the systemic mycelia of *Puccinia suaveolens* are very numerous, reddish, highly scented, nectar-excreting, and attractive to insects and, furthermore, that they expel great numbers of pycnidiospores and are provided not only with periphyses but also with flexuous hyphae. It would indeed be surprising if these pycnidia, which are so well developed, were functionless.

In 1940, Brown and I at Winnipeg obtained evidence that the pycnidia of the systemic mycelia of *Puccinia suaveolens* are actually functional. Our experiments were made with the pycnidia of the systemic mycelium described in the preceding Section. This mycelium, it will be recalled, bore pycnidia only and appeared to be unisexual.

Nectar taken from pycnidia of systemic mycelia on other plants was applied to the pycnidia of the shoot that bore pycnidia only at one particular spot on one leaf.

After a few days primary uredospore sori developed at the spot on the leaf where the nectar had been applied and nowhere else. This experiment was carried out three times in succession and always with the same result: uredospore sori appeared only at the spots where the nectar from the pycnidia of other infected plants had been applied.

Doubtless, in the experiments just described, the mixing of the nectar enabled (+) pycnidiospores to unite with (−) flexuous hyphae of (−) pycnidia, or (−) pycnidiospores to unite with (+) flexuous hyphae of (+) pycnidia, with the result that the haploid cells of the proto-uredospore sori became diploidised and then gave rise to diploid primary uredospores.

The experiments of Brown and myself just described indicate that the pycnidia of the systemic mycelia of *Puccinia suaveolens* are functional. Under natural conditions, the mixing of the nectar of (+) and (−) systemic mycelia must often be effected by insects.

Kursanov imagined that, in proto-uredospore sori, the only migration of nuclei causing diploidisation was from one cell to the next. This was an error. The experiments of Brown and myself indicate that a nucleus of one sex can enter a flexuous hypha and migrate down the flexuous hypha, through the wall of the pycnidium, through intercellular mycelium, and into a proto-uredospore sorus where its descendents by nuclear division are able to diploidise the haploid "fertile" cells and so make the production of diploid uredospores possible. In a proto-uredospore sorus of *Puccinia suaveolens* the "fertile" cells, like the "fertile" cells in a proto-aecidium of *P. helianthi* or *P. graminis*, are all (+) in sex or all (−), and the only ways in which they can be diploidised and come to contain a pair of conjugate nuclei is by the migration to them: (1) of (−) or (+) nuclei derived from pycnidiospores of opposite sex; or (2) of (−) or (+) nuclei derived from neighbouring mycelia.

Experiments Proving that Systemic Mycelia may arise from Uredospores under Natural Conditions.—Already, in an earlier Section, in treating of the artificial introduction of *Puccinia suaveolens* into Manitoba, three experiments were described in which it was shown that the inoculation of wild thistles in the summer with uredospores resulted, in the following spring, in the development of several thistle shoots containing systemic mycelium and bearing thousands and tens of thousands of orange pycnidia followed by primary uredospores. Since the first two of these experiments were made in a locality from which *P. suaveolens* had previously been completely absent and where, therefore, there had never been any teleutospores that could germinate and give rise to basidiospores, it is clear that the systemic mycelia in the spring thistle-shoots must have been derived from the uredospores sown during the previous summer. Some of the uredospores may have fallen upon and caused the infection of young shoots of old thistles, just appearing above the ground, or of the leaves of seedling thistles. We may suppose that the mycelium then grew down the young stems, entered roots, and so made its way into adventitious root-buds that became the aerial shoots in the next spring.

Uredospores and the Persistence of Puccinia suaveolens.—As already stated, the teleutospores of *Puccinia suaveolens* germinate and give rise to basidiospores and the mycelium derived from basidiospores causes the formation of haploid pustules on thistle leaves. After diploidisation, one would expect such pustules to produce uredospores. Doubtless this actually happens although, for the present, it has not been verified by actual observation. For the persistence of *P. suaveolens* from year to year, however, it may be that the production and germination of basidiospores is not of very great importance.

Since (1) Creeping Thistles bearing a systemic mycelium of *Puccinia suaveolens* are fairly common and may persist for many years, since (2) the systemic mycelium on such thistles produces on new shoots great numbers of uredospores annually, since (3) it has been shown that uredospores deposited on seedling thistles and on root-buds of old established thistles may germinate and give rise to new systemic mycelia, and since (4) it has been proved experimentally that, under natural conditions in the open, thistles can readily become infected with systemic mycelium derived from uredospores, it would appear that *P. suaveolens*, even if its teleutospores did not germinate and give rise to basidiospores, could persist indefinitely.

CHAPTER XI

THE GENUS GYMNOSPORANGIUM

Introductory Remarks — The Genus Described — Heteroecism — Geographical Distribution — Species in Central Canada — Cytology and Heterothallism — Appearance and Origin of Cornute Aecidia — Proto-aecidia and their Development into Cornute Aecidia — Two Groups of Species, one with Cupulate the other with Cornute Aecidia — Cornute Aecidia and Puff-balls — Hygroscopic Movements of the Peridium of Cornute Aecidia — Non-violent and Violent Discharge of Aecidiospores in Rust Fungi in general — Evolution of Gymnosporangium — Pores and Pore-plugs in the Aecidiospores of *Gymnosporangium ellisii* and other Rust Fungi.

Introductory Remarks—The genus Gymnosporangium, owing to the peculiar appearance of its teleutospore sori and aecidia and by reason of the economic importance of certain of its species, has long attracted attention, and the gelatinous teleutospore-sorus stage of the species now known as *Gymnosporangium clavariiforme* was described and illustrated by Micheli[1] as long ago as 1729 (Fig. 109).

Today, if the vicissitudes of nomenclature had not intervened, Gymnosporangium would be known as Puccinia. Micheli[2] called what we now know as *Gymnosporangium clavariiforme*, *Puccinia non ramosa*, the generic name having been chosen by him in honour of T. Puccini, a Florentine physician and teacher. As Micheli described and illustrated this fungus so well that its identity is unmistakable, *Puccinia* became the first genus of Uredinales to be definitely established and *Puccinia non ramosa* became the species on which the genus Puccinia was founded. However, in the words of Arthur[3]: "The name Puccinia was afterwards bandied about by various writers without regard to earlier usage, as was the habit among botanists of the time and for many years thereafter, and finally became settled upon the rusts that bear more or less resemblance to *Puccinia graminis*." We

---

[1]P. A. Micheli, *Nova Plantarum Genera*, Florentiae, 1729, Tab. XVII, Fig. 1.
[2]*Ibid.*
[3]J. C. Arthur *et al.*, *The Plant Rusts (Uredinales)*, New York, 1929, p. 36.

owe the name Gymnosporangium to R. A. Hedwig[1] who first employed it 1805.

The genus Gymnosporangium, regarded as a whole, is *macrocylic* (having pycnidiospores, aecidiospores, teleutospores, but no uredo-spores), *heteroecious* (alternate hosts: chiefly Juniperaceae and Mal-aceae), and *heterothallic*. No short-cycled species are known, but G. *nootkatense* differs from all its fellows in having uredospores in addition to all the other spore-forms. A single species, G. *bermudianum*, is autoecious. The tele-utospores are bicellular, as in Puccinia; but the pedicels of the teleuto-spores, unlike those of Puccinia, become more or less evidently gelati-nised, and the conspic-uous gelatinous teleuto-spore sori of such species as G. *sabinae* (Fig. 5, p. 33), G. *clavariiforme* (Fig. 70, p. 202, and Fig. 109), and G. *juniperi-virginianae* (Volume III, Figs. 207-210, pp. 514-517) on Juniper branches are well-known objects. The pycnidia are large and deep red, and they are provided not only with flexuous hyphae but also with well-developed

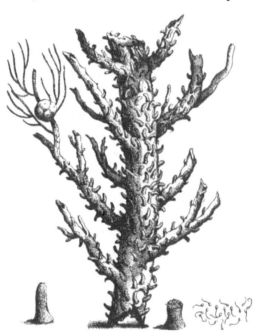

FIG. 109.—*Puccinia non ramosa*, now known as *Gymnosporangium clavariiforme*, as illus-trated by Micheli in his *Nova plantarum genera* in 1729. It shows the gelatinous teliospore sori scattered over the trunk and branches of *Juniperus sabina*, with one healthy branch and cone at the upper left. Below, from left to right: a whole sorus, a truncated sorus, and some teleutospores with their pedicels.

slenderly conical ostiolar Trichomes resembling in appearance those of Puccinia. These trichomes arise at intervals all over the pycnidial wall (Fig. 41, p. 137) and are therefore *paraphyses*. Except in a very few species the aecidia are horn-like. The walls of the aecidiospores are often very thick (Figs. 116, 117, 118, and 122), a characteristic which may possibly be correlated with the fact that the aecidiospores retain their vitality for a long period, germinate better after being subjected to

[1]R. A. Hedwig, 1805.

cold, and are able to germinate and infect their hosts some weeks or months after they have been produced.[1]

Heteroecism.—Heteroecism was discovered in Gymnosporangium almost as soon as in Puccinia; and the story of how *Gymnosporangium sabinae* was found to have alternate hosts is scarcely less interesting than the story of how *Puccinia graminis* was found to pass from grass-plants to Barberry bushes and from Barberry bushes back again to grass-plants.

In Europe *Gymnosporangium sabinae* parasitises Pear-trees (*Pyrus communis*) and Junipers (*Juniperus sabina*).   As far back as 1837, it had been held by Eudes-Deslongchamps[2] that a Juniper tree which grew in his garden was causally connected with the Rust disease of the neighbouring Pear trees.   He therefore had the Juniper cut down and removed.   In the following winter, after the leaves of the Juniper had died and fallen away, he observed that the central branches had been infected with a Gymnosporangium.   During the next summer, the Pear-trees, which for some years previously had always been badly infected with Rust, remained free from the disease.   Eudes-Deslong-champs therefore came to the conclusion that the Rust of the Pear-tree had been caused by the Gymnosporangium.

The connexion between *Gymnosporangium sabinae* on *Juniperus sabina* (Fig. 5, p. 33) and *Roestelia cancellata* on *Pyrus communis* (Fig. 4, p. 32) was proved experimentally beyond doubt by Ørsted in 1865, a few months after de Bary had shown that *Puccinia graminis* is heteroecious.   Ørsted[3] sowed the basidiospores of the Gymnosporangium on Pear leaves.   The experiment was crowned with success;

---

[1]T. Fukushi ("Studies on the Apple Rust caused by *Gymnosporangium yamadae* Miyabe," *Journ. Coll. Agric. Hokkaido, Imp. Univ.*, Vol. XV, 1925, pp. 287-294) showed that exposure to low temperatures increased the germinability of the aecidio-spores of *Gymnosporangium yamadae*; H. E. Thomas and W. D. Mills ("Three Rust Diseases of the Apple," *Cornell Univ. Agric. Exp. Sta.*, Mem. CXXIII, 1929, pp. 14-15) obtained similar results with aecidiospores of *G. juniperi-virginianae* and *G. clavipes* (*G. germinale*); and P. R. Miller ("Pathogenicity of Three Red-Cedar Rusts that Occur on Apple," *Phytopathology*, Vol. XXII, 1932, pp. 723-740), working with *G. juniperi-virginianae*, found that aecidiospores held under natural conditions germinated most readily between January and April.   D. E. Bliss ("The Pathogenicity and Seasonal Development of Gymnosporangium in Iowa," *Iowa Agric. Exp. Sta.*, Research Bull. No. CLXVI, 1933, pp. 337-392) has confirmed the previous work in respect to the effect of low temperatures.

[2]*Vide* H. Klebahn, *Die wirtswechselnden Rostpilze*, Berlin, 1904, pp. 331-333.

[3]A. S. Ørsted, "Vorläufige Berichterstattung über einige Beobachtungen, welche beweisen, dass *Podisoma Sabinae*, welches auf den Zweigen von *Juniperus Sabina* wächst, und *Roestelia cancellata*, welche die Blätter der Birnbäume angreift, derselben Pilzart sind," *Bot. Zeit.*, Vol. XXIII, 1865, pp. 291-293.   For other references *vide* Klebahn.   English translation by F. Currey in *Journ. Roy. Hort. Soc.*, n.s., Vol. I, 1866, pp. 84-86.

for Ørsted had the satisfaction of seeing the Pear leaves become infected and of observing the rust pustules develop spermogonia (pycnidia) about ten days after the spores had been sown, and aecidia subsequently.

Ørsted, by the cultural method, proved: not only (1) the connexion between *Roestelia cancellata* on the Pear and *Gymnosporangium sabinae* on *Juniperus sabina*; but also (2) the connexion between *Roestelia cornuta* on the Mountain Ash or Rowan Tree (*Sorbus aucuparia*) and *Gymnosporangium aurantiacum* on *Juniperus communis*,[1] and (3) the connexion between *Roestelia lacerata* on the Hawthorn (*Crataegus oxyacantha*) and *Gymnosporangium clavariiforme* on *Juniperus communis*.[2]

Since Ørsted's day, other workers have added greatly to our knowledge of the alternate hosts of Gymnosporangium, with the result that we now know that of the nearly fifty species included in the genus only one is autoecious, namely, *G. bermudianum*, while all the others are heteroecious.

Gymnosporangium bermudianum (Farlow) Earle is found in the United States of America along the coast of the Gulf of Mexico, further inland in the State of Georgia, and also in the Bahama Islands and in Bermuda. It is therefore a species with a very restricted geographical distribution. It parasitises four species of Juniper: · *J. virginiana*, *J. barbadensis*, *J. bermudiana*, and *J. lucayana*.[3]

Gymnosporangium bermudianum was first described by Farlow[4] in 1887. He found the aecidia on galls on the Red Cedar (*Juniperus virginiana*) in Bermuda; and, not observing any teleutospore sori, he did not recognise the fungus as a Gymnosporangium and accordingly named it *Aecidium bermudianum*. Galls bearing both aecidia and teleutospore sori were distributed by Semour and Earle[5] in 1872: they described the fungus as a *Gymnosporangium* and remarked "Teleutospore stage following aecidial stage on the same galls in the spring." In 1896, Underwood and Earle[6] noted that *Gymnosporangium ber-*

---

[1]A. S. Ørsted, "Indpodningsforsög, hvorved det bewises, at der finder et Generationsskifte sted mellem den paa Enens Grene snyltende Baevrernst (*Podisoma juniperinum*) og den paa Rönnens Blade voxende Hornrust (*Roestelia cornuta*)," *Oversigt K. danske Vidensk. Selsk. Forhandl.*, 1866, pp. 185-196, Pl. III and IV.

[2]A. S. Ørsted, "Über *Roestelia lacerata* (Sow.) nebst Bemerkungen über die anderen Arten der Gattung *Roestelia*," *Bot. Zeit.*, Bd. XXV, 1867, pp. 222-223.

[3]*Vide* Kern's monograph and Thurston's paper, cited below.

[4]W. G. Farlow, "Aecidium on *Juniperus virginianum*," *Botanical Gazette*, Vol. XII, 1887, pp. 205-207.   [5]Cited from Thurston's paper.

[6]L. M. Underwood and F. S. Earle, "Distribution of *Gymnosporangium* in the South," *Botanical Gazette*, Vol. XXII, 1896, pp. 255-258.

mudianum "unlike its congeners, produces its aecidial and teleutosporic stages on the same host, from the same gall, and in all probability from the same mycelium." Kern,[1] in 1911, described the species more fully and published a photograph showing the cylindrical-cornute aecidia projecting from a gall on a Juniper. In 1923, Thurston,[2] with the help of the microtome, investigated *G. bermudianum* cytologically and, in one of his illustrations he shows a section through a gall, which includes an aecidial sorus and also a teleutosorus. Thus the fact that *G. bermudianum* is autoecious has become well established.

Geographical Distribution. — The geographical distribution of Gymnosporangium species in relation to their hosts was discussed by Kern[3] in his monograph on Gymnosporangium in 1911. In more recent years Crowell[4] has devoted his attention to this subject and, in 1940, he embodied the results of his study in a paper provided with forty-three maps showing the geographical distribution of Gymnosporangium species and of the associated alternate hosts.

Crowell's conclusions were as follows. The species of Gymnosporangium are found in the northern hemisphere only and occur most abundantly in the temperate zone. Three species (broadly conceived), namely, *Gymnosporangium aurantiacum* Chev., *G. clavariiforme* (Jacq.) DC., and *G. juniperinum* (L.) Mart. are found in all the three continents mainly concerned, *i.e.* Europe (including the Mediterranean coast of Africa), Asia, and North America. In addition, in each of these three continents there is a distinctive endemic set of Gymnosporangium species. The genus contains about 48 species of which 33 occur in North America, 15 in Asia, and 6 in Europe (including the three tricontinental species in each record).

In respect to their distribution, according to Crowell,[5] the North American species of Gymnosporangium are divisible into two groups: (A) species that occupy all potential territory covered by the coincident ranges of their alternate hosts, *e.g. G. libocedri* and *G. nootkatense*; and (B) species that occupy less than all potential territory. The latter

[1]F. D. Kern, *A Biologic and Taxonomic Study of the Genus Gymnosporangium*, Bulletin of the New York Bot. Garden, Vol. VII, No. 26, 1911, p. 475 and Plate CLIII, Fig. 43.

[2]H. W. Thurston, "Intermingling Gametophytic and Sporophytic Mycelium in *Gymnosporangium bermudianum*," *Botanical Gazette*, Vol. LXXV, 1923, pp. 225-248, Fig. 4 in the text.

[3]F. D. Kern, *A Biologic and Taxonomic Study of the Genus Gymnosporangium*, Bulletin of the New York Bot. Garden, Vol. VII, No. 26, pp. 397-400.

[4]I. H. Crowell, "The Geographical Distribution of the Genus Gymnosporangium," *Canadian Journal of Research*, Vol. XVIII, C, 1940, pp. 469-488, with 43 maps in the text.   [5]*Ibid.*, pp. 482-488.

group may be subdivided as follows: (1) species that are confined by the range of their primary telial host, *e.g. G. juniperi-virginianae* and *G. juvenescens*; (2) localised species that are confined within a portion of the coincident ranges of their alternate host, *e.g. G. corniculans* and *G. externum*; and (3) widely distributed species that are not limited in their range by either alternate host-group, namely, the three tri-continental species, *G. aurantiacum*, *G. clavariiforme*, and *G. juniperinum*.

Species in Central Canada.—In central Canada, I have found and personally identified the following five species of Gymnosporangium:

(1) *G. aurantiacum*. O, I on *Pyrus (Sorbus) americana*; III on *Juniperus communis*. At Minaki (western Ontario). O, I, also at Kenora (western Ontario). This species was observed by Bisby[1] on Mountain Ash leaves as far north as Norway House at the north end of Lake Winnipeg.

(2) *G. clavariiforme*. O, I on *Amelanchier alnifolia*; III on *Juniperus communis*. At Minaki.

(3) *G. clavipes*. O, I on *Amelanchier alnifolia*; III on *Juniperus communis*. At Victoria Beach, on the east side of Lake Winnipeg; also at Minaki.

(4) *G. corniculans*. O, I on *Amelanchier alnifolia* probably, since these stages were near to III forming galls on *Juniperus horizontalis*. At Pine Ridge (north of Winnipeg). III at Saskatoon, on banks of the South Saskatchewan River.

(5) *G. juvenescens*. O, I on *Amelanchier alnifolia* probably, since these stages were near to III forming witches brooms on *Juniperus horizontalis*. At Pine Ridge. III at Saskatoon, on banks of the South Saskatchewan River.[2]

By cultural methods I succeeded in transferring the last four species from Juniper bushes to the leaves of *Amelanchier alnifolia*.[3]

Cytology and Heterothallism.—In 1896, Sappin-Trouffy[4] described the nuclear condition throughout the life-cycle of *Gymnosporangium sabinae* and *G. clavariiforme* and also in the aecidium of *G. aurantiacum* his *G. juniperinum*), and he showed that it was similar to that of

---

[1]G. R. Bisby *et al.*, *The Fungi of Manitoba and Saskatchewan*, National Research Council of Canada, Ottawa, 1938, p. 64.

[2]Other species of Gymnosporangium mentioned by Bisby as occurring in Manitoba and Saskatchewan are: *G. betheli*, *G. globosum* (?), and *G. nelsoni*.

[3]*Cf.* in Chapter II, p. 92.

[4]P. Sappin-Trouffy, "Recherches histologiques sur la famille des Urédinées," *Le Botaniste*, T. V, 1896, pp. 59-244.

Puccinia graminis and other long-cycled Rust Fungi. Sappin-Trouffy's observations were confirmed in 1904 by Blackman[1] in investigations made on *Gymnosporangium clavariiforme* and *Phragmidium violaceum.* "In this cycle," said Blackman, "the mature teleutospore is uninucleate and gives rise to four uninucleate sporidia, from which a mycelium is developed with the nuclei arranged singly, usually in separate cells. The spermatia produced on this mycelium are uninucleate, but in the young aecidium the nuclei become paired (forming binucleate cells) and divide together in very close association. This paired condition is then persistent throughout the rest of the life-cycle (aecidiospores, uredospores, and mycelia produced from them) up to the formation of the teleutospores, which in the young state are binucleate, but when mature become uninucleate by the fusion of the two paired nuclei." In 1915, Kursanov[2] gave an account of the nuclear condition of the aecidial rudiments and aecidia of *Gymnosporangium juniperinum* (his *G. tremelloides*) and of *G. aurantiacum*, and he showed that the aecidial rudiments are at first uninucleate but later give rise to binucleate hyphae which produce binucleate aecidiospores and peridial cells.

Owing to difficulties in obtaining early stages, Sappin-Trouffy, Blackman, and Kursanov were not able to investigate how, in the aecidium of Gymnosporangium species, the so-called fertile cells pass from the uninucleate to the binucleate condition. However, the theoretical solution of this problem became possible as soon as Craigie in 1927 had proved by experiment that *Puccinia graminis* and *P. helianthi* are heterothallic and it was realised that long-cycled Rust Fungi in general are heterothallic.

That all the heteroecious species included in the genus Gymnosporangium are heterothallic, with the basidiospores giving rise to (+) haploid mycelia and (−) haploid mycelia, may be inferred from the fact that all of the four species so far tested, namely, *Gymnosporangium juniperi-virginianae*, *G. globosum*, *G. haraeanum*, and *G. clavipes*, have proved to be heterothallic.

Experimental proof of heterothallism was obtained: (1) for *Gymnosporangium juniperi-virginianae* by Hanna (1929, recorded in this volume) and Miller (1932); (2) for *G. globosum* by Miller (1932); (3) for

[1]V. H. Blackman, "On the Fertilization, Alternation of Generations, and General Cytology of the Uredineae," *Annals of Botany*, Vol. XVIII, 1904, pp. 323-373.

[2]L. Kursanov, "Morphological and Cytological Researches on the Uredineae (Russian)," *Sci. Mem. Nat. Hist. Imp. Univ. Moscou*, No. XXXVI, 1915, 228 pp., 6 plates. French translation: "Recherches morphologiques et cytologiques sur les Urédinées," *Bull. Soc. Nat. Moscou*, n. ser. (1917), T. XXXI (*i.e.* XXX), pp. 1-129, 4 plates, 1922.

G. haraeanum by Kawamura (1934); and (4) for *G. clavipes* by myself (recorded in this volume).[1]

In *Gymnosporangium juniperi-virginianae*, Hanna observed that simple pustules formed on Apple leaves as a result of infection by single basidiospores remain sterile, while certain compound pustules formed by the coalescence of two simple pustules give rise to aecidia; and, in the same fungus, Miller demonstrated that the formation of aecidia can be induced by applying mixed nectar containing both (+) and (−) pycnidiospores to the upper surface of (+) or (−) pustules. The mixed nectar technique was also employed in demonstrating hetero-thallism by Miller in *G. globosum* and by Kamamura in *G. haraeanum*. In *G. clavipes*, after mixed nectar had been applied to some simple pustules, actual fusions of a (−) pycnidiospore with a (+) flexuous hypha or of a (+) pycnidiospore with a (−) flexuous hypha were observed by myself, and seven such fusions are illustrated in Fig. 80 (p. 239).

While there is every reason to suppose that all the ordinary species of Gymnosporangium that produce pycnidia and aecidia on a dicotyledonous host and teleutospores and basidiospores on a coniferous host are heterothallic, the sexual condition of *Gymnosporangium bermudianum*, the only known autoecious species of the genus, is uncertain. No one has reported pycnidia and it has therefore been assumed by Arthur, Thurston, and others that pycnidia are not included in the life-history.

Jackson[2] has supposed that the basidiospores of *Gymnosporangium bermudianum* suddenly became possessed of the power to infect Junipers and thus the haploid phase was transferred to the Juniper, so that now both haploid and diploid phases are on one and the same coniferous host. If we suppose that the basidiospores produced on Juniper galls infect Juniper leaves and that pycnidia are never produced on the galls, then we may infer that *G. bermudianum* is homothallic; for, in the Uredinales, the twenty-one species that have been proved to be heterothallic have basidiospores that infect host-plants and give rise to pycnidia, and the four species that have been proved to be homothallic have basidiospores that infect host-plants but do not give rise to pycnidia or give rise to pycnidia that are non-functional and vestigial.[3]

---

[1]References to the work of these authors is made in connexion with the list of heterothallic Rust Fungi given in Chapter I, and details of some of the experiments have been recorded in Chapter III.

[2]H. S. Jackson, "Present Evolutionary Tendencies and the Origin of Life Cycles in the Uredinales," *Mem. Torrey Bot. Club*, Vol. XVIII, 1931, p. 73.

[3]*Vide* Chap. VI.

The assumption that the basidiospores of *Gymnosporangium bermudianum* produced on Juniper leaf-galls are able to infect Juniper leaves and so form fresh galls has yet to be proved and may not be correct. In other Gymnosporangia, *e.g. G. juniperi-virginianae*, it is the aecidiospores that infect the Juniper leaves and stimulate the formation of galls, and it is possible that gall-formation by *G. bermudianum* is entirely due to infection of leaves by the aecidiospores.

Thurston[1] attempted to infect seedling trees of *Juniperus barbadensis*, 6-18 inches high, raised in a greenhouse, with the basidiospores and aecidiospores of *Gymnosporangium bermudianum*, unfortunately without any success; and therefore, by experimental means, he was unable to throw any light on the life-history of the fungus.

Thurston's experiments ought to be repeated with variations in the greenhouse conditions or, better still, on Juniper trees growing under natural conditions; for there is every reason to suppose that, in the open, galls are formed as a result of infection of leaves by aecidiospores.

On examining stained microtome sections of galls containing *Gymnosporangium bermudianum*, Thurston observed that the aecidia are produced on uninucleate mycelium and the teleutospore sori on binucleate mycelium. Binucleate mycelium was found in the galls underneath the uninucleate, and it was noticed that the uninucleate mycelium, after it had given rise to aecidia, appeared to be moribund or dead.

What is the origin of the two kinds of mycelium, uninucleate and binucleate, in a gall of *Gymnosporangium bermudianum*? In an attempt to answer this question, Thurston[2] suggested that the Juniper is first infected by aecidiospores, as is the case with all other Gymnosporangia, and that the gall thus resulting is reinfected by basidiospores from germinating teleutospores. Thus, in a gall, the binucleate mycelium is derived from the germ-tubes of the binucleate aecidiospores, while the uninucleate mycelium is derived from the germ-tubes of the uninucleate basidiospores. This theory, which involves a double infection of each gall, seems too complicated and but little likely to be true.

Thurston assumed that a binucleate aecidiospore or uredospore can give rise exclusively to a binucleate mycelium; but this we now know, on the basis of experimental and cytological investigations made on *Puccinia minussensis*[3] and *P. suaveolens*,[4] is not always true. Thus,

[1] H. W. Thurston, "Intermingling Gametophytic and Sporophytic Mycelium in *Gymnosporangium bermudianum*," *Botanical Gazette*, Vol. LXXV, 1923, pp. 231-232.

[2] *Ibid.*, p. 245.

[3] *Vide* Chapter VI.

[4] *Vide* Chapter X.

if one infects a Blue Lettuce plant (*Lactuca pulchella*) with binucleate aecidiospores or binucleate uredospores of *P. minussensis*, the binucleate mycelium so produced, on passing into the leaves, usually becomes de-diploidised, so that haploid uninucleate hyphae come into existence, and it is on this uninucleate mycelium that the pycnidia and proto-aecidia arise. Subsequently, the proto-aecidia become diploidised by the entry into them of nuclei of opposite sex, and so binucleate aecidia come to be developed.

In the light of the facts just presented, we may suppose that, in *Gymnosporangium bermudianum*, when infection of a Juniper leaf has taken place as a result of the germination of an aecidiospore, the germ-tube produces in the gall a binucleate mycelium which sooner or later undergoes partial de-diploïdisation, with the result that in a normal gall of large size there comes into existence near the surface a mass of uninucleate mycelium which gives rise to aecidia and, deeper within the gall, a mass of binucleate mycelium which eventually produces the overwintering teleutospore sori. On this theory, which it would be easy to test if once galls could be obtained as a result of sowing aecidiospores, the necessity for supposing that a gall is infected first by one or more aecidiospores and then by one or more basidiospores is avoided.

If we assume that the basidiospores of *Gymnosporangium bermudianum* do not infect Juniper leaves and therefore do not cause gall-formation and also that the uninucleate mycelium in the galls is always formed as a result of de-diploidisation of a binucleate mycelium developed from the germ-tubes of one or more aecidiospores, it is conceivable that *G. bermudianum* is heterothallic in a limited sense. Already, in a discussion of the sexual process in *Puccinia suaveolens* and *P. minussensis*,[1] the phenomena of de-diploidisation and of re-diploidisation by self-diploidisation have been discussed, and these phenomena may also take place in the galls of *Gymnosporangium bermudianum*. We may suppose that the uninucleate mycelium in a gall, formed by de-diploidisation of a bisexual, binucleate mycelium derived from the germ-tube of an aecidiospore, consists of (+) branches and (−) branches, both of which form haploid proto-aecidia; and, further, we may suppose that the (+) proto-aecidia are diploidised by (−) nuclei which migrate to them from (−) hyphae and that the (−) proto-aecidia are diploidised by (+) nuclei which migrate to them from (+) hyphae.

If the basidiospores of *Gymnosporangium bermudianum* do not infect Juniper leaves, then the basidia and basidiospores, which are known from Thurston's observations to be actually produced by the

[1] *Vide* Chapters VI and X.

teleutospores,[1] are either now functionless or they are still able to infect a dicotyledonous host and give rise to pustules bearing pycnidia and aecidia. The latter possibility seems improbable, for no Rust Fungus is known that regularly produces aecidia on both of the alternate hosts. Nevertheless, attempts should be made to infect the leaves of Rosaceous trees and shrubs, such as Amelanchier and

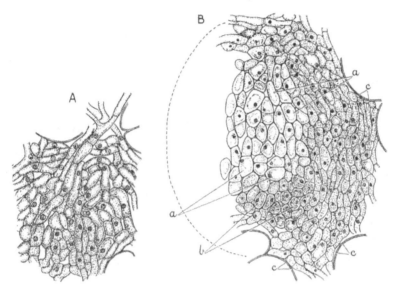

FIG. 110.—*Gymnosporangium juniperium* (Kursanov's *G. tremelloides*) in an Apple leaf. A very young (+) or (−) proto-aecidium (Kursanov's *primordium*) composed of a weft of hyphae some of which are directed from below upwards and are continued as a vegetative mycelium. B, an older proto-aecidium: oval in form; *a a*, sterile cells (corresponding to the pseudoparenchyma of a proto-aecidium of *Puccinia graminis*); *b*, fertile cells, full of protoplasm and destined to be diploidised; *c c*, host-cells. That the proto-aecidium in both drawings is haploid is indicated by the uninucleate condition of the hyphal cells. Copied and lettered by A. H. R. Buller, from Kursanov's *Recherches* (1917). Magnification, 340.

Crataegus, and other possible hosts; for, if positive results should be obtained from such experiments, our ideas concerned with the life-history and sexuality of *G. bermudianum* would have to be revised.

Assuming that *Gymnosporangium bermudianum* has repeating aecidiospores on Junipers, then we may conclude that this species has come to resemble certain Cronartium species in which the aecidiospore stage repeats itself on Pines. Among these species, listed in the form-genus Peridermium, are *Peridermium pini* on *Pinus sylvestris* in

[1]H. W. Thurston, *loc. cit.*, p. 231.

Europe,[1] Meinecke's Peridermium on *Pinus alternuata*, *P. contorta*, *P. jeffreyi*, *P. ponderosa*, and *P. radiata* in the western United States,[2] the so-called *Woodgate Peridermium* on *Pinus sylvestris* in New York State,[3] and a Peridermium on *Pinus banksiana* in New York State and Massachusetts.[4]

Appearance and Origin of Cornute Aecidia.—The appearance of mature *cornute* (horn-like) or *roestelioid* (beak-like) aecidia character-istic of so many species of Gymnosporangium has already been well shown: for *G. aurantiacum* (Rostrup's *G. juniperinum*) on the back of a leaflet of *Pyrus aucuparia* in Fig. 3 (p. 31); and for *Gymnosporangium sabinae* on the back of a leaf of *Pyrus communis* (Pear), with two aecidia enlarged, in Fig. 4 (p. 32).

As Arthur,[5] in treating of the aecidia of Gymnosporangium has remarked: "The cornute or roestelioid aecium when mature and un-ruptured has a horn-like form, and is usually deeply colored, due to the dark colored spores within. It is definite in form, with its long axis perpendicular to the leaf surface, and never without a peridium. The primordium is ovoid, instead of globoid as in the cupulate form (Richards,[6] 1896; Kursanov,[7] 1915). It usually arises deep in the mesophyll tissues of its host. The hymenium is composed of long and slender basal cells pressed closely together" (Fig. 115).

Proto-aecidia and their Development into Cornute Aecidia.— Kursanov[8] (1915-1922) investigated the development of cornute

---

[1]G. Haack, "Der Kienzopf (*Peridermium pini* (Willd.) Kleb.). Seine Ueber-tragung von Kiefer zu Kiefer ohne Zwischenwirt," *Zeitschr. Forst. u. Jagdw.*, Bd. XLVI, 1914, pp. 3-46; also H. Klebahn, "*Peridermium pini* (Willd.) Kleb. und seine Uebertragung von Kiefer zu Kiefer," *Flora*, Bd. CXI—CXII, 1918, pp. 194-207.

[2]E. P. Meinecke: "*Peridermium harknessii* and *Cronartium quercuum*," *Phyto-pathology*, Vol. VI, 1916, pp. 225-240; "Faculative heteroecism in *Peridermium cerebrum* and *Peridermium harknessi*," *Phytopathology*, Vol. X, 1920, pp. 279-297; "Experiments with Repeating Pine Rusts," *Phytopathology*, Vol. XIX, 1929, pp. 327-342.

[3]H. H. York, "Inoculations with Forest Tree Rusts," *Phytopathology*, Vol. XXVIII, 1938, pp. 210-212.

[4]M. A. McKenzie, "Experimental Autoecism and other Biological Studies of a Gall-forming Peridermium on Northern Hard Pines," *Phytopathology*, Vol. XXXII, 1942, pp. 785-798.

[5]J. C. Arthur, *The Plant Rusts (Uredinales)*, New York, 1929, p. 132.

[6]H. M. Richards, "On Some Points in the Development of Aecidia," *Proc Amer. Acad. Arts and Sci.*, Vol. XXXI, 1896, p. 265.

[7]L. Kursanov, "Morphological and Cytological Researches on the Uredineae (Russian)," *Sci. Mem. Nat. Hist. Imp. Univ. Moscou*, No. XXXVI, 1915, 228 pp., 6 plates. French translation: "Recherches morphologiques et cytologiques sur les Urédinées," *Bull. Soc. Nat. Moscou*, n. sér. (1917), T. XXXI (*i.e.* XXX), pp. 1-129, 4 plates, 1922.   [8]*Ibid.*, pp. 30-36.

aecidia. In the cornute aecidium of *Gymnosporangium juniperinum* (Kursanov's *G. tremelloides, Roestelia penicillata*), which develops on Apple leaves, Kursanov found that the proto-aecidium (his primordium) is formed from hyphae which grow toward the epidermis (Fig. 110, A). By division of the hyphae into uninucleate, more or less isodiametric cells, a plectenchyma arises as a small ellipsoidal area within the central region of the proto-aecidium (Fig. 110, B). These cells increase somewhat in size, their protoplasm diminishes in quantity, and their walls undergo gelatinisation. Evidently, they are equivalent to the sterile cells that compose the pseudoparenchyma in a proto-aecidium of a Puccinia. In *Gymnosporangium aurantiacum* (Kursanov's *G. juniperinum, G. cornutum*) on Sorbus leaves the additional cross-walls are not formed and consequently the cells of the plectenchyma remain oblong (Fig. 111). Diploidatision[1] is effected in cornute aecidia deep within the fertile part of the proto-aecidium, and the first binucleate cells appear as members of branched hyphae (Fig. 112). These binucleate hyphae push upwards into the gelatinous mass of disorganised cells of the plectenchyma (Fig. 111) and, as soon as they have grown from one-third to half-way up, their terminal cells form an indistinct hymenium and begin to function as aecidiosporophores. The

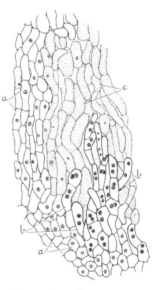

Fig. 111.—*Gymnosporangium aurantiacum* (Kursanov's *G. juniperinum*). Diploidised (+) or (−) proto-aecidium on a Sorbus leaf: *a*, haploid wall-cells; *b*, dikaryotic diploid hyphae pushing upwards; *c*, oblong sterile cells, becoming gelatinous. Copied and lettered by A. H. R. Buller. From Kursanov (1917). Mag. 340.

first cells cut off by the aecidiosporophores are at the centre of the hymenium (Fig. 113). They increase in size, thicken their cell-walls, and so take on the appearance of peridial elements. The central aecidiosporophores give rise to three or four peridial cells, their immediate neighbours fewer, and the marginal aecidiosporophores one only (Fig. 114). Thus there is formed a more or less conical and rigid peridial cap (Fig. 114). The aecidiosporophores below the cap

[1]The term *diploidisation* is my own. Kursanov's work was done prior to the discovery of heterothallism in the Hymenomycetes and Uredinales.

then form chains of aecidiospores in the usual way, and the cylindrical lateral wall of the peridium, as in *Puccinia graminis*, originates from a ring of aecidiosporophores which produce chains of aecidiospores that become metamorphosed into peridial cells (Fig. 115).

Kursanov[1] has shown that in *Gymnosporangium juniperinum* (his *G. tremelloides*), just as in *Puccinia graminis*, the intercalary cells cut off laterally by the peridial wall-cells form a layer of cells between the peridium and the surrounding cells, that becomes gelatinous (Fig. 38, C). Here, too, we may suppose that the jelly acts as a lubricating agent as the long cornute aecidium pushes its way out of the host-leaf.

The hard conical peridial cap of a roestelioid aecidium such as that of *Gymnosporangium aurantiacum* (Fig. 114) is pushed up, as the aecidium grows in length, through the thick overlying host tissues, and it thus functions as an efficient boring organ. The conical peridial cap of *G. juniperi-virginianae* just after the aecidium has bored through the epidermis of its host is illustrated in Fig. 122, D. When a stout peridial cap is present, the peridium of a mature aecidium does not usually open at the apex but ruptures laterally, thereby forming longitudinal slits (Figs. 4 on p. 32, 116, A, on p. 405, and 120 on p. 408).

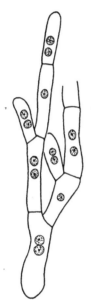

Fig. 112

G. aurantiacum. Branching hypha in a diploidised proto-aecidium. Kursanov (1917). Mag. 750.

Two Groups of Species, one with Cupulate the other with Cornute Aecidia.—In the genus Gymnosporangium, in respect to the structure of the peridium[2] and the mode of liberating the aecidiospores into the air, there are two contrasting groups of species: Group I and Group II.

In Group I, containing relatively few species (four in N. Amer.: *Gymnosporangium nootkatense* (*Aecidium sorbi*), *G. libocedri* (*Aecidium blasdaleanum*), *G. speciosum*, and *G. ellisii*)[3]: (1) the aecidia are cupulate (aecidioid); (2) the outer walls of the peridial cells are thicker than the inner; and (3), as we know from B. O. Dodge's

[1]L. Kursanov, "Über die Peridienentwicklung im Aecidium," *Ber. d. Deutschen Bot. Gesellschaft*, Jahrg. XIV, 1914, pp. 317-327.

[2]F. D. Kern: "The Morphology of the Peridial Cells in the Roesteliae," *Botanical Gazette*, Vol. XLIX, 1910, pp. 445-452, Text-figs. 1 and 2, and Plates XXI and XXII; and "A Biologic and Taxonomic Study of the Genus Gymnosporangium," *Bulletin of the New York Botanical Garden*, Vol. VII, 1911.

[3]For the structure of the peridium of *Gymnosporangium ellisii*, *vide* J. C. Arthur, *Manual of the Rusts in United States and Canada*, Lafayette, U.S.A., 1934, pp. 359-360.

investigation of *G. ellisii*,[1] the aecidiospores are shot away violently into the air. Thus the aecidia of Group I, in their general structure and mode of discharging their spores, exactly resemble the cupulate aecidia of *Puccinia graminis*.

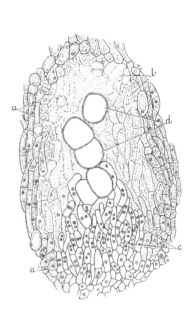

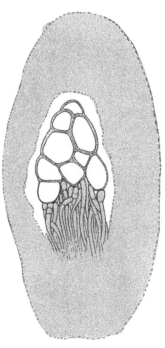

Fig. 113.—*Gymnosporangium aurantiacum*. Young aecidium: *a a*, haploid bounding cells; *b*, sterile haploid cells that have become gelatinous; *c*, dikaryotic diploid cells forming a hymenium; *d*, the first-formed peridial cap-cells. Copied from Kursanov (1917) by A. H. R. Buller. Magnification, 250.

Fig. 114.—*Gymnosporangium aurantiacum*. Aecidiosporophores supporting the cap-cells of the cornute peridium. The dotted area indicates the position of host tissues. Copied from Kursanov (1917) by A. H. R. Buller. Magnification, 190.

On the other hand, in Group II, containing relatively many species including the common ones, (1) the aecidia are horn-like (cornute or roestelioid); (2) the inner walls of the peridial cells are thicker than the outer; and (3) the spores are not violently discharged from the aecidia but form a powdery mass which is gradually extruded through slits and other openings in the peridium with the help of hygroscopic movements of the peridial cells.

[1]B. O. Dodge, "Aecidiospore Discharge as Related to the Character of the Spore Wall," *Journ. Agric. Research*, Vol. XXVII, 1924, pp. 749-756.

Illustrations of Cornute Aecidia.—The long horn-like aecidia of *Gymnosporangium aurantiacum* as seen on leaves of *Pyrus aucuparia* (Mountain Ash) in Europe are well shown in Rostrup's illustration reproduced in Chapter I as Fig. 3 (p. 31), and the lateral slits in two aecidia of *Gymnosporangium sabinae* on a leaf of *Pyrus communis* (Pear)

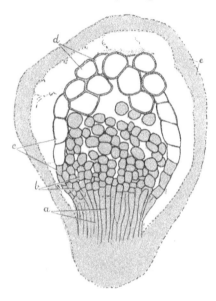

are shown in the same Chapter in another of Rostrup's illustrations reproduced as Fig. 4 (p. 32).

D. E. Bliss,[1] in his studies of the pathogenicity and seasonal development of Gymnosporangium in Iowa, U.S.A., published drawings of the cornute aecidia of *Gymnosporangium globosum*, *G. clavipes*, *G. clavariiforme*, and *G. juniperi-virginianae* which show: the aecidia split down their sides thus permitting of the escape of the aecidiospores; peridial cells with their inner ornamented walls much thicker than their smooth thin outer walls; and some isolated aecidiospores with their thick walls seen in optical section. These illustrations have been reproduced two-thirds of the original size in Figs. 116, 117, 118, and 122.

FIG. 115.—*Gymnosporangium juniperinum* (Kursanov's *G. tremelloides*). Aecidium on an Apple leaf: *a*, aecidiosporophores; *b*, aecidiospores; *c*, wall-cells of peridium; *d*, cap-cells of peridium; *e*, tissues of host. Copied and lettered by A. H. R. Buller. From Kursanov (1917). Magnification, 125.

Fischer's[2] illustration of a cross-section of the peridium of an aecidium of *Gymnosporangium clavariiforme* in which it is well shown that the inner walls of the peridial cells are, contrary to what is found in *Puccinia graminis* (Fig. 38, p. 120, and Fig. 39, p. 121) and *Gymnosporangium ellisii*, much thicker than the outer walls, has been reproduced in Fig. 119, A.

Some photographs of the cornute aecidia of *Gymnosporangium clavariiforme* illustrating their mode of dehiscence, made by Dr. B. O. Dodge and kindly given to the author, are reproduced in Figs. 120 and 121.

[1]D. E. Bliss, "The Pathogenicity and Seasonal Development of Gymnosporangium in Iowa," *Iowa Agric. Experiment Station*, Research Bulletin No. CLXVI, 1933, pp. 337-392.

[2]E. Fischer, *Die Uredineen der Schweiz*, Bern, 1904, Fig. 275, p. 383.

Cornute Aecidia and Puff-balls.—In their mode of liberating spores, the aecidiosporocarps of the cornute species of Gymnosporangium may be compared with the basidiosporocarps of such Puff-balls as Lycoperdon and Calvatia.

In Puff-balls, the basidia do not shoot away their basidiospores. Instead, the basidium-bodies, along with most of the tramal cells, are dissolved by autodigestion, and then the basidiospores dry up and form

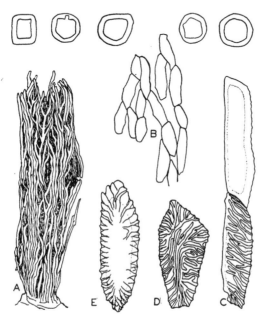

FIG. 116.—*Gymnosporangium globosum. A*, an aecidium, which was cyclindric, dehiscing: its upper part is split and it has become reticulate below. The aecidiospores gradually escape as a powder through the apertures. B, a group of peridial cells in face view, the cells are attached more securely at ends than at sides. C, two peridial cells in side view- the thickest wall of each cell is directed toward the interior of the aecidium; ridge-like papillae cover the inner and side walls. D, inner face view of a peridial cell. E, outer face view of a peridial cell. F, aecidiospores showing their thick walls. Drawings by D. E. Bliss (1933), arranged by the author. Magnification: A, 42; B, 144; C, D, and E, 434.

a light-brown powder, loosely held among the capillitium fibres. Finally, the peridium breaks away above or falls away over a large area of the sporocarp, thus permitting the basidiospores to be carried off by the wind when this is blowing with sufficient strength. It may well be that the Puff-balls have been derived from hymenomycetous ancestors in which, as happens now in all Hymenomycetes, the spores were violently discharged from their sterigmata and so were projected into the outer air as soon as they were ripe and whilst they were still turgid.

So, too, we may suppose that, in Gymnosporangium, the species of Group II, *e.g.* *G. aurantiacum* and *G. clavariiforme*, in which the aecidiospores escape as a powder through the cleft peridium have been derived from ancestors resembling the species of Group I, *e.g.* *G. ellisii*, in which the aecidia were not cornute but cupulate and the aecidiospores were discharged violently into the air singly and in succession.

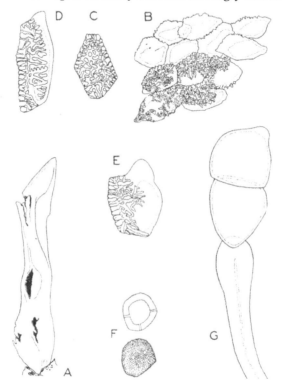

Fɪɢ. 117.—*Gymnosporangium clavipes* (*G. germinale* of Bliss). A, an aecidium: coarse cracks have formed down the sides, allowing the escape of the aecidiospores. B, a group of peridial cells. C, a peridial cell as seen from within the aecidium. D and E, peridial cells in side view; the thicker vertical cell-wall faces the interior of the aecidium. F, two aecidiospores, one in optical section showing three germ-pores, the other showing verrucose papillae on its surface. G, a teleutospore with its clavate pedicel. Drawings by D. E. Bliss (1933) arranged by the author. Magnification: A, 16; B, 167; C, D, E, 332; F, 336; G, 596.

Hygroscopic Movements of the Peridium of Cornute Aecidia.— The roestelioid aecidia of Gymnosporangium grow relatively slowly and dehisce only after they have attained their full length; and they take much longer to discharge their spores than do such cupulate aecidia as those of *Puccinia graminis* and *P. helianthi*.

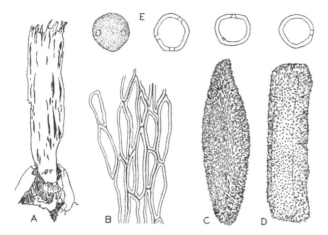

Fig. 118.—*Gymnosporangium clavariiforme*. A, an aecidium, the top portion of which has been shattered; the peridium shows vertical rents through which aecidio-spores can escape; a few aecidiospores are shown on the peridium near its base. B, a group of peridial cells. C, the inner face of a peridial cell as seen from the interior of the aecidium. D, a peridial cell in side view; the thicker vertical wall faces the interior of the aecidium. E, aecidiospores, one showing surface markings, the others in optical section showing germ-pores. Drawings by D. E. Bliss (1933) arranged by the author. Magnification: A, 16; B, 157; C, D, E, 418.

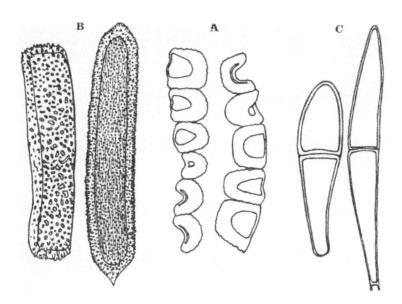

Fig. 119.—*Gymnosporangium clavariiforme*. A, cross-section of a peridium, showing that the thick walls of the peridial cells face toward the interior of the aecidium. B, two peridial cells, one (on left) in side view, the other showing its inner face. C, two teleutospores. From Fischer's *Die Uredineen der Schweiz* (1904).

In 1910, Kern[1] remarked that when he placed dry peridial cells of certain Gymnosporangium species having roestelioid aecidia, *e.g.* those of *G. juniperi-virginianae*, in water they became bent inwards (Fig. 122, E) and that, when they were allowed to dry, they straightened out again (Fig. 122, A). The hygroscopic movements of the peridial ribbons of *G. juniperi-virginianae* were described by Lloyd and Ridgway[2] in 1910, Reed and Crabill[3] in 1915, and Bliss[4] in 1933. All these writers agree in stating that in this species the peridial strips are

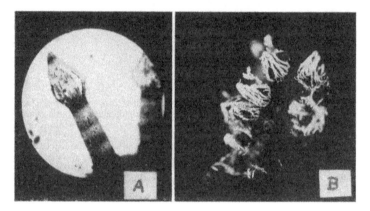

FIG. 120.—*Gymnosporangium clavariiforme* on *Amelanchier alnifolia*. Dehiscence of the cornute aecidium. A, early stage: below the cap-cells the peridium is bulging outwards and becoming lacerate. B, late stage: several aecidia in which the peridium has become lacerate to the base. The aecidiospores are not shot out of the aecidium but escape gradually as a powder. Photographed by B. O. Dodge. Magnification: A, about 20; B, about 6.

strongly recurved when dry but close over the mouth of the aecidium when wet. Lloyd and Ridgway, one of whose illustrations showing a closed aecidium has been reproduced by Arthur,[5] have suggested that the closing inwards of the peridium in wet weather is advantageous in that it prevents the escape of the spores in wet weather; but, possibly, the closing in is only an extreme form of the hygroscopic movements which tend to work the aecidiospores gradually out of the aecidium.

[1]F. D. Kern, 1910, *loc. cit.*, p. 449.

[2]F. E. Lloyd and C. S. Ridgway, "Cedar Apples and Apples," *Alabama Department of Agric.*, Bull. XXXIX, 1911, pp. 12-14, Figs. 8-10. Same article in *Annual Report of Alabama Department of Agric.*, 1911, pp. 11-27.

[3]H. S. Reed and C. H. Crabill, "The Cedar Rust of Apples caused by *Gymnosporangium juniperi-virginianae* Schw.," *Virginia Agric. Experiment Station*, Bull. IX, 1915, pp. 48-49.

[4]D. E. Bliss, *loc. cit.*, p. 2, in a description of his Fig. 2.

[5]J. C. Arthur, *The Plant Rusts (Uredinales)*, New York, 1929, Fig. 74, p. 135.

This view has been expressed by Reed and Crabill[1] who remark for *G. juniperi-virginianae*: "The peridial strand is very sensitive to moisture. When it becomes dry it curls outwards, opening wide the aecidium. Simply breathing on the open aecidia is sufficient to close them immediately. This alternate opening and closing of the aecidia in the presence of moisture or dryness aids in the expulsion of the aecidiospores."

For *Gymnosporangium clavariiforme* (Figs. 120 and 121) Dodge[2] has remarked: "The aecidiospores of this species, which are comparatively long-lived, collect in large quantities in the cancellate peridia and are allowed to escape through the hygroscopic action of the peridial cells."

FIG. 121.—*G. clavariiforme*, two aecidia dehiscing: stage intermediate between A and B in Fig. 120. Photographed by B. O. Dodge. Magnification, about 20.

The curious warts and ridges that roughen the inner walls of the peridial cells of cornute species of Gymnosporangium have a characteristic pattern for each species and are therefore helpful to taxonomists (Fig. 116, C-E; Fig. 117, C-E; Fig. 118, C, D; Fig. 122, E). Not improbably they have some functional significance. By increasing the area of the surface of the inner walls, they may promote the absorption and loss of water vapour from the peridium as weather conditions change and so make the hygroscopic movements of the peridium more sensitive and frequent, thus assisting the escape of the aecidiospores.

Many plants that produce a large number of organs of dissemination (spores or seeds) have arrangements which secure that these organs shall be gradually and not simultaneously set free. Thus in Mycetozoa, Liverworts, and Puff-balls, elaters or capillitium threads are present from which the spore-dust in the opened sporocarps disentangles itself only with difficulty and with the assistance of the hygroscopic movements of the threads; and, in the capsules of certain Orchids,[3] on the inside of the split pericarp, there are elater-like hairs which help in the dispersal of the seed-dust. The part played by these

[1]H. S. Reed and C. H. Crabill, *loc. cit.*

[2]B. O. Dodge, *loc. cit.*, p. 753.

[3]E. Pfitzer, in Engler and Prantl, *Pflanzenfamilien*, Teil II, Abteilung 6, Orchidaceae, 1889, p. 73.

elaters, capillitium threads, and pericarp hairs in Mycetozoa, Liver-worts, Puff-balls, and Orchids seems to be played in the cornute species of the Rust-genus Gymnosporangium by the cancellate or lacerate peridium.

A precise experimental investigation of the factors which cause the splitting of the peridium of dehiscing cornute aecidia in the first place and an exact determination of the extent of the hygroscopic movements of the opened peridia during changes in moisture conditions of the surrounding air are desirable.

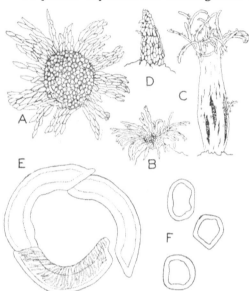

FIG. 122.—*Gymnosporangium juniperi-virginianae.* A, an open aecidium containing many aecidiospores. In this position the strips of peridial cells are turned outwards and allow the aecidiospores to escape; but, when wet, they curl inwards and close the mouth of the aecidium. B, an aecidium from which most of the aecidiospores have escaped. C, side view of a mature aecidium: strings of peridial cells are splitting downwards and thus allowing the aecidiospores to escape. D, the pointed tip of a young aecidium that has pierced the epidermis of the leaf. E, three peridial cells in side view showing inward curvature when wet. F, aecidiospores showing their thick walls. Drawings by D. E. Bliss (1933), arranged by the author. Magnification: A, 40; B, C, and D, 28; E, 346.

As pointed out by Fischer,[1] there are some Pucciniae of the *P. hieracii* type, *e.g.* *P. variabilis*, *P. willemetiae*, *P. podospermi*, *P. crepidis*, and *P. tragopogonis* which, like the roestelioid species of Gymnosporangium, have the inner walls of the peridial cells thicker than the outer. It is desirable that an attempt should be made to determine how in these Pucciniae the thickening of the inner walls of the peridial cells affects the dehiscence of the peridium and the process of spore-discharge.

Non-violent and Violent Discharge of Aecidiospores in Rust Fungi in General.—Among the Uredinales the roestelioid species of Gymnosporangium do not stand alone in liberating their aecidiospores in the form of a spore-powder; for one finds a similar means of spore-discharge

[1] E. Fischer, *Die Uredineen der Schweiz*, Bern, 1904, pp. L, 202-215.

in the genus Cronartium. Thus, in *Cronartium ribicola*, the bladdery aecidia do not discharge their aecidiospores violently, but as dust.[1] A cloud of aecidiospore-dust produced by striking a White Pine tree bearing ripe aecidia of *C. ribicola* with a stick is shown in a photograph reproduced as Fig. 227 in Volume III (p. 263).

At present the violent discharge of aecidiospores is known only in the following genera: Puccinia,[2] Uromyces,[3] Gymnosporangium (*G. ellisii*),[4] and Gymnoconia (*G. peckiana*).[5] Investigations on the mode of liberation of the aecidiospores of other genera of the Pucciniaceae and of the genera of the Melampsoraceae other than Cronartium have not as yet been undertaken. It is possible that such investigations might bring to light facts of taxonomic significance.

Evolution of Gymnosporangium.—Jackson,[6] in his discussion of the evolutionary tendencies of the Uredinales concluded that, like *Gymnosporangium nootkatense* at the present day, the primitive forms of Gymnosporangium "bore uredinia and cupulate aecia." My own researches also suggest that the pycnidia of the primitive forms of Gymnosporangium were provided with pycnidial periphyses, cell-organs which appear to be entirely lacking in the Melampsoraceae, in Phragmidium, and in Gymnoconia, but are present in Puccinia, Uromyces, and Endophyllum.

It seems probable that the genus Gymnosporangium was evolved from a Rust resembling a long-cycled species of Puccinia, for: (1) one species of Gymnosporangium, *G. nootkatense*, still has uredospores in addition to all the other spore-forms; (2) the aecidia of *G. ellisii*, *G. nootkatense*, and a few other species of Gymnosporangium are cupulate and discharge their spores violently into the air, as in Puccinia; and (3) Gymnosporangium species have long pointed ostiolar trichomes projecting from the ostioles of their pycnidia, resembling those in Puccinia.

The characters in which the genus Gymnosporangium has become specialised are: (1) the mycelia producing teleutospores are systemic and long-persistent in leaf-galls or in the bark of shrubs or trees; (2) the pedicels of the teleutospores are more or less evidently gelatinised;

---

[1]H. Klebahn, *Die wirtswechselnden Rostpilze*, Berlin, 1904, pp. 15-16.

[2]and[3]For a review of the work on aecidiospore-discharge in Puccinia and Uromyces by Zaleski, Craigie, and Dodge and for the author's own observations *vide* these *Researches*, Vol. III, 1924, pp. 5.2-559.

[4]B. O. Dodge, "Aecidiospore Discharge as Related to the Character of the Spore Wall," *Journ. Agric. Research*, Vol. XXVII, 1924, pp. 749-756.

[5]B. O. Dodge, *loc. cit.*, pp. 753-754. Also these *Researches*, Vol. III, 1924, p. 554.

[6]H. S. Jackson, "Present Evolutionary Tendencies and the Origin of Life Cycles in the Uredinales," *Mem. Torrey Bot. Club*, Vol. XVIII, 1931, p. 73.

(3) except in one species only, the production of uredospores has been eliminated; (4), the paraphyses are developed at intervals between the pycnidiosporophores (*cf.* Fig. 41, p. 137), instead of being periphyses at the mouth of the pycnidium as in Puccinia (*cf.* Fig. 40); (5) the aecidium in most species is cornute and has its aecidiospores dispersed as a powder by the wind; and (6) the aecidiospores are thick-walled, retain their vitality for a long time, and germinate better after being subject to cold.

Jackson[1] has suggested that the loss of the uredospore stage in the life-history of Gymnosporangium is correlated with the fact that the diploid mycelium which bore the uredospores became normally over-wintering or perennial.

Gymnosporangium bermudianum, as we have seen, has the distinction of being the only known autoecious species of its genus. One may ask: how did it become autoecious? To this question Jackson[2] has given the following answer: the species from which *G. bermudianum* was derived, like other Gymnosporangia, was heteroecious with pycnidia and aecidia on one or more species of Malaceae or other dicotyledonous plants and its teleutospore sori on Junipers; then the basidiospores acquired the ability to infect the host which bore them, *i.e.* a Juniper; and thus the haplophase and the diplophase in the life-history came to be established on one and the same coniferous host-plant. If this solution of the problem is the correct one, there is the possibility that the basidiospores which suddenly acquired the ability to infect Juniper trees at the same time lost the ability to infect any species of Malaceae: the mutation which brought to the basidiospores new powers of infection may have resulted in the destruction of the old powers.

Another solution of the problem of the evolution of *Gymnosporangium bermudianum* will now be suggested. One may suppose that the binucleate mycelium in a Juniper leaf-gall, produced as a result of infection of a leaf by one or more aecidiospores derived from a dicotyledonous host, suddenly acquired the power of becoming partially de-diploidised, with the result that proto-aecidia were formed on the uninucleate (+) and (−) branches of the mycelium, diploidisation of the proto-aecidia was effected by self-diploidisation (due to the migration of (−) nuclei to the (+) proto-aecidia and of (+) nuclei to the (−) proto-aecidia), and aecidia came to be developed on the leaf-gall. In this way mycelia derived from aecidiospores came to develop aecidiospores on its uninucleate branches before producing teleutospore

[1] *Loc. cit.*
[2] *Loc. cit.*, p. 95.

sori on its binucleate branches. If this theory is correct then the haplophase in a leaf-gall did not originate as a result of basidiospores acquiring the power of infecting Juniper leaves and losing the power of infecting the dicotyledonous host.

The phenomenon of de-diploidisation, recognised by the author[1] and discussed by him in 1941, occurs in the Hymenomycetes, Uredinales, and Ustilaginales. It must have been an important factor in the evolution of such Pucciniae as *Puccinia suaveolens* and *P. minussensis*, and it may well have been that its sudden introduction into the life-history of *Gymnosporangium bermudianum* was a first step in the passing of that species from heteroecism to autoecism.

Pores and Pore-plugs in the Aecidiospores of Gymnosporangium ellisii and other Rust Fungi.—B. O. Dodge[2] has investigated the development of pores in the walls of aecidiospores, more particularly in *Gymnosporangium ellisii* (his *G. myricatum*) which produces its aecidia on the leaves of *Myrica carolinensis*, the Bayberry. Dodge found that the aecidiospores of *Gymnosporangium ellisii* are each provided with some six or seven germ-pores. At the point where a germ-pore is to be developed, the spore-wall thickens over a small area in such a way that there is formed a little ball-like body or *pore-plug* which is separated from the rest of the spore-wall. The germ-pore becomes evident as soon as the plug is dislodged. The formation of a pore and its plug begins with about the fourth aecidiospore in the chain and is completed when the spore reaches maturity. In an aecidiospore-deposit formed as a result of the violent discharge of the spores from an aecidium one can see the pore-plugs as tiny discoid bodies, some still clinging to the spores and others that separated from the spores at the moment of discharge or on the spores striking the substratum.

Dodge[3] found pore-plugs similar to those of *Gymnosporangium ellisii* in *Puccinia podophylli*. On a single aecidiospore of this species there appear to be three or four pores and as many plugs. Here, again, the pore-plugs can be seen in aecidiospore-deposits, some still clinging to the spores and others lying free. Ingold,[4] who does not seem to have been acquainted with Dodge's investigations, shows pore-plugs attached to discharged aecidiospores of *P. pulverulata* but with-

---

[1]A. H. R. Buller, "The Diploid Cell and the Diploidisation Process in Plants and Animals, with Special Reference to the Higher Fungi," *Botanical Review*, Vol VII, 1941, pp. 411-414.

[2]B. O. Dodge, *loc. cit.*

[3]B. O. Dodge, "Expulsion of Aecidiospores by the Mayapple Rust, *Puccinia podophylli* Schw.," *Journ. Agric. Research*, Washington, Vol. XXVIII, 1924, pp. 923-926.

[4]C. T. Ingold, *Spore Discharge in Land Plants*, Oxford, 1939, Fig. 34, p. 73.

out any remark as to their nature. I, myself, have observed pore-plugs in aecidiospore deposits of *P. graminis*, and some plugs of this species can be seen attached to the sides of aecidiospores in a spore-chain in Fig. 37, B (p. 118).

Dodge also investigated *Gymnoconia peckiana* (his *G. interstitialis*), the long-cycled Orange Rust on Rubus, but found that in this species pore-plugs on aecidiospores are not developed.

Persistent pore-plugs are present in *Gymnosporangium ellisii*, *Puccinia podophylli*, but not in *Gymnoconia peckiana*; yet, in all three species the aecidiospores are violently discharged. Dodge, having noted these facts, discussed their significance. He[1] said: in *Gymnosporangium ellisii*, the aecidiospores in the aecidium are densely packed together and the pore-plugs indent the elastic walls, so that one can readily understand how the plugs function in providing a more effective mechanism for spore discharge; and, concerning *Puccinia podophylli*, he[2] said: the plugs serve as fulcrums against which the walls react as the spore is set free. Finally, he[3] remarked: "although the long-cycled Orange Rust, Gymnoconia, on Rubus, does not form such pore plugs, still the spore is discharged with considerable force, proving the plug, although serving a useful purpose in this connexion, must act only in an accessory capacity in dislodging the spores in the case of the Bay-berry Rust."

In the chains of aecidiospores of *Gymnosporangium ellisii*, *Puccinia graminis*, etc., the pores and pore-plugs are formed on the *lateral* walls of the spores and not the upper or lower walls in the line of flight where, for mechanical reasons, one might expect to find them if they were to be effective in aecidiospore discharge. The pore-plugs are a by-product resulting from the formation of germ-pores and, as Dodge has shown, aecidiospore-discharge can take place in their absence. Whether or not, when they are present, they really serve to make the discharge process more effective than it would be in their absence is a question that seems to be not yet finally decided.

[1]B. O. Dodge, "Expulsion, etc.," *loc. cit.*, p. 923.
[2]*Ibid.*, p. 926.   [3]*Ibid.*, p. 923.

CHAPTER XII

THE GEOLOGICAL TIME DURING WHICH THE PYCNIDIA
OF THE UREDINALES ATTAINED THEIR PRESENT FORM
AND FUNCTION, WITH SOME REMARKS ON THE
EVOLUTION OF OTHER ENTOMOPHILOUS FUNGI

Introduction — Rock-systems — Land Plants of the Past — Cambrian and Ordovician Periods — Silurian Period — Devonian Period — Carboniferous and
Permian Periods — Triassic Period — Jurassic Period — Cretaceous Period —
Tertiary and Quaternary Periods — Insects of the Past — Flowers and Insects —
Fungi and Insects — Fungi dependent on Insects and the Time of their Evolution — Geological Time of Evolution of the Pycnidia of the Uredinales — Time
of Evolution of the Phallales, etc. — Conclusion.

Introduction.—The Uredinales are obligate parasites and, therefore,
their evolution must have been bound up with the evolution of their
host-plants. Furthermore, since the Uredinales at the present day
are entomophilous in that the initiation of the sexual process in isolated
(+) and (−) pustules is largely dependent on the visits of insects, the
specialisation of the pycnidia as mechanisms to be worked by insects
must have been delayed until nectar-sucking insects had come into
existence. Therefore, before treating of the evolution of the Rust
Fungi, it will be necessary to review briefly our knowledge of the evolution of land-plants and of insects.

Rock-systems.—For reference, the rock-systems making up the
geological record and the eras during which they were formed may be
set out as follows:

Quaternary . } Cainozoic era
Tertiary . . }

Cretaceous . ⎫
Jurassic . . } Mesozoic era
Triassic . . ⎭

Permian . . ⎫
Carboniferous |
Devonian . |
Silurian . . } Palaeozoic era
Ordovician . |
Cambrian . ⎭

Pre-Cambrian . . . Pre-Cambrian era

Land Plants of the Past.—It was the appearance of land plants with aerial stems and leaves that made possible the evolution of the Basidiomycetes and, in particular for our present considerations, the evolution of the Uredinales. Here therefore, will be briefly summarised some of the chief factors relating to the appearance and spread of terrestrial vegetation during the successive periods of the earth's geological history. This summary is based chiefly on the information given in Seward's *Plant Life Through the Ages* (1931).

Cambrian and Ordovician Periods.—Up to the present, the rocks of the Cambrian and Ordovician periods have not yielded a single trace of terrestrial vegetation that can with certainty be interpreted as such. Seward examined the whorled stem-like structures known as *Protoannularia*, obtained from the Skiddaw Slates (Ordovician) and convinced himself that the affinity of these fossils could not be determined. Protoannularia, instead of being related to the Carboniferous land-plant *Annularia*, may be nothing more than an alga. Although "we know nothing of Ordovician terrestrial vegetation, not even whether it existed," it may well be that in the Ordovician period the land was receiving from the sea or from fresh water "the germs of a vegetation which was soon to clothe the continents with a new mantle."

Silurian Period.—Evidence that land plants existed in the Silurian period is afforded by fossil stems bearing small leaves, obtained from beds of Upper Silurian age in Victoria, Australia. Moreover, a "slender axis, 50 mm. long, bearing a lateral branch and covered with short linear appendages" described by Halle from Silurian beds in the island of Gothland, "may perhaps be a fragment of a land plant"; and very similar specimens have been found in Upper Silurian beds in England. It would appear that, in the Silurian period, land plants had already come into existence, were undergoing a steady evolution, and were taking up their stations far and wide on what had formerly been vast stretches of barren ground.

Devonian Period.—As proved by fossil evidence, in the Devonian period land plants of varied form and size grew thickly together and formed a definite flora. Among the more famous fossil plants are those which have been obtained from a Middle Devonian marsh at Rhynie in Aberdeenshire, Scotland. The record of the rocks shows that among the Devonian land plants were: *Psilophytales* (Psilophyton, Rhynia, Hornea, Asteroxylon, etc.); *Lycopodiales* (Lepidodendron, Protolepidodendron, Leptophloeum, etc.); *Protoarticulatae* (Calamophyton, Hyenia); *Articulata* (Pseudobornia, Sphenophyllum); *Filicales* (Asteropteris); *Pteridospermae and other gymnosperms* (Archaeopteris,

Sphenopteris, Dadoxylon, etc.); and *Plantae incertae sedis* (Palaeopitys, etc.).

In *Outlines of Historical Geology* (1941), Schuchert and Dunbar[1] place the Rhynie bed in the Lower rather than in the Middle Devonian, and they say: "The oldest known deposit of well-preserved land plants is that of the Lower Devonian at Rhynie, Scotland, which seems to be a silicified portion of a bog. Here the plants are of several species, but all are small and structurally as simple as a vascular plant could be. Once adapted to land life, the vascular plants evolved rapidly, and before the close of the Devonian were represented by ferns, seed ferns, scouring rushes (calamites), scale trees, and precursors of the conifers (cordaites). Petrified logs and stumps in the Middle and Upper Devonian strata of the Catskill delta in New York record clearly the oldest known forests, with stumps more than 2 feet in diameter."

There can be but little doubt that Fungi and Bacteria were main factors in the destruction of the tissues of dead land plants of Devonian times. So far as Fungi are concerned, Seward remarks: "The tissues of Rhynia and of the other vascular plants contain in their cells innumerable tubular threads, which often occur as tangled clusters swelling into vesicles that developed into well-protected resting spores. The thick-walled spores are often invaded by still more delicate threads of other fungi, and these in turn produced spherical reproductive cells within the cavities of the larger spores. Some of the fossils exhibit a strong family likeness to such common fungi as Saprolegnia, Pythium, and Peronospora, genera familiar to all botanical students and, under other names, well known as garden pests. Several forms are described under the comprehensive term *Palaeomyces*."[2]

Carboniferous and Permian Periods.—In the Permo-Carboniferous period land plants abounded; and, today, their remains in the form of coal are mined. Among the groups of plants which were well represented in the Permo-Carboniferous period are: *Equisetales* (Calamites, Asterophyllites, Annularia, Equisetites, etc.); *Sphenophyllales* (Sphenophyllum); *Lycopodiales* (Lepidodendron, Sigillaria, Stigmaria, Lycopodites, Selaginellites); *Filicales* (Botryopteris, Etapteris, Psaronius); *Pteridospermae* (Pecopteris, Alethopteris, Lonchopteris, Odontopteris, Neuropteris, Sphenopteris, Lyginopteris, etc.); *Cycadophyta* (Dionites, Pterophyllum); *Ginkgoales* (Baiera, Saportaea); *Cordaitales* (Cordaites,

[1]Schuchert and C. O. Dunbar, *Outlines of Historical Geology*, Ed. 4, in *Outlines of Geology* by Longwell, Knopf, Flint, Schuchert and Dunbar, Ed. 2, 1941, pp. 177-178.

[2][E. J. Butler, *Trans. Brit. Mycol. Soc.*, Vol. XXII, 1939, pp. 274-301, points out that some of these fungi resemble the present day species of Rhizophagus.]

Pitys, Dolerophyllum); *Coniferales* (Araucarites, Volzia, Dicrano-phyllum, etc.).   As yet, however, there were no Angiosperms.

Triassic Period.—In the Triassic period the groups of plants characteristic of the Permo-Carboniferous period continued to exist; but their representatives were much changed.   Says Seward: "If we contrast the Keuper and Rhaetic floras with those which preceded them we become conscious of very great differences; the plant-world in the Permian period was dominated by *Calamites, Lepidodendron,* sigillarias, a host of pteridosperms and several kinds of gymnosperm trees such as *Cordaites;* here and there were a few plants of another type, the advance guard of a Mesozoic host.   Before the end of the Triassic period the vegetation had put on a new dress; the dominant genera were of a much more modern type and most of them bore no obvious marks of direct relationship to the trees in the Palaeozoic forests."

Jurassic Period.—In the Jurassic period, plants took on a still more modern appearance.   The Cycadophyta became a dominant group, and Conifers flourished.   The view that Angiosperms existed in the Jurassic period is supported by the discovery in rocks of Jurassic age of: impressions of leaves, often far from perfect; the fossil fruits of *Caytonia* and *Gristhorpia,* associated with foliage known as *Sagenopteris;* an oval leaf-like structure resembling a petal; certain seed-like bodies; and some fossil wood characterised by wood vessels.

Cretaceous Period.—The Cretaceous period presents us with the dawn of a new era so far as vegetation is concerned.   As Seward says: "in the earlier part of the Cretaceous period there was added to the world's carpet of vegetation a new design and a new pattern."   The Angiosperms had arrived; and, in the Lower Cretaceous rocks of Greenland, we are provided with fossil leaves hardly distinguishable from the foliage of dicotyledonous trees existing at the present day. The Angiosperms increased in number and variety during the Cretaceous period so that in the latter half of the period the plant-world "differed in no outstanding feature from that of our own time." Among the fossil leaves obtained from Cretaceous rocks may be mentioned those of *Platanus, Magnoliaephyllum, Cinnamomites, Menispermites, Artocarpus, Juglans, Salix, Myrica, Ficus, Dalbergia, Bauhinia, Sterculia,* and *Eucalyptus.*   In the Cretaceous period the "Ferns and Gymnosperms which had long occupied a dominant place became subordinate groups in a company consisting chiefly of Flowering Plants."

Tertiary and Quaternary Periods.—The Tertiary floras agree very closely with those that exist today; but it appears that, in the early part of the Tertiary period, the arborescent Angiosperms were pre-dominant over the herbaceous. The first stage of the Tertiary, the Eocene, has yielded fossil evidence of the existence of Palms (Nipa, Sabal, Phoenix), Typha, Statioides, Magnolia, Liriodendron, Cinna-momum, Viburnum, Nelumbium, Betula, Alnus, Quercus, etc. The recent Quaternary period had floras which were identical or almost identical with those which clothe the earth at the present moment.

Insects of the Past.—It is of importance to note that no remains of insects are known in the earlier palaeozoic rock system, *i.e.* in the Cambrian, Ordovician, and Silurian.

Devonian Period.—Imms[1] (1931), in reviewing the geological history of insects, states that in the Devonian: "the presumed insect remains are extremely small and fragmentary and have been found in flakes of Rhynie Chert from the Middle Old Red Sandstone of Scotland. The most complete remains consist of portions of the head-capsule, mouth-parts and antennae of the species *Rhyniella praecursor* Hirst and Maulik, which is regarded by Tillyard as being an early Col-lembolan. The other fossils consist of mandibles only, and the species to which they have been relegated, viz. *Rhyniognatha hirsti* Till., may possibly have been a Thysanuran. If, as appears very probable, Rhyniella is a species of true Apterygota, it is remarkable that no evidence of the existence of such insects has so far been revealed in intervening rocks until the Oligocene period." Thus the first known fossil insects appear to have been wingless.

Carboniferous Period.—Imms[2] says: "Remains of insect life are found in comparative abundance during this period, and the earliest undoubted examples of the class occur in the lower layers of the Upper Carboniferous of North America. This early fauna consisted almost entirely of Palaeodictyoptera, Protorthoptera (including Protoblat-toidea), Protodonata, Megasecoptera, and Blattidae. With the exception of the last mentioned all these groups are extinct today—they are ancestral types from which more recent types have been derived."

In the United States of America the Carboniferous Period is divided into the earlier Mississipian Period and the later Pennsylvanian Period, and it is in the Pennsylvanian rocks that numerous remains of insects have been found. Schuchert and Dunbar say:[3] "The Pennsylvanian insects belong mostly to primitive extinct orders, and many of them were of notable size. The greatest of all insects, either fossil or

[1]A. D. Imms, *Recent Advances in Entomology*, Philadelphia, 1931, p. 75.
[2]*Loc. cit.*
[3]C. Schuchert and C. O. Dunbar, *loc. cit.*, p. 188.

modern, was a dragon-fly-like form, *Meganeuron*, found in the Coal Measures of Belgium with a wing spread of 29 inches."

Permian Period.—According to Imms,[1] all the ancient Carboniferous order of insects extended into Permian times. Permian insects in considerable numbers and variety have been found in Europe, North America, and Australia. Schuchert and Dunbar[2] remark: "In the Permian there was, in general, a reduction in size and rapid specialization, leading to the origin of several of the modern orders, notably those that embrace the may-flies, the wasps, the true dragon-flies, and the beetles. A single locality in the Early Permian near Elmo, Kansas, has yielded about 10,000 specimens, many of which retain color markings and microscopic 'hairs' on the wings."

Mesozoic Era.—In the Mesozoic Era (Triassic, Jurassic, and Cretaceous Periods) insects became more and more numerous and varied, and today they are represented by hundreds of thousands of species.

Australian rocks of Triassic age have yielded a rich insect fauna, and other Triassic insects have been found in Europe. Among the Triassic insects were modern dragon-flies, true Coleoptera (beetles), and also Blattidae (cockroaches) that were very like those of the Palaeozoic era.

The Jurassic Period, says Imms,[3] "is notable from the fact that all the main orders of insects, with the exception of the Lepidoptera, had come into existence before it terminated."

In the Cretaceous rocks there are only scanty and fragmentary remains of insects, but this is attributed by Imms[4] not to actual scarcity of those animals during that period, but rather to the absence of suitable fresh-water deposits wherein examples might have accumulated and become fossilised; and he remarks: "The comparatively sudden appearance of highly organised Lepidoptera, Aculeata, and cyclorrhapous Diptera in the early Tertiary period suggests that these groups must have been represented by forerunners in the Cretaceous era."

Palaeontological evidence shows that true Diptera (flies) and the earliest true Hymenoptera (Symphyta and Parasitica) came into existence in the Jurassic period. Many of the insects belonging to these orders (our modern flies, ants, bees, and wasps) have mouth-parts which permit of sucking up sweet juice from the nectaries of flowers.

In the Post-Devonian and Pre-Jurassic times there existed insects which may be roughly classed as Dragon-flies, Cockroaches, May-flies,

[1]A. D. Imms, *loc. cit.*, p. 78.

[2]C. Schuchert and C. O. Dunbar, *loc. cit.*, pp. 188-189.

[3]A. D. Imms, *loc. cit.*, p. 79.   [4]*Loc. cit.*, p. 80.

and Grass-hoppers, etc.; but these orders of insects, whose descendents have survived to the present day, are not adapted for visiting either flowers or Rust fungi or for sucking up the nectar.

Lepidoptera (butterflies, moths) are present as fossils in Eocene rocks (Cainozoic Era) onwards; but, as Imms[1] says: "it is practically certain that the order is of Pre-Tertiary origin, and it may be added that certain leaf mines of Upper Cretaceous age are very possibly the work of their larvae." Lepidoptera with their long tubular probosides are well adapted for sucking nectar from flowers; but, so far, no moths or butterflies have been recorded as visiting the nectar of the Rust Fungi.

Tertiary Period.—Says Imms[2]: "Many of the Tertiary deposits are extremely rich in insects and they have yielded nearly three-fourths of the known fossil species, almost all modern orders being represented. Taken as a whole the fauna of these times did not differ markedly in composition from that of today—even during the Oligocene, which is relatively early in the Tertiary period, the insects were very much the same as those which prevail at present."

Collembola and Thysanura, wingless insects, are well represented in Baltic amber, and Coleoptera are extremely abundant in Tertiary rocks. Imms[3] says: A number of modern genera of Coleoptera "were existing in Eocene times and more than 430 species of beetles are known from Baltic amber, while an even larger number have been collected from the Miocene of Florissant."

The earliest known Aculeate Hymenoptera (ants, bees, wasps), according to Imms,[4] date from the Eocene period, and he says: "It is evident that they were one of the latest of the great groups to be evolved. . . . Considerably over 300 species of ants are known from Tertiary strata, and among them male, female, and worker castes were differentiated much as today."

Flowers and Insects.—Palaeobotanical and palaeontological facts strongly support the view that the Angiosperms and the insects which visit their flowers and so bring about cross-pollination were evolved in the middle and upper Mesozoic era, *i.e.* when the Jurassic and Cretaceous systems of rocks were being formed.

The dependence of certain Flowering Plants on insects and the dependence of certain insects on Flowering Plants is generally recognised. Many Flowering Plants are so constructed that, without the visits of insects to their flowers, they are unable to set seed; and many insects, including bees, moths, butterflies, and certain flies, unless they

[1]*Loc. cit.*, p. 81.   [2]*Loc. cit.*, p. 81.
[3]*Loc. cit.*, p. 82.   [4]*Loc. cit.*, p. 82.

can obtain nectar or nectar and pollen grains from flowers, starve to death.   Among the classical instances of mutual interdependence of Flowering Plants and insects may be cited: (1) Yucca (Liliaceae) and Pronuba (a moth); (2) the Lady Slipper Orchid and bees; and (3) Curcurbitaceae, in which there are separate staminate and pistillate flowers, and bees.

Fungi and Insects.—Certain Fungi have become dependent on insects for: (1) the dispersal of their spores; or (2) acts which correspond to the pollination of flowers and are a necessary preliminary to the initiation of the sexual process.   Among the Fungi which are thus dependent the following may be cited:

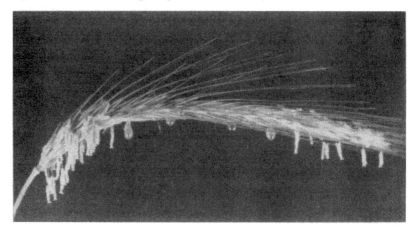

FIG. 123.—*Claviceps purpurea* in the Sphacelia stage on an ear of Rye (*Secale cereale*), showing four pale-yellow, sweetish, odoriferous, gelatinous, conidia-containing drops of nectar hanging below as many infected ovaries.   The young ovaries were inoculated with mycelium, derived from ascospores, growing within tiny blocks of nutrient agar.   Cultures and photograph made by A. M. Brown at the Dominion Rust Research Laboratory, Winnipeg.   Natural size.

(1) The Phallales, *e.g. Phallus impudicus* and *Mutinus caninus*. The gleba deliquesces; by its position in space, its form, its colour, and its strong odour it makes itself attractive to dung-flies, etc.   The insects are rewarded for their visits by sugar[1] dissolved in the glebal fluid and, in their turn, they disperse the spores and so are of service to the fungus.

(2) Claviceps, a genus of Pyrenomycetes, exemplified by *Claviceps purpurea*, the pathogen which causes the disease known as the Ergot

[1] E. Ráthay and B. Haas, "Über *Phallus impudicus* und einige Coprinus-Arten nach Wiesner," *Sitzungsber. d. Mathem.—Naturwiss. Klasse d. K. Akad. der Wiss. in Wien*, Bd. LXXXVII, 1883, pp. 18-24.

of Rye. Here the fungus, in its Sphacelia stage, produces great numbers of minute conidia (Fig. 124) enveloped in mucilaginous drops which are exuded from the flowers and hang from the inflorescences (Fig. 123). These drops have an unpleasant odour, and contain sugar. Flies, attracted by the scent, visit the drops, suck up the sugary liquid, carry away the conidia, and deposit the conidia on healthy flowers. The conidia soon germinate and their germ-tubes enter the ovaries. Thus, with the help of insects, the fungus comes to infect new host-plants.

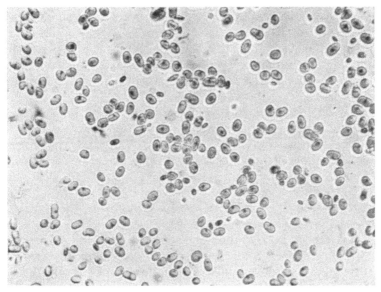

FIG. 124.—Conidia produced in the Sphacelia stage of *Claviceps purpurea* and set free in a drop of nectar hanging below an infected ovary of an ear of Rye (*Secale cereale*, *cf*. Fig. 123). Photographed by A. M. Brown. Highly magnified: the conidia are about 6-7μ long.

(3) Certain species of Sclerotinia, a genus of Discomycetes, *e.g. Sclerotinia vaccinii* which parasitises the Cowberry, *Vaccinium vitis-idaea*. The life-history of *Sclerotinia vaccinii* was elucidated by Woronin.[1] In the spring, on the twigs and leaves of its host, the fungus produces monilioid hyphae which break up into a snow-white or slightly yellow pollen-like mass of loose cells (gonidia). This dust has a strong smell of almonds and attracts flies and bees. The insects carry the pseudo-pollen to the stigmas of the Cowberry flowers. Here

[1]M. Woronin, "Über die Sclerotinienkrankheit der Vaccinieen-Beeren," *Mém. de l'Acad. Imp. des Sci. de St. Petersbourg*, Sér. 7, T. XXXVI, 1888, pp. 1-49, Taf. I-X.

the fungus cells germinate. The germ-tubes then grow down the stigmas and the mycelium infects the ovary. The ovary becomes sclerotic. Next spring the mummified berries give rise to stalked apothecia which discharge numerous ascospores, and these spores serve to infect the twigs and leaves of healthy Cowberry plants.

(4) Certain species of Coprinus, a genus of Hymenomycetes, *e.g.* *Coprinus lagopus* which commonly occurs on horse dung. The haploid mycelia of *C. lagopus*, derived from single basidiospores, produce an abundance of oidiophores bearing oidia enclosed in a mucilaginous drop. As shown by experiments made by Brodie,[1] flies carry (+) oidia to (−) mycelia derived from (−) basidiospores and (−) oidia to (+) mycelia derived from (+) basidiospores, with the result that the transported oidia germinate, the germ-tubes or young mycelia so produced unite with the mycelium of opposite sex, and thus there is initiated a sexual process which culminates in the production of diploid basidiocarps.

(5) Certain species of Ascobolus, a genus of Discomycetes, *e.g.* *Ascobolus stercorarius*. Experiments made on this species by Miss Dowding[2] have shown that mites and flies, by transporting (+) or (−) oidia from a mycelium of one sex to a mycelium of opposite sex, play an important part in initiating the sexual process in haploid mycelia derived from ascospores.

In *Coprinus lagopus* and *Ascobolus stercorarius*, the initiation of the sexual process is dependent on insect visits only when the (+) and (−) mycelia are isolated from one another. It often happens that a dung-ball contains many basidiospores or ascospores of these species. In that case young (+) and (−) mycelia produced from the spores come into contact with one another and diploidise one another without insect intervention. If all insects were exterminated, it may well be that *Coprinus lagopus* and *Ascobolus stercorarius* would continue to flourish.

(6) The Uredinales. Here, as already shown, whenever the haploid pustules of a heterothallic Rust Fungus are well separated from one another, insects are indispensible for transporting the pycnidiospores from a (+) pustule to a (−) pustule, or from a (−) pustule to a (+) pustule, and thus creating the conditions requisite for the initiation of the sexual process. A fly at work on a Barberry leaf bearing haploid pustules of *Puccinia graminis* has been illustrated in Fig. 13 (p. 68).

[1]H. J. Brodie, "The Oidia of *Coprinus lagopus* and their Relation with Insects," *Annals of Botany*, Vol. XLV, 1931, pp. 315-344.

[2]E. Silver Dowding, "The Sexuality of *Ascobolus stercorarius* and the Transference of the Oidia by Mites and Flies," *Annals of Botany*, Vol. XLV, 1931, pp. 621-637.

It should be emphasised that, in a heterothallic Rust species, under natural conditions, separate haploid pustules are the rule and compound pustules formed by the fusion of two simple pustules of opposite sex are comparatively rare. Only when this is borne in mind can one fully realise how important for a heterothallic Rust the visits of insects actually are.

There seems every reason to suppose that the primitive Rusts were long-cycled and heterothallic and that the present-day microcyclic species, such as *Puccinia malvacearum*, which are homothallic, have no pycnidia, and do not require the visits of insects, are the degenerate progeny of long-cycled, heterothallic species.[1] The relations of Rust Fungi with insects must therefore be considered as of very ancient origin.

In four of the six cases listed above, namely, the Phallales, *Claviceps purpurea*, *Sclerotinia vaccinii*, and the Uredinales, it is obvious that we have relations with insects resembling in their essentials those between flowers and insects. In all four cases the insects are attracted by a special scent and the fruiting structures are so placed that they are readily accessible to insect visitors. Colour enhances the attractiveness or means of recognition of the fruit-bodies of the Phallales, the pseudo-pollen of *S. vaccinii*, and the pycnidia of the Rust Fungi, while the conidia-bearing drops of *Claviceps purpurea* hanging from the ears of Rye and other grasses glisten in the sunlight and are yellowish. In the Phallales, Claviceps, and the Rust Fungi, the liquid in which the spores are embedded is known to contain sugar, while in *Sclerotinia vaccinii*, where there is no liquid, the disjuncted cells of the monilioid chains resemble pollen-grains and as such appear to be regarded by bees and other insects which are attracted to them.

In two of the cases listed above, namely, *Coprinus lagopus* and *Ascobolus stercorarius*, the oidia are produced on the surface of fresh horse-dung balls by mycelia which lack both colour and scent. The insects which visit these mycelia and transport the oidia are doubtless attracted by the appearance and odour of the dung-balls. In our Coprinus and Ascobolus, therefore, the colours and odours which attract insects are provided not by the fungi themselves but by the substratum on which the fungi grow.

Fungi dependent on Insects and the Time of their Evolution.—In view of the fact that in the Phallales, *Claviceps purpurea*, *Sclerotinia vaccinii*, and the Uredinales, insects are indispensible for enabling the fungi to flourish, it would seem logical to conclude that these fungi

[1]For the evidence supporting this conclusion *vide* H. S. Jackson, "Present Evolutionary Tendencies and the Origin of Life Cycles in the Uredinales," *Memoirs Torrey Bot. Club*, Vol. XVIII, 1931.

could not have existed in their present condition before the insects upon which they now depend had been evolved. Thus we are provided with a clue as to the earliest geological period during which the fungi in question became specially adapted for insect visits.

Geological Time of Evolution of the Pycnidia of the Uredinales.— An attempt will now be made to indicate the geological period during which the pycnidia of the Uredinales attained their present form and function.

No fossil Rust Fungi are known and therefore palaeomycology cannot give us any assistance in our task.

Ráthay[1] identified 135 species of insects which visited the pycnidia of various Rust Fungi, and his analysis shows that they consisted of:

1. Diptera........chiefly flies........... 47.40		
2. Hymenoptera....chiefly ants........... 23.70		per cent.
3. Coleoptera......flower-visiting beetles... 22.96		
4. Hemiptera, etc...accidental visitors...... 5.94		
100.00		

Ráthay realised that the Diptera, Hymenoptera, and Coleoptera which visit the pycnidia of the Rust Fungi are insects which visit flowers. Even the beetles which suck the nectar of Rusts are pollen-eaters and honey-lickers, largely dependent upon flowers.

It is generally agreed that such flowers as those of Cypripedium, Salvia, Rhododendron, Althaea, etc., must have been evolved in the presence of insects and after insects had begun to visit them. Similarly, it may be admitted that the pycnidial pustules of heterothallic Rust Fungi, *e.g.* those of *Puccinia graminis*, must have attained their present condition in the presence of insects and after insects had begun to visit them. The position of the pycnidia, their colour, strong odour, and sweet nectar, and the flexuous hyphae which protrude into the nectar (also in Puccinia, Uromyces, Gymnosporangium and Endophyllum the very red pointed periphyses), may all be regarded as special adaptations which enable a Rust Fungus to employ insects for mixing the nectar of its (+) and (−) pustules and thus for establishing conditions which permit of the initiation of the sexual process.

It is possible that, in the earliest stages of the evolution of the Uredinales, the pycnidiospores behaved like ordinary conidia in that they germinated and produced haploid mycelia which infected the host-plants. If the pycnidiospores of the Uredinales once had this

---

[1]E. Ráthay, "Untersuchungen über die Spermogonien der Rostpilze," *Denkschrift d. Kais. Akad. d. Wissensch.*, Wien, Bd. XLVI, 2 Abt., 1883, pp. 1-51. An extended review of Ráthay's work has already been given in this volume in Chapter I.

non-sexual function, they may well have lost their power of germination when the pycnidia underwent their final stages of evolution and became adapted for insect visits.

The two families of the Uredinales existing at the present day, namely, the Melampsoraceae and the Pucciniaceae, resemble such angiospermous families as the Orchidaceae and the Papaveraceae in that they are entomophilous. In being entomophilous the Uredinales stand in marked contrast with the Ustilaginales with whose sexual processes insects have nothing whatever to do.

Since no insects are known in the Cambrian, Ordovician, and Silurian rocks, and since only traces of wingless insects are known from Devonian rocks, it is improbable that the pycnidia of the Rust Fungi, as we know them today, were evolved in Pre-Devonian times.

As we have seen, Grass-hoppers, Cockroaches, Dragon-flies, May-flies, etc., were present in the Carboniferous and Permian periods but such insects are not adapted for visiting flowers and the pycnidia of Rust Fungi and sucking up the sweet nectar. Beetles also occurred in the Permian period, but the flower-visiting and rust-visiting beetles of today may have become adapted to nectar-licking in Mesozoic or Tertiary eras. Ráthay's beetles that visit Rusts, except an occasional weevil, are flower-visiting forms. Our present knowledge of fossil insects, therefore, suggests that insects suitable for working the pycnidial mechanism of the Rust Fungi did not exist in the Palaeozoic era.

As we have seen, the Diptera and Hymenoptera, to which orders belong flies and ants that today visit flowers and the pycnidia of the Rust Fungi, came into existence in the Jurassic period, while the Lepidoptera which visit flowers but not Rust Fungi date from the beginning of the Tertiary and were probably evolved in the Jurassic period.

It would appear that, whilst flowers and flower-visiting insects were evolving together during the Jurassic and Cretaceous Periods, the Uredinales became adapted to use certain of the flower-visiting insects for their own welfare. Apparently, at that time, the pycnidia of the Rust Fungi came to resemble flowers in respect to colour, scent, and the provision of nectar and thus, like flowers, gained the help of insects in creating physical conditions suitable for the initiation of the sexual process.

Time of Evolution of the Phallales, etc.—*Phallus impudicus* and *P. ravenelii* have white stipes and green glebae. In *Mutinus caninus* the stipe is white and the gleba red. Other Phallales, *e.g. Aseroe rubra*, *Clathrus cancellatus*, and *Dictyophora indusiata* more or less resemble flowers in their peculiar forms and colours. Also, the Phallales give out very pungent and disagreeable odours. By their form, colours, and odours the fruit-bodies of the Phallales are attractive to insects, particularly to dung-flies. I myself have seen not only dung-flies but also wasps visiting the gleba of *Phallus ravenelii*. Using the same arguments as have been applied to the Rust Fungi, it may be concluded that the Phallales were probably evolved after the Diptera and Hymenoptera had come into existence. Assuming this to be true, one may conclude further that the Phallales were evolved not earlier than the Jurassic, probably in the Jurassic and Cretaceous Periods and, if not so soon, then in the Tertiary Period.

Similarly, too, it may be argued that the Sphacelia stage of Claviceps in its present form and the pseudo-pollen of certain Sclerotineae were evolved in Jurassic or Post-Jurassic times.

Conclusion.—Having regard to the geological history of insects, one may conclude that the pycnidia of the Rust Fungi as we know them today, as well as the sporophores of the Phallales, the Sphacelia stage of Claviceps, and the pseudo-pollen of certain Sclerotineae, all as known today with their adaptations for insect visits, were evolved:

(1) later than the Cambrian, Ordovician, Silurian, and early Devonian periods;

(2) at the earliest possible time in the middle Devonian;

(3) possibly in the Carboniferous and Permian periods when there were plenty of insects, mostly belonging to extinct orders, which presumably were capable of visiting the infected leaves; and

(4) more probably, in the middle and upper Mesozoic era (Jurassic and Cretaceous periods), at a time when flies, ants, wasps, bees, and probably also moths and butterflies were spreading over the earth and nectar-excreting and scent-bearing flowers of the Angiosperms were evolving.

GENERAL INDEX

Abies balsamea, inoculated, 249
   ,,   ,,   in Manitoba and Saskatchewan, 197
Abies lasiocarpa, and aecidiospores, 195
Abies pectinata, and *Stellaria nemorosa*, 195
   ,,   ,,   and witches' brooms, 194
Abnormal, F₃ cultures, of *Puccinia graminis*, 254
   ,, pycnidia, 149
Abscission, from the oidiophore, 304
Absence, of *Puccinia suaveolens* from central Canada, 351-354
Adams, J. F., on nuclear migration, 49
Aecia, and Telia, in pycnial cup, 263
Aecidia, and atmospheric conditions, 123
   ,, and paired nuclei, 46
   ,, and pycnidia, successive formation of, 326-334
   ,, and sexual process, 53, 111
   ,, and unredospore sori, 258
   ,, belated in simple and compound pustules, 62
   ,, cupulate, 109, 402
   ,, dehiscence of, in saturated air, 117
   ,, dehiscence, mode of, 404
   ,, formed spontaneously, 250
   ,, genetic composition of, 235
   ,, haploid, 293
   ,, hornlike in Gymnosporangium, 390, 400, 403-409
   ,, in compound pustules, 275
   ,, of *Puccinia graminis*, 113-124
   ,, of *Uromyces caladii*, 76
   ,, on underside of leaf, 113
   ,, on uninucleate mycelium, 397
   ,, preparation for production of, 103
   ,, second year after pycnidia, 327
   ,, small, 335
   ,, split down sides, 404
   ,, three-dimensional studies of, 317
Aecidiospore chains, in mature aecidium, 118, 235
Aecidiospore-discharge, distance of, 118
   ,,   ,, in damp air, 122

Aecidiospore-discharge, violence of, 119, 403, 406, 410-411
Aecidiospore stage, repeats on Pines, 399
Aecidiospores, and basal cells, 278
   ,, and female cells, 51
   ,, and peridial cells, 401
   ,, and pore-plugs, 123
   ,, as vestigial structures, 208, 272
   ,, develop basidia, 368
   ,, displace pseudoparenchyma, 116
   ,, first generation of, 292
   ,, germ-tubes of, 398
   ,, infect Graminaceous hosts, 4
   ,, infect Juniper leaves, 397
   ,, infect only Ribes, 322
   ,, in teleutospore sori, 247
   ,, liberated as powder, 406, 410
   ,, longevity of, 324, 390-391, 412
   ,, metamorphosed into peridial cells, 402
   ,, number of, on Barberry leaves, 122
   ,, on Euphorbia leaves, 368
   ,, produce aecidiospores, 369
   ,, sown near haploid pustules, 270
   ,, sown on *Pinus sylvestris*, 337
   ,, sown on Rhizome bud of *Lactuca pulchella*, 286
   ,, uninucleate in belated aecidia, 63
   ,, uninucleate in *Kunkelia nitens*, 291
   ,, uredospores, and binucleate mycelia, 397
   ,, wind dispersal of, 123, 265, 322
Aecidium, a diploid organ, 109
   ,, and moisture, 121
   ,, and nuclear association, 77
   ,, and pressure surface, 123

Aecidium, and Roestelia, 11
,,   conversion of Proto-aecidium
          into, 115-124
,,   cornute in Gymnosporangi-
          um, 412
,,   in transverse section, 317
,.   on *Euphorbia virgata*, 31
,.   on *E. virgata*, lacks sucrose, 33
,,   ruptures epidermis, 117
,,   under wet and dry conditions,
          122
Aecidium bifrons, and *Aconitum koelleanum*, 12
Aecidium clematidis, cuprous oxide from
nectar of, 31
Aecidium leucospermum, nuclear   condi-
                        tion of aeci-
                        diospores, 64
,,   ,,   nuclear division,
                        in aecidia, 44
,,   ,,   on *A n e m o n e
                        nemorosa*, 367
Aecidium magelhaenicum, lack of dex-
                        trose and
                        laevulose, 33
,,   ,,   nectar of pyc-
                        nidia, 38
,,   ,,   nectar taste-
                        less, 32
Aecidium oenotherae, on *Onagra biennis*,
21
Aecidium punctatum, and   *Tranzschelia
                        pruni-spinosae*,
296
,,   ,,   finger-like hyphae,
                        227
Aerial shoots, and rust infection, 286
,,   · ,,   and spore sequence on, 385
,,   ,,   and systemic mycelium of
               Puccinia suaveolens, 352
,,   ,,   die in autumn, 370
Ajreker, S. L., and Parandekar, S. A.,
on aecidiospores, 262
Allen, Miss R. F., and *Puccinia triticina*,
78
,,   ,,   ,, and pycnidium for-
                  mation, 134
,,   ,,   ,, and transfer of pyc-
                  nidiospores by in-
                  sects, 169-170
,,   ,,   ,, on emergent hyphae,
                  227
,,   ,,   ,, on fusion of spermatia
                  with paraphyses, 79
,,   ,,   ,, on heterothallism in
                  Puccinia graminis,
                  125
,,   ,,   ,, on migrating nuclei,
                  223
,,   ,,   ,, on multinucleate cells,
                  342
,,   ,,   ,, on *Puccinia malva-
                  cearum*, 244

Allen, Miss R. F., on receptive hyphae,
314
,,   ,,   ,, on spermatial nuclei,
                   228
Althaea rosea, and basidiospores, 245
Amelanchier alnifolia, and Gymnospor-
                        rangium, 91-92,
                        158, 204
,,   ,,   and *Gymnosporangi-
                        um juvenescens*,
                        158
,,   ,,   inoculated leaves
                        of, 202
,,   ,,   inoculation of, 92
,,   ,,   living and pickled
                        leaves of, 203
Ames, L. M., on *Pleurage anserina*, 8
Andrus, C. F., on fusion of trichogynes
              with pycnidiospores,
              314
,,   ,,   on haploid pustules, 227
,,   ,,   on hyphae as trichogynes,
               227
,,   ,,   on sex in Uromyces, 77
Anemone quinquefolia, and *Puccinia
rubigo-vera* var. *agropyrina*, 91
Anemone ranunculoides, and *Aecidium
leucospermum*, 293
Angiosperms, absence of, 418
,,   petals of, 190
,,   pollen grains of, 233
Annual changes, in *Puccinia minussensis*
and *P. suaveolens*, 307
Antheridia, of Discomycetes and Pyreno-
           mycetes, 9
,,   of Laboulbeniales, 8-11
Aphidae, Coccidae and Chermidae, 30
Apothecia, discharge of ascospores in,
424
Appendix, origin and meaning of the
terms Haploid and Diploid, 308-312
Apple, and *Gymnosporangium juniperi-
virginianae*, 91
Apple leaves, inoculated, 205
Arabinose, in nectar of Gymnosporangium species, 35
Arthur, J. C., on *Kunkelia nitens*, 290
,,   ,,   on *Melampsorella cerastii*,
               195
,,   ,,   on sexual reproduction,
               52-53
,,   ,,   on treating pycnidial exu-
               date, 74
,,   ,,   pycnium and pycniospores,
               3
Arthur, J. C., and Kern, F. D., on
witches'-broom, 195
Articulata, Filicales, and Pteridospermae, 416
Artificial, introduction of *Puccinia
suaveolens* into central Canada, 351-
354
Ascobolus, a genus of Discomycetes, 424

Ascobolus carbonarius, antheridial conidium of, 10-11

Ascobolus magnificus, single and mated mycelia of, 54

Ascobolus stercorarius, flies carry oidia, 423-424

Ascomycetes, and discovery of Pycnidia, 5

„ and heterothallism, 54

„ and Uredinales, spermatia of, 3

Ascospores, infect Cowberry plants, 424

Ashworth, Miss D., killed pycnidia, 287

„ „ „ observations on Melampsoridium and Melampsora, 213-214

„ „ „ on hyphae and spermatial nuclei, 78

„ „ „ on *Puccinia malvacearum*, 243

Axillary buds, on the rhizome, 354

„ „ on vertical rhizomes, 374

Bache-Wiig, Sara, on a systemic Fungus Parasite, 369-370

Bachus, M. P., on *Neurospora sitophila*, 8

Backmann, F. H., and sexual theory, 6

Bailey, D. L., on inoculation methods, 261

„ „ on *Puccinia helianthi*, 260

Balsam Fir, and witches' brooms, 198

„ „ pycnidia on leaves, 250

Baltic amber, and wingless insects, 421

Barberry, and flies on infected Barberry, 24

„ and proto-aecidia, 110

„ and size of pustule, 99

„ illustration, 70

„ inoculated with basidiospores, 3

„ inoculated with sporidia, 67, 96-98

„ leaves boiled, 111

„ mature leaf resistance, 98

„ prepared for inoculation, 97

Barberry bushes, and *Puccinia graminis*, 216

Barclay, A., on *Uromyces cunninghamianus*, 261

Bark, covering the pycnidium, 331

„ ruptured by pressure, 191

Basal cells, alone are truly diploidised, 302

„ „ enlarged in Proto-aecidium, 115

„ „ exhausted by aecidiospore development, 117

„ „ fuse in pairs, 334

„ „ of the Proto-aecidia, cell-fusions in, 341-343

Basal stroma, and single pycnidium, 249

Basidia, become synkaryotic, 298

Basidiospore, nucleus divides, 263

Basidiospores, and Flax, 93

„ and haploid mycelium, 100

„ and haploid oidia, 307

„ and heterothallic species, 396

„ and mutation, 412

„ and *Rhamnus alnifolia*, 93

„ and White Pine, 283

„ are short lived, 324

„ become uninucleate, 294

„ begin to germinate, 366

„ binucleate, 263

„ carried by wind, 405

„ developed asymetrically, 298

„ derived from localised mucelia, 365-367

„ discharged, 366

„ dry up, 405

„ escape, 261

„ from overwintered teleutospores, 376

„ infect Juniper, 395, 412

„ infect White Pine, 324

„ on a needle, 322

„ on first foliage leaves, 366

„ on Flax seedlings, 193

„ on *Helianthus annuus*, 270

„ on *Jasminum grandiflorum*, 263

„ on leaves of *Euphorbia sylvatica*, 292

„ on seedling thistles, 361

„ produce teleutospores, 241

„ violently discharged, 298

Basidium, and nuclear divisions, 307

„ from aecidiospore, 307

„ two-celled, 291

Basidium-bodies, dissolved by autodigestion, 405

Batho, George, on Canada Thistle in Manitoba, 353

Bauch, R., on *Camorophyllus virgineus*, 295

Baur, E., and sexual theory, 6

Beck, W. A., and Redman, R., on Plant Pigments, 142

Bensaude, Mlle., and heterothallism in *Coprinus lagopus*, 56-57

Berberis vulgaris, and *Puccinia graminis*, 26

„ „ and uredospore sori, 253

„ „ waxy epidermis of, 164

Bessey, E. A., and aecial primordium, 107

Betts, E. M., on sex in *Ascobolus carbonarius*, 56

Bibliography, of Appendix, 312

Binucleate, and uninucleate mycelia, 378

Binucleate, basidial cells and basidio-spores, 281
,, basidiospores, 262
Binucleate condition, and lateral cell-fusion, 282
,, ,, of systemic myceli-um in stems and roots, 375-377
Binucleate mycelium, and basal cells, 263
,, ,, de-diploidised, 398
,, ,, in species of Rusts, 281
,, ,, in stems and roots, 377
,, ,, partially de-diploi-dised, 412
,, ,, to uninucleate hy-phal branches, 378
Biological Significance, of Pycnidia on Systemic Mycelia, 383-384
Bisby, G. R., and *Melampsorella cerastii*, 197
,, ,, on absence of *Puccinia suaveolens*, 351
Black Chaff, a disease of wheat, 150
Blackman, V. H., and binucleate con-dition, 48
,, ,, and receptive cell, 76
,, ,, on *Phragmidium vio-laceum*, 22
,, ,, on sexual nature of pycnidia, 23
Blackman, V. H., and Fraser, Helen, on sexuality of Uredinales, 52
Blakeslee, A. F., and heterothallic Mu-corineae, 54
Bliss, D. E., on seasonal development of Gymnosporangium, 404
Blister Rust, disease of White Pines, 321-324
,, ,, from Canada to United States, 324
,, ,, introduced into Canada and United States, 322
,, ,, mode of infection, 324-326
,, ,, of Asiatic origin, 321
Blueberry patches, at Minaki, Ontario, 249
Blue Lettuce, Creeping Thistle, and systemic mycelia, 335
Botrytis cinerea, and *Hypochnus solani*, 302
Boys and Girls, eat nectar, 157
Branch, cortex and cambial region, 335
Brefeld, O., on germination of spermatia in Rust spp., 18-19
,, ,, on pycnidia as sexual organs, 19
,, ,, on sexuality in Higher Fungi, 18
Brodie, J. H., experiments with oidia, 424

Brodie, J. H., on *Collybia velutipes*, 306
,, ,, on pycnidial exudate, 167
Brown, A. M., diploidisation experiments of, 88
,, ,, discovered Mode III of diploidisation, 268-272
,, ,, on compound pustules, 180
,, ,, on diploidisation, 269-270
,, ,, on life-history of *Puccinia minussensis*, 284-288
,, ,, on *Phragmidium specios-um*, *272*
,, ,, on Phragmidium, Uro-myces and Puccinia, 85
,, ,, on *Puccinia helianthi*, 236, 260
,, ,, on *P. xanthii*, 246
,, ,, on *Uromyces fabae*, 257-258
Brown, A. M., and Craigie, J. H., on monosporidial pustules, 250
Brush, of periphyses and flexuous hy-phae, 176-177
Bud-scales, develop stomata, 374
Buffer cells, deteriorate, 135
Buller, A. H. R., on haploid phenome-non, 294-295
,, ,, on significance of conju-gate nuclei, 58-59
,, ,, presidential address, R.S. of Canada, 321
Buller, A. H. R., and Brown, A. M., on uredospores as the origin of *Puccinia suaveolens*, 346-347
Buller phenomenon, investigated, 305
Bunning, E., on pigment in *Pilobolus kleinii*, 141

Caeoma, a form-genus, 290
,, and proto-caeoma, 109
,, on *Poterium sanguisorba*, 31
Caeoma nitens, on *Rubus villosus*, 25
Caeomoid aecidia, 208
,, ,, of *Gymnoconia pecki-ana*, 290
Calamagrostis canadensis, teleutospores on straw, 93
Calamites, and Lepidodendron, 418
Caley, Dorothy M., on homothallic Pyrenomycetes, 55
Calyptospora goeppertiana, and *Abies bal-samea*, 250
,, ,, is heteroeci-ous, 249
Cambrian, and Ordovician Periods, 416
Canada thistle, or *Cirsium arvense*, 350
Canadian Pacific Railway, first train, 353
Cane sugar, molar solution of, 164
Canker, and bladdery aecidia, 329
,, completes ripening, 322
,, pycnidia precedes aecidia, 327

Canker, zone of discoloration and new pycnidia, 329

Canker Formation, and incubation period, 326

Cankers, and drops of nectar, 190

„ delayed aecidial formation in, 327

„ on *Pinus caribaea*, 93

„ simple and compound, 337

„ that have produced aecidia, 339

Capitulum, and systemic mycelium, 357

„ of a pistillate plant, 355

„ of a staminate plant, 355

„ receptacle of, 354

Carboniferous and Permian periods, 417-418

Carboniferous and Permian periods, and insects, 427

Carboniferous period, and insect remains, 419-420

Carleton, M. A., on *Puccinia helianthi*, 260

„ „ on spore germination in Uredineae, 20

Carotin, and Flowering Plants, 141

„ and Pilobolus, 141

Carotinoid pigment, in periphyses, 141

„ „ in Uredinales, 141

„ „ of pycnidiospores, 141

Carpogonium, a female organ, 6

Caytonia, and Gristhorpia, fossil fruits of, 418

Cedar-apple, galls of, 206

Cell-fusions, and diploidisation of a Proto-aecidium, 124-131

Cells, fuse in pairs, 341

Cellulae disseminulae, are pycnidiospores, 132

Central arch, of peridium, 119

Cerastium arvense, and Picea spp., 197

Chains, of binucleate aecidiospores, 334

Channel, pycnidiospore to flexuous hypha, 239

Christensen, J. J., on insects and pycnidial nectar, 172-173

Christman, A. H., and *Phragmidium speciosum*, 48-49

„ „ and *P. speciosum* and *Uromyces caladii*, 23

„ „ on fusion cells, 23

„ „ on origin of caeomoid aecidia, 52

Chromosomes, affect cell protoplasm, 297

„ full complement of, 301

Cell-fusions, in Basal cells of Proto-aecidia, 341-343

„ „ in teleutospore sori, 280

„ „ of haploid cells, 264

Cirsium arvense, and odour of *Puccinia suaveolens*, 26, 287

Cirsium arvense, and pycnidia, 29

„ „ and systemic mycelia of *Puccinia suaveolens*, 370-373

„ „ and *Taraxacum officinale*, 349

„ „ bearing ripe pycnidia, 39

„ „ distribution of, 350-351

„ „ growth and reproduction, 354-356

„ „ horizontal and vertical roots, 354

„ „ host of *Puccinia suaveolens*, 349

„ „ illustrations, 345, 363

„ „ inoculated with *Puccinia suaveolens*, 351-352

„ „ in pastures and waste places, 344

„ „ introduced into Canada, 350

„ „ is dioecious, 354

„ „ seeds germinate, 372

„ „ shoots hoed, 374

„ „ sprayed with uredospores, 359

„ „ spread of, 350

Clamp-connexions, and conjugate nuclei, 295

Claviceps, a genus of Pyrenomycetes, 422

Claviceps purpurea, and conidia-bearing drops, 425

„ „ conidia germinating, 165

„ „ illustrations, 422-423

„ „ insects and conidia, 42

„ „ Sphacelia stage, 35

Clinton, G. P., and McCormick, Florence A., experiments of, 325

Clinton, G. P., and McCormick, Florence A., inoculation experiments, 25

Clitocybe illudens, illustration, 284

Coleoptera, and Blattidae, 420

„ and pycnidia, 36

„ in Eocene period, 421

Coleosporium campanulae, paired nuclei in uredospores, 43

Coleosporium euphrasiae, nuclear division in teleutospores, 44

„ „ nuclear fusion in teleutospore, 44

Coleosporium helianthi, pycnidia of, 82

Coleosporium pinicola, *Colyptospora goeppertiana*, and vestigial pycnidia, 248-250

„ „ is homothallic, 282

Collema pulposum, sexual process in, 11

Colley, R. H., and Cytology of *Cronartium ribicola*, 329

„   „   on flexuous hyphae, 338

„   „   on multiple fusions, 343

Collybia velutipes, oidia of, 307

Colour, attractive to flies, 141

„   of pycnidiospores, 141

Comparison of systemic mycelia, 334-337

Component, (+) and (−) mycelia, 340

Compound canker, initiation of diploidisation in, 340-341

Compound drops, of nectar, 161

Compound pustules, on Barberry, 173

„   „   produce aecidia, 73, 396

„   „   produce uredospore sori, 258

„   „   remain haploid, 258

Conidia, germinate, 423

Coniferous host, and teleutospores, 396

Conifers, precursors of, 417

Conjugate division, of associated nuclei, 264

Conjugate nuclei, as sister nuclei discarded, 279

„   „   in aecidiosporophores, 124

„   „   in Hymenomycetes, 298

„   „   in perennial mycelium, 369

„   „   in young aecidiosporophores, 343

Conjugate pairs, of (+) and (−) nuclei, 300

Conjugation, of nuclei, 22

Conversion, flexuous hyphae into periphyses, 185-186

Co-operation, of nuclei, 314

„   pairs of cells in *Puccinia graminis*

Coprinus, a genus of Hymenomycetes, 424

Coprinus lagopus, a haploid strain of, 295

„   „   and aerial oidiophores, 299

„   „   and *Ascobolus stercorarius*, 425

„   „   and flexuous hyphae, 132

„   „   and *Puccinia graminis* as types for comparison, 298-300

„   „   and *Puccinia helianthi*, 305

„   „   cultivated by Hanna, 295

„   „   differs from *Puccinia graminis*, 299

„   „   diploidisation not localised, 303

Coprinus lagopus, diploidisation process in, 58

„   „   four sexual groups, 73

„   „   groups of basidiospores, 304

„   „   haploid and diploid mycelia, 269

„   „   haploid generations of, 296

„   „   insects carry oidia of haploid mycelium, 303

„   „   mating experiments, 303

„   „   mycelia and oidia of, 424

„   „   mycelia of opposite sex, 277

„   „   nuclear migration in mycelium, 221

Coprinus Rostrupianus and *C. radians*, 61

Coprinus species, and homothallic Rusts, 306

Cornu, M., and pycnidiospores, 17

Cornute aecidia, and Puff-balls, 405-406

„   „   development from Protoaecidia, 400-402

„   „   hygroscopic movements of peridium, 406-410

„   „   illustrations of, 404

„   „   of *Gymnosporangium juniperinum*, 401

„   „   of two groups of species, 402-403

Cotter, R. W., on hybridization in *Puccinia graminis*, 216

Cotton-blue, and lacto-phenol, 112

Cotyledons, inoculated with basidiospores, 372

„   inoculated with uredospores, 371

Cracked bark, and excreted nectar, 331

Craigie, J. H., and fusions, 215

„   „   and monosporidial pustules, 65

„   „   experiments described, 67-71

„   „   experiments with pycnidiospores, 231

„   „   mixed pycnidial nectar, 66

„   „   Modes, I and II of diploidisation, 266

„   „   on aecial development, 107

„   „   on caged flies, 170

„   „   on flexuous hyphae, 4

„   „   on heterothallic Rusts, 65

„   „   on heterothallic species of Puccinia, 395

„   „   on heterothallism, 60-62

„   „   on nectar of *Puccinia graminis*, 158

„   „   on *P. helianthi*, 330

Craigie, J. H., on pycnidia, 4
   ,,   ,, pycnidiospores fuse with flexuous hyphae, 80-81
   ,,   ,, sex distribution in Puccinia spp., 72-73
   ,,   ,, sex in Puccinia spp., 60
   ,,   ,, unions discovered, 227
Creeping Thistles and *Puccinia suaveolens*, 165
   ,,   ,, at Banbury, England, 360
   ,,   ,, observed in pastures, 370
   ,,   ,, readily infected, 349
Creeping Thistle leaves, and young pycnidia, 379
Creeping Thistle Rust, a brachy-form, 344
Cretaceous period, and Angiosperms, 418
Cretaceous rocks, and scanty remains of insects, 420
Critical remarks, concerning Mode I, 272-279
Cronartium, evolutionary advance of, 336
Cromartium species, and flexuous hyphae, 190-192
Cronartium cerebrum, and living pycnidia, 190
   ,,   ,, pycnidia of, 93
Cronartium fusiforme, and flexuous hyphae, 191
   ,,   ,, and living pycnidia, 190
   ,,   ,, cankers on *Pinus caribaea*, 93
   ,,   ,, examined pycnidia of, 92
Cronartium quercuum, excrescences on trunk, 157
   ,,   ,, in Japan, 332
Cronartium ribicola, aecidiospores discharged as dust, 411
   ,,   ,, and *Endophyllum euphorbiae-sylvaticae*, 336
   ,,   ,, and *Pinus strobus*, 321, 329
   ,,   ,, and *Puccinia graminis*, 338
   ,,   ,, illustrations, 323-335
   ,,   ,, is heterothallic, 88, 332, 338
   ,,   ,, nectar and pycnidiospores, 22
   ,,   ,, nectar sweet, 332
   ,,   ,, odour of pycnidia, 28
   ,,   ,, on *Pinus monticola*, 326
   ,,   ,, systemic mycelium of, 283, 326

Cross-diploidisation, De-diploidisation, Self-diploidisation, and Re-diploidisation, 382-383
Cross-diploidisation, in *Puccinia minussensis*, 287
   ,,   ,, in Uredinales, and Hymenomycetes, 288
Cross-fertilization, Plant and Animal Kingdom, 288
Crossing, between different strains, 384
Cross-wall, in the oidium, 306
Crowell, I. H., on Gymnosporangium, 393
Crozier unions, 218
Crustacea, and underground caverns, 250
Cryptomycina pteridis, and Sara Bache-Wiig, 369-370
   ,,   ,, attacks Bracken Fern, 369
   ,,   ,, resembles Rusts, 369
Crystals, in dried nectar, 158
Cullen, E. O., and *Anemone nemorosa*, 369
Culture media, and pycnidiospores, 232
Cultures, by J. C. Arthur, 195
Cummins, G. B., *Puccinia sorghi* is heterothallic, 74
Cunningham, G. H., and *Puccinia suaveolens*, 359
Cuprous oxide, and nectar of Rust species, 31
Cupulate aecidia, of *Puccinia graminis*, 405
   ,,   ,, of Puccinia spp., 406
Cuticle, not ruptured, 250
Cuticular collar, and ostiolar band, 129
Cytological background, function of the pycnidia, 43-53
Cytological work, review of Periods, I, II, III, 313-317
Cytologist, and sexual process, 320
Cytology and heterothallism, in Gymnosporangium, 394-400
Cytoplasm, and nucleus of pycnidiospore, 221
   ,, flows from cell to cell, 319
   ,, moving rapidly, 128
Cytospora chrysosperma, and Black Poplar, 150
Czechinsky, N. I., on effect of Rust, 357-358

DANDELION, and DeBary, 349
   ,, parasitism by *Puccinia suaveolens*, 349-350
Dangeard, P. A., and Sappin-Trouffy, P., on cytology of the Uredinales, 44
Darbyshire, D. V., and sexual theory, 6
Darrell, L. W., on stem rust in Iowa, 98-99

Darwin, F., and *Pteris aquilina*, 35

Davis, R. J., on *Ophiobolus graminis*, 55

Dead preparations, and living Fungi, 317-320

Dearness, John, found infected shoots, 351

DeBary, A., and aecidial primordium, 106

„ „ and developing aecidia, 226

„ „ and paraphyses, 133

„ „ and *Puccinia graminis*, 3

„ „ and *P. tragopogonis*, 106

„ „ and sex in Uredineae, 76

„ „ and species of Rusts, 15

„ „ deprecated sexual process, 226

„ „ experiments of, 362

„ „ inoculation of Barberry, 3

„ „ on *P. graminis*, life history, 43

„ „ on *P. suaveolens*, 349

„ „ on pycnidia and aecidia, 15

„ „ on pycnidia and pycnidiospores, 16

„ „ ⁻on pycnidiospores, 38

„ „ on scent of Evening Primrose, 27

„ „ on spermogonia, significance of, 7

„ „ on *Uromyces fabae*, 257

De-diploidisation, and Self-diploidisation, in *Puccinia minussensis*, 284-288

„ in Puccinia spp., 413

„ defined, 306

„ depends on stimulus, 386

„ of bisexual, binucleate mycelium, 398

„ Re-diploidisation, Self-diploidisation, and Cross-diploidisation, 382-383

Defect, of Homothallism, 288-289

Degeneracy, of pycnidia, 251

Derx, H. G., on *Penicillium lutem*, 55

Development, of a pycnidium of *Puccinia graminis*, 134-136

Devonian Period, fossil evidence, 416-417

„ „ insect remains of, 419

Dextrose and laevulose, in nectar of *Gymnosporangium sabinae*, 157

Dextrose, in nectar of *Gymnosporangium sabinae*, 33

Dickinson, S., on monosporidial mycelia of Ustilaginales, 58

Dicotyledonous host, pycnidia and aecidia, 396

Dikaryophase, in teleutospore sori, 248

Dikaryotic and synkaryotic diploid cells, 297-298

Dikaryotic cells, in proto-aecidia, 233

Dikaryotic diploid cells, in *Puccinia graminis*, 298

Dikaryotic hybrids, of Hymenomycetes and Rusts, 298

Dikaryotic nuclei, 311

Dillenius, J. J., and Hedwig, J., on pycnidia of Lichens, 5-6

Diploid and haploid mycelia, fused, 270

Diploid, in Crystallography, 309

„ origin and meaning of, 308-312

Diploid cell, defined by Buller, 297

Diploid hyphae, from perennial mycelium, 381

Diploid mycelium, and haploid branches, 286, 382

„ „ partially de-diploidised, 378, 382

Diploidisation, and Mode IV, 279

„ and nuclear migration, 286

„ by cross-diploidisation, 382

„ by fusion, and by pycnidiospore, 340

„ by pycnidiospores, 382, 383

„ completed, 301

„ effected, 380

„ fusion of cells in pairs, 338

„ in cornute aecidia, 401

„ initiated, 276, 380

„ initiation in a compound canker, 340-341

„ initiation in a simple canker, 338-340

„ last stages, 126

„ of a haploid mycelium, 320

„ of haploid branches, 288

„ part of the sexual process, 264

„ progressive, 276-277

„ term introduced, 300

„ without pycnidiospores, 274

Diploidise, a haploid mycelium, 301

„ basal cells, 127

Diptera, and pycnidia, 36

„ Hymenoptera and Coleoptera, 171

Discomycetes, Pycnidia and Spermogonia of, 7-8

Displacement cells, displaced, 116

Dodge, B. O., and Neurospora, 7-8

„ „ on *Coleosporium pinicola*, 281-282

„ „ on *Gymnosporangium ellisii*, 402-403

„ „ on *Kunkelia nitens*, 290-291

„ „ on sexuality of Uredinales, 25-26

Dodge, B. O., pores and pore-plugs in aecidiospores, 413-414

Dodge, B. O., and Gaiser, L. O., on *Kunkelia nitens*, 290

Dominion Rust Research Laboratory, 246, 273

Dowding, Eleanor S., and Buller, A. H. R., on nuclear migration, 128

Dried seeds, vitality retained, 373

Drop of nectar, 152-164

Drops of water, on Barberry, 162

EARLY PERMIAN, many specimens of, 420

Eastman's Opaque, 274

Economic value of, *Puccinia suaveolens*, 359

Ectosphere, and stationary nuclei, 316

Edgerton, C. W., on (+) and (−) strains in Pyrenomycetes, 54

Elater-like hairs, and seed-dust dispersal, 409

Electric lamps, in Greenhouse, 360

Elongated cell, in *Puccinia graminis*, 231

Emerging hyphae, and pycnidiospores, 226-229

Emmons, C. W., and Ascocarps in Penicillium, 56

Endophyllum euphorbiae-sylvaticae, a haploid strain of, 296

Endophyllum euphorbiae-sylvaticae, and cuprous oxide, 31

Endophyllum euphorbiae-sylvaticae, and the cytologist, 291

Endophyllum euphorbiae-sylvaticae, and *Puccinia suaveolens*, 368

Endophyllum euphorbiae-sylvaticae, lacks dextrose and laevulose, 33

Endophyllum euphorbiae-sylvaticae, lacks fermentable sugar, 35

Endophyllum euphorbiae-sylvaticae, lacks sucrose, 33

Endophyllum euphorbiae-sylvaticae, nectar tasteless, 32

Endophyllum euphorbiae-sylvaticae, on *Euphorbia amygdaloides*, 367

Endophyllum euphorbiae-sylvaticae, perennial mycelium of, 41

Endophyllum euphorbiae-sylvaticae, sexual process imperfect in, 289

Endophyllum sempervivi, pycnidia not required, 287

Entomological background, and function of pycnidia, 26-43

Eocene period, ants and other insects, 421

Epidermal cells, and emerging hyphae, 226-229

Epipycnidial papilla, unbroken, 194

Equisetales, and Sphenophyllales, 417

Eradication, of Ribes, 324

Ergot, of Rye, 422-423

Euphorbia amygdaloides, and *Endophyllum euphorbiae-sylvaticae*, 36

Euphorbia amygdaloides, and *Lactuca pulchella*, 370

Euphorbia amygdaloides, and pycnidia, 27

Euphorbia cyparissias, and pycnidia, 27

Euphorbia seedlings, and basidiospores, 370

Euphorbia species, infected, 41

Euphorbia virgata, and pycnidial pustules, 39

Evolution, of insects, and fungi, 425-426

    „   of Gymnosporangium, 411-413

    „   of land plants and insects, 415

    „   of Phallales, etc., 427-428

    „   of pycnidia, Geological Time of, 426-427

Examination, of needle bundles, 325

Experimental background, function of the pycnidia, 53-59

External conditions, effect on uredospore and teleutospore production, 359-360

FAIRY RING, and fruit-bodies, 295

Farlow, W. G., and *Aecidium bermudianum*, 392

Faull, J. H., and *Galyptospora goeppertiana*, 249

    „   „   on *Milesia polypodophila*, 29

Fauna, of Carboniferous period, 419

Fehling's solution, and experiments of Rathay, 32, 40

Ferns, Gymnosperms, and flowering plants, 418

Fertile cells, all (+), or all (−), 387

    „   „   in proto-aecidium, 320

Fertilisation, and conjugation, 301

Fimentaria fimicola and vacuoles, 128

Fisher, E., on *Gymnosporangium clavariiforme*, 404

Flax stems, and teleutospore sori, 193

Flexuous hypha, a receptive hypha, 131

    „   „   and basal cell, distance between, 234

    „   „   co-operates with pycnidiospore, 233

    „   „   not a trichogyne, 131

Flexuous hyphae, and pycnidiospores, 5, 143

    „   „   bent, curved or twisted, 217

    „   „   branching, 182

    „   „   break down and disappear, 131, 224

    „   „   conversion of mature periphyses into, 185-186

    „   „   described, 131

    „   „   differ from periphyses, 176-177

    „   „   discovery in Uredinales, 75-84

    „   „   failure to distinguish, 316

Flexuous hyphae, from mouths of pyc-
    nidia, 18
,,   ,,   fully grown, 180
,,   ,,   fuse with pycnidio-
    spores, 75-84, 217-
    222, 229
,,   ,,   fusions outside host-
    leaf, 180
,,   ,,   in Cronartium species,
    190-192
,,   ,,   in Gymnosporangium
    species, 178-180,
    200-207
,,   ,,   in living condition, 176
,,   ,,   in rust species, 90,
    214
,,   ,,   in *Melampsorella cer-
    astii*, 194-198
,,   ,,   in Milesia, 198-199
,,   ,,   in Phragmidium, 207-
    209
,,   ,,   in *Puccinia graminis*
    and *P. helianthi*,
    180-185
,,   ,,   in pycnidia, 379
,,   ,,   in Rust species, 90
,,   ,,   in *Transchelia pruni-
    spinosae*, 210
,,   ,,   long-lived, 229
,,   ,,   longer than pycnidio-
    sporophores, 331
,,   ,,   methods of observa-
    tion, 173-176
,,   ,,   never unite, 183
,,   ,,   new observations on,
    187-190
,,   ,,   non-septate in *Puc-
    cinia graminis*, 234
,,   ,,   observed in Melamp-
    soridium and Me-
    lampsora, 213-214
,,   ,,   observed in twenty
    species of Rusts, 187
,,   ,,   of Melampsoraceae
    and Pucciniaceae,
    214
,,   ,,   of *Melampsora lini*,
    192-194
,,   ,,   tips protrude into air,
    178
,,   ,,   treated in detail, 173-
    177
,,   ,,   turgid, and flaccid, 83
,,   ,,   union with pycnidio-
    spores, 215-240
Flies, access to Rust pustules, 237
,, and haploid rust pustules of
    Puccinia graminis, 24
,, carry conidia, 423
Florets, become infected in rusted thistle,
    357
,,   open in succession in staminate
    plant, 355

Florets, open simultaneously in pistillate
    plant, 355
Flowering plants, and insects, interde-
    pendence of, 422
,,   ,,   depend on insects, 421
Flowers, and odour of Rust Fungi, 168
Forms, of localised mycelia, 362
,,   of systemic mycelia, 362-363
Fort, Margaret, on pointed periphyses,
    226
Fossil insects, wingless, 419
Fossil leaves, from Greenland, 418
Fossil plants, from Middle Devonian, 416
Fossil, fruits and wood, 418
,, Rust Fungi unknown, 426
Fossils, from Cretaceous rocks, 418
,,   likeness to common fungi, 417
Fraser, W. P., and Ledingham, G. A., on
    Puccinia coronata, 251
Fromme, F. D., on peridial cells, 119
,,   ,,   on *Melampsora lini*, 342
Fruits, bore aecidia, 201
Fruit-bodies, haploid and diploid, 295
Fukushi, T., on the apple Rust, 391
Function of the Pycnidia, experimental
    background,
    53-59
,,   ,, ,,   ,,   entomological
    background,
    26-43
,,   ,, ,,   ,,   observations
    and specu-
    lation on, 12-
    26
Functions, of the periphyses, 136-143
Fungi and insects, 422-425
Fungi, Bacteria, and destruction of dead
    plants, 417
,, dependent on insects, 425
,, earliest geological period of, 426
Fungi Imperfecti, conidia extruded from
    pycnidia, 150
,,   ,,   extrusion of spores, 146
Fungus material, source of, 360-361
Fusion cells, multinucleate, 342
Fusions, diploid with haploid mycelium,
    Mode III, 267-272
,,   hypha-to-peg, 220
,,   in *Coprinus lagopus*, 317
,,   in *Puccinia graminis*, 217, 222
,,   in *P. helianthi*, 235-236
,,   in Puccinia species, 224
,,   methods of observing, 216-217
,,   pycnidiospores and flexuous
    hyphae, 190, 217-222
,,   time to complete, 222-223

GALL, and binucleate mycelium, 398
,, and double infection, 397
Gall mycelium, is uninucleate, 337
Galls, bearing aecidia and teleutospore
    sori, 392
Gasteromycetes, and flies, 141

Gelatinous cell-walls, of pycnidiospores, 145

Gelatinised flexuous hyphae, 211, 238

Genera and species, flexuous hyphae in, 214

Geographical distribution, of Gymnosporangium, 393-394

Geological Time, and evolution of pycnidia of Uredinales, 426-427

Germination, of teleutospores, 93, 294

Germ-tube, becomes a basidium, 289

Germ-tubes, become bicellular, 262

,, ,, enter leaf-parenchyma, 336

,, ,, grown down stigmas, 424

,, ,, in opposite directions, 366

,, ,, interdigitate with hyphae, 315

Ginkgoales, and Cordaitales, 417

Glass slide, and drop of nectar, 191

Gleba, deliquesces, 422

Gregory, H. P., on pycnidial and aecidial stages, 91

Grove, W. B., on systemic infections, 346

,, ,, on pycnidiospore function, 25

Growth and reproduction, of Cirsium arvense, 354-356

Growth curvatures, and pycnidiospores, 239

Guard-cells, increase dimensions, 358

Gwynne-Vaughan, Dame H. C. I., and Barnes, B., on pycnidiospore function, 26

Gymnoconia peckiana, is systemic on Rubus, 199

,, ,, illustration, 199

,, ,, pore-plugs not developed in, 414

,, ,, pycnidial elements discussed, 199-200

Gymnosperm trees, as Cordaites, 418

Gymnosporangium, and Eudes-Deslongchamps, 391

,, and Origin of Cornute Aecidia, 400

,, and Puccinia, ostiolar hairs of, 133

,, cytology and heterothallism of, 394-400

,, endemic species of, 393

,, evolution of, 411-413

,, flexuous hyphae in species of, 200-207

,, geographical distribution of, 393-394

,, heteroecism of, 391-393

,, heterothallic species of, 240

Gymnosporangium, in Europe and North America, 393-394

,, introductory remarks to, 389-391

,, is macrocyclic, 390

,, one species is autoecious, 392

,, perennial mycelium of, 412

,, primitive forms of, 411

,, specialized characters of, 411-412

,, species in Central Canada, 394

,, species of, 390-397, 406

Gymnosporangium aurantiacum, illustrations, 31, 401-403

Gymnosporangium aurantiacum, on Juniperus communis, 392

Gymnosporangium aurantiacum, on Pyrus aucuparia, 400

Gymnosporangium bermudianum, and Juniper, 392

Gymnosporangium bermudianum, basidiospores of, 398

Gymnosporangium bermudianum from heteroecious to autoecious, 413

Gymnosporangium bermudianum is autoecious, 393, 396

Gymnosporangium clavariiforme, and cancellate peridia, 409

Gymnosporangium clavariiforme, and flexuous-hyphae, 178-180

Gymnosporangium clavariiforme, and Phragmidium violaceum, 395

Gymnosporangium clavariiforme, illustrations, 202, 390, 407-409

Gymnosporangium clavariiforme, pycnidial nectar of, 32-33

Gymnosporangium clavariiforme, systemic mycelium of, 283

Gymnosporangium clavipes, and Amelanchier alnifolia, 238

Gymnosporangium clavipes, and G. clavariiforme, 201

Gymnosporangium clavipes, a simple pustule of, 396

Gymnosporangium clavipes, illustrations, 239, 406

Gymnosporangium clavipes, is heterothallic, 88, 240

Gymnosporangium clavipes, pycnidial elements of, 238-240

Gymnosporangium ellisii, and Myrica carolinensis, 413

Gymnosporangium ellisii, and Puccinia graminis, 414

Gymnosporangium ellisii, and P. podophylii, 123

Gymnosporangium ellisii, pores and pore-plugs in aecidiospores of, 413-414

Gymnosporangium globosum, and *G. haraeanum*, 396

Gymnosporangium globosum, and Klebahn, H., 22

Gymnosporangium globosum, colour of pycnidia, 26

Gymnosporangium globosum, illustration, 405

Gymnosporangium globosum, is heterothallic, 206

Gymnosporangium juniperinum, and *G. sabinae*, 157

Gymnosporangium juniperinum, and ripe pycnidia, 41

Gymnosporangium juniperinum, and sugar from pycnidial nectar, 33

Gymnosporangium juniperinum, and visiting insect species, 37

Gymnosporangium juniperinum, illustrations, 120, 399, 404

Gymnosporangium juniperinum, on *Sorbus aria*, 40

Gymnosporangium juniperinum, pycnidial exudate of, 30

Gymnosporangium juniperinum, pycnidial nectar reduces Fehling's solution, 30-32

Gymnosporangium juniperi-virginianae, and flexuous hyphae, 178-180

Gymnosporangium juniperi-virginianae, and *Juniperus virginiana*, 92

Gymnosporangium juniperi-virginianae, and pycnidia, 136

Gymnosporangium juniperi-virginianae, illustrations, 137, 410

Gymnosporangium juniperi-virginianae, on *Pyrus malus*, 25

Gymnosporangium juniperi-virginianae, opening of aecidia, 409

Gymnosporangium juniperi-virginianae, pycnidiospores failed to germinate, 232

Gymnosporangium juvenescens, illustrations, 159, 204

Gymnosporangium nootkatense, uredospores of, 390, 411

Gymnosporangium peckiana, illustrations, 199, 200

Gymnosporangium sabinae, and sugar in pycnidial nectar, 35

Gymnosporangium sabinae, illustrations, 32, 33

Gymnosporangium sabinae, on Pear-trees and Junipers, 391

Gymnosporangium sabinae, on *Pyrus communis*, 400

Gymnosporangium sabinae, pycnidial nectar of, 32-33, 38

Gymnosporangium sabinae, pycnidial nectar reduces Fehling's solution, 31

Gymnosporangium species, alternate hosts of, 394

Gymnosporangium species, and violent spore discharge, 411

Gymnosporangium species, are heterothallic, 240, 395, 396

Gymnosporangium species, on Pyrus, 404

HAACK, G., on repeating aecidia, 336

Hagborg, W. A. F., on *Puccinia bardanae*, 87

Hanging drops, of culture media, 232

Hanna, W. F., on diploidised mycelia in *Coprinus lagopus*, 64

„   „   on *Gymnosporangium juniperi-virginianae* 204, 396

„   „   on nuclear association, in *Puccinia graminis*, 79

„   „   on osmotic pressure, 160

„   „   on primordia of *P. graminis*, 107

„   „   on pycnidia and aecidia, 113

„   „   on pycnidiospore germination, 232

„   „   on pycnidiospores of *P. graminis*, 24, 315

„   „   on young aecidia of *P. graminis*, 234

Hanna, W. F., and Popp, W., on hybridization of Oat Smuts, 58

Haploid, and compound pustules, 425

„   and diploid oidia, 306

„   origin and meaning of, 308-312

Haploid aecidia, unknown in *Puccinia helianthi*, 63

Haploid and diploid, origin and meaning of, 308-312

Haploid branches, of two kinds, 285

„   „   re-diploidised, 286

Haploid fruit-bodies, of heterothallic Hymenomycetes, 295

Haploid hyphae, from diploid mycelium, 386

Haploid mycelium, diploidised, 4, 87

„   „   from an oidium, 304

Haploid pustules, and diploid mycelium, 236

„   „   changed to diploid, 70

„   „   formation of protoaecidia in, 98-115

„   „   in Petri dishes, 87

„   „   on *Helianthus annuus*, 273

„   „   on *Puccinia graminis*, 229

„   „   spontaneous change, cause unknown, 70

Haploid pycnidia, and haploid sori, 386

Haplophase, and diplophase, 412

„   transition to diplophase, 126

Hawthorn leaves, and pycnidial lesions, 206

Hayden, Miss Ada, on viability of thistle seeds, 356

Hedgecock, G. G., on the genus Cronartium, 190

Hedwig, R. A., and Gymnosporangium, 390

Helianthus annuus, inoculated with basidiospores, 235

   „   „   introduced into Europe, 259

Hemi, Greek prefix for *half*, 311

Hemiptera, pycnidia visited by, 36

Henderson, I. F., and M. A., on derivation of *id*, 308

Herbaceous perennials, and establishment of systemic mycelia, 367-370

Heteroecious and autoecious heterothallic rusts, 267

Heteroecious species, and self-propagating haploid strains, 294-296

Heteroecism, in Gymnosporangium, 391-393

Heterokarotised mycelia, fusion of, 301

Heterothallic Coprinus species, 288

Heterothallic Hymenomycetes, 60

Heterothallic Rusts, list of, 84-85

Heterothallic species, of Gymnosporangium, 200-201

Heterothallic Uredinales, and sexual process, 314

Heterothallism, and cytology of Gymnosporangium, 394-400

   „   discovery in Uredinales, 53, 59-65

   „   of *Cronartium ribicola*, 338

Hibiscus esculentus, and Glomerella, 54

Higgins, B. B., on *Mycosphaerella tulipifera*, 8

Higher Fungi, and plant and animal cells, 297

Hollyhocks, at Kew, 243

Holton, C. S., on sporidial fusion in Ustilago, 223

Homologous chromosomes, 297

Homothallic Rusts, and diploidisation, Mode IV, 279-283

   „   „   and Hymenomycetes, 279

   „   „   produce no pycnidia, 279

Homothallic species, from heterothallic species, 288

   „   „   of Coprinus, produce no oidia, 243

   „   „   of Puccinia, 246

   „   „   of Rust Fungi, 84-88

Homothallism, a defective sexual process, 288-289

Honey, Dr., photograph of blister rust, 332

Honey-dew, of *Claviceps purpurea*, 164

Hooke, Robert, in Micrographia, 207

Horizontal roots, and adventitious buds, 354, 356

Horns of pycnidiospores, in *Puccinia graminis*, 148-151

Hunter, Lillian, on *Coleosporium pinicola*, 249

   „   „   on flexuous hyphae in Melampsoraceae, 82

Hybrid aecidiospores, of *Puccinia graminis tritici*, 235

Hybridisation, and multiple fusions, 234-235

Hygroscopic action, of peridia, 408, 409

Hygroscopic movements, of peridium in cornute aecidia, 406-410

Hymenial cells, and pycnidiosporophores, 178

Hymenium, composed of palisade layer, 329

Hymenomycetes, and Discomycetes, 133

   „   and migrating nuclei, 319

   „   and sexual process, 243

   „   and Uredinales, 298, 307

   „   Dikaryotic and Synkaryotic diploid cells in, 297-298

   „   haploid and diploid strains of, 295

   „   heterothallic species of, 295

   „   importance of nuclear migration, 126

Hymenoptera, and Rust pycnidia, 36

Hyphae, are short lived, 229

   „   as functioning trichogynes, 81

   „   grow into stomatal apertures, 77-78

   „   in staminal filaments, 357

   „   monilioid, 423

   „   of palisade layer, 124

   „   react to stimulus, 219

   „   with blunt tips, 228

Hyphal fusions, and nuclear exchange, 277

   „   „   in heterothallic Rusts, 317

   „   „   three-dimensional network, 183

Hypocotyl, and cotyledons, 356

IMMS, A. D., on geological history of insects, 419

Incipient cankers, on young trees, 326

Incubation period, 326-327

Infected plant, and conditions of light, 360

Infected thistles, and healthy ones, 344

Ingold, C. T., on aecidiospores of *Puccinia pulverulata*, 413

Initiation of sexual process, Modes I, II, III, IV, V, discussed, 300-307

Inoculation, with pipette, 205

Insects, absence of, 286, 292
,, and compound canker, 340
,, and *Cronartium ribicola*, 327
,, and fertilisation process in Rusts, 17
,, and flowers, 421-422
,, and fungi, 422-425
,, and isolated pustules, 266
,, and *Puccinia graminis*, 38
,, and Rust Fungi, 169-171, 426
,, attracted by odour of fungi, 425
,, disseminate pycnidiospores, 36, 152, 169
,, fungi dependent on, 425-426
,, in Palaeozoic rock system, 419
,, list of, 37-38
,, mix pycnidial nectar, 380
,, remove pollen, 355
Intercalary cells, and aecidiospores, 119-120
,, ,, as disjunctor cells, 119
,, ,, elongated, 334
,, ,, in *Gymnosporangium juniperinum*, 402
Intercellular hyphae, and mycelial mat, 329
Intercellular mycelium, is heterokarotised, 302
Introduction to evolution of Uredinales, 415
Isodiametric cells, and plectenchyma, 401

JACKSON, D. D., and botanic terms, 309
Jackson, H. S., and Mains, E. B., on *Puccinia triticina*, 29
Jackson, H. S., and uninucleate teleutospores, 294
,, ,, on ancestors of Rust Fungi, 88
,, ,, on evolutionary tendencies, 243
,, ,, on *Gymnosporangium bermudianum*, 412
,, ,, on *Kunkelia nitens*, 290
,, ,, on life cycles in Uredinales, 241
Jasminum grandiflorum and *Uromyces hobsoni*, 261-263
Johnson, T., on forms of *Puccinia graminis*, 93
,, ,, on hybridisation in *P. graminis*, 216
Juniper, and Gymnosporangium, 391
Juniperaceae, and Malaceae, 390
Juniperus barbadensis, and *Gymnosporangium bermudianum*, 397
Juniperus communis, and Gymnosporangium spp., 202
,, ,, and *Lactuca pulchella*, 283
,, ,, and teleutospore sori, 238

Juniperus horizontalis, and Gymnosporangium spp., 203
Juniperus virginiana, and parasite of, 43
Jurassic and Cretaceous periods, insects in, 427
Jurassic period, orders of insects in, 420

KAMAMURA, R., on *Gymnosporangium haraeanum*, 170
Kamei, S., on flexuous hyphae in Milesia, 83
Karsten, H., on Laboulbeniales, 9
Keeping, E. S., on *Gelasinospora tetrasperma*, 319
Kern, F. D., on cornute aecidia, 393
,, ,, on dry peridial cells, 408
Keuper and Rhaetic floras, 418
Kirby, R. S., on *Ophibolus graminis*, 55
,, ,, on stem rust in Iowa, 98-99
Kirchhoff, H., on condidia of *Claviceps purpurea*, 164-165
Klebahn, H., on aecidial development, 226
,, ,, on *Cronartium ribicola*, 22
,, ,, on inoculation of *Picea excelsa*, 28
,, ,, on inoculation of *Pinus strobus*, 324
,, ,, on wefts of hyphae and pycnidiospores, 76
Kniep, H., on heterothallism in Ustilaginales, 57
,, ,, on *Schizophyllum commune*, 57
Kny, L., on tropical orchids, 152
Kuhn, J., on *Claviceps purpurea*, 35
Kunkelia nitens, a microcyclic form, 290
,, ,, and Rubus species, 289
,, ,, life-history of, 290
Kursanov, L., and *Gymnoconia peckiana*, 199
,, ,, on *Aecidium punctatum*, 293
,, ,, on Gymnosporangium, 395, 402
,, ,, on nuclear migration in aecidia, 49
,, ,, on peridium in Puccinia and Uromyces, 119
,, ,, on sexual theories, 52
,, ,, on the developing aecidium, 124
,, ,, on *Uromyces ficariae*, 280-281

LABOULBENIALES, antheridia of, 8-11
,, identified as fungi, 9
,, on *Brachinus crepitans*, 9
,, spermatia and trichogynes of, 7-11
Lachmund, H. G., field observations of, 325

Lachmund, H. G., on incubation of *Cronartium ribicola*, 326

Lacto-phenol and cotton blue, 193, 204

Lactuca pulchella, a herbaceous perennial, 284

,, ,, and *Puccinia minussensis*, 26

,, ,, inoculated with uredospores, 398

Laevulose, in nectar of *Gymnosporangium sabinae*, 33

Lamb, I. M., on *Puccinia phragmitis*, 211, 223-224

Land plants, and evolution of Uredinales, 416

Lathyrus venosus, and *L. ochroleucus*, 257

Lechmere, E., on mycelium of *Tuberculina maxima*, 327

Lehmann, E., Kummer, H., and Dannenmann, H., on Black Stem Rust, 84

Lepidoptera, and Jurassic period, 427

,, fossils in Eocene rocks, 421

Leveille, J. H., on scent of *Puccinia tragopogonis*, 27

Levine, M. N., on hybridisation in *Puccinia graminis*, 216

Lichens, ascogonium and trichogyne of, 6, 227

,, carpogonia discovered by Stahl, 6

,, pycnidia and pycnidiospores of, 5-6

,, sexual process initiated in, 6

,, spermogonia and spermatia of, 3, 14

Lindan, G., on sexual theory of Stahl, 6

Lindfors, T., on binucleate young aecidia, 25

,, ,, on origin of paired nuclei, 52

Link, H. F., on *Puccinia suaveolens*, 27

,, ,, on *Uredo obtegens*, 347

Linum, species of, 192

Living Fungi, and dead preparations, 317-320

Living pycnidia, of Cronartium species, 330

Lloyd, F. E., and Ridgway, C. S., on closing of peridium, 408

Lloyd, F. E., and Ridgway, C. S., on *Gymnosporangium juniperi-virginianae*, 43

Localised mycelia, from basidiospores, 365-367

,, ,, from uredospores, 364-365

Lower Devonian, well-preserved plants in, 417

Ludwigs, K., on *Puccinia malvacearum*, 280

Lumen, of an epidermal cell, 315

Lycopodiales, and Filicales, 417

Magnus, P., on *Caeoma obtegens*, 348

Maire, R., on *Endophyllum sempervivi*, 47

Malva rotundifolia, and *Puccinia malvacearum*, 245

Massee, G., on *Puccinia poae*, 18

Material, prepared for inoculation, 93-96

McAlpine, D., on function of pycnidiospores, 23

McCubbin, W. A., field observations of, 324-325

McKenzie, M. A., on Peridermium in New York, 400

Meganeuron, in Coal Measures, 420

Meinecke, E. P., on *Cronartium harknessii*, 337

,, ,, on Peridermium, 400

Meiosis, and haploid nuclei, 295

Melampsoraceae, and flexuous hyphae, 82

,, and Pucciniaceae, 189 427

,, and pycnidia, 189

,, species of, 189, 335

Melampsora abieti-capraearum, flexuous hyphae of, 82

Melampsora euphorviae-dulcis, lacks dextrose and laevulose, 33

Melampsora farinosa, nuclear fusion in teleutospores of, 44

Melampsora larici-capraearum, pycnidia and flexuous hyphae of, 213-214

Melampsora larici-epitea, pycnidial stage of, 27

Melampsora lini, and flax, 192

,, ,, and *Puccinia sorghi*, 315

,, ,, flexuous hyphae of, 192-194

,, ,, illustration, 194

,, ,, is heterothallic, 192

,, ,, pycnidia of, 193

Melampsoridium betulinum, pycnidia subcutilar, 213

Melampsorella, on Abies and Picea, 196

Melampsorella cerastii, and witches' brooms, 92, 166, 194

Melampsorella cerastii, illustrations, 196-198

Melampsorella cerastii, pycnidia and flexuous hyphae of, 194-198

Melhus, I. E., on stem rust in Iowa, 98-99

Mendelian dominance, and haploid mycelia, 297

Mesophyll cells, of the host leaf, 161

Mesozoic era, Angiosperms and insects of, 420-421

Meyen, J. F., on pycnidia and aecidia, 12

Miche, H., and Phanerogams, 50

Micheli, P. A., on *Gymnosporangium clavariiforme*, 389

Microcyclic rusts, evolved from long-cycled, 247, 257

,, ,, with pycnidia, 246-248

Microcyclic rusts, without pycnidia, 241-246

Mid-rib, and proto-aecidia, 285

Milesia, flexuous hyphae in, 198-199

Milesia intermedia, pycnidia and aecidia of, 198

Milesia polypodophila, and *Abies balsamea*, 158

Milesia scolopendrii, on *Abies alba*, 82

Miller, P. R., and nectar-mixing, 206

„   „   on *Gymnosporangium juniperi-virginianae*, 204-205

Mitra, M., on wheat bunt in India, 168-169

Modes of initiating the sexual process, 266-272

Moreau, Mme, on Endophyllum, 63, 291

„   „   on *Uromyces ficariae*, 280

Morphogenic and chemotropic stimuli, 220

Mountain Ash, or Rowan Tree, 392

Mucilage, and pycnidiospores, 157

Mucilage and sugars, in nectar, 155

Mucilaginous drop, sucked up by flies, 304

Mucor mucedo, (+) and (−) strains of, 296

Multiple fusions and hybridisation, 234-235

Mycelia, fuse and exchange nuclei, 340

„   of opposite sex, meet, 266, 300

„   second-generation, 364

„   spontaneous change in, 62

Mycelium, and healthy bark, 322

„   becomes systemic, 384

„   contains conjugate nuclei, 284

„   continues haploid for years, 334

„   (dikaryotic) diploid, 385

„   diploid in stems, 376

„   from uredospores, 375, 382

„   gametophytic generation of, 375

„   grows from rhizomes, 285

„   grows into young leaf, 378

„   haploid and unisexual, 339

„   in roots of old thistle plants, 374

„   in terminal bud, 372

„   intermingling of two kinds, 376

„   of a herothallic rust, 277-278

„   of opposite sex, 132

„   old in abnormal pustule, 254

„   localised and systemic, 335

„   not de-diploidised, 374

„   perennates in rhizomes, 336

„   perennates in roots, 377

„   produces odour, 28

„   remains sterile, 336

„   sporophytic generation of, 375

Mycena, species of, 296

Mycetozoa, Liverworts, and Puff-balls, 409

NÄGELI, C. V., on idioplasm, 308

Nectar, and periphyses, 138

„   applied to pycnidia, 387

„   as adhesive, 156

„   attractive to insects, 154

„   colour in different species, 153

„   diluted with water, 148, 216

„   drops fuse, 144

„   due to osmosis, 39

„   eaten by boys and girls, 157

„   experiments with heated, 70

„   exuding from ostioles, 84

„   filtered and unfiltered, 74

„   in summer and winter, 161

„   is antiseptic, 164-165

„   mixed by insects, 387

„   mixed on detached leaves, 216

„   neutral, 33

„   odour of, 26, 338

„   osmotic pressure of, 160

„   pycnidiospores removed from, 74-75

„   surface tension of, 163

„   sweet and tasteless in Rust species, 26, 31-32, 157

„   term introduced by Craigie and Buller, 157

„   the drop of, 113, 152-164

„   transferred from pustule to pustules, 72

Nectar-drops, and refracted light, 154

Nectar mixing, and appearance of fusions, 222-223

„   „   and natural conditions, 380

Neurospora sitophila, fruiting bodies of, 8

Neurospora tetrasperma, dwarf spores in ascus of, 56

„   „   hybrid perithecia of, 55

Newton, Margaret, on hybridisation in *Puccinia graminis*, 216

„   „   on *Uromyces betae*, 351

Newton, Margaret, and Johnson, T., on abnormal races, 293

Newton, Margaret, and Johnson, T., on diluting nectar, 172

Newton, Margaret, and Johnson, T., on selfed strains of *Puccinia graminis*, 253

Newton, Margaret, and Johnson, T., on uredospore colour, 142

Newton, Margaret, Johnson, T., and Brown, A. M., on hybridisation, 278

Niederhauser, J. S., on *Puccinia menthae*, 85

Nomenclature, vicissitudes of, 389

Non-septate hyphae, 213

Non-violent and violent discharge of aecidiospores, 410-411

Nuclear association, discussion of, 281, 341-342

,, ,, (+) and (−) nuclei, 4, 388

Nuclear division, and conjugate mates, 127

,, ,, in Uredinales, 44

Nuclear fusion, in rust species, 44

Nuclear migration, and association, 264

,, ,, and diploidisation of a proto-aecidium, 124-131

,, ,, from vegetative cells, 51

,, ,, in proto-uredospore sori, 380

,, ,, time required, 234, 319

,, ,, unnecessary, 282

Nuclei, and septal pores, 127

,, at opposite ends of cells, 377

,, become paired, 277-278

,, conjugate pairs of, 45, 216

,, diploid in uninucleate Rusts, 296

,, fuse in teleutospore, 22, 46

,, illustration, 46

,, in young teleutospores, 45

,, migration of, 126, 277, 316, 317

,, of pycnidia and proto-aecidia, 265, 341

,, unequal distribution of, 377

Nucleus, drawn out, 128

,, enters flexuous hyphae, 387

,, passage to proto-aecidium, 233-234

Ochrospora ariae, on Anemone nemorosa, 369

Odour, not detected by Link, 347

,, of Puccinia monoica, 166

,, of pycnidia, 90

,, of Rust Fungi, 165-169

,, of Tilletia spp., 168

,, unpleasant, 423

Oidia, and pycnidiospores, 299

,, germinate, 303

,, give rise to haploid mycelia, 307

Oil-drops, extruded from pycnidia, 152

,, ,, in masses of pycnidiospores, 151-152

Old pycnidia, no flexuous hyphae in, 177

Oligocene period, and insects, 419

Olive, E. W., on binucleate hyphae, 364

,, ,, on sterile primordia, 107

,, ,, on Uromyces rudbeckiae, 293

Olive, L. S., on Pucciniastrum hydrangeae, 83

Oliveira, B. D., on Uromyces fabae, 271

Ophiobolus cariceti heterothallism in, 55

Ophiobolus graminis, sexuality of, 55

Optimum temperature, for uredospore germination, 359

Orange Rust, on Rubus, 414

Orchidaceae, and Papaveraceae, 427

Orchids, and insects, 152

,, capsules of, 409

Ordovician period, vegetation unknown in, 416

Orsted, A. S., sowed basidiospores of Gymnosporangium, 391

Orton, C. R., experiments of, 364

Osmotic pressure, of infected and healthy plants, 358

,, ,, of nectar and sugar solution, 144, 159, 160

,, ,, reduced, 147

Ostiolar filaments, 193

Ostiolar trichomes, are paraphyses, 390

,, ,, Pucciniaceae, 190

,, ,, of Gymnosporangium juvenescens, 203

,, ,, or ostiolar bristles, 134

,, ,, wanting in Cronartium ribicola, 330

Overwintered teleutospores, germinate, 373

Ovule, and pollen-tube, 233

Oxalis corniculata, infected, 93

Pady, S. M., on Witches' brooms, 195

Palisade cells, and parenchyma, 358

Palisade layer, of aecidiosporophores, 116

Panaeolus solidipes, pilei of, 155

Paraphyses, and periphyses, 412

,, as receptive hyphae, 315

,, in Rust Fungi, 133, 137

,, scattered, 137

Parenchymatous cells, longitudinal sections of, 378

Pear-leaves, become infected, 392

Peg, contact with spore, 219

Penicillium luteum, is heterothallic, 55

Pennsylvanian rocks, insect remains in, 419

Perennating mycelium, of Puccinia minussensis, 287

,, ,, of P. suaveolens, 384

,, ,, in Euphorbia rhizone, 368

Peridermium, on Pinus banksiana, 400

Peridermium cerebroides, repeating aecidia of, 337

Peredermium pini, on Pinus sylvestris, 336, 399

Peridermium pini var. acicola, illustration, 45

Peridermium pini var. acicola, nuclear division in, 44

Peridial cap, of Gymnosporangium juniperi-virginianae, 402

Peridial cells, hygroscopic movements of, 403
  „   „   warts and ridges, 409
Peridial strand, sensitive to moisture, 409, 410
Peridial ribbons, in *Gymnosporangium juniperi-virginianae*, 408
Peridium, and epidermis, 116, 118
  „   function of, 124
  „   hygroscopic movements of, 406-410
       406-410
  „   splitting of, 410
Peripheral hyphae, become diploidised, 303
Periphyses, and drops of nectar, 75, 129
  „   and flexuous hyphae, 136, 178, 185-186
  „   and light, 142
  „   and pycnidiospores, 223-226
  „   and their function, 136-143
  „   are absent in some genera, 161
  „   described, 129
  „   immersed, 138
  „   in Puccinia, 133, 137
  „   number of, 136
  „   or ostiolar trichomes, 315
  „   unicellular, 139
  „   wanting in Melampsorella, 198
Permian and carboniferous periods, 417-418
Permian period, and insects, 420
Persoon, D. C. H., on *Aecidium cancellatum*, 11
  „   „   on *Puccinia suaveolens*, 27
  „   „   on systemic mycelium, 362
  „   „   on *Uredo suaveolens*, 347
Peturson, B., on *Uromyces betae*, 351
Petrified logs, of Catskill delta, 417
Phallales, evolutionary time of, 427-428
  „   odour of, 428
  „   Uredinales, and insects, 425
Phallus, stipes and glebae of, 427
Phallus impudicus, and *Mutinus caninus*, 422
Phragmidium, and Rosaceae, 207
Phragmidium species, flexuous hyphae in, 207-209
Phragmidium potentillae, aecidiospores of, 92
  „   „   and *Potentilla bipinnatifida*, 209
Phragmidium rubi, nuclear fusion in teleutospores of, 44
Phragmidium sanguisorbe, lacks periphyses, 40

Phragmidium sanguisorbe, nectar tasteless, 32
Phragmidium speciosum, flexuous hyphae in, 208-209
  „   „   illustrations, 207-209
  „   „   is heterothallic, 51, 85
  „   „   nuclear migration in, 127
  „   „   on *Rosa blanda*, 208
  „   „   spore sequence of, 208
Phragmidium violaceum, aecial spore bed of, 48
  „   „   aecidia of, 51, 379
  „   „   nuclear migration in, 22, 48
  „   „   sterile cells of, 76
Picea canadensis, and *Melampsorella cerastii*, 198
  „   „   *P. mariana*, and Serastium sp., 197
Picea engelmanni, and aecidiospores, 195
Pierson, R. K., on *Cronartium ribicola*, 80, 190
  „   „   on flexuous hyphae, 330
  „   „   on fusions, 215
Pigment, in pycnidia, 153
Pinus caribaea, infected seedlings of, 190
Pinus clausa, branches of, 191-192
Pinus monticola, in British Columbia, 326
Pinus strobus, infection experiments with, 22
Pinus virginiana, and *Coleosporium pinicola*, 248
Pistillate colonies, produce seed, 356
Pisum sativum, and *Vicia faba*, 257
Pitcher Plants, and insects, 141
Plant list, of Permo - Carboniferous period, 417
Plowright, C. B., and germination of pycnidiospores, 18
  „   „   failed to infect old plant, 368
  „   „   on basidiospores, 336
  „   „   on *Circium arvense*, 356
  „   „   on insects and pycnidia, 42
Poirault, G., and Raciborski, M., on nuclear division, 44
Pollen-grains, and disjuncted cells, 425
  „   „   oidia and pycnidiospores, 233
Populus deltoides, and Glomerella, 54
Pore, of upper and lower cell, 366
Pore-plugs, and germ pores, 413
  „   „   are by-products, 414

Pores and pore-plugs, in aecidiospores of Rust Fungi, 413-414
  „   „   „   „ on lateral walls, 414
Post-Devonian and Pre-Jurassic, insects of, 420-421
Poterium sanguisorba, and *Phragmidium sanguisorbae*, 40
Prillieux, E., on pycnidiospore germination, 20
Primary and secondary root, 375
Primary and secondary uredospores, 361-362
Primary hypha, degenerates, 134
Primary uredospore sori, bear few teleutospores, 365
  „   „   „ of long-cycled Rusts, 379
  „   „   „ where nectar applied, 387
Primary uredospores, produced, 383
Privet, inflorescence of, 27
Proboscides, pycnidiospores adhere to, 146
Production of pycnidia, primary uredospores and teleutospores, 377-382
Proto-aecidia, and aecidiospores, 278
  „   „ and cell-fusions in basal cells, 341-343
  „   „ and light, 110
  „   „ and proto-uredospore sori, 345
  „   „ and pycnidia, 109
  „   „ and veinlets, 113
  „   „ apex and base of, 104
  „   „ conversion into aecidia, 115-124, 334, 400-402
  „   „ diploidisation of, 124-131, 339-340, 398
  „   „ fertile cells of, 106
  „   „ flexuous hyphae, and nuclear passage to, 233-234
  „   „ formation in haploid pustule, 98-1 5
  „   „ haploid structures, 100, 132
  „   „ in Rust species studied, 89-93
  „   „ of *Cronartium ribicola*, 332
  „   „ of *Puccinia graminis*, 111
  „   „ proposed name, 108
  „   „ pseudoparenchymatous cells of, 105
Proto-uredospore sori, and diploid hyphae, 381
  „   „   „ are functionless, 381
  „   „   „ suppression of, 381
Protoannularia, from Skiddaw Slates, 416
Protoplasm, continuity of, 234

Protoplasmic bridle, and epidermal wall, 315
Pseudoparenchymatous cells, function of, 106
Psilophytales, Lycopodiales, and proto-articulatae, 416
Pteridospermae, and Cycadophyta, 417
Puccini, T., and *Puccinia non ramosa*, 389
Puccinia and Uromyces, pycnidial elements of, 210-212
Puccinia, first genus of Uredinales, 389
Puccinia species, and walls of peridial cells, 410
  „   „ haploid mycelium of, 280
Puccinia adoxae and *Uromyces scillarum*, 280
Puccinia asarina, nuclei in teleutospores of, 44
Puccinia asteris, uni- and binucleate cells of, 282
Puccinia bardanae, sexuality of, 87
Puccinia buxi, nuclear fusion in, 44
Puccinia caricis, nuclear condition of, 44
Puccinia coronata, colour of pycnidia, 26
  „   „ cuprous oxide from nectar of, 31
  „   „ heterothallic to homothallic, 251
  „   „ homothallic and heterothallic varieties of, 247
  „   „ illustration, 252
  „   „ nectar tasteless, 32
  „   „ nuclear fusion in, 44
  „   „ pycnidial nectar scent of, 168
  „   „ receptive hyphae of, 78
  „   „ vestigial pycnidia of, 250-251
Puccinia coranata avenae, annual species of, 89
  „   „   „ fusions in, 237-238
  „   „   „ illustration, 237
  „   „   „ living preparations of, 211
  „   „   „ the Oat Rust, 237
Puccinia coronata elaeagni, and *Elaeagnus commutata*, 250
Puccinia coronata elaeagni, homothallic long-cycled, 252
Puccinia coronata elaeagni, illustration, 251
Puccinia coronata elaeagni, vestigial pycnidia of, 210, 251
Puccinia cryptotaeniae, binucleate condition of, 283

Puccinia falcariae, cuprous oxide from
  nectar of, 31
  „   „   lacks dextrose, laevu-
               lose, and sucrose,
               33
  „   „   nectar tasteless, 32
Puccinia fusca, cuprous oxide from nectar
  of, 31
  „   „   lacks dextrose and laevu-
               lose, 33
  „   „   nectar tasteless, 32
  „   „   pycnidia of, 282
Puccinia glumarum, absent from Manitoba, 351
Puccinia graminis, aecidia, uredospore and
               teleutospore   sori,
               29, 101
  „   „   aecidiospore discharge,
               122
  „   „   and Barberry, 3, 96-
               98
  „   „   and *Coprinus lagopus*,
               264, 298-300
  „   „   and *Gymnosporangium
               ellisii*, 404
  „   „   and odour of Rust
               Fungi, 165-169
  „   „   and periphyses, 136
  „   „   and Phanerogams, 299
  „   „   and *P. helianthi*, 235,
               266, 320
               and *P. helianthi*, flexu-
               ous hyphae de-
               scribed for, 80
  „   „   and *P. helianthi*, illus-
               tration, 121
  „   „   and *P. helianthi*, mono-
               sporidial pustules
               of, 75
  „   „   and *Uromyces rud-
               beckiae*, 294
  „   „   cell fusions in, 50
  „   „   colour of pycnidia, 26
  „   „   conjugate nuclear di-
               vision initiated, 265
  „   „   crossing   physiologic
               forms of, 71
  „   „   cuprous oxide from
               nectar of, 31
  „   „   development of a pyc-
               nidium in, 134-136
  „   „   dimension of peri-
               physes of, 139
  „   „   diploidisation effected
               in, 125, 278
  „   „   emergent hyphae of,
               79
  „   „   evolutionary view of,
               257
  „   „   experiments with pyc-
               nidiospores of, 22
  „   „   flexuous hypha and
               pycnidiospore of, 84

Puccinia graminis, flexuous hyphae in,
               180-195, 210
  „   „   flexuous hyphae grow
               towards pycnidio-
               spores, 240
  „   „   horns of pycnidio-
               spores produced,
               148-151
  „   „   hybridisation of va-
               rieties, 125
  „   „   hybrids of, 215-216
  „   „   illustrations, 4-5, 47,
               64-68, 71-74, 94-100,
               103-105,   108-130,
               139-140,   145-148,
               150-151,   154-155,
               163, 174, 179-185,
               218-221,   254-256,
               399
  „   „   insects intermix nectar
               of, 303
               introduction to fusions
               in, 215-216
  „   „   is heterothallic, 51,
               64
  „   „   Jaczewski on, 24
  „   „   list of insects visiting,
               37
  „   „   may become autoeci-
               ous, 255
  „   „   nectar dries, 156
  „   „   nectar mixed, 67
  „   „   nectar of, and atmos-
               pheric conditions,
               162
  „   „   nectar of pycnidia, 74
  „   „   nectar tasteless, 32
  „   „   nuclear condition of,
               44
  „   „   nuclear fusion in, 44
  „   „   nuclear migration in,
               127
  „   „   on Barberry, 113, 362
  „   „   odour of pycnidia, 28
  „   „   origin of periphyses,
               177
  „   „   periphyses described
               for, 137-139
  „   „   (+) and (−) mycelia
               fuse, 301
  „   „   pore-plugs of, 414
  „   „   proto-aecidium de-
               scribed, 109
  „   „   pycnidia of, develop,
               101
  „   „   pycnidia of, stained,
               112
  „   „   pycnidial pustules of,
               426
  „   „   pycnidiospore germi-
               nation, 20, 24
  „   „   pycnidiospore size,
               217

Puccinia graminis, pycnidiospores do not germinate, 231
,, ,, pycnidium development in, 135-136
,, ,, scent of the pycnidia, 29, 167
,, ,, sexual process initiated in, 71
,, ,, stimulates hypertrophy, 99
,, ,, "stylospores" germinate in water, 23
,, ,, surface hyphae of, 228
,, ,, the diplophase and monokaryophase, 313
,, ,, two sexual groups of, 73
,, ,, uninucleate uredospores of, 293
,, ,, young aecidium of, 119
Puccinia graminis avenae, teleutospore germination, 95
Puccinia graminis tritici, strains of, 253
Puccinia graminis tritici, teleutospore germination, 94
Puccinia graminis tritici, "white" race of, 142
Puccinia glumarum, and *Uromyces betae*, 351
Puccinia grindeliae, and *Grindelia squarrosa*, 246
,, ,, is homothallic, 246
Puccinia helianthi, and Coprinus species, 305
,, ,, and natural conditions, 272
,, ,, and periphyses, 136
,, ,, and *P. graminis*, localized mycelia of, 380
,, ,, and *P. graminis*, (+) and (−) mycelia of, 277
,, ,, and *P. graminis*, nectar heated, 70
,, ,, autoecious Rust, 269
,, ,, association of pycnidia with uredospore sori, 259-261
,, ,, compound pustules of, produce aecidia, 180
,, ,, crossing of strains of, 62
,, ,, diploid and haploid mycelia of, interact, 305
,, ,, diploidised haploid pustules of, 365
,, ,, flies and pycnidiospores, 143
,, ,, flies mix nectar, 69

Puccinia helianthi, flexuous hyphae in, 180-185
,, ,, germination of teleutospores, 95-96
,, ,, illustrations, 61, 81, 102, 121, 145, 176-177, 236, 267-270, 273-276
,, ,, interfertility of four strains of, 236
,, ,, is heterothallic, 62
,, ,, monosporidial and compound pustules of, 60
,, ,, multiple fusions in, 236
,, ,, nectar mixed, 67
,, ,, observations on fusions in, 235-236
,, ,, oil drops of, 151
,, ,, protected from insects, 62
,, ,, pycnidial odour of, 168
,, ,, pycnidiospore and flexuous hyphae fused, 80
,, ,, resting period of teleutospore, 96
,, ,, sexual process initiated in, 71
,, ,, short-cycled, 260
,, ,, size of pustule, 99
,, ,, two sexual groups of spores in, 73, 305
Puccinia hieracii, and *Taraxacum officinale*, 349
Puccinia liliacearum, nuclear division in, 44
Puccinia malvacearum, illustrations, 242, 244
,, ,, is homothallic, 425
,, ,, lacks pycnidia, 171
,, ,, The Hollyhock Rust, 241
Puccinia menthae, nuclear fusion in, 44
Puccinia micro-forms, are homothallic, 86-87
Puccinia minussensis, a perennial species, 89
,, ,, and *P. helianthi*, 288
,, ,, and *P. suaveolens* 306, 382, 397
,, ,, number of periphyses in, 136
,, ,, De-diploidisation and Self-diploidisation in, 284-288

Puccinia minussensis, illustrations, 150, 153, 156, 175, 284-285
,,   ,,   mycelium de-diploidised, 285
,,   ,,   mycelium hibernates, 166
,,   ,,   odour of pycnidia of, 26
,,   ,,   oil drops of, 151
,,   ,,   on *Lactuca pulchella,* 334, 367
,,   ,,   sexual process, discussion of, 398
,,   ,,   systemic mycelium of, 283
Puccinia monoica, aecidia of, 92
,,   ,,   and systemic mycelium, 166
Puccinia non ramosa, illustration, 390
Puccinia obscura, nectar and indigo-carmine, 43
Puccinia obtegens, erroneous designation of, 347-348
,,   ,,   usage in England, 348
Puccinia phragmitis, is heterothallic, 211
Puccinia pimpinellae, cuprous oxide from nectar of, 31
Puccinia poarum, cuprous oxide from nectar of, 31
,,   ,,   fertilisation process in, 128
,,   ,,   nectar tasteless, 3
,,   ,,   nuclear condition of, 44
Puccinia podophylii, and *Podophyllum peltatum,* 376
,,   ,,   pore-plugs of, 413
Puccinia prostii, micro-form with pycnidia, 248
Puccinia rubogo-vera, conidia of, 47
,,   ,,   ,,   cuprous oxide from nectar of, 31
,,   ,,   ,,   nectar tasteless, 32
,,   ,,   ,,   pycnidiospore germination in, 20-21
Puccinia sorghi, and flexuous hyphae, 212
,,   ,,   is heterothallic, 225
Puccinia suaveolens, and *Cirsium arvense,* 334
,,   ,,   and climatic factors, 354
,,   ,,   and *P. podophylli,* 375
,,   ,,   as parasite of Dandelion, 349-350
,,   ,,   cuprous oxide from nectar of, 31
,,   ,,   derived from eu-form, 345
,,   ,,   distribution of, 350-351

Puccinia suaveolens, economic value of, 359
,,   ,,   English strain of, 361
,,   ,,   experimental investigations on, 347
,,   ,,   failure to observe haploid pustules in, 365
,,   ,,   infected plants distinguished, 357
,,   ,,   illustration, 373
,,   ,,   in Europe, 344
,,   ,,   introduced into central Canada, 351-354
,,   ,,   investigated cytologically, 346
,,   ,,   is heterothallic, 380, 386-388
,,   ,,   its effect on structure and physiology of host, 356-359
,,   ,,   local haploid pustules of, 367
,,   ,,   mycelium in root-system of, 356
,,   ,,   name discussed, 347-348
,,   ,,   natural conditions of, 367
,,   ,,   nectar of pycnidia, 38
,,   ,,   nectar tasteless, 32
,,   ,,   nuclear migration in, 50
,,   ,,   odour of pycnidia, 26-28, 70-71, 165
,,   ,,   parasite of *Cirsium arvense,* 348
,,   ,,   persistence of, 388
,,   ,,   prevents flowering, 356
,,   ,,   produces teleuto-spores, 361
,,   ,,   proto-uredospore sori of, 381
,,   ,,   pycnidia and uredo-spore sori of, 380
,,   ,,   pycnidia numerous in, 387
,,   ,,   sexual process, discussion of, 344-347, 398
,,   ,,   spread by uredo-spores, 352
,,   ,,   systemic mycelia established in *Cirsium arvense,* 370-373
,,   ,,   systemic mycelium of, 362, 372, 375-377

Puccinia suaveolens, teleutospores detachable in, 366
,, ,, uredospore inoculations, 371
,, ,, uredospores between bud-scales, 374
Puccinia sylvatica, cuprous oxide from nectar of, 31
,, ,, lacks dextrose, sucrose, and laevulose, 33
,, ,, reduces Fehling's solution, 32
Puccinia tragopogi, cuprous oxide from nectar of, 31
,, ,, lacks dextrose, sucrose, and laevulose, 33
,, ,, nectar tasteless in, 32
Puccinia tragopogonis, illustration, 105
,, ,, scent of, 27
Puccinia triticina, pycnidial odour of, 29
Puccinia violae, cuprous oxide from nectar of, 31
Puccinia xanthii, and *Xanthium commune*, 241
,, ,, cell fusions in, 282
,, ,, illustrations, 245-246
Pucciniastrum americanum, pycnidia and flexuous hyphae of, 83
Pucciniae, flexuous hyphae in, 224
Puff-balls, and cornute aecidia, 405-406
,, ,, basidiospores not shot away, 405
Pustules, hand sections of, 206
,, ,, sterile and fertile, 202
,, white in colour, 255
Pycnia, colour of, 70
,, function investigated, 67
Pycnial nectar, intermixing eliminated, 274
Pycnidia, amphigenous and subepidermal, 193
,, and aecidia, closely associated, 13
,, and aecidia, successive formation of, 326-334
,, and aecidia on Malaceae, 412
,, and fine weather, 172
,, and flexuous hyphae, 194-198, 390
,, and primary uredospores, 352, 384
,, and proto-aecidia, 111, 378-379, 398
,, and pycnidiospores of Lichens, 5-6
,, and spermogonia of Discomycetes and Pyrenomycetes, 7-8
,, and stylospores, 7
,, applanate type, 249
,, are functional, 77

Pycnidia, as male organs, 5
,, colour and scent, 344, 379
,, colour in Uredinales, 26
,, conspicuous, 169
,, cytological background, 43-53
,, directed outwards, 176
,, discovered in Ascomycetes, 5
,, discovered in Lichens, 5
,, discovery in Uredinales, 5, 11-12
,, effect of rain on, 171-173
,, entomological background of, 26-43
,, evolution of, in Uredinales, 426-427
,, followed by pycnidia, 327
,, formation in haploid pustule, 98-115
,, function of, 12-26, 42, 387
,, improbable in Pre-Devonian periods, 427
,, in four rows, 198
,, in *Gymnoconia peckiana*, 199-200
,, in light and darkness, 41
,, in species studied, 89-93
,, lack odour, 29
,, nectar and pycnidiospores, 4
,, not degenerate, 43
,, number and position of, 102, 111
,, of *Gymnosporangium clavipes*, 238-240
,, of Rust species, 90, 91, 194, 200-209
,, of *Uromyces hobsoni*, 261-263
,, of systemic mycelia, 287, 383-384
,, on both sides of leaf, 371
,, on Hemlock, 83
,, on perennial mycelium, 292
,, presence or absence of, 252-253
,, present form of, attained, 426
,, produced annually, 383
,, proto-aecidia, and aecidia, 285
,, remain active, 135
,, rudimentary, 255
,, seen by insects, 102
,, species of insects, visiting, 4, 43
,, subcortical, 329
,, subcuticular and subepidermal, 196
,, suppressed, 247
,, sweet exudate of, 23
,, transverse sections of, 192
,, unisexual, 387
,, uredospores and teleutospores of *Puccinia graminis*, 253-257
,, use to *P. sauveolens*, 383
,, weakened in inbred strains, 253

Pycnidial brush, and periphyses, 142
Pycnidial development, begins, 134
Pycnidial function, discovery in Uredinales, 65-75
Pycnidial nectar, and insects, 62
,, ,, of systemic mycelia, 380
Pycnidial nuclei, and flexuous hyphae, 254
Pycnidial production, and warm weather, 327
Pycnidiospore nucleus, passage to protoaecidium, 233-234
Pycnidiospores, abnormal, 24, 191
,, and budding process, 21
,, and cotton-blue, 145
,, and emerging hyphae, 226-229
,, and filamentous hyphae, 338
,, and flexuous hyphae, 383, 387
,, and gametophytic hyphae, fusions, 228
,, and germ-tubes of, 79
,, and oil-drops, 151-152
,, and papilla, 219-220
,, and pointed periphyses, 223-226, 316
,, and pycnidiosporophores, 143-148, 330
,, asexual spores, 23, 25
,, attached by ends, 240
,, carried by flies, 4, 17
,, colour and aroma of, 304
,, discovery in Uredinales, 11-12
,, do not infect host plant, 24
,, equivalent to oidia, 243
,, extruded through ostiole, 129, 142
,, formation stops, 135
,, fuse with flexuous hyphae, 59, 217-222, 229, 237, 320
,, incapable of germination, 229-233
,, in mass, 144
,, lack germ-tubes in Puccinia graminis, 224
,, not vestigial structures, 48
,, oidia, and pollen grains, 233
,, oozing from pycnidium, 146
,, production in P. graminis, 148-151, 191
,, (+) and (−), 4, 75
,, scattered in nectar, 74
,, spindle shaped, 239

Pycnidiospores, union with flexuous hyphae, 215-240
,, swallowed by insects, 24
,, variable length, 231
,, without protoplasm and oil-drops, 218
Pycnidiosporophores, and the pycnidiospores, 143-148, 330
,, described, 129, 135
,, of Cronartium fusiforme, 191
Pycnidium, described, 329
,, development of, in Puccinia graminis, 134-136
,, exposed to air, 331
,, pycnium and pycniospores, 3
,, ruptures epidermal cells, 12
,, structure of, 129
,, subcuticular and lacking periphyses, 210
Pycniospores, as conidia, 60
Pycnium, and pycnidium, 132
,, derivation of, 132
,, entomological solution of, 66
,, two types of hyphae protrude from, 80
Pyrenomycetes, and Discomycetes, 7
,, and heterokaryotisation, 302
,, and periphyses, 133
,, and species of Neurospora, 302
,, facultatively heterothallic, 55
,, pycnidia and spermogonia of, 7-8
,, sexual function in, 7
Pyreonema confluens, and Pleurage anserina, 319
,, ,, and streaming of cytoplasm, 318
Pyrus communis, and Juniperus sabina, 391

Quaternary and Tertiary Periods, 418-419
Quintanilha, A., on Buller Phenomenon, 304-305

Radicle, develops into tap root, 356
Rain, effect of on pycnidia, 171-173
Ranunculus ficaria, and young pycnidium, 12
Ráthay, E., found nectar sweet, 31
,, ,, on insects visiting Rusts, 4, 37-38, 426
,, ,, on nectar of Rust Fungi, 38, 157
,, ,, on pycnidia of Rust spp., 27
,, ,, spermogonia are male organs, 42

Ráthay, E., tests with Fehling's solution, 30

Rebentisch, J. E., on aecidial peridium, 11

Receptive hyphae, in *Puccinia graminis*, 227

,, ,, misnomer, 75

Recessive mutations, and abnormal characters, 255

Re-diploidisation, 382-383

Reed, H. H., and Crabill, C. H., on pycnidiospore germination, 25

Reed, H. H., Crabill, C. H., and Bliss, D. E., on movements of peridial ribbons, 408

Refractive contents, of pycnidiospores, 217

Respiration of sick and healthy plants, 358

Rhamnus, and *Puccinia coronata avenae*, 91, 237-238

Rhamnus cathartica, and *Puccinia coronata*, 26

,, ,, inoculation of, 237

Rhizome, of *Lactuca pulchella*, 286

Rhizomes, and aerial shoots, 284

Rhyme, "The Creeping Thistle," 361

Rhyniella praecursor, remains of, 419

Rhyniognatha hirsti, mandibles of, 419

Rice, Mable A., on Puccinia spp., 227

,, ,, on receptive hyphae, 82

Richards, H. M., on emergent hyphae, 76

,, ,, on peridium, in a Puccinia, 118-119

,, ,, on *Uromyces caladii*, 226

Rock-systems, list of periods and eras of, 415

Roestelia cancellata, on *Gymnosporangium sabinae*, 392

,, ,, on *Pyrus communis*, 391

,, ,, on *Sorbus aucuparia*, 392

Roestelia lacerata, on *Crataegus oxyacantha*, 392

Roestelioid aecidia, in Gymnosporangium, 200, 400

,, ,, grow slowly, 406

Root-buds, of *Cirsium arvense*, experiments with, 373-374

,, ,, inoculated with uredospores, 385

Rosen, F., on *Uromyces pisi*, 44

Rose Rust, on cultivated Rose, 91

Rostrup, F. G. E., on second generation mycelia, 364

Rouget, A., on Laboulbeniales, 9

Rust Fungi, aecidial primordia of, 78

,, ,, aecidiospore discharge in, 410-414

,, ,, aecidiospores, pores and pore-plugs in, 413-414

Rust Fungi, and Hymenomycetes, 304

,, ,, and imperfect sexual process, 289-294

,, ,, and insects, 169-171, 426

,, ,, and ostiolar trichomes, 134

,, ,, comparison of Modes of initiating sexual process in, 300-307

,, ,, flexuous hyphae observed in, 188-189, 210-212, 214

,, ,, haploid pustules of, 424-425

,, ,, haploid strains of, 296

,, ,, heterothallic and homothallic species of, 84-88

,, ,, in pycnidial stage, 187

,, ,, insects mixing nectar of, 426

,, ,, macrocyclic Rusts, 88

,, ,, modes of initiating sexual process in, 59, 184, 264-296

,, ,, odour of pycnidial stage, 165-169

,, ,, pycnidia are functional, 71, 248

,, ,, pycnidia, discovery of, 5

,, ,, pycnidiospores fuse with hyphae of opposite sex, 79-80

,, ,, without pycnidia, 93, 241

Rust mycelium, and central pore in each septum, 127

Rust pustule, cut with hand razor, 173

Rust species, that are brachy-forms, 346

SAPPIN-TROUFFY, P., on binucleate aecidiospores, 124

Sappin-Trouffy, P., on *Endophyllum euphorbiae-sylvaticae*, 289

Sappin-Trouffy, P., on nuclear condition of Gymnosporangium, 394-395

Sappin-Trouffy, P., on *Puccinia graminis*, 45

Savile, D. B. O., on diploidising process in Rusts, 318

,, ,, on fusions in Pucciniaceae, 225

,, ,, on nuclear structure, 212

,, ,, on *Puccinia sorghi* and *Uromyces fabae*, 316

,, ,, pores in septa of Rust Fungi, 318

Scented flowers, and pollination, 167

Schaffnit, E., on scent of aecidiospores and uredospores, 28

Schizophyllum commune, four sexual groups of, 73

Schleiden, M. J., rejected pycnidia as male organs, 13-14

Schuchert, and Dunbar, C. O., on Rhynic beds, 417

Schmitz, F. K. J., on *Coleosporium campanulae*, 43

Sclerotinia vaccinii, and pseudo-pollen, 425
   ,,   ,,   and *Vaccinium vitis-idaea*, 423
Sea-anemones, and pycnidia, 136
Secondary uredospores, and teleutospores, 362, 384-386
Sections, mounted in water, 217
Seed-beds, for cereals and weeds, 353
Seedling thistles, infected, 361
   ,,   ,,   inoculated with basidiospores, 366
   ,,   ,,   inoculated with uredospores, 372, 384
Self-diploidisation, 382-383
   ,,   ,,   of *Puccinia minussensis*, 286
   ,,   ,,   in *P. minussensis*, 284-288, 340
   ,,   ,,   in rusts having a systemic mycelium, 283-284
Self-propagating haploid strains, derived from heteroecious species, 294-296
Selkirk Settlers, arrival of in Manitoba, 353
Senecio jacobaea, and *Puccinia schoeleriana*, 43
Septa, absent in flexuous hyphae, 183
Septal Pores, and diploidisation of protoaecidium, 124-131
   ,,   ,,   in mycelium of Higher Fungi, 318
Septum, described by Buller, 318
Sexual process, and presence or absence of pycnidia, 252-253
   ,,   ,,   in diploid mycelium, 365
   ,,   ,,   in Discomycetes and Pyrenomycetes, 6
   ,,   ,,   inoperative in *Kunkelia nitens*, 291
   ,,   ,,   in *Puccinia graminis* and *P. helianthi*, 289
   ,,   ,,   modes of initiating in Rust Fungi, 264-296
   ,,   ,,   of a heterothallic Rust, 264
   ,,   ,,   of *P. suaveolens*, 344-347
   ,,   ,,   spermatia and carpogonia, 6
Sexual processes, compared, 300
Shear, C. L., and Dodge, B. O., on Neurospora species, 56
Shirai, M., states nectar eaten by boys in Japan, 332
Shoots, infected with perennial mycelium, 374-375
Short-cycling, in *Uromyces fabae*, 257-258
   ,,   ,,   occurs infrequently, 261
Silurian beds, in England and Gothland, 416
Silurian Period, land plants existed in, 416

Simple and compound cankers of *Cronartium ribicola*, 337-340
Simple and compound pustules, observed by Hanna, 204
Simple pustules, more numerous than compound, 267
   ,,   ,,   remain sterile, 396
Smith, Gilbert M., on *Puccinia graminis*, 274-275
Soppit, H. J., on *Aecidium leucospermum*, 369
Sorbus aria, and *Gymnosporangium juniperinum*, 41
   ,,   ,,   and ants observed by Rathay, 30
Spaulding, P., on infection of young bark, 325
Species, with flexuous hyphae, 214
Species studied, in Uredinales, 89-93
Spermatia, are uninucleate, 395
   ,,   as sex cells, 25-26
   ,,   attached to trichogyne, 6
   ,,   fuse with trichogynes, 78
   ,,   germination in honey and sugar, 21
   ,,   in Discomycetes and Pyrenomycetes, 7
   ,,   of Laboulbeniales, 8-11
Spermatial nuclei, entrance into hyphae, 314
Spermatiophores, in pycnidia of Uredinopsis spp., 83
Spermogonia, and spermatia, 3, 7, 131
   ,,   as male organs, 6, 17
   ,,   not reduced organs, 22
   ,,   of Discomycetes and Pyrenomycetes, 7-8
Sphacelia stage, in Ergot of Rye, 422-423
Spores, discharged, 117, 403, 405
   ,,   form powdery mass, 403
   ,,   resting, observed by Seward, 417
Spore-chains, turgid in moist atmosphere, 122
Sporidia, fusion in Ustilago spp., 223
   ,,   of opposite sex, 61
   ,,   of same sex, 61
   ,,   sexual distribution of, 73
Stahl, E., sexual theory of, supported by Baur, 6
Stakman, E. C., on hybridisation in *Puccinia graminis*, 216
Staminate, and pistillate plants, 354
Stättler, H., and sexual theory, 6
Sterile cells, and pseudoparenchyma, 401
Stevens, Edith, on *Gymnosporangium juniperi-virginianae*, 263
Stigma, and pollen grains, 233
Stinkhorn Fungus, and flies, 70
Stomata, increase in number of, 358
Stomatic cleft, used as ostiole, 194
Strasburger, E., invented terms haploid and diploid, 309

Structure and physiology of host, effect of *Puccinia suaveolens* on, 356-359

Successive formation of pycnidia and aecidia, 326-334

Sugar, identified by Fehling's solution, 30

,, in the glebal fluid, 422

Sunflowers, inoculated with basidiospores, 65, 98

,, killed by mysterious disease, 259

Sunflower leaf, illustration, 69

,, ,, pycnidia situated on upper side of, 101

,, ,, rust pustules and fly, 65

Sunflower rust, see *Puccinia helianthi*

Sunlight, pycnidia viewed in, 147

Sydows, P., and H., and *Puccinia obtegens*, 348

Synkaryotic and dikaryotic diploid cells, 297-298

Synkaryotic diploid nucleus, 311

Synkaryotic hybrids, of Higher Plants, 298

Syracuse dishes, and culture media, 232

Systemic infections, resulting from uredospore, 346

Systemic mycelia, comparison of, 334-337

,, ,, forms of, in *Puccinia suaveolens*, 362-363

Systemic mycelium, and annual self-diploidisation of, Mode V, 283-384

,, ,, and pycnidia, 363

,, ,, bearing pycnidia only, 386

,, ,, behaviour of, 362-363

,, ,, binucleate condition of, in stems and roots, 375-377

,, ,, established in *Cirsium arvense*, 370-373

,, ,, established slowly, 336

,, ,, failed to de-diploidise, 385

,, ,, from aecidiospores and uredospores, 334

,, ,, from uredospores of *Puccinia suaveolens*, 378

,, ,, in axillary buds, 371

,, ,, in herbaceous perennials, 367-370

,, ,, in terminal buds, 371-372

,, ,, in thistle roots, 374-375

,, ,, in thistle seedlings, 372

Systemic mycelium, may arise from uredospores, 388

,, ,, of *Cronartium ribicola*, 335, 340

,, ,, pycnidia, primary uredospores, and teleutospores, 377-382

,, ,, pycnidia on, and biological significance of, 383-384

,, ,, secondary uredospores and teleutospores, 384-386

Tannic acid, and ferric chloride, 318

Taylor, R. E., on mycelium and pycnidia, 386

Teleutospore sori, on Junipers, 412

,, ,, overwintering of, 398

,, ,, rudiments of, 280

Teleutospore stage, of Gymnosporangium spp., 92

Teleutospores, among uredospores, 379

,, and secondary uredospores on systemic mycelia, 384-386

,, and uninucleate sporidia, 395

,, and uredospores, effect of external conditions on, 359-360

,, formation of, 360

,, from England, 365

,, germinated in usual manner, 361, 366

,, in exhausted aecidia, 262

,, kept in refrigerator, 94, 366

,, liberate basidiospores, 324

,, nuclear condition of, 46, 395

,, on Corn, 93

,, on May-apple, 376

,, on petals of florets, 357

,, pedicles gelatinised, 390, 411

,, prepared for germination, 93-96, 366

,, primary uredospores, and production of pycnidia, 377-382

Terminal bud, and systemic mycelium of *Puccinia suaveolens*, 372

Terminology, discussion of, 131-140

Tertiary, and Quaternary periods, 418-419

Tertiary deposits, and insect fossils, 421

Thalictrum dioicum, and *Puccinia rubigovera* var. *agropyrina*, 91

Thatcher, F. S., on osmotic pressure, 160

Thaxter, R., experiments with pycnidiospores, 17-18

Thaxter, R., on *Gymnosporangium obosum*, 18

Thecopsora padi, inoculation of Spruce Trees, 28

,,　　　　,,　odour without pycnidia, 28

Thirumalacher, M. J., on *Uromyces hobsoni*, 262-263

Thistles, in laboratory garden at Winnipeg, 352

,,　inoculation of, 388

Thistle roots, systemic mycelium in, 374-375

Thistle rust, source of, 360-361

Thistle shoots, come up in spring, 335

,,　　　,,　number infected, 352

Thurston, H. W., cytological investigations of, 393

,,　　　,,　failed to infect Junipers, 397

Trachyspora intrusa, and *Triphragmium ulmariae*, 50

,,　　　,,　perennial mycelium of, 50

Tranzschelia pruni-spinosae, flexuous hyphae observed, 210

Tranzschelia pruni-spinosae, on *Hepatica acutiloba*, 210

Trees, girdled and killed by blister rust, 322

Triassic age, rich insect fauna of, 420

Triassic period, vegetation in, 418

Trichogynes, of Rust Fungi, 23

Trichomes, and flexuous hyphae, 134

Trinucleate, and uninucleate cells, 376

Triphragmium ulmariae, binucleate condition of, 51

Triphragmium ulmariae, nuclear fusion in young teleutospores of, 44

Triphragmium ulmariae, nuclear migrations and artifact in, 50

True, R. P., on Woodgate Peridermium, 337

Tuberculina maxima, may prevent diploidisation, 329

,,　　　,,　parasite of *Cronartium ribicola*, 327

Tuberculina persicina, a parasite of Rust pycnidia, 21

Tulasne, L. R., on Discomycetes and Pyrenomycetes, 7

UNDERWOOD, L. M., and Earle, F. S., on *Gymnosporangium bermudianum*, 392-393

Unger, F., on *Aecidiolum exanthematum*, 11

,,　　,,　on periphyses, 11

,,　　,,　on *Rhamnus catharticus*, 12

Uninucleate aecidiospores, uredospores, and teleutospores, 293

Uninucleate cells, degeneration and death of, 379

Uninucleate cells, fertilised in *Phragmidium violaceum*, 379

,,　　　,,　fuse in pairs, 280

Uninucleate hyphae, originate by de-diploidisation, 378

Uninucleate mycelium, and proto-aecidium, 124

Uninucleate mycelium becomes moribund after aecidia, 397

Uninucleate mycelium, by de-diploidisation of binucleate, 398

Uninucleate oidia, on dikaryotic mycelium, 307

Uninucleate Rusts, derived from Heteroecious species, 294-296

Upper Silurian age, in Australia, 416

Uredinales, aecidia without pycnidia, 17

,,　and Ascomycetes, 5

,,　and flexuous hyphae, 75-84, 187

,,　and Hymenomycetes as related groups, 298

,,　and sexual process, 53

,,　are entomophilous, 415

,,　comparison of modes of initiating sexual process in, 264-296

,,　contrast with Ustilaginales, 427

,,　dikaryotic and synkaryotic diploid cells in, 297-298

,,　discovery o :heterothallism in, 59-65

,,　function of pycnidiospores, 8

,,　geological time and evolution of pycnidia, 426-427

,,　heterothallic and homothallic species of, 252-253

,,　nuclear migration in, 49

,,　odour of pycnidial stage, 165

,,　pycnidia and pycnidiospores discovered in, 11-12

,,　pycnidial function discovered in, 65-75

,,　sexual process imperfect in, 289

,,　species studied, 89-93

,,　their evolution, 415, 426

Uredinia, and primary uredinia, 359

,,　contain teliospores, 360

Urediniospores, formation inhibited, 360

Uredinopsis, and Milesia, 335

Uredo Caeoma-nitens, spermatia of, 20

Uredospore pustules, localised, 371

Uredospore sori, after pycnidia, 379

,,　　　,,　are hypophyllous, 379

,,　　　,,　associated with pycnidia in *Puccinia helianthi*, 259-261

,,　　　,,　primary and secondary, 364

,,　　　,,　pycnidia wanting in, 385

Uredospore sori, with pycnidia, 261
Uredospores and teleutospores, effect of external conditions on, 359-360
Uredospores and teleutospores, on Barberry, 253-255
Uredospores, and the persistence of *Puccinia suaveolens*, 388
,, carried by wind, 265, 372
,, generations of, 265
,, genetically identical, 381-382, 384
,, germinating on seedling thistles, 362
,, localised mycelia derived from, 364-365
,, on Currants and Gooseberries, 324
,, on old infected thistle plants, 372
,, primary and secondary, 361-362
,, production of eliminated, 412
,, retain vitality, 261
,, sown on rhizome buds, 370
,, sown on Tragopogon species, 349
,, sown previous summer, 388
,, systemic mycelia may arise from, 388
,, washed by rain to root-system, 370
Uromyces betae, absent from Manitoba, 351
,, ,, nuclear fusion in young teleutospore of, 44
Uromyces dactylis, cuprous oxide from nectar of, 31
Uromyces fabae, and *Pisum sativum*, 160
,, ,, haploid pustules diploidised, 270
,, ,, illustration, 258
,, ,, is heterothallic, 225
,, ,, short-cycling in, 257-258
Uromyces geranii, nuclear fusion in young teleutospore of, 44
Uromyces glycyrrhizae, and Olive, E. W., 375
Uromyces hobsoni, and its pycnidia, 261-263
,, ,, teleutospore germination in, 262
Uromyces perigynius, on *Rudbeckia laciniata*, 211
Uromyces pisi, cuprous oxide from nectar of, 31
,, ,, illustration, 34
,, ,, lacks dextrose, laevulose, and sucrose, 33
,, ,, nectar slightly sweet, 32
,, ,, nuclear condition in, 44

Uromyces pisi, perennial mycelium on *Euphorbia cyparissias*, 41
Uromyces rudbeckiae, a variation of, 293
Uromyces scirpi, and *Cenanthe crocata*, 226
Uromyces sp., illustration, 105
Uromyces trifolii hybridi, haploid pustules diploidised, 270
Uromyces trifolii hybridi, illustration, 271
Uromyces trifolii hybridi, is heterothallic, 85
Ustilaginales, Uredinales, and insects, 169, 427
Ustilago, interspecific hybridisation of, 57-58

Vaccinium vitis-idaea, and *Abies pectinata*, 249
Vacuolar pressure, and cytoplasm, 318
Vandendries, R., and Martens, P., on *Pholiota aurivella*, 306
Van Tieghem, P., on sexual theory, 6
Vaucheria, oogonium of, 77
Vegetative mycelium, and flexuous hyphae, 184
,, ,, and uninucleate cells, 332
Vegetative nuclear division, in fusion cell, 290
Veins, and xylem in cross section of a leaf, 358
Vestigial pycnidia, in *Coleosporium pinicola*, and *Colyptospora goeppertiana*, 248-250
,, ,, in *Puccinia coronata*, 250-252
Vestigial stamens, in Salvia, 250
Viability of thistle seeds, collected at Winnipeg, 356
Virgin cankers, mycelium of, not diploidised, 339
Vize, J. E., on *Uromyces hobsoni*, 261
Von Jaczenski, A., on pycnidiospore germination, 23
Von Tavel, F., and Brefeld, O., on morphology of Fungi, 20
Von Tubeuf, C., inoculated White Pine, 324-332
Wahrlick, W., discovery of pores in Higher Fungi, 318
,, ,, on septal pores and protoplasmic bridges, 50, 126-128
Waino, E. A., on sexual theory, 6
Walker, Ruth, on micro-forms of Puccinia, 282
Walls, of pycnidiospores become gelatinous, 330
Water, absorbed by pycnidiospore walls, 150
,, added to pycnidial nectar, 147

Waters, C. W., on uredospores and teleutospores, 359

Wehmeyer, L. E., on perithecia from single ascospores, 55

Weir, J. R., and Hubert, E. E., sowed aecidiospores on Stellaria, 195

Weismann, A., on germ plasm theory, 308

Welsford, E. J., on nuclear migration in Uredinales, 49

Werth, E., on *Puccinia malvacearum*, 280

White Pine, Blister Rust Disease of, 321-324

Wild Lettuce, bearing pycnidia, 90

Willows, male catkins of, 27

Wilson, W. P., on Nectaries, 39

Witches' broom, illustration, 196

Witches'-broom Rust, on Picea, 195

Witches' brooms, of *Gymnosporangium juvenescens*, 92

    ,,   ,,   on Vaccinium, 249

Whetzel, H. H., Jackson, H. S., and Mains, E. B., on *Puccinia podophylli*, 376

White flies and Thrips, exclusion of, 72

  ,,   ,,   ,,   ,, mix pycnial nectar, 63

Woodgate Peridermium, on *Pinus sylvestris*, 337, 400

Woronin, M., on life history of Sclerotinia sp., 423

  ,,   ,, on *Puccinia helianthi*, 259

YEASTS, and culture media, 230

York, H. H., and Woodgate Rust, 336-337

Xanthium commune, rust infections on, 246

Xanthomonas translucens f. sp. *undulosa*, illustration, 149

ZICKLER, H., on *Bombardia lunata*, 8

Zodiomyces, and Ceratomyces, 9

Lightning Source UK Ltd.
Milton Keynes UK
UKHW030614210722
406167UK00006B/620

9 781487 598167